D0079351

CAMBRIDGE STUDIES IN ECOLOGY

The ecology of bird communities

Volume 1

Foundations and patterns

Also in the series

Hugh G. Gaugh, Jr *Multivariate analysis in community ecology*

Robert Henry Peters *The ecology implications of body size*

C.S. Reynolds *The ecology of freshwater phytoplankton*

K.A. Kershaw *Physiological ecology of lichens*

Robert P. McIntosh *The background of ecology: concept and theory*

Andrew J. Beattie *The evolutionary ecology of ant–plant mutualisms*

F.I. Woodward *Climate and plant distribution*

Jeremy J. Burdon *Diseases and plant population biology*

Janet I. Sprent *The ecology of the nitrogen cycle*

H. Stolp *Microbial ecology: organisms, habitats and activities*

R. Norman Owen-Smith *Megaherbivores: the influence of very large body size on ecology*

The Ecology of Bird Communities

Volume 1

Foundations and Patterns

JOHN A. WIENS

Professor of Ecology
Department of Biology
Colorado State University
Fort Collins, Colorado

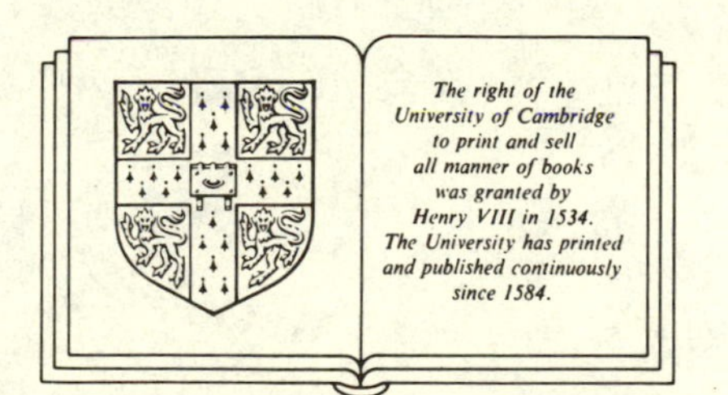

CAMBRIDGE UNIVERSITY PRESS

Cambridge

New York Port Chester

Melbourne Sydney

Published by the Press Syndicate of the University of Cambridge
The Pitt Building, Trumpington Street, Cambridge CB2 1RP
40 West 20th Street, New York NY 10011, USA
10 Stamford Road, Oakleigh, Melbourne 3166, Australia

First published 1989

Printed in Great Britain at the University Press, Cambridge

British Library cataloguing in publication data

Wiens, J A
The ecology of bird communities.
Vol. 1: Foundations and patterns
1. Birds. Ecological communities
I. Title
598.2′52′47

Library of Congress cataloguing in publication data

Wiens, John A.
The ecology of bird communities / John A. Wiens.
p. cm. – (Cambridge studies in ecology)
Includes bibliographies and indexes.
Contents: v. 1. Foundations and patterns – v. 2. Processes and variations.
ISBN 0 521 26030 2 (v. 1). – ISBN 0 521 36558 9 (v. 2)
1. Birds–Ecology. I. Title. II. Series.
QL673.W523 1989
598.2′5–dc19 88-38331

ISBN 0 521 26030 2

SE

For Ann, David and Kyra

Contents

Preface

This book and its companion volume represent a personal statement about the ecology of bird communities – what they are, what we know about them, and what we need to know. They are also about how avian community ecology has been practiced as a science – how we have gone about gaining our knowledge of bird communities and how logical and methodological considerations affect the certainty we can attach to that knowledge. Because studies of birds have contributed a good deal to the foundation of contemporary community ecology and because concerns about logic, methodology, and epistemology are central to any science, I believe that the themes and viewpoints I develop are relevant to community ecology well beyond the somewhat artificial boundaries dictated by my focus on birds. My topic is really community ecology as it has been practiced on birds rather than bird communities *per se*.

Avian community ecology is a complex, multifaceted discipline that is enriched by controversy. I have written these volumes partly in an attempt to examine the complexity of communities and to probe the dimensions of the controversies, but partly also out of a simple enjoyment of the subject. I have directed my comments particularly toward advanced undergraduates and graduate students with interests in avian community ecology or, more broadly, in birds or in ecology, for I feel that they are in the best position to put my comments into practice or to challenge my views. I hope that my colleagues – practicing ecologists – will also find much to interest (or outrage!) them.

My objective when I began this project was to provide a critical assessment of the current state of affairs in avian community ecology. I rapidly discovered that this required that many studies and areas of investigation be developed in considerable detail. It also led to excursions into aspects of the philosophy of science and the history of community ecology that provide essential perspectives on current thinking in this discipline. When I finished, it was apparent that the manuscript would result in a book of such formidable size (and cost!) that only the most dedicated students would be likely to read it. Accordingly, it was decided to publish the work in two volumes. Although these volumes are a closely integrated set, each has a particular focus. In this volume, I consider why avian community ecologists ask the questions that they do and how they have gone about answering them. The questions have generally involved both 'what' (pattern) and 'why' (process) components. Most of this volume is devoted to a critical evaluation of the patterns of bird communities. In the companion volume, I consider how these patterns have been interpreted in terms of causal processes, how the operation of those processes has been determined, and how the patterns and our efforts to discern and understand them are influenced by the complexity and variability of natural environments.

It is clear that we know rather less about bird communities than we thought we knew. Some patterns that we have accepted as real are equivocal, while the generality of others may be quite limited. The ercsion of our certainty about these patterns, however, has its rewards: we can now see more clearly what needs to be done in order to pose and answer questions (or test hypotheses) in community ecology with greater rigor, so that when answers are obtained we may have greater confidence in their being correct. Community ecology is a difficult science, but it is not impossible. I hope that the views I develop will not discourage readers but will inspire them to design and conduct appropriate field studies and to generate useful theory for dealing with the complexity that, after all, *is* nature.

I should say what this book is not. It is not a review or synthesis of all the literature dealing with bird communities. Instead, I have used selected examples that, for one reason or another, seemed to me appropriate. It is not a book about the biogeography of bird communities. Community ecology grades gradually into biogeography, but biogeography in the broad sense, although certainly relevant to community ecology, is somewhat removed from the more local scale at which I believe investigations of community ecology are most fruitful. Most importantly, it is not a book about community theory. There are lots of concepts presented, but very little formal theory *per se*. There are few equations in the book. Others (e.g. Roughgarden 1979, Pielou 1974, 1977, Vandermeer 1981, May 1981) do that well, and I am not a theorist (although I could be called a conceptualist).

I think that the time is appropriate for a clear, critical examination of our observations of nature, to determine where both our field studies and our development of theory should be directed. I hope that this book and its companion volume will provide the stimulus and guidance for these developments.

Acknowledgments

I contemplated writing a book on bird communities for many years, but the opportunity and the impetus to stop contemplating and start writing came during a sabbatical leave from the University of New Mexico that I spent as a Fulbright Senior Scholar in Australia in 1984–85. Peter Sale, Tony Underwood, Rich Bradley, Charles Birch, and other members of the School of Biological Sciences at the University of Sydney provided support, good discussions, and southern hemispheric insights that broadened and sharpened my thinking about communities. At the University of New Mexico, Don Duszynski and Mary Alice Root provided assistance and understanding as I continued to write, and faculty, students, and administators at Colorado State University tolerated my absent-minded preoccupation as I completed the project. My graduate students, Gary, Luke, Hanni, Rachel, and Tasha, provided a refreshing diversity of viewpoints about communities for me to think about..

Several of my colleagues found themselves volunteering to read chapters in varying stages of development: Ian Abbott (Chapters 4, 5, and 8), Rauno Alatalo (11), Dave Bradley (3), Rich Bradley (2, 6), Joel Cracraft (2), Jared Diamond (4), Luke George (11), Nick Gotelli (4, 5, 6), Peter Grant (7), Yrjö Haila (2), Fran James (7), Kerry Kilburn (7), Terry Root (4, 5, 6, 8), John Rotenberry (10), Tony Underwood (2, 3), Bea Van Horne (1–6), and Jerry Verner (3). Their comments were unfailingly helpful and deeply appreciated, and of course I followed all of their suggestions to the letter. Jean Ferner read the entire manuscript in its initial draft, kindly pointing out my lapses in style and my affronts to the English language. John Birks did the same for the penultimate draft; thanks to Jean, his job was relatively easy. I owe a special debt of gratitude to each. Sylvia Sullivan handled the subediting at Cambridge University Press, and was extremely perceptive in finding small inconsistencies that no one else had noticed and quite gracious in pointing them out.

Authors usually thank members of their family for putting up with them during the writing of a book. I never understood what they were talking about. It seems, however, that there have been moments during the writing of this book when my behavior was not entirely normal, when I seemed detached from what was going on about me, or when I even was not a joy to be around. Thanks, Bea, for your understanding and support.

PART I

The foundation of avian community ecology

A scientific discipline operates on a foundation composed of three factors: its history, its logical and theoretical structure, and its methodology. These factors indicate what is known in the discipline and how that knowledge was gained, how practitioners of the science think, and what procedures are available and acceptable. They determine what sorts of questions are interesting or appropriate, how the questions should be framed, whether or not answers can be obtained, and how much confidence should be placed in those answers. The three factors are interrelated: history exerts a powerful influence on the theory and thinking that are current in a discipline at any time, and these in turn often dictate what sort of research is likely to produce findings that will contribute to further progress. The research traditions that develop in this way both determine and are constrained by the methods that may be used to gather observations or conduct experiments. Developments in methodology and thinking and the accumulation of facts comprise the history on which subsequent activities in the discipline build.

In order to understand and evaluate how science is done in avian community ecology and what is known or thought about bird communities, we must consider these elements of its foundation. In the three chapters that follow I develop this background.

1

The development of avian community ecology

Community ecology is concerned with identifying the patterns that characterize natural assemblages of species, understanding what has caused these patterns, and determining how general they are. Birds have been especially popular with ecologists investigating communities, perhaps because birds are diurnal and often conspicuous, their behavior can be documented with relative ease, and their distribution, natural history, and systematics are generally well known. As a consequence, a tremendous amount of information has been published on bird communities, and studies of such communities have contributed in major ways to the conceptual and theoretical framework of community ecology as a whole (Wiens 1983, Karr 1983). We know a lot about bird communities, or at least we think we do.

One of the themes of this book, however, is that we may not know as much as we think about bird communities. Knowledge and thinking in any science should be subjected to periodic re-examination. In this chapter I begin such an evaluation by considering how communities are defined, how recent views of communities developed, and how controversy has emerged about these views.

What is community ecology about?

The best way to determine the scope of operation of a discipline is to ask how it is formally defined. Dictionaries or introductory ecology textbooks provide an array of such definitions, and they cover a broad range of meanings. Whittaker (1975), for example, considered a community to be 'an assemblage of populations of plants, animals, bacteria, and fungi that live in an environment and interact with one another, forming together a distinctive living system with its own composition, structure, environmental relations, development, and function'. At the other extreme, MacArthur (1971: 190) advocated a more relaxed definition: 'any set of organisms currently living near each other and about which it is interesting to talk'.

If one surveys the spectrum of community definitions, several features consistently emerge (Table 1.1). The co-occurrence of individuals of several species in time and space is central to all definitions. Most also stress the importance of interactions (indeed, interdependencies) among these populations. Communities are often viewed as possessing a dynamic stability that tends toward an equilibrium composition. Moreover, this composition may be repeatable or constant under similar environmental conditions, allowing the recognition of community 'types'. Some authors stress the emergent properties of communities, attributes of structure (e.g. species diversity) or functioning (e.g. energy flows) that are found only at this level of organization.

To some extent, the different definitions of communities are consequences of different objectives. Plant ecologists, dealing with spatially fixed assemblages, often emphasize the description of such associations and their changes over successional time; animal ecologists, confronted with mobile, behaviorally active organisms, often stress the importance of interactions and functional relationships among species. Some workers define communities in terms of habitat units (e.g. rocky intertidal communities; Menge 1976), others by life-form categories (e.g. tree or herbaceous understory communities), others by taxonomy (e.g. bird or lizard communities). Any of these approaches can be rationalized by persuasive arguments. MacArthur, for example, justified his restriction of attention to co-occurring birds rather than, say, birds and butterflies by emphasizing the importance of finding general patterns and noting that 'there are many of these if we confine our attention to birds or butterflies, but no one has ever claimed to find a diversity pattern in which birds plus butterflies made more sense than either one alone' (1972: 176).

MacArthur's emphasis was on the operational value of a community definition rather than its abstract soundness. The difficulty of dealing operationally with the apparently enormous and complex web of interactions among populations, however, led Andrewartha and Birch (1954, 1984) to conclude that community ecology 'is not likely to give rise to a satisfactory, or even to any, general theory about the "distribution and numbers of animals in nature"' (1954: 4). It is nonetheless clear that the notion of the community must have some operational value, for it has provided a framework for a tremendous number of studies. As Schoener (1986a) has observed, community ecology is one of the most alluring (but also among the most tumultuous) of ecology's subdisciplines. For the remainder of this book I will accept the operational utility of talking about bird communities as assemblages of individuals of several species that occur together.

Table 1.1. *Common elements of definitions of ecological communities, and some problem areas related to them*

Element	Problem
1. Contain several species	Degree of inclusiveness
2. Species co-occur in time and space	Scale of study
3. Populations interact	Nature of interactions; non-interactive sets of species
4. Assemblage is stable; properties self-regulating	Definition of equilibrium; nonequilibrium conditions
5. Composition repeatable in similar environments	Determination of boundaries
6. Possesses emergent properties in structure and functioning	Are emergent properties statistical consequences of individual components?

The historical roots of contemporary avian community ecology

Initial development

In the early part of this century, previous thinking about ecological communities crystalized into two sharply differing views, which, with different dressings, continue to dominate our thinking today (McIntosh 1985, Underwood 1986a). On the one hand, Clements (1916) argued that communities were discrete, repeatable assemblages of species that were so closely integrated in their functioning that the community as a whole possessed properties paralleling those of individual organisms. At the other extreme, Gleason (1917, 1926) held that communities lacked internal organization or integration, being no more than fortuitous coincidences of the ranges of species that responded to environmental conditions quite independently of one another. The opposing views were strongly held and sharply argued, but the Clementsian view took hold and dominated community ecology for several decades (McIntosh 1967, 1970, 1975, 1982; Richardson 1980, Worster 1977). A major ecology textbook of the period (Allee *et al.* 1949) aggressively promoted Clements' view while scarcely mentioning Gleason.

For various reasons, the conceptual edifice of Clementsian communities began to crumble during the later 1940s and early 1950s (McIntosh 1975, 1985). Disillusioned with the rigid inflexibility of community classification, the strained analogies with 'supraorganisms', and the exaggerated emphasis on climate as the determinant of community patterns, many ecologists studying plant communities began to consider Gleasonian alternative. This trend was bolstered by the development of mathematical techniques such as ordination and gradient analysis, which often demonstrated quantitatively the lack of interdependence in species' distributions (Whittaker 1953, Curtis 1959). Some elements of the Clementsian view, however, were transformed into principles of ecosystems (Odum 1969) and continue to play a dominant role in research in that discipline (Simberloff 1982).

How did all of this affect studies of bird communities? Not very much, really. Rather few such studies were conducted prior to the 1950s. Some studies (e.g. Kendeigh 1934, 1945, Twomey 1945, Fawver 1947) were conducted within the general Clementsian framework, but most often the associations of various bird species with life zones, vegetation formations, or biotic communities (biomes) were simply described (e.g. Palmgren 1930, Lack 1933, Pitelka 1941, Odum 1950). Birds were not considered to form communities in their own right but rather were superimposed on communities that were climatically or vegetationally defined. The studies of Bond (1957) and Beals (1960) of assemblages of breeding birds in Wisconsin woodlands, however, were conducted in the context of a Gleasonian view, and both clearly documented that the bird species were distributed on gradients of forest vegetational composition largely independently of one another.

The transformation of avian community ecology

During the late 1950s and early 1960s, however, animal community ecology underwent a transformation and revitalization, largely as a result of the work on birds by David Lack and, especially, Robert MacArthur. Lack's views on bird communities took form during the early 1940s and were expressed most clearly in his studies of shags and cormorants (1945b), raptors (1946), and Galápagos finches (1945a, 1947). He emphasized the importance of interspecific competition in structuring communities through its effects on the evolutionary development of ecological and morphological differences among co-existing species. The hypotheses that he derived from his observations were intuitive and qualitative rather than mathematical, and were not designed to be especially testable (Ratcliffe and Boag 1983). Lack apparently was influenced little by either the Clementsian or Gleasonian view, but instead drew his inspiration from the theory of competition that emerged from Gause's (1934) laboratory experiments and Mayr's (1942) views on speciation (Ratcliffe and Boag 1983, Kingsland 1985). Lack turned his attention to the mechanisms of population regulation in the late 1940s and did not return to considerations of community ecology until two decades later (Lack 1971, 1976).

The proximate impetus for the transformation of avian community ecology came from the work of MacArthur and his colleagues. Drawing on themes developed by Hutchinson (1951, 1957) and influenced by Lack's studies, MacArthur looked at bird communities in new and different ways. MacArthur visualized community patterns in neat and simple theories, usually accompanied by mathematical formulations. As had Lack, he derived his concepts more often from population ecology than from Clementsian or Gleasonian community ecology. He shared with Clementsians, however, a firm belief that communities were organized, integrated assemblages, not simply independent collections of species that happened to co-occur because of ecological similarities. Both MacArthur and Lack, then, emphasized the importance of interactions and interdependents among species in producing structured community patterns.

The effect of this work on community ecology was profound. From a somewhat biased perspective, Cody and Diamond (1975: vii) observed that

> when this era began in the 1950's, ecology was still mainly a descriptive science. It consisted of qualitative, situation-bound statements that had low predictive value, plus empirical facts and numbers that often seemed to defy generalization. Within two decades new paradigms had transformed large areas of ecology into a structured, predictive science that combined powerful quantitive theories with the recognition of widespread patterns in nature.

Kingsland (1985: 176) suggested that, together with Hutchinson, MacArthur brought mathematical methods that 'had been lingering on the fringes of ecology squarely into the mainstream of the science'.

Characteristics of the MacArthur approach

What was so powerful about the MacArthur approach to community studies? To some degree, it became popular because it filled a void. Animal community ecology in the 1950s was stagnating, and the MacArthur approach posed a large number of new and interesting questions and provided a model for answering them. MacArthur's studies were a special impetus to investigations of bird communities. With the benefit of hindsight, we can discern from MacArthur's own writings (e.g. MacArthur and Connell 1966, MacArthur 1971, 1972) and those of his students and colleagues (e.g. Cody 1974a, Cody and Diamond 1975, Fretwell 1975, Diamond 1978, Pianka 1981, Hutchinson 1978; see also Ricklefs 1975, Wiens 1976a, 1977a) some of the basic features of this approach (modified from Wiens 1983):

1. There was an emphasis on the importance of detecting and explaining patterns in nature. 'To do science,' MacArthur wrote, 'is to search for repeated patterns, not simply to accumulate facts' (1972: 1). Observations that failed to form clear and repeated patterns were not interesting.
2. The most interesting and important patterns and theories were those that possessed *generality*: 'science should be general in its principles' (1972: 1). General patterns were most likely to have a general explanation, whereas little progress could be expected if scientists became excessively preoccupied with the details of patterns that characterized only specific situations and that, therefore, were likely to have a specific explanation of limited generality. As MacArthur (1968: 159) noted, 'these kinds of general events are only seen by ecologists with rather blurred vision. The very sharp-sighted always find discrepancies and are able to say that there is no generality, only a spectrum of special cases.'
3. The patterns of communities were usually held to be consequences of *deterministic* processes. Although chance did enter into some of MacArthur's formulations (e.g. island colonization, MacArthur and Wilson 1967; the occupancy of patchy habitats, Horn and MacArthur 1972), the primary emphasis was upon a strictly deterministic relationship between cause and effect.
4. Most community patterns were a result of *competition* between species in the face of limited resources. Such interactions lead to the exclusion of marginally adapted species from communities, leaving the assemblages optimally structured by the ecological spacing among their members. As Cody and Diamond (1975: 5) put it, 'it is natural selection, operating through competition, that makes the strategic decisions on how sets of species allocate their time and energy; the outcome of this process is the segregation of species along resource–utilization axes.'
5. Communities, and therefore the patterns they expressed, were presumed to be at or close to an *equilibrium* determined by resource limitations. Environments might vary in time or space, but population changes, driven by competition, would rapidly adjust the structure of the community to the changed conditions, preserving the general patterns. A corollary of this assumption was that suitable habitat was fully occupied by individuals and that the community as a whole was 'saturated', containing the optimal number and kinds of species for the resource conditions.

Kingsland (1985) suggested that, in deliberately selecting problems so as to minimize the complicating effects of past history, MacArthur emphasized populations and communities in apparent equilibrium.

6. *Optimality thinking* was prevalent. The patterns of populations and communities were considered to be products of natural selection, acting primarily through competition, and as a consequence were optimal. Thus, Cody (1974b: 1156) observed that 'all important processes and patterns in biology are products of natural selection', which is 'a mechanism that maximizes fitness . . .; it leaves only the best-adapted or optimal phenotypes for our inspection'.
7. Investigations were guided by *theory*, which provided predictions about the patterns of communities that could be tested with observations. Theory thus determined which measures or observations were appropriate and pointed field or laboratory work in specific directions. MacArthur (1972) observed that it is not only important to distrust theory until it is confirmed by facts but also that one should not put too much faith in facts until they have been confirmed by theory, and May (1976) lamented that the wide variation among observations 'tended to discourage theoretical activity'.
8. The most appropriate and efficient way to discern patterns and test theories was through *comparisons* of different situations in nature. Such 'natural experiments' capitalized on the variation that occurs naturally in the environment: differences or similarities between islands and mainland, between different elevational zones on a mountainside, between habitats at different latitudes, and so on. Active experimental manipulation of species or their resources was considered unnecessary and possibly confusing (Diamond 1978, 1983, Cody 1974b).
9. Often, theories were tested or patterns explained by the use of *selected examples*. Observations that agreed with theoretical expectations especially well were used to confirm the theory, while contradictory or less confirmatory observations received little attention or were simply not mentioned. Attempts to define patterns or test the match of observations to predictions through rigorous statistical procedures were infrequent.
10. The *methodology* of documenting patterns or testing the predictions of theory generally received little attention and often lacked

rigor. Although much of the theory was expressed in mathematical terms, field data were frequently gathered following no particular design or sampling protocol, and resource conditions were often simply inferred rather than measured. Moreover, observation was usually restricted to a few variables of interest, those that were 'already identified with environmental variables for which competition takes place, as indicated by resultant displacement patterns' (Cody 1974a). MacArthur (1971: 190) expressed a similar view when he wrote that 'there is no such thing as a "correct" or "incorrect" measure of diversity or stability or anything else. The virtue of a measure resides solely in the neatness of the relations constructed from it.'

Perhaps the most evident manifestation of the importance of the transformation in ecology precipitated by MacArthur's work is in the questions that community ecologists began to ask in the 1960s. When Beals (1960) analyzed the bird communities of the Apostle Islands in Lake Superior, for example, he focused on ordinating the species and only noted in passing that the number of species present decreased as island size became smaller. A few years later, the species–area relationships among these islands and their relations to colonization and species–turnover patterns would more likely have been the targets of such an investigation. Community ecologists increasingly asked questions such as: What factors determine the limits to the similarity of coexisting species? How does the diversity of bird communities relate to the structural diversity of their habitats? What happens when a species competes with several other species? What determines the number of species in a habitat? On an island? In a continent? Are there regularities in the size distributions of species in communities? Do communities in similar environments converge in structure? Are there 'rules' that describe the process of community assembly? Such questions were new and interesting and generated considerable enthusiasm. They formed the framework for new kinds of community studies, and the characteristics of the 'MacArthur approach' noted above prescribed a protocol for this research.

The MacArthur approach as a paradigm

Elsewhere (Wiens 1983) I have argued that this approach displayed many of the attributes of a Kuhnian paradigm (Kuhn 1970a). It presented a coherent, unified tradition of scientific research that defined which problems were interesting and merited study and which were not. It formed the intellectual and conceptual foundation for the education and training of students in the discipline. And it discouraged alternative explanations or views of communities (e.g. Van Valen and Pitelka 1974, Haila 1986). There is little question that the centerpiece of this research tradition, the assertion that competition is the process that produces community patterns, was widely accepted. Ricklefs (1975), for example, stated that 'few ecologists doubt that competition is a potent ecological force or that it has guided the evolution of species relationships within communities,' and Diamond (1979) observed that 'now, few ornithologists question the widespread role of competition'. Giller (1984) reiterated this assessment. Schoener (1983a), however, has argued that ecology has probably never had any true paradigms in the Kuhnian sense, and there is considerable dispute among philosophers of science about the value of Kuhn's thesis (e.g. Lakatos and Musgrave 1970, Gutting 1980). Kuhn himself has had some difficulty in defining exactly what a paradigm is (Kuhn 1970b).

Nonetheless, considering community ecology within Kuhn's framework of scientific progress may be instructive. In particular, Kuhn envisages a phase of 'normal science' that follows the widespread adoption of a paradigm or research tradition in a discipline. In periods of normal science, most research activity is directed toward a restricted set of problems or questions that, according to the paradigm, are of greatest importance. Other questions are ignored or actively suppressed. Using the methods and approaches of the paradigm, scientists make rapid progress in answering their questions, to some degree because only confirmatory evidence is considered. The discipline seems vigorous, and there is widespread agreement among scientists. These traits characterized a good deal of community ecology during the period from 1960 to 1975. The questions of interest centered about species packing in communities, resource partitioning, diversity, geographical variations in community structure, and patterns in island biogeography, and there was rapid progress in the development and enrichment of community theory and in the generation of evidence supporting the theories (e.g. Cody and Diamond 1975; see Haila 1986). Our apparent understanding of communities grew considerably.

Although the MacArthur approach came to dominate investigations of animal (especially vertebrate) communities in North America, its power was not felt everywhere with equal strength. Jackson (1981) suggested that the basic elements of the body of 'niche theory' developed during the 1960s were already well understood by many plant ecologists early in the century. Be that as it may, many contemporary plant ecologists simply ignored MacArthur's work, although as interest in plant population biology developed, MacArthur's contributions to population theory were rapidly embraced. Insect ecologists likewise did not rush to adopt MacArthur's teachings (but see Price 1975), perhaps because their efforts focused so strongly on population dynamics (e.g. Clark *et al.* 1967). It took some time before MacArthur's thinking began to influence animal community studies conducted in Europe or Scandinavia (e.g. Herrera 1977, 1978a, Ulfstrand 1976, Järvinen and Väisänen 1976, Järvinen 1978), where a tradition of quantitative analyses of species interactions, species–area relationships, and environmental heterogeneity had developed much earlier (A. Palmgren 1915–17, 1922; P. Palmgren 1930, Merikallio 1951; see Ranta and Järvinen 1987). Many British animal ecologists, who had never shown much affection for community studies, continued to focus on intensive studies of populations. In Australia, ecology developed as a discipline later than in North America or Europe, and much of this development was fostered by applied problems (L.C. Birch, personal communication). The theory that was the rage in North America was thus of little interest to many Australian ecologists, who were deeply immersed in detailed studies of population problems.

The emergence of controversy

Part of Kuhn's scenario of the history of science is that, during periods of normal science, observations will appear that conflict with the views and predictions of the prevailing paradigm. Initially, such anomalies may be ignored, but, as they accumulate, some workers may begin to question the veracity of the paradigm. Controversy appears. Eventually, dissatisfaction with the prevailing paradigm may become widespread – it no longer provides a satisfactory research tradition for the discipline, and it may be rejected in favor of a new or developing paradigm that does a better job of explaining the anomalous observations and pointing toward productive research. A 'scientific revolution' has occurred.

Beginning in the mid-1970s, some ecologists began to express doubts about the view of communities espoused by the 'MacArthur school.' By the early 1980s, controversy was widespread, fueled not only by an increasing volume of observations that did not fit the predictions of current theory but also by concerns about the logical procedures and methods followed (Lewin 1983a, b). The elements of this controversy will be considered in some detail throughout this book, but they should be noted briefly now.

The chief debate was over the centerpiece of the MacArthur/Lack view, the role of competition in determining community patterns. Some workers (e.g. Connell 1975, 1980, Wiens 1977b, 1983, Strong *et al.* 1979), failing to find much evidence of competition, suggested that its role had been vastly overplayed, while others (e.g. Diamond 1978, 1979, Schoener 1982, 1983b,

Giller 1984) staunchly defended its importance. But other features of the MacArthur approach were involved in the controversy as well:

1. Are the well-ordered community patterns really what they seem to be? No one disputed the importance of searching for patterns – to do otherwise would make science a dull and sterile activity – but there was widespread disagreement about the reality of particular patterns. Thus, Simberloff and his colleagues (Simberloff 1978, Simberloff and Boecklen 1981, Connor and Simberloff 1978) contended that some of the 'patterns' that others had found could not be distinguished from those generated by random processes. Other patterns, such as some species–abundance distributions, seemed to be no more than statistical artifacts (May 1975).
2. Are communities really in equilibrium? Price (1980), Connell (1975), and I (1974a, 1977b, 1984a) suggested that often they were not, at least in the sense demanded by theory.
3. Are the patterns of communities general, and can we therefore expect realistic theory to have much generality? Increasingly detailed studies of communities indicated that patterns often had multiple causes and varied in ways not accounted for by theory, casting doubt on the value of general models (Underwood and Denley 1984, Quinn and Dunham 1983).
4. Are communities structured by entirely deterministic processes? Some of the challenges to prevailing community notions emphasized the importance of stochastic effects. Chance, far from being only meaningless 'noise', might play a major role in determining community patterns.
5. What is the role of theory in ecology? A feeling developed in some quarters that theory had so dominated community ecology that it had blinded us to observations. Optimality thinking and adaptationist arguments were challenged on other fronts (Maynard Smith 1978, Gould and Lewontin 1979). Roughgarden (1983), however, suggested that theory was largely responsible for progress in science and tenaciously defended its importance to ecology.
6. How does one document the patterns of communities and determine what processes cause them? The spatial scale on which communities were studied, for example, was usually determined arbitrarily, largely for logistic reasons, but studies at different scales often yielded quite different patterns (Wiens 1981a, Wiens *et al.* 1987). Assertions that resources were limiting or that a given

process had produced an observed pattern were challenged as being illogical or unjustified. The basic tenets of the comparative methods were also subjected to critical scrutiny (Wiens 1984a, Connell 1980, Clutton–Brock and Harvey 1979).

These areas of controversy go deeper than simple disagreements about the views or procedures of workers following the MacArthur tradition, however. They strike at the heart of how one defines communities by identifying problems associated with each of the elements of a conventional community definition (Table 1.1.). Thus, skepticism about equilibrium affects the notion that a community can be defined as stable or self-regulating, while difficulties associated with the scale of investigation may affect whether a given set of species does or does not co-occur and whether that combination of species is repeatable. If competition is not the primary process operating in communities, do other sorts of population interactions occur, or may community members co-occur but not interact at all? If some of the patterns of communities are nothing more than artifacts of statistics or of the scale on which they are studied, they can scarcely represent 'emergent properties' of the community. And how inclusive should a community be? Should it be restricted only to those species that do interact, or should it include a wide diversity of co-occurring organisms – birds, butterflies, spiders, rabbits, slugs – with little potential for direct interaction?

Controversy has turned community ecology into a much more complex and difficult discipline than it was during the halcyon days of 1960–75. This controversy has brought new excitement and challenge to the discipline, however – far from being disruptive (although it is to some), it can stimulate new thinking, rid the discipline of rigid and dogmatic thinking, and point toward new directions for research.

General themes of this book

The elements of this controversy penetrate all aspects of contemporary community ecology, and they thus play a dominant role in my treatment of community patterns. There are several general themes that are especially important, which I note briefly here.

Community composition and boundaries

How does one determine the boundaries of a community, both in terms of the species that are considered and the temporal and spatial domain? 'Bird communities' are delimited by taxonomic boundaries, which impose a certainly artificiality on them. In practice, it is rare that *all* of the birds in an area are considered in a community study; as a matter of convenience or necessity, attention is usually restricted to some subset of birds (e.g. warblers, small passerines, raptors, nectarivores). The time and space boundaries of community studies are also usually arbitrary, and this may affect the degree to which the community is open or closed to influences from elsewhere.

Variation and communities

Environments vary in both time and space, and the effects of these variations on observations and theories of communities have not always been appreciated. Temporal variations in resource levels, for example, may affect the degree to which a community matches the equilibrium assumptions of theory. The equilibrium status of a community affects not only how one interprets community patterns but how one conducts research, as different theories and research designs may be needed depending on this status. The configuration of habitat patches in a landscape or of resource patches in a habitat may have important effects on both community patterns and processes because of what transpires across boundaries between patches (Wiens 1986, Wiens *et al.* 1985).

Individuals and species also vary. Individuals within different local populations may differ in their patterns of habitat selection depending on local densities, the availability of habitat types in an area, geographic variation in the ecological responses of the species, or the complex of other species that is present. A species is not an ecological constant, and determining its community role or position by averaging over samples that are geographically widespread may obscure this important variation. Individuals and populations may also vary in the degree to which they respond to environmental changes or in the time lag of their responses. These differences, in turn, influence the probability that the community will approach an equilibrium with its proximate environment.

Scale and its consequences

Our ability to detect community patterns and to determine what may have caused them is strongly affected by the scale of space or time at which they are viewed. Patterns that emerge at the scale of local communities may vanish at a broader scale, where different patterns suddenly appear. Which are likely to be real or important? The problems of determining the proper scale on which to study a community are exacerbated by differences between species in longevities and in lifetime home ranges. Any local assemblage therefore includes species that respond to environmental variations in time and space on quite different scales.

Resources

Much of our theorizing about communities requires assumptions about the nature, availability, or quality of resources such as food, shelter, habitat, or nest sites. 'Resources' are by no means easy to define (contrast Wiens 1984b with Andrewartha and Birch 1984), much less to measure. As a consequence, discussions about resources in the community literature have often been fuzzy and inferential. If the operation of a mechanism such as competition is dependent on limitation in resource supplies, that limitation should be documented before one uses competition as an explanation of community patterns with any degree of certainty. If one wishes to use measures like population density as an index of habitat quality, some independent measure of habitat quality and a determination of its relation to population density are first necessary.

How should community ecology be done?

These themes all relate to how we gain knowledge in community ecology, to how the science is structured. If it is true that many community investigations have suffered from inadequate attention to methodology or logical procedures (Wiens 1983), then a high priority should be placed upon rectifying this neglect. Observations and the patterns that emerge from them are sensitive to the procedures used to gather them. Experiments should be an integral part of research designed to unveil the causes underlying patterns, but experiments that are not carefully designed may be worse than useless in that they create a false sense of confidence in the results simply because an 'experiment' was done. Research on communities may be more productive if it tests alternative hypotheses, but those hypotheses must be framed in a way that makes their predictions separable and truly testable. Full consideration should be given to the possible influences of chance effects on community patterns. Such stochastic factors may not mesh with our preconceptions of well-ordered Nature, but they are nonetheless real and may be quite important. Finally, although we wish to retain a focus on communities as ecological units of interest, we must realize that the patterns we see at that level are consequences of how individual organisms relate to their environment and to one another. The details of their behaviour and physiology may thus be important and should be considered.

My approach to developing a perspective on avian community ecology as a science relies heavily on the development of specific examples. Although I often focus on flaws or inadequacies in these studies, this should not be taken as condemnation of the work or of the scientists who conducted it. One must always remember that studies should be evaluated in their proper historical context. A study conducted in the early 1960s would (I hope) not be done the same way today – we have new techniques available, better insights about communities, and a keener sense of how science should be done. A study conducted by, say, W.A. Mozart in 1965 might, with the benefit of hindsight, provide good examples of various logical, methodological, or interpretational flaws. This does not necessarily mean that the study was ill-conceived at the time it was done, that Mozart was a naive or incompetent scientist, or that the study did not contribute to scientific progress. Neither does it mean that Mozart's more recent work is tainted by the flaws of the earlier study. Science and scientists grow and mature. Probably few of the people whose work figures prominently in this book would unhesitatingly embrace statements they made 20 or 10 or perhaps even 5 years ago. To hold a scientist forever responsible for inadequacies in earlier work is to ignore the context in which the work was done and deny the capacity to learn and mature.

2

Determining pattern and process: the logical structure of community ecology

Communities are interesting to ecologists, as MacArthur said, because of their patterns. Once detected, a pattern is something that to most ecologists cries out for an explanation – what process or processes caused the pattern? To document the existence of a particular pattern and determine the associated processes, however, is not so simple as many ecologists have thought. Patterns are derived from observations, which are our closest approximation of 'facts'. If the observations are made using flawed procedures, these 'facts' are suspect. If the derivation of patterns from the observations or the development of process explanations do not follow proper logical pathways, neither the patterns nor their interpretations can be trusted. Unless we are certain that the patterns we detect are accurate representations of nature, there is really nothing to be explained. Elaborating process explanations of such patterns simply compounds the uncertainty. Detecting community patterns and understanding their relationships to processes therefore requires a clear understanding of the logical structure of scientific investigation, which is my focus in this chapter. It also requires attentiveness to the operational structure of gathering and analyzing observations, which I will consider in the following chapter.

Pattern and process

In ecology, a *pattern* is a statement about relationships among several observations of nature. It thus connotes a particular configuration of properties of the system under examination. Neither observations nor patterns, however, are free of biases. Sale (1984) has noted that we screen or filter observations for patterns at a most elemental level, as our sensory abilities have evolved for detecting patterns. As a consequence, what we see or hear is a simplified, more ordered approximation of reality. As Grant (1977: 298) has observed, 'pattern, like beauty, is to some extent in the eye of the beholder'.

Superimposed in this sensory filtering of observations is an intellectual filtering that also influences pattern detection. What we perceive is determined by our mental preconceptions about what there is, and these preconceptions are closely bound to the language we use to express and organize our perceptions (Haila 1986). The specialized jargon used in a scientific discipline simply enhances the way language categorizes and constrains our preconceptions of nature. The patterns we see are strongly influenced by our culture.

Culture also affects pattern detection through its worldview, its preconceptions about how nature is. Western science is conducted within the context of a worldview that seeks *order* in nature (Wiens 1984a, Simberloff 1982, Dayton and Oliver 1980, Greene 1981). In accordance with this preconception, we find most interesting those patterns that express some regularity, and 'pattern' is often equated with the expression of an aspect of the apparent orderliness of life (e.g. Eldredge and Cracraft 1980). Because they are repeatable, ordered relationships also permit predictions to be made, and there is thus a pragmatic emphasis on such patterns as well.

Relationships among observations need not express regularity, however, in order to represent patterns. A scatter of points on a graph is a pattern, even though it may display no statistically evident regularity or order and may offer no predictions. The statement that there is no statistically significant pattern among such observations is not a statement of no relationship among the points but only a conclusion that the relationship is not *ordered* in some specific fashion. Admittedly, regular, repeatable relationships are more interesting than less regular patterns, and they should probably remain the primary focus of our investigations. To emphasize only those patterns that display apparent order and neglect those that do not, however, may foster unwarranted confidence in the orderliness of nature and leave many important patterns unexplored. Our challenge is to detect patterns in observations and to understand them, whether or not regularity in the patterns is immediately apparent.

A pattern, once established, represents something to be explained. Explanation rests on an understanding of *processes*, of what causes the pattern to be as it is. A process is thus synonymous with a cause, the operation of some factor or factors that produce a particular relationship among observations. A process is not simply something happening. If a community increases in diversity through time, for example, the diversity change or species enrichment is not a process but rather a change in pattern. Processes such as immigration or extinction are what brought about the diversity increase. Pattern and process are often confused. Thus, Thomson (1980:

720) considered numerical responses and niche shifts to be responses to competition, but he referred to them as processes that produced community patterns: 'both processes have equivalent end results, i.e. both affect community structure, species diversity, etc.' Numerical responses and niche shifts are population-level patterns that may contribute to community-level patterns, but the process causing them, in Thomson's view, is competition. The distinction between a pattern and a process is not always clear-cut, as all patterns contain the effects of processes. Moreover, what we call a process at one level (e.g. extinction) may be recognized as a pattern at some other level (e.g. the change in abundance from >1 to 0). Despite the conceptual difficulty of dichotomizing patterns and process, it is important to distinguish between them operationally, for a failure to recognize pattern detection and process explanation as different phases of scientific activity has led to fuzzy thinking and incorrect conclusions, as we shall see.

There is, by definition, an inevitability in the sequence: cause (process) → effect (pattern). Because of this, there is a time lag between the operation of a process and the appearance of or change in a pattern. Current patterns are consequences of past or ongoing processes. In attempting to define cause–effect relationships, we usually assume that this time lag is sufficiently short that it can be ignored or that past processes operating in the same way that they do now (the assumption of uniformitarianism). If processes operate in an episodic fashion with considerable time lags, however, drawing links between process and pattern may be difficult. This difficulty is especially apparent in attempts to test evolutionary or adaptive arguments (e.g. Gould and Lewontin 1979, Brady 1982, Sober 1984), but it besets a good deal of community ecology as well (Connell 1980).

The philosophical foundation: hypothesis testing

How does one structure an investigation in order to detect pattern or determine process? Part of the task involves observing and measuring accurately, according to a proper design (see Chapter 3). Investigations must also have a proper logical structure. Insofar as the philosophy of science provides guidance in this matter, it deserves careful consideration. There exists a wide array of philosphical positions on how science should be or has been done, ranging from the rigid inductive framework of Logical Positivism (e.g. Carnap 1923; see Rosenberg 1985) or the strict falsification program of Popper (1959, 1962) through the more flexible positions of Lakatos (1978) and Laudan (1977), to the relaxed, 'anything goes' approach advocated by Feyerabend (1975). As a consequence, an ecologist can probably find a philosophical position to legitimatize whatever approach he takes. Roughgarden (1983: 583) suggests that science is rarely done following any 'formal rules'; rather, we develop knowledge by making a 'convincing case' for a position on the basis of experience and 'common sense'. What is convincing or common sense to one person may not be to another, of course. More importantly, common sense is what accords with preconceptions that are widely shared in a discipline, its 'conventional wisdom' (Strong 1983). Adopting an approach based on common sense increases the likelihood that only evidence that accords with this 'wisdom' will be considered.

It is not my intention here to review philosophies of science (see Hull 1974, Suppe 1977, Sarkar 1983, Rosenberg 1985) or to advocate any specific position as being 'best'. Nor do I wish to develop any tightly structured, formal philosophical framework that must be followed in order to practice 'good' science. Some degree of formalism in the logical operations of science nonetheless seems essential.

Most ecologists now purport to be testing hypotheses when they conduct their research, so I will emphasize the logical structure of hypothesis testing. This does not mean, however, that testing hypotheses is the only way to do science and gain knowledge. In order to develop enough basic understanding of an ecological system that tests are focused on realistic and sensible hypotheses, the system must generally be investigated in a less formalized fashion first. This may be through the formulation of preliminary models or 'working hypotheses', educated guesses or speculations that serve to guide research but are not directly testable (Thornhill and Alcock 1983). Exploratory data analysis, in which data are subjected to various forms of pattern analysis to see what emerges rather than to formal statistical hypothesis tests (Tukey 1977), may also be an important activity (James and McCulloch 1985). Such analysis may help to determine which of several working hypotheses merits formal testing. These less formalized approaches are *not* atheoretical – in fact, the observations are usually gathered with a particular theory in mind, even if explicit tests of the predictions of that theory may not be immediately forthcoming. To claim that research is not scientific unless it involves testing of hypotheses is absurd. Indeed, hypothesis testing in the absence of the necessary background provided by natural-history studies is likely to be a sterile and meaningless activity.

Deriving hypotheses and predictions

At some point, knowledge of a system is sufficient that provisional models or working hypotheses can be restated as formal hypotheses. Hypotheses are statements that take the general form 'if A, B, C, and D hold, then X follows'. They represent conjectures about nature, questions or propositions that, for a specific situation, have not yet been answered. The value of hypotheses lies in their capacity to provide clear answers to specific questions, but to do so the hypotheses must be testable and the tests must be fair. I have outlined the generalized logical procedures for testing hypotheses in Fig. 2.1; after discussing this framework, I will illustrate it with a specific example.

Hypotheses may develop by two pathways: inductively, from observations and patterns, or deductively, from theoretical constructs. In practice, this distinction is far from absolute, for there is a continual interplay between observations/patterns and theory in science. Although it is true that theory at times springs from bold, intuitive leaps, most often its development is prompted by observations of patterns or problems posed by them. On the other hand, theory often determines what sorts of observations are made or what kinds of patterns are sought (e.g. Haila and Järvinen 1982). Grant (1977) has suggested that, after Lack embraced the theory that competition produces ecological segregation among species, he forced observations to fit that framework.

In order to be testable, hypotheses must offer clear predictions. Often neither the general theory nor a more restricted hypothesis is very specific or unambiguous, but the predictions (which are what we test) must be precise and must logically follow from the hypothesis. There is more behind a prediction, however, than a logical connection with a hypothesis. In order to generate specific, restricted predictions, assumptions must be made; more often than not, these are left unstated. Many of these assumptions are contained in other theories and are usually accepted without challenge (Kuhn 1970a, Lakatos 1978, Dunbar 1982). Brown (1981), for example, noted how predictions of the general theory of competition in ecology are not only dependent upon the assumptions of that theory but also upon the premises of other background theories, such as the Lotka–Volterra model of population interactions. Changing these assumptions in biologically realistic ways may lead to totally different predictions (e.g. Abrams 1976, 1983). Moreover, any prediction contains a *ceteris paribus* (all other things being equal) clause, which effectively permits one to ignore a host of other variables or effects (Brady 1982). The more that such detail is contained in the predictions instead of sheltered in the background assumptions, however, the clearer tests of the predictions will be.

It is at this stage that one of several violations of logic may appear. Suppose that one has gathered a number of observations and from them detected a pattern. A hypothesis is developed to explain the pattern.

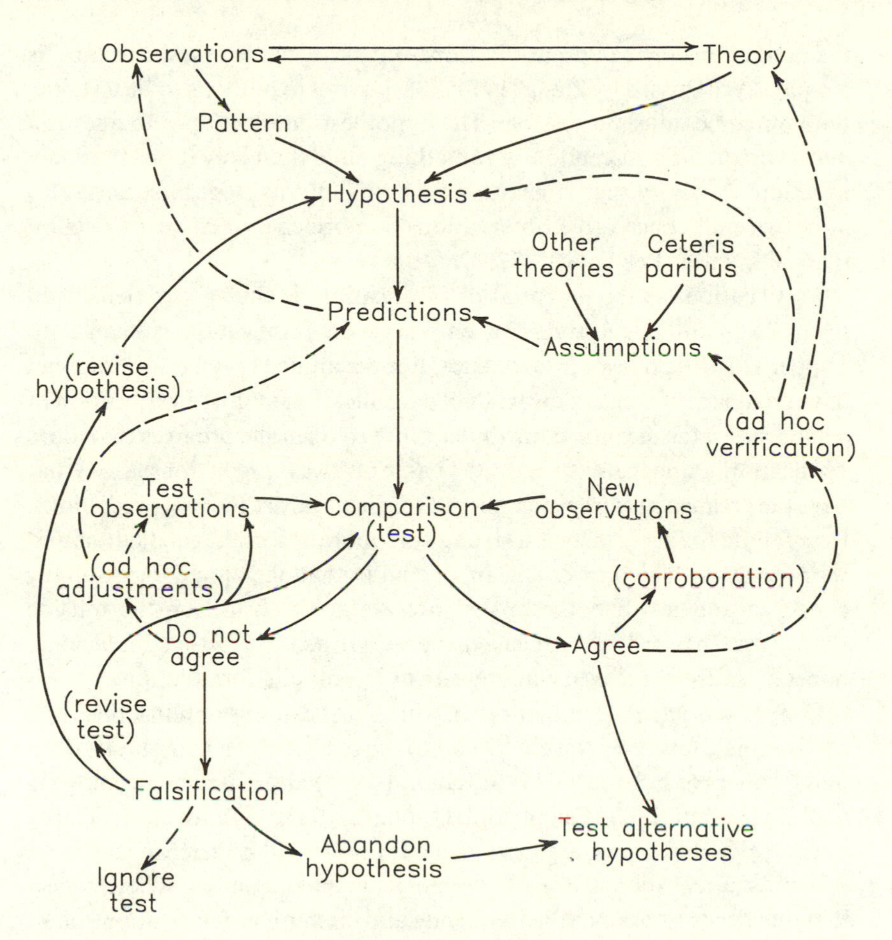

Fig. 2.1. Flow chart of the steps involved in hypothesis-testing. Solid lines show acceptable paths; dashed lines indicate logically flawed operations.

Predictions are generated and then compared with the original observations. They agree, and the original observations and pattern are then considered 'evidence' that 'proves' the explanation (hypothesis). Because the same observations were used to generate and to test the hypothesis, this agreement is not surprising. The 'test' depends on circular logic, however, and is totally invalid.

Testing and falsification

What, then, is a test of the predictions? Testing rests on a comparison of the predictions with some set of new observations. If the hypothesis is constructed properly, two results are possible: the observations agree with the predictions or they do not [a third outcome, neither 'yes' nor 'no', is eloquently discussed by Pirsig (1974: 288)]. For a hypothesis to be testable, both outcomes must be possible. The hypothesis must be open to question and its predictions susceptible of something other than only support or only rejection. A theory that is guaranteed by its internal logical structure to agree with all conceivable observations is more characteristic of religion than of science (Lewontin 1972, Dayton 1979).

Falsification, or the potential of falsification, is thus a key element in scientific hypothesis-testing. An emphasis on falsification is central to Popper's (1959) philosophy of science. Popper argued that, because one can have greater confidence in the truth of a falsification than of a verification of a hypothesis, falsification contributes more to scientific progress than does verification. Popper also stressed the value of 'risky' predictions: those that have the greatest *a priori* probability of falsification are the best predictions, for a failure to falsify them is a strong corroboration of the predictions and the hypothesis. This does not mean, of course, that one should intentionally construct and test hypotheses that are blatantly absurd on the outside chance that they will not be falsified and we will all be surprised. Intellectual honesty is also a necessary ingredient in hypothesis formulation.

Popper's original formulation of this falsificationist philosophy was rigid: a single falsifying test could lead to the rejection of a hypothesis. Both philosophers (e.g. Lakatos 1978, Laudan 1977) and practicing ecologists (e.g. Simberloff 1982, Quinn and Dunham 1983, Hilborn and Stearns 1982, Haila 1982) have argued (correctly) that rigid adherence to such a scheme is unrealistic and likely to detract from understanding. They suggest that one needs to balance the frequency and strength of falsifications of an hypothesis against the frequency and strength of prior corroboration of its predictions. Popper himself relaxed his initial position in later writings (1962, 1983). Some philosophers, however, abandoned Popper's falsificationist emphasis entirely (see Suppe 1977, Bronowski 1977), leading ecologists such as Diamond (1986a) to question the value of the approach in scientific investigations. Although it is true that a rigid adherence to a strict falsificationist philosophy can lead one to discard useful hypotheses on the basis of trivial exceptions, a more balanced approach to falsification provides a valuable logical structure for scientific investigations.

One outcome of a test, then, is falsification: the observations do not agree with those predicted. What can one conclude from this? The most obvious conclusion, of course, is that the predictions were incorrect and the hypothesis is therefore false. Several other explanations of such a result are possible, however (see Clutton-Brock and Harvey 1979, Brady 1982, Jarman 1982, Hilborn and Stearns 1982):

1. The observations are faulty. They may have been gathered in a biased manner, such that negative evidence was somehow favored and confirmatory evidence disregarded. The observations may have been incomplete or insufficient, perhaps because some key variables were not measured. This leads to an incorrect documentation of pattern and a failure of the comparison. A test based on such observations is of course unfair, but this may not become evident until the test is actually made (if then). Such problems with observations can apply either to the test observations or to the original observations on which the hypothesis was founded, or to both. If the test observations are faulty, the test must be revised and conducted again; if the problems are with the original observations, the hypothesis itself must be revised (Fig. 2.1).
2. The observations are accurate but the result is a statistical artifact. There is always a possibility that the results may not agree with the predictions as a consequence of chance and sampling variation. This is likely to lead to an unjustified rejection of the hypothesis (a Type I error), especially if there has been little replication in the test. There are other ways in which the behavior of samples or sets of numbers can produce apparent patterns that are mathematical or statistical artifacts; May (1975) has considered some of these in detail. In either instance, revisions of the test are necessary.
3. Patterns in the results may be affected by other variables or causes. Interacting variables that were not controlled for in the test comparison or inappropriate variables that were included in the observations may act as 'noise', obscuring the relationship being tested. The pattern examined by the hypothesis may have more than one cause and, if the hypothesis and its predictions focused on only one of these, the expected pattern might fail to appear. These problems stem from violations of underlying assumptions of the hypothesis. Revisions of both the hypothesis and the test are required.
4. The comparison itself is flawed. Perhaps the predictions and the test observations are both logically and empirically correct, but they are compared in an inappropriate manner (e.g. an incorrect application of multivarate statistics), such that the expected pattern simply cannot be demonstrated. Perhaps the predicted pattern is tested with a heterogeneous data set that contains observations

that are not comparable but are nonetheless compared (Andrewartha and Birch 1984). This calls for a revision of the test.

5. The assumptions may be violated. When a test of predictions is conducted, everything that is behind those predictions is part of the test, including assumptions derived from other theories and the *ceteris paribus* clause. If one tests a hypothesis that assumes ecological equilibrium by using observations from a system that is clearly nonequilibrial, for example, the failure to match the predicted patterns may reflect problems with the underlying assumptions and inappropriateness of the test rather than a failure of the hypothesis itself. Revision of the test is needed, but a hypothesis so heavily laden with assumptions that clear tests are difficult also merits reconsideration.

   The *ceteris paribus* assumption is often especially suspect, and this leads to another sort of logical transgression that is frequent in community ecology. Faced with an apparent falsification of a cherished hypothesis (a failure to document ecological differences or 'niche partitioning' among similar, syntopic species, for example), one may argue that the test and the falsification are invalid, as other things were not equal (e.g. the species may partition niches on some other, unmeasured dimension). As the domain of possible unmeasured variables is quite large, such *ad hoc* practices effectively make the hypothesis untestable by sheltering it from falsifications.

6. The logic is faulty. Each of the steps between the original observations or theory and the comparison of test observations with predictions in Fig. 2.1 involves a logical operation. If one has erred, for example, in deducing the hypothesis from a general theory or in deducing predictions from the hypothesis, falsification is a likely consequence. Revision of the hypothesis and review of the logic used in each step are required.

Clearly, a falsification may result from much more than a simple failure of the hypothesis being tested. These additional reasons for falsification all reflect poor design of either the hypothesis or the test (or both). If a test is carefully structured, the influences of confounding variables will be minimized, observations will be accurate and complete, the comparison will be valid, all assumptions will be clearly stated and matched in the test, and so on. Few tests in ecology are so carefully designed, however, and errors or biases may affect the test even if one *is* careful. Unless one considers these other possible reasons for a falsification, there is a real danger of being misled by such a result.

There is a third abuse of logical procedures that may occur at this stage (Fig. 2.1). Finding that one's observations do not agree with the predictions, one may be tempted to alter either the observations or the predictions in an *ad hoc* manner to bring them into closer agreement. This may be done, for example, by adding another term to an equation, by changing a model from a linear to a nonlinear form, or by removing 'outlier' data points to change a nonsignificant relationship into a significant one. Sometimes such procedures are justified, and part of the process of testing hypotheses includes changing the hypothesis or predictions to accommodate apparent anomalies. Such revised hypotheses then require fresh tests, however. To modify the data or the predictions and then to treat the modified set as a 'test' and 'verification' of the hypothesis and theory in the absence of additional, independent tests is logically invalid. Nonetheless, it is done (see Wiens 1983 for examples).

Corroboration

What if the observations of a test agree with the predictions? This is usually taken as evidence that the hypothesis is correct or, at the very least, corroborated. Our confidence in the relationships expressed in the hypothesis is bolstered, especially if the test was a 'risky' one. Such seemingly confirmatory results, however, are subject to the same sorts of alternative interpretations that apply to falsifications (items 1–6 above). Thus, just as faulty observations can mislead one into falsification, so also can they produce apparent agreement with expectations. Inadequate sampling can contribute to this, by fostering Type II statistical errors. The comparison used in the test or the logic used in deriving the hypothesis or predictions may be flawed in such a way that the likelihood of confirmation is enhanced. The predicted results may be due to other totally different causes, and our conclusions regarding the hypothesis under test will thus be misguided. Elements of untestability hidden among the underlying assumptions may also lead only to confirmatory results. As a consequence of these factors, corroboration may not be as compelling as it might appear. Further tests, structured so as to deal with such confounding factors, are required.

There is yet another logical error that often accompanies apparent corroborative tests (Fig. 2.1). Provided with observations that match the predictions of some hypothesis, one may conclude not only that the predictions have been verified, but that the hypothesis, its underlying assumptions, or the general theory as a whole have been proven correct as well. Finding ecological differences among coexisting species, for example, one might conclude that such differences are a result of competition, which created the niche partitioning that permits the species to coexist. Such conclusions are not warranted by the test but are inferences and assertions only. The problem is that the observations (differences among the species) do match the predicted pattern, but the role of the postulated process (competition) in producing the pattern has itself not been tested. Burdon-Sanderson drew attention to this problem in his inaugural address to the British Association for the Advancement of Science in 1893: 'To assert that the link between *a* and *b* is mechanical, for no better reason than that *b* always follows *a*, is an error of statement, which is apt to lead the incautious reader or hearer to imagine that the relation between *a* and *b* is understood, when in fact its nature may be wholly unknown' (1893: 467). To a large extent, this logical trap is a consequence of an emphasis on verification rather than falsification of hypotheses. Andrewartha and Birch (1984) term this the 'error of the misplaced premise', noting that the explanation or theory seems so plausible that its truth may be accepted with absolute certainty.

Testing alternatives

One way to reduce the likelihood of errors in falsification or corroboration is to make the hypothesis and predictions clear and concise, minimize their reliance on underlying assumptions, and conduct the test following rigorous procedures. It is also important to consider alternative hypotheses that address the same general question but do so using different sets of assumptions or premises and thus lead to different predictions. This method of 'strong inference' (Chamberlin 1965, Platt 1964) may be particularly appropriate when one suspects that a pattern is the result of multiple causes or the interaction of several variables (but see Haila 1982, Hilborn and Stearns 1982, Quinn and Dunham 1983). This approach is represented in tests of statistical hypotheses by contrasting a null hypothesis of no effects of a treatment or relationship among variables with one or more alternative hypotheses that offer specific predictions of patterns. Ideally, these tests are constructed so that falsification of the null hypothesis leads logically to provisional acceptance of the alternative hypothesis. In ecology, however, it is often difficult to devise null and alternative hypotheses that are logically complementary to each other (Connor and Simberloff 1986) and the null and alternative hypotheses must therefore be tested separately.

Ecologists have also given particular attention to conceptual null hypotheses or neutral models, which are conjectures that a particular process of interest has *not* contributed causally to the observations under study (e.g. Caswell 1976, Strong 1982, Connor and Simberloff 1983, 1986, Simberloff 1983a, Strong *et al.* 1979, Harvey *et al.* 1983). Such conceptual null hypotheses contrast with the approach of examining data for consistency with the predictions of hypotheses that contain specific causal processes in the absence of any knowledge of what values such data would take were those processes not operating (Connor and Simberloff 1986). Roughgarden (1983) has argued in favor of this approach, noting that null hypotheses are 'empirically empty' because they contain no biological processes that produce the predicted patterns. We learn nothing by falsifying such hypotheses, he claims, and they therefore do not represent viable alternatives.

At the opposite pole, Strong *et al.* (1979) have argued that, because the null hypothesis portrays what might be expected in the absence of a particular process, it has logical priority over other hypotheses. Whether or not the null hypothesis has logical priority or is simply a good starting point for investigations (Strong 1983, Simberloff 1983a), it seems clear that some sort of hypothesis based on the absence of key process assumptions should be among the alternatives considered when evaluating a specific hypothesis. Contrary to Roughgarden's view, conceptual null hypotheses *are* formulated relative to some specific causal explanation; consequently, they are not empirically empty, so long as they are correctly formulated. Both the value of null hypotheses in ecology and the procedures used in testing them have generated considerable controversy (e.g. Connor and Simberloff 1984a, Gilpin and Diamond 1984a), which I consider further in Chapter 4.

Concluding testing

How does one know when to stop cycling through the loops of Fig. 2.1 and to conclude that a hypothesis has been satisfactorily falsified or corroborated? When this occurs is a matter of judgment, based on 'convincing evidence' and 'common sense' (Roughgarden 1983). The danger, as Kuhn (1970a) so clearly shows, is in stopping prematurely in allowing one's preconceptions to equate 'convincing evidence' with the 'right' results. In a sense, testing of a hypothesis is never really completed. It is appropriate to move on to other questions, however, when, on the basis of sound logic, good design, appropriate methods, and 'risky' tests, the predictions have been clearly falsified, repeatedly corroborated, or have failed in a way that leads to a revision of the hypothesis.

Hypothesis testing and normal science

If Kuhn (1970a) is correct in his view of how science actually operates, it is important to know if the form or use of hypothesis testing differs during different phases of activity in a discipline. Do individuals engaged in 'normal science' test hypotheses in the same fashion as those embroiled in controversy or crisis? Both Kuhn (1970c) and Popper (1970) have addressed this issue; not surprisingly, they reach different conclusions. Popper does not dispute that some evidence of 'normal science' exists. He regards an individual engaged in such activity, however, as 'poorly taught': 'he has been taught in a dogmatic spirit; he is a victim of indoctrination. He has learned a technique which can be applied without asking for the reason why.' Popper suggests that, because one can always undertake critical comparisons of competing theories, using the methods of bold conjectures and criticisms that he regards as central to hypothetico-deductive science, *good* scientists do not suffer from the symptoms that Kuhn describes.

Kuhn, on the other hand, suggests that during periods of 'normal science' most individuals are engaged in 'puzzle solving', that is, in answering questions that are posed within the context of the prevailing paradigm. Hypothesis testing in a falsificationist spirit is usually not part of this endeavor (contra Fretwell 1972) because the emphasis is not on strong or 'risky' tests or falsification but rather on verification of variations of the paradigm. Because they take current theory for granted, individuals 'are freed to explore nature to an esoteric depth and detail otherwise unimaginable. Because that exploration will ultimately isolate severe trouble spots, they can be confident that the pursuit of normal science will inform them when and where they can most usefully become Popperian critics' (Kuhn 1970b: 247). Kuhn views formal hypothesis testing as 'the strategy appropriate to those occasions when something goes wrong with normal science, when the discipline encounters crisis'.

Neither of these views is entirely correct. Whether labelled as such or not, many scientists do undertake a form of hypothesis testing in their normal activities, whether or not controversy exists. There is no doubt, however, that interest among ecologists in philosophy and concern over the appropriate logical structure of scientific investigation were expressed much more strongly beginning in the late 1970s than they were during the previous two decades, when symptoms of Kuhnian 'normal science' were more apparent in community ecology.

Testing hypotheses: an example

To explore some aspects of how hypotheses of pattern and process are actually developed and tested in community ecology, it is helpful to consider a specific example. Using the format of Fig. 2.1, I have indicated some of the major features of what might be termed the 'size–ratio hypothesis' in Fig 2.2 (see Wiens 1982, Simberloff and Boecklen 1981, Roth 1981, Simberloff 1983b). This idea is evaluated further in Chapter 7.

We begin with Hutchinson's (1959) original observation that bill lengths within several sets of congeneric, sympatric bird species differed by an average ratio of 1.3. Hutchinson's ratio value was derived inductively, but the observation was made theoretically explicit by Hutchinson and MacArthur (1959). The theory was subsequently enriched through general competition theory and, especially, limiting similarity theory (e.g. MacArthur and Levins 1967, May and MacArthur 1972). This work generates the hypothesis that competition imposes a limit to the similarity of sizes of coexisting, ecologically similar bird species. A specific, testable prediction of this hypothesis is that such species will differ in bill lengths by an average ratio of 1.3; this also implies that several coexisting species will be spaced along a bill-length spectrum by constant ratios of 1.3.

Several assumptions are contained in this prediction. Because it is derived in part from general competition theory, we assume that food resources are limiting to the species and that the species use similar resources (i.e. they are indeed potential competitors). Implicitly, it is also assumed that the community is fully packed with species. More specifically, one must assume that the food-size resource spectrum is continuous and continuously available to the species (MacArthur 1972) and that food-resource use is a direct function of bill length. Finally, there is, of course, the *ceteris paribus* clause.

The predictions have been tested by measuring the bill lengths within sets of similar coexisting species and calculating the ratios. Additional tests have involved determinations of the frequency distributions of size ratios over a large number of sets of species (e.g. Schoener 1965, 1984, Roth 1981, Simberloff and Boecklen 1981). It turns out that for many of these tests the prediction has not been upheld: the size ratios neither average 1.3 nor are they constant. Does this represent a clear falsification of the hypothesis? Perhaps, but other explanations come to mind. The tests might be flawed in various ways, for example. Perhaps the comparison has included species that are not really ecologically similar, obscuring any pattern that might exist among the subset of similar species (e.g. Bowers and Brown 1982, Grant and Abbott 1980). Bill length was measured because it was consid-

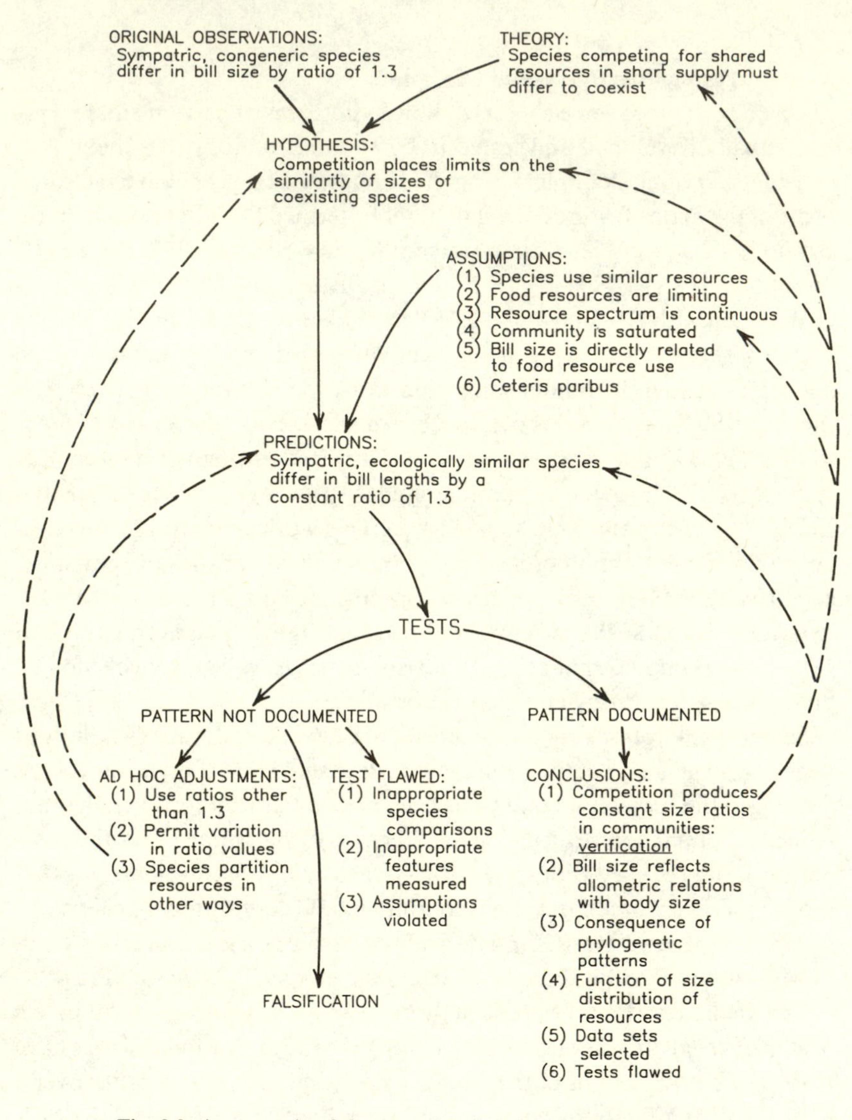

Fig. 2.2. An example of the operations involved in testing the 'size–ratio' hypothesis. The background of this hypothesis and the tests are developed more fully in Chapter 7. The arrangement of the diagram follows that of Fig. 2.1.

ered the variable most relevant to capturing prey of different sizes, but perhaps bill shape is more important than length alone (e.g. James 1982) or perhaps allometric relationships with body size impose constraints on variation in bill lengths (Clutton-Brock and Harvey 1979). Perhaps one or several of the underlying assumptions has not been met, and the tests are therefore invalid. These possibilities all reflect inappropriate tests of the hypothesis, but they may nonetheless affect the results that have been obtained.

Clearly, there may be valid reasons for concluding that the lack of agreement between the prediction and the test observations is not a strict falsification of the hypothesis. A less logical route to preserving the hypothesis involves *ad hoc* adjustments of various sorts. Thus, size–ratio values other than 1.3 (e.g. 1.05, 1.1, 1.2, 1.4, 2.0) have been considered to be consistent with the hypothesis by various investigators (Simberloff 1983b). Maiorana (1978) reconciled observations of substantial variation in ratio values within a sequence with the hypothesis by arguing that larger species in a sequence should be expected to exhibit greater variance in resource use than smaller species and thus require wider spacing in order to coexist. Others have suggested that the apparent falsification simply means that the species partition resources in ways other than by food size.

In some studies, reasonable agreement of observations with the prediction has been found. What conclusions can be drawn from this corroboration? The most obvious is that the prediction is upheld in this instance: the expected pattern, in fact, was found. But other possibilities exist. If bill length varies allometrically with body size, the pattern may reflect something having to do with body size rather than bill length alone. If the comparison has been restricted to a single phylogenetic lineage, the size patterns may be related to the phylogenetic history of the lineage rather than to the proximate ecology of the species (Wiens 1982). The regular spacing of species along the size gradient may be a consequence of the distribution of nodes and gaps along a discontinuous resource–size spectrum. In some instances, the pattern may apply only to a selected data set within a larger group of suitable data (Simberloff 1983b). Finally, the test itself may be flawed by the lack of any statistical analysis to determine whether the observations actually fit the predictions or just seem to.

If the results of the test agree with the predictions, however, such alternatives usually are not considered. The findings are regarded as verifying the hypothesis and may be used to support the correctness of the underlying assumptions as well (e.g. bill length *is* directly related to food-resource use). Frequently, the consistency of the observations with the predictions leads to the conclusion that the process contained in the general theory, competition, is indeed the cause of the observed pattern and has been demonstrated to occur. The pattern is 'explained'. This conclusion is not only logically unwarranted, but it may stifle further, more specific tests (e.g. coupling body-size differences to differences in foraging efficiency and thence resource use; Werner 1984). Such inferential 'proof' of a process underlying an observed pattern rests on the presumption that nothing else is likely to have produced that pattern, but this may not be so. The pattern *may* result from competition, to be sure, but it may also arise from other quite different processes (Wiens 1982, Simberloff 1983b). Given this, no conclusions about process from the documentation of a particular pattern are warranted.

Pattern hypotheses and process hypotheses

The temptation to explain patterns by inferences about processes, as in the above example, arises because the hypothesis contains statements about *both* pattern and process. Thus, even though the specific prediction relates only to pattern, it is a small step to accept the close coupling of process with pattern contained in hypothesis. In fact, most ecological hypotheses contain the proposition that a pattern (X) has been produced by a process (Y). They state (either explicitly or implicitly) that if Y occurs, then X will follow. Such causal 'if–then' statements contain two parts, which really represent *separate* hypotheses to be tested: (a) does X occur? and (b) if so, was it a consequence of Y? A failure to separate the process and pattern portions of a hypothesis in tests generally leaves the process buried within the underlying assumptions and available to 'testing' only by assertion and inference.

In pattern hypotheses we ask whether a particular relationship holds among a number of observations. Patterns are usually detected by comparisons, and the so-called 'comparative method' (Clutton-Brock and Harvey 1979, Jarman 1982, Thornhill and Alcock 1983) provides a framework for testing pattern hypotheses. The actual tests of such hypotheses, however, are often statistical. We may hypothesize, for example, that there is a linear relationship between two variables and then use linear regression and correlation statistics to test this hypothesis. James and McCulloch (1985) and Connor and Simberloff (1986) have emphasized the distinction between these statistical or 'empiric' hypotheses and what they call 'scientific', 'research', or 'theoretic' hypotheses, which are statements about processes. Statistical (pattern) hypotheses are tested by inductive inference, by sampling from clearly defined populations to evaluate pattern predictions. The emphasis is on developing reliable *predictions*, and such tests address theory only to the extent that a theory generates unique and unambiguous predictions. Popper (1959, 1962, 1983; Popper and Miller 1983) explicitly disavowed a legitimate role for inductivism in scientific investigations, and statistical hypotheses, which are founded firmly on inductive tests, are therefore not 'proper' Popperian hypotheses, even though the emphasis is still upon falsification. This, of course, does not diminish their usefulness in scientific investigations.

The documentation of a regular, repeatable pattern reveals something requiring explanation: what *caused* the pattern? Hypotheses about processes are considerably more difficult than pattern hypotheses to frame and test. This is because a process must be isolated in the test to reveal its effects upon pattern; *ceteris paribus* must be assured. Some workers (Alexander 1979, Mayr 1982, Thornhill and Alcock 1983, Thornhill 1984) have argued that, with appropriate comparisons, the comparative method can be used to examine causal forces as well as to detect patterns. Appropriate comparisons are those that, because of their number and diversity, are likely to randomize and therefore control the influence of other variables on the result (Thornhill 1984). If it were possible to achieve complete randomization of other variables in a comparison, this might be true, but intercorrelations among variables make such randomization of effects unlikely.

Increasingly, experiments are suggested as the best way to test process hypotheses and determine causality. In experiments, critical confounding variables are controlled rather than presumed to be randomized and the variable of interest is manipulated. For example, if one hypothesizes that the habitat niche breadth of species A is restricted by the presence of a competing species B, the expansion of A's habitat niche when B is experimentally removed should be clear evidence of B's competitive effects when it is present. Such experiments and their interpretations, however, are not without problems (Hilborn and Stearns 1982, Quinn and Dunham 1983, Hurlbert 1984, Bender *et al.* 1984, Underwood and Denley 1984, Underwood 1986b), some of which are considered in Chapter 3. Moreover, in many instances logistic constraints or other considerations (e.g. endangered species or habitats) make experimental manipulations impractical. In these situations, one must rely on using carefully planned, nonmanipulative comparisons to test process hypotheses (Diamond 1986a).

In the end, of course, experiments are simply another form of comparison, in which the investigator establishes the domain of the comparison rather than leaving this up to Nature. The results of experiments are patterns, not processes. Thus, although experiments may isolate the *effects* of processes, they rarely test process explanations directly. One must still rely on inference to reach conclusions about process. Experiments may foster a greater confidence in such causal inferences than does the comparative method, but they do not assure the truth of such inferences. As Andrewartha and Birch (1984: 222) have observed, 'there is no epistemological magic by which a precise probability (far less a certainty) can be attributed to a causal explanation'.

Conclusions

The philosophical and logical foundation of posing questions or testing hypotheses is important here because it provides a necessary framework for evaluating and designing investigations of bird communities. Unfortunately, not all investigators have been attentive to these matters, and this casts doubt upon the conclusions they have reached. Ecologists should recognize that the detection of pattern and the documentation of underlying process are conceptually separate and generally sequential phases of scientific investigation. Both pattern detection and process documentation are best done by following a flexible hypothetico-deductive procedure in which hypotheses that are derived from observations and/or theory are used to generate specific, testable predictions. 'Testable' means that the predictions, and thus the hypothesis, must be susceptible to falsification or corroboration. Interpreting the outcome of such tests, however, is not a simple matter, for a variety of factors other than the apparent truth or falsity of the hypothesis can contribute to the outcome. Caution and insight are required in reaching the conclusion that a particular test falsifies or corroborates the hypothesis. Pattern hypotheses may be tested by direct, nonmanipulative comparisons, but this 'comparative method' is of limited value in testing process hypotheses, for which experiments may be more appropriate. Conclusions about processes drawn from experimental tests, however, still rely on inference. A well designed experiment may increase our confidence in such conclusions, but it cannot guarantee certainty.

Opportunities for committing logical errors exist at several points in this hypothetico-deductive framework. Most apparent are those in which (a) *ad hoc* changes are made in the hypothesis or its predictions in order to fit the observations, which are then used as 'proof' of the hypothesis, or (b) the agreement between observations and predictions is used as a basis for inferring that the underlying assumptions or the process explanation have been confirmed. Neither is an acceptable practice.

3

The importance of methodology

The detection of a pattern or test of a hypothesis can be no better than the data on which it is built. An impeccable experimental design cannot remedy poor observations, clever statistical analyses cannot overcome poorly designed tests, and logically deduced predictions cannot save a logically flawed hypothesis. How one determines which variables to measure, the way to measure them, the design of observational or experimental tests of hypotheses, and the way to analyze and interpret results are critically important to the conduct or evaluation of any scientific work. Lack of attention to the details of methodology can seriously impair the value of an otherwise worthwhile study.

The methods of an investigation are determined by its objectives. Clear specification of objectives is thus essential, and this is one of the benefits of following a hypothetico–deductive approach. A well-framed hypothesis offering specific predictions defines the objectives of an investigation and thus guides the development of a sound methodology. Unfocused description with no hypothesis in mind does not. Such research is not only aimless, but it is more likely to be methodologically flawed, especially if data gathered with no objective in mind using a particular method are later applied to some specific objective or *post-facto* hypothesis test for which that method is inappropriate.

Objectives differ among studies, of course, and there is thus no ideal, all-purpose methodology that can be applied uniformly in research on bird communities. If one wishes to test a hypothesis relating the number of bird species present on an island to island size, for example, it may be more important to survey a large number of islands than to record the abundance of each species on each island with great precision. Single-visit censuses in which only the presence or absence of species is noted may be appropriate (Järvinen and Väisänen 1981). If one intends to test hypotheses of species turnover rates on those islands, however, greater care must be taken to ensure that *all* of the species present on an island at a given time are recorded in order to avoid error associated with 'pseudoturnover' (Lynch and Johnson 1974, Jones and Diamond 1976; see Volume 2). If one's interest is in 'density compensation' (changes in the density of one species in the presence or absence of another species), simple enumeration of the presence or absence of another species will not do, and accurate estimations of population densities in the islands are required. The patterns revealed by qualitative measures such as presence or absence may differ substantially from those derived from quantitative measures (e.g. abundance) of the same systems (Haila *et al.* 1987).

Measurements and methods are thus determined by one's objectives. Our objectives, however, may be strongly influenced by our preconceptions of nature. If one believes natural systems to be more or less equilibrial, for example, an appropriate methodology for examining avian niche relationships may involve the measurement of only a few key variables (the resource dimensions that account for partitioning), and valid patterns may be derived from comparisons of single samples of a number of locations (as conducting repeated samplings of each would presumably yield much the same results). If one imagines nature to be nonequilibrial or variable, on the other hand, many resource variables must be measured (as different niche dimensions may be important to different species in the same or different places at different times). Repeated, replicate sampling of locations will be necessary (in order to avoid comparisons among locations sampled at different phases of variation) (Wiens 1981b, 1984a).

The methodology of community ecology is a complex discipline in itself, involving mensuration, sampling theory, experimental design, and statistical analysis. It is not my intention in this chapter to review these areas. Rather, I wish to draw attention to some especially critical aspects of gathering data and testing hypotheses, note some potential pitfalls and problems, and indicate in general how these might be avoided or minimized.

Measurement of variables

Determinations of patterns are sensitive to *how* measurements are taken. The hope that sophisticated statistical analyses may somehow heal methodologically ailing observations is futile; if anything, such analyses only magnify and compound initial errors, while giving the measures a false aura of precision (Johnson 1981).

There are actually several elements to the 'how' of measurement: how samples are taken in space or time or how many are taken (the realm of sampling theory), how one goes about gathering the measures (sampling procedures), and how one decides what to measure, at what level of precision. Here I consider chiefly the latter two; several excellent treatments of sampling theory in ecology are available (e.g. Southwood, 1978, Caughley 1977, Tanner 1978, Burnham *et al.* 1980), and these should be required reading for anyone contemplating quantitative studies of bird communities.

Censusing birds

The interplay between the objectives of a study and the methodology employed is especially clear in the estimation of species' abundances. Determining patterns of variation in species richness, for example, requires only information on the presence or absence of species, whereas assessing population fluctuations demands estimates of absolute densities. The methods most appropriate to obtain such kinds of information differ (Table 3.1). Avian ecologists have generally paid little attention to these differences, however; consequently, they often have unwarranted confidence in the density estimates they calculate (Verner 1985).

Census methods. Most interesting community patterns relate in one way or another to the abundances or relative abundances of species (Table 3.1), and some reliable method of estimating abundances is therefore essential. Avian ecologists have devoted much effort to the development of such techniques, but, because there are many sources of variation in the estimates, censuses often contain errors much greater than those encountered in many other scientific measurements (Mayfield 1981). Some indication of the magnitude of error is provided by comparisons of census results obtained using different methods. Franzreb (1981a, b), for example, surveyed the birds breeding in a mixed-coniferous forest in Arizona with a spot-mapping procedure (Williams 1936), a variable-width strip transect method (Emlen 1971), and a fixed-width transect (Kendeigh 1944). When the variable-strip transect was used, density estimates obtained for Western Flycatchers, Golden- and Ruby-crowned Kinglets, and Yellow-rumped and Red-faced Warblers were greater and those of Hermit Thrushes were lower than estimates derived from the spot-map procedure (Table 3.2). The fixed-width transect estimated greater densities of Golden-crowned but not Ruby-crowned Kinglets in relation to the spot-map results. Total community density estimates were considerably lower for the fixed-strip transect. All three procedures tallied the same number of species, but they differed in which species those were.

It is clear from this example that the use of different methods yields results

Table 3.1. *The relationship between the objectives of a study and the scale of measurement and level of sampling detail required.*

The nominal scale involves recording only the presence or absence of species, the ratio scale information on the relative abundances of the species, and the absolute scale estimates of the true abundances.

	Measurement scale needed			Applicable level of sampling detail			
Study objective	Nominal	Ratio	Absolute	List	Unadjusted count	Adjusted[a] count	Census
Biogeography	X			XX[b]	X	X	X
Species richness	X			XX	X	X	X
Frequency	X			XX	X	X	X
Annual population trends		X			XX	X	X
Seasonal population trends		X				XX	X
Successional trends (intra- and interspecific)		X				XX	X
Habitat suitability or preference		X				XX	X
Density-dependent and density-independent effects		X				XX	X
Species diversity		X				XX	X
Population fluctuation			X				XX
Trophic dynamics			X				XX

Notes:

[a] Adjusted counts compensate for differences in area sampled and for differences in detectability of birds in different habitat and seasons. They consequently result in an *estimate* of density. Unadjusted counts do not permit an estimate of density, but properly designed counts that control for observer differences, phenological changes, and other such variables should deliver abundance measures on the same ratio scale for the same species in the same or structurally comparable habitats.

[b] XX shows the least costly (hence preferred) sampling level for the indicated study objective. An X in this section shows a more costly way to the same study objective.

Source: From Verner (1985).

Table 3.2. *Densities (individuals per 40 ha) and relative densities (%, in parentheses) of the most abundant species in an Arizona coniferous forest as recorded by three census procedures.*

Species	Spot-map	Variable-strip transect	Fixed-strip transect[a]
Broad-tailed Hummingbird (*Selasphorus platycercus*)	30.8 (3.6)	20.8 (2.5)	9.6 (1.9)
Northern Flicker (*Colaptes auratus*)	25.6 (3.0)	18.3 (2.2)	11.0 (2.1)
Western Flycatcher (*Empidonax difficilis*)	48.7 (5.6)	71.6 (8.6)	32.9 (6.4)
Mountain Chickadee (*Parus gambeli*)	58.9 (6.8)	64.8 (7.8)	37.0 (7.1)
Pygmy Nuthatch (*Sitta pygmaea*)	25.6 (3.0)	27.2 (3.3)	20.6 (4.0)
Brown Creeper (*Certhia americana*)	51.3 (6.0)	46.4 (5.6)	34.3 (6.6)
Hermit Thrush (*Catharus guttatus*)	76.9 (8.9)	42.8 (5.1)	48.0 (9.2)
Golden-crowned Kinglet (*Regulus satrapa*)	30.8 (3.6)	51.4 (6.2)	45.2 (8.7)
Ruby-crowned Kinglet (*Regulus calendula*)	74.4 (8.6)	88.6 (10.6)	49.3 (9.5)
Yellow-rumped Warbler (*Dendroica coronata*)	89.8 (10.4)	136.5 (16.3)	87.7 (16.9)
Red-faced Warbler (*Cardellina rubrifrons*)	25.6 (3.0)	40.3 (4.8)	15.1 (2.9)
Dark-eyed Junco (*Junco hyemalis*)	51.3 (5.9)	66.6 (8.0)	43.8 (8.4)
Total density	865.9	835.6	519.3
Number of species	35	35	32
Exclusive species[b]	3	0	0
Missing species[c]	2	3	3

Notes:
[a] Strip width = 30 m.
[b] Species recorded only by this method.
[c] Species recorded by other methods but not by this method.
Source: After Franzreb (1981a, b).

that are both quantitatively and qualitatively different. In this case, none of the census procedures was detailed enough to provide a reliable estimate of absolute densities, so we cannot determine which results are most likely to be correct. Telleria and Garza (1981) used intensive mapping surveys as a standard for comparison in a Spanish oak woodland, and determined that single fixed-width transects recorded 66% of the species and 88% of the individuals present. In subalpine forests in the Sierra Nevada mountains of California, DeSante (1986) compared estimates obtained by variable circular-plot (VCP) surveys with those derived from intensive spot-mapping of territories coupled with monitoring of active nests. With minimal effort devoted to the VCP, general community parameters were estimated fairly well but the determination of densities and relative densities of most species departed substantially from the 'actual' values (Table 3.3). A three-fold increase in the number of sampling stations increased the accuracy of relative density estimates obtained by the VCP method, although the absolute density estimates still contained many errors. DeSante concluded that the accuracy of the VCP method is likely to be low for species with low population densities, large territories, high mobility, or ventriloquial vocalizations and in habitats that are dense and structurally complex. DeSante's study was conducted in a very heterogeneous area, and it is therefore not too surprising that counts at a small number of VCP stations failed to estimate densities as well as intensive mapping of the entire area. Verner and Ritter (1985) reported a similarly dismal performance of the VCP method in mixed oak-pine woodlands in California. In more homogeneous and open coastal scrub habitat in California, the errors in VCP density estimates were much lower (DeSante 1981), and the method also produced more accurate estimates in desert riparian and desert scrub habitats in Arizona (Szaro and Jakle 1982) and in oak-hickory forests in South Carolina (Hamel 1984). The level of accuracy of this (and other) census procedures clearly differs among species and among habitats.

The increased accuracy obtained in intensive surveys based on mapping, color-marking of individuals, and nest monitoring is bought at a price of the considerable effort involved. In many situations it is probably not possible to obtain absolute values of species' densities, and for most objectives it is probably not worth the effort. Hamel (1984), for example, suggested that the VCP method might provide estimates comparable in accuracy to those from spot-mapping censuses in 20% of the time, and Burnham *et al.* (1985) demonstrated that line transect procedures are generally more efficient than strip transects, especially when the detectability of individuals drops off rapidly with increasing distance from the transect line. Reed (1985) con-

Table 3.3. *Estimates of breeding densities of several species in a California subalpine forest obtained by the variable circular-plot (VCP) procedure in relation to the 'actual' densities determined by intense mapping and nest monitoring.*

	Territory density/48 ha[a]		
Species	'Actual'	VCP[b]	% Error
Dark-eyed Junco[c]	23.1 (17.5)	10.0 (9.1)	−56.7
Cassin's Finch (*Carpodacus cassinii*)	20.6 (15.6)	16.8 (15.3)	−18.5
Dusky Flycatcher (*Empidonax oberholseri*)	16.4 (12.4)	19.7 (17.9)	+20.3
Yellow-rumped Warbler[c]	15.0 (11.4)	19.4 (17.6)	+29.3
Chipping Sparrow[c] (*Spizella passerina*)	8.5 (6.4)	1.6 (1.5)	−81.0
Mountain Chickadee	11.9 (9.0)	9.7 (8.8)	−18.3
Hermit Thrush[c]	7.8 (5.9)	12.9 (11.7)	+65.3
White-crowned Sparrow[c] (*Zonotrichia leucophrys*)	3.0 (2.3)	1.6 (1.5)	−46.9
White-breasted Nuthatch (*Sitta carolinensis*)	2.0 (1.5)	1.1 (1.0)	−47.1
Golden-crowned Kinglet[c]	1.0 (0.8)	0.7 (0.6)	−35.1
Ruby-crowned Kinglet[c]	0.8 (0.6)	1.3 (1.2)	+59.2

Notes:
[a] Relative densities in parentheses.
[b] Based on 48 stations.
[c] Counts of singing males only.
Source: Modified from DeSante (1986).

cluded that a 'flush-mapping' procedure (Wiens 1969) was faster and more accurate in defining sparrow territories than the spot-map method.

Different census procedures provide relative estimates of density of varying degrees of accuracy. In order to obtain data for the comparisons that will yield community patterns and test hypotheses, the degree of accuracy of the method must be relatively constant and its accuracy must not be density-dependent, species-dependent, habitat-dependent, or anything-else-dependent. This requires that one consider sources of error such as differences between observers, species, locations, and sampling regimes.

Effects of observers. If two observers of equivalent experience simultaneously census the avifauna at the same location with the same method, they will not necessarily obtain the same results. During a 4.5-h survey period in an area in England, for example, one observer recorded 40

Common Snipe (*Gallinago gallinago*), 520 Wood Pigeons (*Columba palumbus*), 86 Starlings (*Sturnus vulgaris*), and 40 Dunnocks (*Prunella modularis*). A paired observer recorded 62 snipe, 1285 pigeons, 20 Starlings, and 22 Dunnocks. They recorded nearly the same total number of species (57 vs. 55), although each saw species that the other did not (Lack 1981). In another study, four observers of varied experience differed significantly in the number of territorial clusters they recorded during spot-mapping censuses (O'Connor 1981). Verner and Ritter (Verner 1985) found similar between-observer differences in spot-mappings of territory boundaries of a single species (Fig. 3.1). Spot-map surveys produce plottings of observation points that require interpretation to derive a density estimate. O'Connor had three trained individuals interpret these maps independently and found a high degree of consistency in their estimates. Best (1975), on the other hand, had individuals with rather little experience interpret territory mappings; their density estimates varied considerably.

Such between-observer differences arise for a variety of reasons – differences in attention span, ability to estimate distances, skill in identification, experience with a given method, or hearing ability (Berthold 1976, Ralph and Scott 1981, DeJong and Emlen 1985). The latter difference is especially troublesome, as it differentially biases the censusing of species with high-frequency vocalizations in counts that use auditory cues. Thus, apparent population decreases in such species over a series of annual censuses taken by the same individual could reflect a loss of hearing ability by the observer rather than actual population changes (Faanes and Bystrak 1981). Kepler and Scott (1981) show that standardized pre-census training of observers can reduce observer bias substantially.

Effects of species characteristics. The behavior of an individual affects its conspicuousness to an observer and therefore its probability of being counted in a survey. Waterfowl species, for example, differ in their affinities for open water versus closed vegetation, and this affects the ease with which they can be detected in aerial versus ground surveys (Broome 1985). Most censuses of terrestrial passerines rely to a considerable extent on registrations of singing individuals, but song frequency varies for a number of reasons, producing variations in the detectability of individuals. The most obvious source of variation is due to species: some species sing loudly and often from exposed positions, others softly from hidden sites, and still others make only crude noises or do not sing at all. Moreover, because most census techniques have been developed in north-temperate areas, they have a bias toward the behavior of north-temperate bird species, in many of

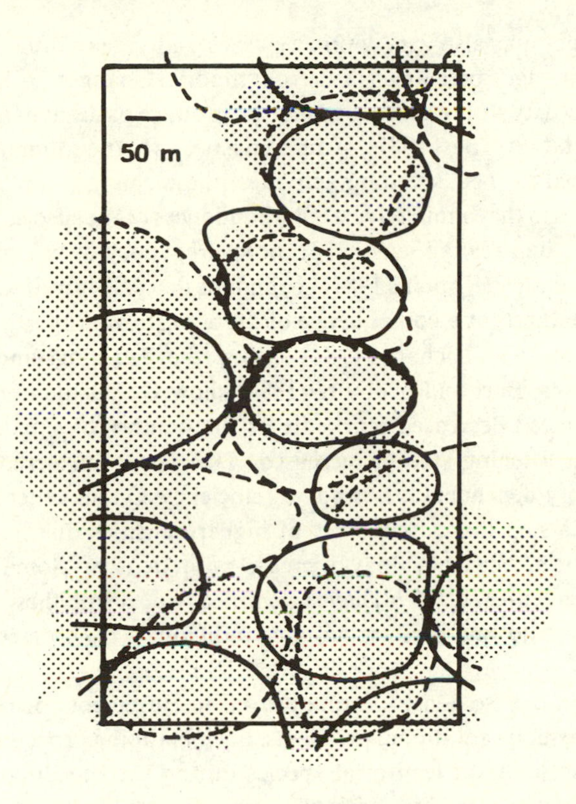

Fig. 3.1. Territory mappings of Bewick's Wrens (*Thryomanes bewickii*) in a 33-ha oak-pine woodland plot in California. The shaded areas represent total mappings; solid and dashed lines indicate the boundaries determined by each of two observers independently using the spot-mapping method. After Verner (1985).

which only males sing and do so from prominent locations in fixed territories during a well-defined breeding season. Many species in other parts of the world are not so well-behaved. Recher (1981), for example, has pointed out that not all species in Australian heath communities sing, while in other species both sexes are equally vocal. There is considerable song mimicry among species, fewer than half of the species are territorial, and breeding is generally asynchronous, with both breeders and nonbreeders being present in an area at any time. A census procedure based solely on singing registrations would fail to provide an accurate record of the composition of such a community or of the densities of the species present. Similar problems occur in tropical areas, exacerbated by the density and height of the vegetation. In such areas, alternative census procedures (e.g. mist nets) may be used, but they have their own special biases and limitations (Karr 1981).

Song frequency may also vary within a species. Individuals differ according to their sex, age, breeding status, or 'mood'. Changes in levels of reproductive activity within a local population produce patterns of seasonal change in detectability (Best 1981). If different species in the community are not synchronized in their breeding activities, these changes can produce seasonal changes in the estimated relative abundances of the species that are more apparent than real (Wiens 1983). Song frequency also varies with time of day or weather (Robbins 1981a, b) and the detectability of song may vary between habitats as a consequence of the acoustical properties of the habitats (Morton 1975, Richards 1981). Using recorded songs under controlled conditions, Bart and Schoultz (1984) showed that the number of individuals detected decreased markedly as the number of singing birds audible from a listening station increased. This vocal interference may produce a density-dependent loss in survey efficiency. On the other hand, if individual birds sing more frequently at higher densities due to greater social stimulation, efficiency may increase with density. Some census methods (e.g. Emlen 1971, 1977a) have attempted to deal with these sources of variation by adjusting direct counts of individuals by a detectability index value.

Techniques that base density estimates on measurements of breeding territories are beset by additional problems. Such mapping procedures will work, of course, only on territorial species during the breeding period. Territories may be mapped by a variety of procedures – direct observation of marked individuals, flushing individuals, recording locations of apparent boundary disputes, plotting the locations of unmarked singing males over several successive visits to the area – but all have drawbacks (Berthold 1976, Oelke 1981). Frequently, territory mappings obtained by different procedures do not agree. Ferry *et al.* (1981), for example, found that individual home ranges determined by capture–recapture of marked birds of five species averaged 2–12 times larger than the estimates of their territory sizes obtained by mapping, and Zach and Falls (1979) noted similar disparities between the sizes of territories defined by foraging behavior versus singing behavior. Sample sizes in both of these studies were too small to generate much confidence in the quantitative conclusions, but the qualitative disparities remain.

Effects of location. Errors or biases may also be introduced into surveys by a failure to standardize features of the locations in which the work is done. If one conducts two surveys in a habitat of the same vegetational composition, one in a homogeneous mixture of the plants and the other in a patchy

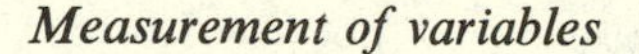

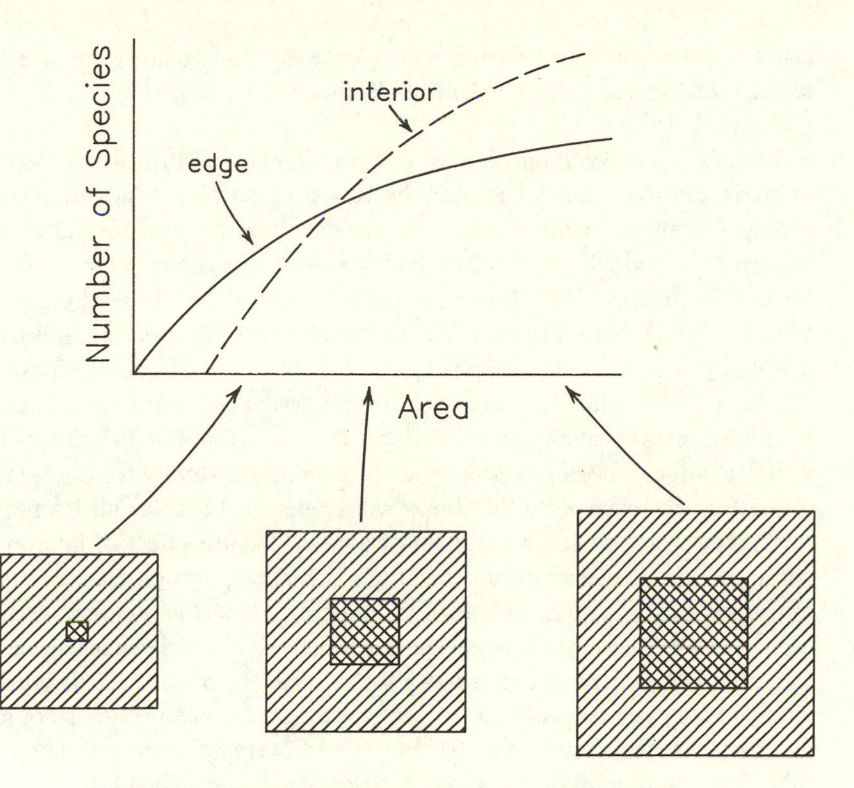

Fig. 3.2. The effects of increasing the area of a census plot on the relative areas of edge and interior habitat (below) and on the rates of accumulation of 'edge' and 'interior' bird species in the census (above).

mosaic, the results are likely to differ. The presence or absence of species in an area and the density levels they reach also depend on the size of the area (see Chapter 5), and censuses taken in areas of different sizes will therefore yield quite different results (Engstrom 1981). Part of this effect of area is associated with the influences of plot or habitat edges, which may be especially important in fragmented landscapes. Different bird species respond differently to habitat edges, and a small plot has proportionately much more edge than a large plot. Species preferring edges will thus be favored in small plots, whereas both edge and interior species may occur in larger plots (Fig. 3.2; see Volume 2) (Whitcomb *et al.* 1981, Ambuel and Temple 1983). As a consequence, census plots or transects that contain differing proportions of habitat edge will be biased by this edge effect, and surveys should be designed to sample edge and interior habitats in relation to their occurrence. Helle and Helle (1982) documented that such differences in survey configuration may have a large effect on density estimates of forest birds on islands in the Gulf of Bothnia.

Effects of sampling procedures. Even if one attempts to control such sources of error, census estimates may still be biased by flaws in conducting the surveys. In survey methods that rely on determinations of distances to observed individuals (e.g. VCP, line transects; Reynolds *et al.* 1980, Burnham *et al.* 1980, 1981; Burnham and Anderson 1984), accurate estimation of distances is essential – a 25% underestimate of distance in the VCP procedure, for example, may produce a 78% overestimate in density (DeSante 1986). Many methods require observers to record birds either from stationary locations or along linear transects, and the duration of a visit to a point or the rate at which the observer moves along a transect may also affect the probability that birds will be detected (Scott and Ramsey 1981). Increasing the time duration of a count has the effect of inflating counts of mobile species (as more of them will move through the area in a longer period; Granholm 1983), while rendering counts of sedentary, cryptic species more accurate (as they are more likely to be recorded in a longer time interval). Fuller and Langslow (1984) conducted counts in fixed-radius circular areas over periods of 5–20 min per station. Using the 20-min count as a standard, they found that 50% of the birds were recorded in the first 5 min, 70% in a 10-min survey, and 87% after 15 min. They found that count duration had relatively little effect on estimates of general community parameters, and suggested that counts of more than 10 min duration per station 'are wasteful of field effort which could be used to improve other aspects of sampling' (1984: 202). The extent to which this is true, of course, depends on the objectives of the study.

Census estimates may be even more strongly influenced by the number of samples taken. As one increases the number of counts on a plot or transect, the variance in density estimates will decrease and the density values will stabilize. The relationship between sampling intensity and variance in density differs between species, however, so that a sampling regime sufficient to obtain a satisfactory estimate of the density of one species may be woefully inadequate for another. Sampling intensity also affects the total number of species recorded – doubling the number of surveys of an area will add some new species to the community list, but further doublings will add progressively fewer new species. There is obviously a point of diminishing returns on the investment of greater sampling intensity, but this point differs between habitats. Most often, decisions regarding sampling adequacy are made for logistical rather than statistical reasons. Clearly, however, census estimates of population densities or community composition based on only one or two surveys are likely to be only coarse approximations of reality and may contain biases favoring some species and discriminating against others.

Censusing methodology is obviously critical to any sort of community investigation. One might conclude from this catalog of errors and biases in census estimations that achieving *any* degree of accuracy is an impossible dream. Many of these problems, however, can be dealt with once they are recognized (see Ralph and Scott 1981, Verner 1985). Attempts to standardize census methodology (e.g. Svensson and Williamson 1970, Järvinen and Väisänen 1975, 1977, 1981) have met with only moderate success. Careful attention to the details of estimating numbers in bird populations is essential if the community patterns derived from such estimates are to have credibility. Measurements taken using *ad hoc* or ill-defined techniques applied in an inconsistent fashion to areas of unknown but different sizes by different observers do not form a foundation for any sort of meaningful community analysis.

Observing and quantifying behavior

Perhaps because the behavior of birds is such a conspicuous feature of their biology, behavioral data have received considerable attention in studies of niche patterns and resource use in bird communities (see Chapter 10). Observations of bird behavior may indeed be relatively easy to obtain in many situations, but they are sensitive to several sources of bias, and patterns should not be derived from the observations without consideration of these biases.

Behavioral observations are used in community analyses to determine how individuals of a species use resources, primarily food and habitat. By recording the details of foraging movements, where they occur, and how much time is spent in foraging and in specific locations, inferences about the use of food resources may be made. By observing how the behavioral activity budget of individuals is allocated among habitats or microhabitats, the importance of various habitat features may be quantified. The basic observations are thus usually some quantified expression of the amount or proportion of time spent in different activities and/or in different locations.

Probably the most serious source of error in such observations is due to visibility bias. One cannot observe a bird without seeing it, which requires that it be discovered and that observations be gathered before it is again lost from view. Errors may be introduced through biases in either the discovery or the loss of individuals (Bradley 1985). Individuals are more likely to be discovered when they are singing in open microhabitats, for example, than when they are foraging or sitting quietly in dense cover. Flying individuals or birds engaged in rapid foraging movements may be lost from view more readily than birds foraging slowing and deliberately. Such biases alter the frequencies at which behaviors and microhabitats appear in the observations from their actual frequencies in nature. Different methods of recording behavioral observations are differentially sensitive to these biases (Wiens 1969, Wagner 1981, Morrison 1984, Bradley 1985). If one records only the behavior or position of a bird when it is initially seen, for example, conspicuous behaviors and exposed microhabitats will be overemphasized in relation to their occurrence in an activity budget obtained through continuous observation of individuals (Table 3.4). Continuous or sequential scan-sampling observations (Altmann 1974, 1984) circumvent the discovery bias that affects single observations by overwhelming it with a large number of subsequent observations in each behavioral sequence. This bias may be minimized further by omitting the first few samples of an individual's behavior following its discovery (Wiens, Van Horne, and Rotenberry 1987). The method still may be sensitive to biases accompanying the loss from sight of individuals under observation (Wagner 1981), but these are generally less important than discovery biases. The important point is that behavioral observations obtained using different methods are not directly comparable and are likely to yield different patterns. Altmann (1984), Bradley (1985), and Bakeman and Gottman (1986) have considered these methodological biases in detail.

The choice of a method of behavioral observation depends in part on the objectives of an investigation. If one wishes to obtain information on the frequency with which various behaviors occur, sampling the behavior of individuals at intervals may be appropriate. Documenting the duration of particular behaviors or the sequence of different behaviors, on the other hand, may require some form of continuous observation. The accuracy of interval sampling of behaviors, however, is sensitive to the relationship between interval length and activity duration. If intervals are long relative to the rate of change in behavior of an individual, infrequent but possibly important activities may be missed and more frequent behaviors overemphasized (Tacha *et al.* 1985, Bakeman and Gottman 1986).

Behavioral observations are also affected by variations among individuals or habitats. Individuals within a population may differ in their allocation of time to behaviors or of behaviors to microhabitats as a consequence of their proximate responses to environmental circumstances (e.g. air temperature, appearance of a predator), their phenotype (e.g. early experi-

Table 3.4. *Comparison of patterns of activity budgeting and microhabitat use by Savannah Sparrows (*Passerculus sandwichensis*) in a Wisconsin grassland as derived from continuous observations versus single point–samples of behavior.*

Values are percentages of total observations (continuous observations, $n = 6309$ s; point–sample observations, $n = 307$).

Category	Continuous observations	Point–sample observations
Activity		
Singing	61.3	90.9
Foraging	23.6	3.6
Perching	7.7	3.3
Aggression and display	3.1	2.3
Flying	3.8	0.0
Preening	0.4	0.0
Microhabitat		
Grass	30.4	13.0
Forbs	50.1	50.1
Wire	3.7	22.1
Post	9.6	13.0
Tree	0.0	1.6
Air	6.2	0.0

Source: From Wiens (1969).

ence, age-dependency of behaviors), or their genotype (e.g. polymorphism) (Davies 1982). Observations gathered from only a few individuals in a population, therefore, may not be representative of the population as a whole, especially if those individuals are selected for observation because they are easier to observe. A large sample of behavior from a few individuals is not the same, either biologically or statistically, as samples of behavior from a large number of individuals (Machlis *et al.* 1985). Individuals also behave differently in different habitats (Collins 1983a, b, Leger *et al.* 1983), so observations that are not apportioned properly among the habitats represented in a study may produce a misleading representation of behavior. Say, for example, that individuals of species A are more numerous in dense woodland and species B is more abundant in open savannah. If one obtained behavioral observations on each species only where it is most abundant, the observations might provide information on the behavior of each species in its respective habitats but they would not permit valid comparisons of the behavioral budgets of the two species independent of habitat.

Measuring resources

Tests of most hypotheses in community ecology eventually require some assessment of resources. The frequent absence of resource measurements in community studies may stem, in part, from the widespread acceptance of assumptions about resource limitation, but it also attests to the difficulty of measuring resources (Wiens 1984b).

One must first determine which variables to measure in order to document something about resources. For birds, food and habitat are the most important general resource categories, but what specific measures of 'food' or 'habitat' will be most useful? Usually this decision is made arbitrarily, based upon varying levels of knowledge or intuition about the natural history of the species. It has long been believed, for example, that features of habitat structure are more important to birds than are the details of habitat floristics (e.g. Lack 1933, Svärdson 1949). Most investigations of the relations of bird community patterns with habitat 'resources' have considered various, easily measured features of habitat structure (e.g. height distribution of ground coverage of vegetation). Some studies (e.g. MacArthur and MacArthur 1961, Cody 1968) have used only a few such features, while others (e.g. Emlen 1956, Wiens 1969) have developed elaborate measurement systems involving a large number of variables. Recently, it has become apparent that features of habitat floristics may also be important to birds (Wiens and Rotenberry 1981a, Holmes and Robinson 1981, Rotenberry 1985, Rice *et al.* 1984, Abbott and Van Heurck 1985), and measurement of the coverages of different plant species has become more widespread.

The value of any resource measurement of course depends on how realistic it is biologically. The ease and precision with which we can measure some feature of the environment (e.g. soil pH) should not lead to an arbitrary conclusion that it is an important aspect of habitat resources. Resources should be viewed and defined from the perspectives of the organisms, not our own (Wiens 1969, 1984b, Jarman 1982). Holmes (1981), for example, has suggested that it is necessary to know how birds use their habitats in order to determine which habitat variables should be measured. From his analysis of the foraging behavior and substrate use of forest birds, he determined that different species of broad-leaved trees could be important habitat components in a mixed hardwoods forest. Without a detailed, quantitative knowledge of the natural history of these species, the importance of this distinction would not have been recognized. A structural approach that measured only the presence or coverage of deciduous versus coniferous trees would have failed to portray habitat characteristics realistically.

Presuming that a 'resource' to be measured can be identified correctly, one must also know what aspect of the organism–resource linkage is being measured. Resources occur in the environment in a certain abundance and distribution. Only a part of that pool of resources may be available to organisms at a given time. What we usually see is organisms *using* resources – eating, building nests, seeking cover, and so on. Resource use bears some relationship to resource availability and thence abundance, but a number of factors can affect this relationship (Wiens 1984b). Drawing inferences about the availability of resources to organisms or about resource abundance from observations of resource use alone is thus logically unjustified. Direct measurements of resources must be taken, but it is often difficult to know exactly what these measurements represent. If one measured the number of seeds in each of dozen 10-cm-deep soil cores arranged in a random fashion in a desert habitat, would the results indicate something about resource abundance? Probably not, given the tremendous spatial variance in seed densities in deserts (Reichmann 1981, 1984). Would they portray resource availability to ground-foraging desert birds? Probably not either, as the portion of the samples below 1–2 cm soil depth would be unavailable to most species. Most attempts to measure resource levels are constrained in similar manners. They usually suffer from a lack of variance measures, they record standing crops rather than resource dynamics, and they produce values that bear unknown relationships to actual resource abundance or availability. In order to measure resources in a system, we must consider what constitutes a 'resource' to the organism and the factors that may influence the linkages between the abundance, availability, and use of that resource.

Measurement of resources is subject to the same sorts of methodological problems and biases that plague other forms of observation or sampling. The method used to obtain data, for example, may influence the nature of the data obtained. Taylor *et al.* (1984) measured both structural and floristic habitat variables in a heterogeneous tropical area in northern Australia using (a) systematic placement of sampling stations irrespective of habitat types and (b) random location of stations within defined habitat types (a stratified random design). The vegetation data were then related to data on faunal composition (birds, reptiles, amphibians, and grasshoppers) by correspondence analysis. In the analysis using the stratified random measurements, vegetation structure and lifeform explained most of the variation in faunal composition, but floristics explained almost all of the faunal variation when the systematic sampling data were used (Fig. 3.3). When sites falling in ecotones between habitat types were excluded from the systematic samples, the patterns more closely approximated those obtained using the stratified random design. The differences, in other words, were due to whether or not 'edge' habitats were included in the analysis.

Differences between observers may also affect resource measurements. In a study at several sites in an oak-maple forest in Ontario, four observers independently and repeatedly measured 20 vegetation variables (Gotfryd and Hansell 1985). The observers differed significantly from one another in their measurements of 18 of the variables, and data transformations and use of nonparametric statistics did little to mitigate these observer effects. Principal Component Analysis (PCA) based on the data of each observer separately provided axes that weighted the vegetation variables and portrayed the sites in PCA space quite differently (Fig. 3.4). Discriminant Function Analysis (DFA) was even more sensitive than PCA to the differences among observers. Gotfryd and Hansell concluded that, by ignoring observer-based variability in such measurements, one may reach conclusions that are 'precariously balanced on artifacts, spurious relations, or irreproducable trends' (1985: 224). Block *et al.* (1987) reported similar sensitivity of vegetational habitat characterizations to observer differences. Measurement of resources, like censusing populations, is clearly a tricky business, full of pitfalls for the unwary or careless scientist.

Comparisons and experiments

Comparisons

Even if observations are obtained with minimal measurement error, difficulties may arise when they are compared in order to derive patterns or test hypotheses. In community ecology, the comparisons usually are of observations gathered at a number of locations – the variations among sites serve as the framework for a 'natural' or 'mensurative' experiment (Cody 1974b, Diamond 1983, 1986a, Hurlbert 1984, James and McCulloch 1985). Such 'experiments' avoid the detail and design problems that may beset conventional manipulative experiments (see below) and may be appropriate for investigating phenomena at broad temporal or spatial scales, but control over exactly what is being compared is sacrificed. A series of locations may be presumed to be arrayed along a dimension of variation, such as elevation, vegetation coverage, or the relative densities of potential competitors, that the investigator had in mind when selecting the locations. There is no assurance, however, that the variable of interest represents the only or even the major dimension of variation among the sites. One may use

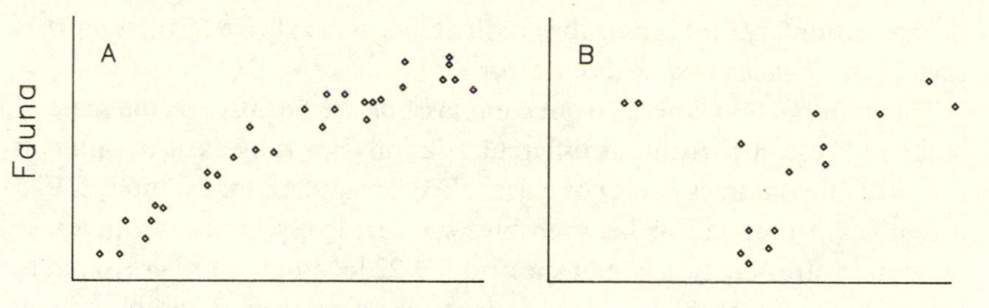

Fig. 3.3. Relationships between the first-dimension vectors derived from correspondence analysis of vegetation measures and faunal surveys in Northern Territory, Australia. A. Faunal composition versus floristics for data derived from systematic sampling; B. the same, using vegetation data obtained by stratified random sampling. For A, $R^2 = 0.96$; for B, $R^2 = 0.40$. Modified from Taylor *et al.* (1984).

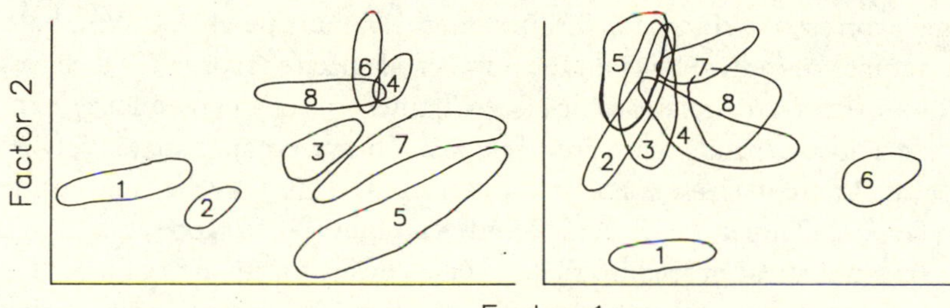

Fig. 3.4. Principal component ordination of vegetation data obtained at seven sites in an Ontario oak-maple forest by two observers working independently. The closed circles in each graph enclose the replicate observations made by that observer at each site. Note that the ordinations derived from the data of the different observers differ markedly in the way sites are arrayed; they also differ in how vegetation features load on these components. From Gotfryd and Hansell (1985).

statistical techniques such as partial correlation analysis or multiple analysis of covariance to disentangle the variable of primary interest from the concurrent effects of other measured variables (e.g. Schoener 1975, Schoener and Toft 1983), but it is naive to expect that other unmeasured and unconsidered factors do not also vary among the locations. In addition, partial correlations will suffer the same instabilities as regression coefficients in the face of multi-colinearity of variables. Because there is usually no replication of points along the gradient, one is also forced to assume that each point is measured without error.

The patterns that emerge from comparisons are sensitive to the scale or scope of the comparison, as different relationships may exist in different subsets of the samples being compared. Erdelen (1984), for example, found a significant correlation between bird-species diversity and an index of vegetation stratification over a spectrum of 22 locations from grassland to forest, but deletion of the three low-stature sites caused the pattern to vanish. We have noted similar scale-dependent changes in patterns in our studies of grassland and shrubsteppe bird communities (Fig. 3.5) (Wiens and Rotenberry 1981a, Wiens 1986, Wiens, Rotenberry and Van Horne 1987). Obviously, what is included in a comparison exerts a considerable influence on the patterns one finds.

In order to be included in a comparison, the observations or samples must be comparable. Observations gathered using different procedures are not, as we have seen. Surveys of bird communities conducted on areas of different size are often compared directly, however, the measurements being expressed in some standardized unit (e.g. pairs per 40 ha). Yet, both the number of individuals and the number of species recorded in a census follow species–area relationships (see Chapter 5). As a consequence, censuses are area-dependent in a nonlinear fashion, and comparisons involving sites of different sizes must be standardized. This may be done using rarefaction (Tipper 1979, James and Rathbun 1981, Kobayashi 1982, 1983), a statistical procedure that, given a census for an area of a certain size, allows one to estimate the number of individuals and species that would be expected to occur had only a smaller portion of that site been surveyed. In this way, censuses from a number of locations may be standardized on the basis of the smallest area sampled. Rarefaction procedures contain several assumptions (the most critical in applications to bird communities is that the spatial distribution of each species in the community is homogeneous; Tipper 1979), but they represent one way to deal with unequal sizes of samples in comparisons. Rarefaction does not compensate for the additional problem of edge effects noted above, except insofar as such effects are contained in the species–abundance distributions.

There is another effect of sample area that is more subtle. Consider the following scenario: We wish to record changes in population density during a series of years on a 10-ha plot that is carefully censused each year. We record considerable temporal variation and, in the process, note that the plot is rarely fully occupied by breeding birds – the plot is not saturated. Given the care we have taken to enumerate completely all individuals on the

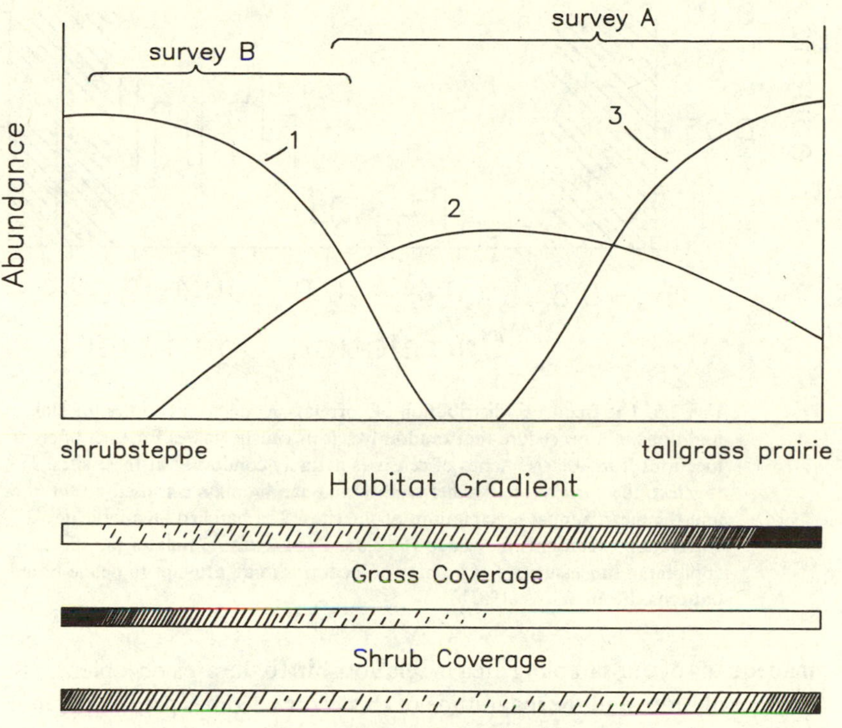

Fig. 3.5. The effects of scale of study on patterns of habitat association. In this schematic diagram, species 1–3 have characteristic distributions on a gradient from shrubsteppe through short- and mixed-grass prairies to tallgrass prairie. Grass coverage, shrub coverage, and overall vegetation height change on the gradient as shown below. In survey A, a large portion of the gradient is sampled but extreme shrubsteppe sites are omited, whereas in survey B only shrubsteppe and a few grass–shrub sites are studied. The species will exhibit different patterns of habitat association in the two surveys. Species 1, for example, will exhibit a strong negative association with grass coverage and a positive association with shrub coverage in survey A but may fail to show either association in survey B. Modified from Wiens and Rotenberry (1981a).

plot, we may believe the yearly density changes to be an accurate representation of true population changes of the species in this habitat and general location. This may be so, but is is also possible that the number of birds occupying a larger area that includes the plot has actually remained constant over this time period and that the density fluctuations we have recorded represent nothing more than reshufflings of individual territories in a sparsely packed habitat (Wiens 1981a). The 'dynamics' reflect the

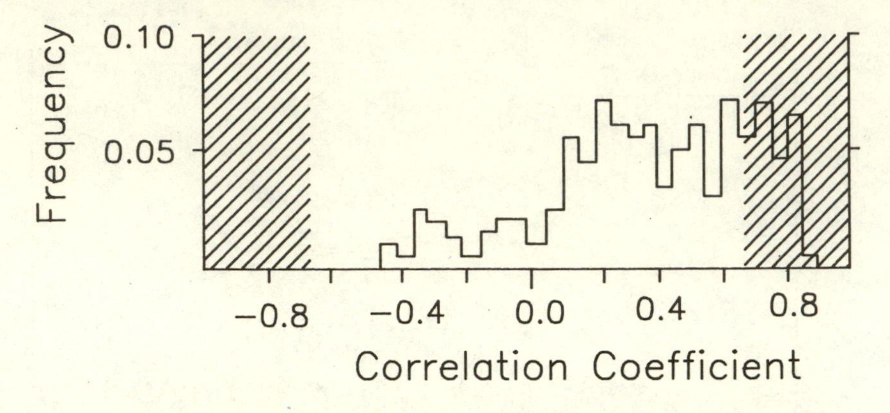

Fig. 3.6. The frequency distribution of correlation coefficient values for 200 iterations of a procedure that randomly selects census values for each of eight locations from a larger series of censuses actually conducted at these sites. In this test, the correlations are between total avian biomass censused on the sites and the mean annual precipitation of the sites. The hatched area indicates correlation coefficients for which $P<0.05$. The exercise simulates the effects of combining single surveys of a series of locations in an attempt to define broad patterns. From Wiens (1981b).

inadequacy of our sampling area in relationship to the area occupied by the real population, and the magnitude of this error becomes greater the more sparsely occupied the habitat. The solution to this problem is to determine the size of sample area on the basis of the biology of the system rather than of convenience or tradition and to replicate plot samples adequately.

Community comparisons are also often based on observations at a number of locations that have been surveyed only once. The validity of this procedure rests on assumptions of community equilibrium. If individual locations vary over time, however, the patterns that one obtains from such comparisons may depend on just when the sites were sampled. To explore these effects, I conducted a simple exercise using a set of 40 census values obtained from eight grassland locations over several years (Wiens 1981b). Each location had been censused several times, and I asked whether significant relationships between total avian biomass and mean annual precipitation might or might not have emerged had each of the sites been sampled only once. The correlation coefficients for 200 iterations of this exercise spanned a broad range of values (Fig. 3.6); roughly 19% of them were statistically significant ($P<0.05$). Thus, if the 'true' pattern in this system was one of increasing biomass with increasing annual precipitation (as indicated by the entire set of 40 samples), this pattern would have been revealed with a probability of only 0.19 using single-sample surveys. Of

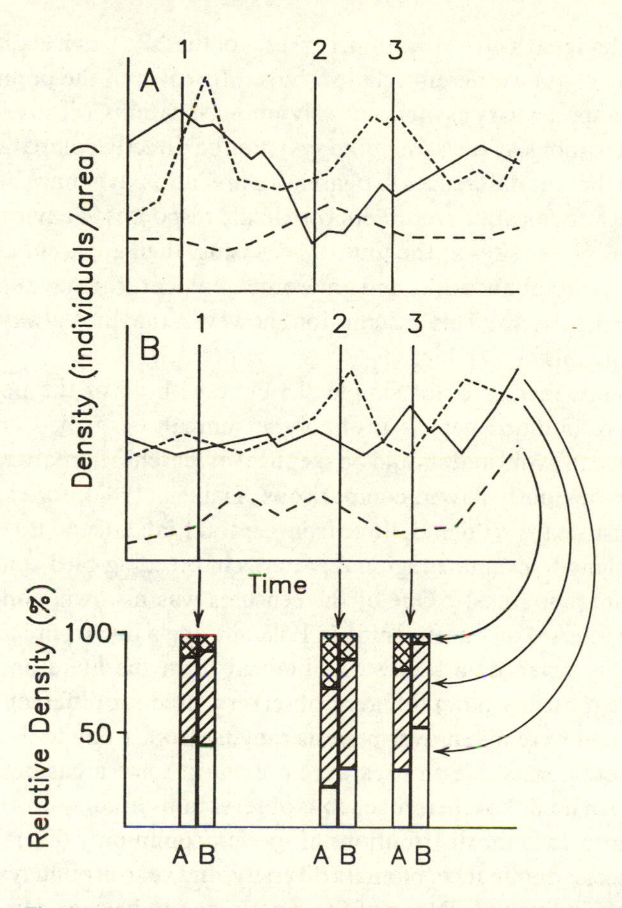

Fig. 3.7. Hypothetical example of density variations of three species (solid, dashed, and dotted lines) at two locations (A and B). Censuses are taken at times 1, 2 and 3 without knowledge of the true dynamics of the populations. Relative densities of the three species at each census are shown at the bottom. Each census reveals a different pattern in the relations between locations A and B, and none captures the temporal behavior of the system.

course, the single-sample surveys in this exercise suffer from a small sample size ($n = 8$). Given this, one might be tempted to ignore statistical tests altogether and simply examine the data for biologically meaningful trends. Because single-sample surveys can produce a broad range of either positive or negative trends (Fig. 3.6), this approach creates the potential for almost any relationship to emerge.

The difficulties associated with deriving patterns from a comparison of single samplings of locations arise largely from their temporal dynamics (Fig. 3.7). If populations of different species are in different phases of population change at a given location, censuses of that site taken at different times will depict quite different relationships. Moreover, if the population dynamics of a species vary asynchronously among locations, censuses taken at different locations at the same time may not be directly comparable. If one assumes that the differences in phases of population dynamics between species or between locations represent proximate responses to environmental conditions on the sites at the time of censusing, then apparent correlations between bird abundances and environmental features may represent real biological patterns. This assumption, however, may not always hold for mobile organisms like birds.

Comparisons, then, are not simple. To cope with all of the potential pitfalls I have outlined here is probably an unrealistic goal. Certainly, however, one can avoid and should be skeptical of searches for patterns that are based on obviously flawed comparisons. Tiainen (1980), for example, used two censuses from Poland, three from central Finland, and three from Finnish Lapland to examine regional patterns in breeding-bird communities of mature pine forests. One of the censuses was his own, conducted during a 2-h interval on one morning in Poland using a line-transect count. The remaining seven data sets were obtained from the literature. They involved five (possibly nine) different observers, used six different survey techniques, and were taken over periods ranging from 1 day to 3 years in areas of different sizes. Nest boxes were present in some areas but not in others. Tiainen used these heterogeneous observations in comparisons that produced patterns in the distributions of species, community diversity and evenness, species dominance, biomass diversity, and year-around residency of species and individuals. None of these patterns can be considered valid on the basis of Tiainen's analysis.

Experimental manipulations

In a manipulative experiment, some variable of interest in a system is altered by the investigator; subsequent changes in the treated system are presumed to be due to the alteration of that variable. Although the laboratory is the traditional domain of experimentation, field experiments have been part of ecological investigations for well over a century (Jackson 1981, Tilman 1982) and, for many of the questions in community ecology, they are the only form of experimentation that is possible. Given the central position of competition in recent community ecology, it is not surprising that most of these experiments have sought evidence of competition rather than anything else. Field experiments on competition have become quite fashionable, as the reviews of Schoener (1983b), Connell (1983), and

Underwood (1986b) clearly show. They are sometimes regarded as the *only* way to document the effects of a process such as competition. In such a climate, there is a danger that the results of poor experiments may be accepted without question if they provide the answers that are sought (Connell 1983).

What constitutes a 'good' or a 'poor' experiments depends on the standards one adopts. In selecting studies for his review of field experiments, Schoener (1983b) excluded those without proper controls, although he admitted to being 'rather liberal' in this judgment. Connell (1983) was more stringent in his criteria for controls and excluded some studies considered suitable by Schoener; in the process, he may have preferentially excluded studies demonstrating competition (Ferson *et al.* 1986). Underwood (1986b), using yet different criteria, evaluated many of the studies cited by both Connell and Schoener. He found

> some studies that were partly incomprehensible, some that were not about competition, some that were not really experiments, and some that were not particularly in the field. Others were lacking in sensible controls and/or replication. Still others were designed to be confounded so that interesting comparisons, and crucial tests of important hypotheses were not possible, were invalid, or only indirectly interpretable (1986b: 243)

Only a third of the 95 studies he evaluated could be considered really useful.

What are the essential ingredients of rigorous experiments in ecology? Assuming that one has framed an unambiguous, logically valid hypothesis, one must first determine what to manipulate and what to measure to determine responses. Dumping an undetermined amount of food into an area and observing how the populations respond, for example, scarcely qualifies as an experiment. Was the food palatable, highly preferred, or perhaps so unnatural that it was not even recognized as food? Were the population responses a consequence of *in situ* reproduction or of emigration? Did the food attract other species, among them predators? In such an exercise, it is difficult to determine exactly what has been manipulated, and measurements of responses may be confounded by unanticipated changes in other variables. An experiment requires a cleaner design if it is to yield unambiguous results.

A well-designed experiment also requires proper control and replication. Controls are necessary to deal with the *ceteris paribus* assumption. They provide a check that tells one whether or not things other than the manipulated variable really were equal and changes in the response measures really were responses to the manipulation rather than changes in the system that occurred independently of the experiment. Designing good controls often involves more than simply comparing the experimental area with some superficially similar area nearby: both the control and treatment areas must be monitored and measured in exactly the same fashion except for the manipulation itself. Ideally, the areas should be identical at the onset of the experiment.

In the field, of course, no two areas are ever identical, and thus it is not likely that a control and a treatment area will differ only in the manipulated variable (Hurlbert 1984, Underwood 1986b, James and McCulloch 1985). Replication provides an essential check on the effects of natural variations in both environments and populations that might by chance produce differences between single control and treatment areas. In many field experiments, 'pseudoreplication' is substituted for true replication: rather than manipulate several independent plots, for example, multiple samples taken from within a single treatment plot are considered 'replicates' (Hurlbert 1984). They are not. Measures of responses to the manipulation are still obtained from a single treatment and are thus sensitive to any chance or deterministic variations in that area that are not directly associated with the manipulation itself (e.g. the plot happens to be within the home range of a predator). Pseudoreplication is a particularly dangerous practice in situations in which there is naturally high spatial or temporal variation. Because of this, attention must also be given to exactly how replicate and control plots are arrayed with respect to one another. Placing several replicated treatment plots at one end of a field and several control plots at the other, for example, creates the possibility that any difference between treatments and controls is due to their position in the field. Placing a treatment plot immediately next to a control creates the possibility of 'spillover' effects between the plots (Underwood 1980, Underwood and Denley 1984). Some form of randomization and interspersion of plot locations is a necessary part of good experimental design (Hurlbert 1984, James and McCulloch 1985).

Replication may be difficult or impossible under some circumstances, and unreplicated experiments may be acceptable when the treatment involves a major perturbation (e.g. clearcutting a forest) or when only a crude estimate of the effects of the treatment is desired. Difficulty is sometimes used as an excuse, however, for the absence of replication in experiments from which firm conclusions have nonetheless been drawn on the basis of subtle effects. A properly designed experiment should include replicated controls and treatments, and the effects must then be gauged by the magnitude of difference between the treatments and any of the controls. In the absence of true replication, one must regard the results of an experiment with considerable caution.

An experimental manipulation by definition imposes an artificial change on the system under study, and there is thus always a 'Heisenberg effect' in experiments: is the response a result of changes in the variable one intended to manipulate or of some aspect of the experimental intervention itself? Placing animals in enclosures many times smaller than their normal home ranges (e.g. Heller 1971) is bound to create aberrations in behavior. Even large enclosures may enhance the spread of diseases or parasites while excluding natural predators, altering mortality sources and perhaps modifying the intensity and directions of species' interactions (Underwood 1986b). Nest boxes have been used in many of the experimental studies of tits and flycatchers in Britain and Europe. Sometimes the experimental design has involved comparisons of areas with and without boxes, but more often boxes have been present in both the treatment and control areas; they are not part of the experiment *per se* but are used because they enable investigators to monitor reproductive behavior and success quite accurately. Their availibility, however, may lead to unrealistically high population densities (e.g. Minot 1981, Alerstam 1985), which are not always taken into account in interpreting the results of experiments.

Experiments are usually conducted within a specified time period, often one determined by logistical or budgetary constraints or the investigator's patience. Among the field experiments included in the reviews of Connell (1983) and Schoener (1983b), for example, relatively few followed the experiment for more than a single post-treatment year. One might feel justified in expecting an immediate response by the organisms to a massive manipulation of the system (but see Wiens and Rotenberry 1985), but most experiments involve less dramatic treatments. There are a variety of reasons not to expect an instantaneous response (Wiens 1986). Many birds, for example, exhibit site tenacity, individuals having previously bred successfully in a habitat or location tending to return there in subsequent breeding episodes despite environmental changes (Hildén 1965). This produces lags in responses, a 'tracking inertia' (Wiens 1984b). It means that clear responses to an experimental manipulation may take several years to materialize and that what may seem to be immediate responses may have nothing to do with the experiment (something unlikely to be recognized in unreplicated experiments). The effects of these time lags will be heightened if the community contains webs of interactions with many indirect effects (e.g. Inouye 1981, Brown *et al.* 1986, Bender *et al.* 1984). Experiments with organisms like birds simply take a long time.

Two general conclusions regarding field experiments are apparent. First, it is probably not worth doing experiments in community ecology unless they can be done correctly. Doing an experiment properly involves paying attention to the key features of experimental design: control, replication, and proper location of plots. It requires that the manipulations be realistic and that the effects of the experiment itself on the organisms be considered. It requires foresight, to determine how long an experiment should be followed and what sorts of indirect effects must also be monitored. And it requires that the results be interpreted with an open mind, unfettered by preconceptions about what 'should' have happened. Second, good experiments rest on good natural history. No matter how much one desires to get on with the business of experimenting with a system, it is simply not possible to design an intelligent experiment without some basic understanding of the organisms and the environment. Moreover, designing a good experiment requires clear objectives, and these also rest on such an understanding of the system. To do experiments without sufficient knowledge of the system is fiddling, not science. It can only serve to perpetuate Underwood's (1986b) assessment of the current state of experimentation: 'appalling'.

Analysis of results

This is not the place to detail all the pitfalls and potential errors that await the ecologist preparing to analyze data – that is what statistics books are for. Nonetheless, several abuses are common enough in the analysis of community data that they merit brief mention.

It is not always true that individuals have a hypothesis clearly in mind when they undertake a study. When observations are gathered in the absence of an *a priori* hypothesis, however, they may be subjected to several forms of 'data-dredging' to produce *ad hoc* tests of *a posteriori* hypotheses derived from the observations (Selven and Stuart 1966, Strong and Simberloff 1981). One may 'snoop' by conducting tests of many possible relationships. Some will be intercorrelated, while others may emerge as significant by chance alone; in either case, the 'patterns' may be artifacts. After examining a large array of variables, one may 'fish' by choosing just those variables that express the desired pattern. Alternatively, one may search through a great many relationships to find some worth testing statistically; this is 'hunting', and failures of the hunt are simply not reported. None of these forms of data-dredging is of itself bad – all represent reasonable ways to explore for patterns. They do not represent valid end-points of an objective data analysis, however, and the probabilities obtained in the statistical tests are of dubious credibility.

More often, statistical analyses are used to test hypotheses that were contained in the initial objectives of a study. Of the wide array of procedures available for doing this, various multivariate analyses have become increasingly prevalent in community studies, perhaps because of their multidimensional parallel with niche theory. Such procedures can be a powerful tool in condensing variation in large, multiple-variable data sets. Each procedure, however, carries with it statistical assumptions, and a failure to meet or at least approximate those assumptions may invalidate the test. Tests of significance in DFA, for example, rest on the assumption of multivariate normality – that values on each of the axes are normally distributed. As Austin (1979) has observed, most ecologists seem to accept normality as a matter of belief: 'the field ecologists fancy that it is a theoretical principle and the theoretical ecologists that it is a field observation'. Ecologists all too rarely heed the statistical requirements of such tests (Johnson 1981, Green 1979, Legrende and Legrende 1983). Even if statistical tests are not done as part of DFA, however, the optimality of the linear discriminant function itself is still dependent on the assumption of multivariate normality with equal covariance matrices. Violation of these assumptions, as though curvilinearity, distorts the distributions of samples in both DFA and PCA multivariate space, which at the very least complicates their interpretation. Some ecologists (e.g. Gauch 1982, Austin 1985) have advocated use of detrended correspondence analysis (DCA) or multidimensional scaling to avoid these problems.

Even if the analyses are sound statistically, errors may be introduced in their interpretation. Simply because a Discriminant Functions Analysis derives a primary multivariate dimension that provides maximum separation among the species in a community does not justify the conclusion that this dimension represents the primary means of niche partitioning in the community. In the same sense, the components derived in PCA cannot properly be considered to be 'niche dimensions' except by means of an arbitrary, operational decision by the investigator. If one plots confidence ellipses about the distributions of samples for different species in multivariate PCA or DFA spaces defined by environmental variables (e.g. Dueser and Shugart 1978, 1979; Noon 1981), the sizes of these ellipses and their positions in the multivariate space are both dependent on sample size (Van Horne and Ford 1982, Carnes and Slade 1982, James *et al.* 1984, Seagle and McCracken 1986). Because of nonequivalence of the axes, equivalent Euclidean distances between species' means in this space also do not have the same probabilistic meaning (James *et al.* 1984, James and McCulloch 1985). Multivariate procedures are attractive analytical tools because they condense similarities among species for a large number of variables into a smaller, more manageable set of derived variables, but both their application to data and their interpretation require careful attention (Birks 1987).

Another difficulty arises when one layers statistical tests (especially multivariate ones) on top of one another. One might subject a matrix of habitat measures to PCA, for example, and find that the first three components accounted for 60% of the variation present in the original matrix. A similar analysis of the densities of bird species at these locations might explain 68% of the variation in the first three components. One might then compare just these three components for the two data sets, using canonical correlation analysis, and find that 20% of this variation is explained in the first canonical correlation. With each step, the resulting model becomes more and more precise, but it deals with less and less of the variation that was present initially. By the time we finish, most of the variation in the system has been left behind, 'unexplained' by the analysis. The obvious end point of this orgy of multivariate layering is an extremely precise statement about almost nothing.

To some degree such excesses are a consequence of a tendency to consider variance as not containing anything of great biological interest. Statistically, it represents 'noise', and the function of most statistical tests is to isolate this noise so that statements can be made about means, regression coefficients, and the like. Variance contains noise from a variety of sources: measurement error, chance variations between samples, and variations due to differences between individuals, species, years, locations, or environmental features. Often some of these will be randomly sampled, providing a basis for statistical inference. Others (especially broad-scale spatial and temporal components) will be fixed at selected points, and the resulting data analysis will be conditional on those chosen points. Statistical inference from a 1-yr study at a few arbitrarily selected sites, for example, permits one to generalize to what might have occurred in imaginary replications of the same study during that year at those locations. Generalizations beyond this would not be data-based but founded on *a priori* assumptions (e.g. equilibrium), because the statistical analysis would not allow one to infer anything about what might happen at other locations or times. In this example, the year in which the study was done has not been randomly chosen and, because variation between years may be considerable, one cannot freely extrapolate to past or future years. Here, years represent a nonsampled component of variance. A considerable portion of the variance that is so easily discarded in statistical analyses may in fact represent such unsampled components and may contain important insights into the dynamics of the community.

Field observations in ecology are often not strictly independent of one another, and this violates the assumption of independence of samples that is basic to all statistical tests. Samples from a 'pseudoreplicated' design, for example, are rarely independent. When multiple observations are made on a small number of individual organisms, the samples obtained from any single individual are not independent but are pseudoreplicated samples. Machlis *et al.* (1985) have discussed this problem in behavioral analysis and have shown that the probability of rejecting a true null hypothesis (Type I error) is almost always substantially greater than the stated alpha level when such pooled samples are used (see also Bakeman and Gottman 1986). Nonindependence of observations is particularly troublesome in the analysis of activity budgets. If one records observations of an individual's behavior at arbitrary time intervals, for example, the observations do not represent independent samples. They are samples extracted from an ongoing stream of behavior in which what the individual does at one moment is strongly influenced by its previous behavior (Altmann 1974, 1984, Morrison 1984, Wiens, Van Horne, and Rotenberry 1987). Adjustments can be made in the analysis to reduce this problem, but they are not easy (see Bradley 1985, Bakeman and Gottman 1986, Wiens, Van Horne, and Rotenberry 1987).

Two related aspects of statistical analysis of observations merit mention. We go to great lengths to avoid Type I errors, setting stringent probability levels for the recognition of significant interactions or effects. We follow the dictum $P < 0.05$ with slavish adherence, often considering any test with greater values of P as lacking biological as well as statistical significance. Setting significance at $P < 0.05$ is a convention adopted by statisticians in the context of experiments with tight controls on extraneous sources of variation. It is often not suitable for the analysis of uncontrolled field observations in a complex and variable world. The attainment of a given significance level is also dependent on sample size – with a large enough sample, almost any test will exhibit significance. Not only should we consider the power of statistical tests (e.g. Toft and Shea 1983, Verner 1983, Rotenberry and Wiens 1985), but we should temper our judgment of 'significance' with biological insight and recognize that, say, $P < 0.15$ may be quite important in some situations.

This adherence to arbitrary statistical standards, combined with the powerful influence of preconceptions about what we *should* see in nature or discover in an experiment, leads to a strong prejudice against negative results. Observations that seem anomalous with respect to some anticipated pattern or results of comparative or experimental tests of hypotheses that are not significant are often discarded or not submitted for publication.

Journal reviewers and editors may not be impressed by studies that 'show nothing', regardless of how carefully the work is done. The consequence, as Connell (1983) has noted, is a literature that may be strongly biased toward reports that are consistent with expectations (which, if one believes Kuhn, are determined by the prevailing paradigm). Negative findings often represent falsifications and thus merit close attention.

Conclusions

Doing community ecology is not easy. A large number of errors and biases can distort patterns and devastate tests of hypotheses. I have summarized some of the major difficulties and ways to minimize their effects in Table 3.5. Overriding these specific problems are three general considerations: 1. Objectives should be determined *before* the methodology of a study is defined. 2. One should be attentive to sources of error and bias and attempt to control for them or measure their effects. 3. One must have the patience to do a study correctly, paying careful attention to the details of methodology, even if this takes considerably longer and involves more effort than the easier, superficial alternative.

Unfortunately, a great many of the studies reported in the literature of avian community ecology are flawed in one way or another by a neglect of these methodological details. Often the flaws are not fatal and require only that one keep them in mind when interpreting the results. There has also been little standardization of methods in these studies, however, even when the objectives are the same. This renders their comparison difficult.

When I initially contemplated writing this book, I though how splendid it might be to draw together a large body of literature, compare the findings, and synthesize some quantitative overview of the 'true' patterns of bird communities from such comparisons. The considerations developed in this chapter quickly dispelled that idea. Quantitative reviews based on the literature are beset by a large number of errors and biases: they are vulnerable to the ways that preconceptions influence the selection of systems to study or hypotheses to test and to the low frequency with which negative findings are reported; they are distorted by the use of different census procedures by different observers on areas of different size; they are weakened by the general lack of measurement of resources or other environmental variables; they are confounded by differing degrees of control or replication or manipulation in experiments; – they are confused by all of the difficulties listed in Table 3.5. For these reasons, reviews that attempt to derive quantitative patterns from a large number of published studies, such as those of Schoener (1983b) or Connell (1983) (see also Schoener 1985a, Hairston 1985, Ferson *et al.* 1986) can be misleading and must be interpreted with great care.

Table 3.5. *Major methodological difficulties in conducting studies in avian community ecology and suggested steps to their solution*

Difficulty	Possible solution
Measurement of variables	
A. Censusing	
1. Different methods yield different results	Select methods most appropriate to objectives; standardize methods within study
2. Observer biases	Use few, trained observers; determine hearing ability
3. Species differ in detectability	Adjust survey design for detectabilities; standardize by phenological state; use techniques appropriate to species
4. Determination of territories	Standardize mapping procedure; adjust for differences in breeding phenologies
5. Differences between locations	Measure features of locations; standardize area and amount of edge
6. Sampling disparities	Standardize duration and number of surveys; conduct adequate number to detect species present
B. Behavioral observation	
1. Visibility bias	Avoid single spot-observations unless adjusted to compensate for visibility; sample behavior sequentially, omitting initial observations and grouping sequences for analysis; sample in all microhabitats
2. Individual variation	Sample many individuals
C. Measuring resources	
1. Which variables?	Knowledge of natural history; quantitative observations of resource use
2. How to measure?	Distinguish among resource abundance, availability, and use; record temporal and spatial variance
Conducting comparisons and experiments	
A. Nonmanipulative comparisons	
1. Effects of unmeasured variables	Measure suspected variables; use partial correlation tests; design comparison to randomize effects
2. Heterogeneous observations	Standardize observation procedures and measurement; use rarefaction or equivalent procedures to adjust for differential sampling
3. Scale-dependency of patterns	Conduct comparisons on same scale; determine scale using biological criteria
4. Single-sample bias; effects of chance	Replicate samples (avoid pseudoreplication)
B. Experiments	
1. Results ambiguous	Check for interaction effects; use clear design; careful determination of what to manipulate; run experiment long enough

Table 3.5. (*cont.*)

Difficulty	Possible solution
2. Temporal variation; *ceteris paribus* violated	Adequate, replicated controls; long-term monitoring
3. Unmeasured plot differences	Replication; randomization and interspersion of plot locations
4. Sampling error	Replication
5. Procedural effects	Use controls; test effects separately
6. Manipulation unrealistic	Use knowledge of natural history when determining manipulation
Analysis of results	
1. Data-dredging	Frame hypotheses before obtaining data; avoid premature conclusions
2. Misapplication of statistical tests	Evaluate assumptions of tests; interpret carefully
3. Interpreting layered tests	Consider unexplained variance
4. Use of variance measures	Examine sources of variation
5. Nonindependence of samples	Test for autocorrelation; adjust sampling design; group samples for analysis
6. Missing real effects because $P > 0.05$	Consider power of tests; evaluate measurement error and sources of variation; determine acceptable P-level on biological grounds prior to analysis; use confidence intervals for the size of an effect rather than significance tests
7. Neglect of negative results	Don't

PART II

The patterns of avian communities

Ecologists can ask two sorts of questions about communities: 'how' questions, which relate to descriptions of their basic patterns, and 'why' questions, which inquire into the processes that produced those patterns. 'Why' questions cannot be asked in the absence of answers to 'how' questions. The study of bird communities, therefore, must begin with a recognition of their patterns.

Community patterns are consequences of the species composition of the community, the distribution, abundance, and morphological and behavioral attributes of those species, and the ways these relate to the environment. Of these factors, the most important determinant of community patterns is the first – which species make up the community. I begin this section by considering how communities might be assembled and then examine several kinds of patterns that are related to the numbers of species in communities and their abundances. Because such communities may be large and contain a heterogeneous mix of species, there may be advantages to considering more homogeneous subsets of species within these communities, and I next describe how such guilds may be defined and recognized. There follow examinations of community patterns defined by the morphological attributes of species, by the broad-scale and local distributional relationships among species, by the ways in which species use resources, and by changes in resource use in time and space. Because many of the patterns bear some relationship to general environmental features, I next consider the possibility that communities occurring in similar environments in widely separated areas might converge in patterns. I conclude this section with an examination of attempts to portray community patterns in terms of energy flow or bioenergetics.

The investigations of bird communities that provide information about these sorts of patterns rarely conclude without offering 'explanations' of the patterns based on the operation of one or more processes. Sometimes these explanations are presented explicitly as speculations, sometimes as formal hypotheses; often they are given as firm conclusions. Rarely are such explanations subjected to separate tests. In almost all instances they rely heavily on inference, usually with little regard for possible alternative explanations. My emphasis in this volume is on patterns, but I also will note, with only brief comments, the process interpretations that investigators have provided for the patterns they report. Evidence relating to the processes underlying bird community patterns will be considered in detail in Volume 2.

4

The assembly of communities

Local communities that are near to one another often differ markedly in species composition, while communities located some distance away may be quite similar. Communities on islands may contain different subsets of the species that are found on an adjacent mainland. What accounts for these variations? To answer this question requires an understanding of how communities have come to be composed of particular species. Communities of birds or other organisms do not suddenly appear as complete units awaiting our study but develop through time as a result of additions and disappearances of species – the dynamics of community assembly.

Factors influencing community assembly

The species composition of a local community is determined by the addition of species through successful colonization and establishment of breeding populations and by the loss of species through local extinction (Fig. 4.1). Colonists come from some broader-scale species pool – the adjacent mainland for an island or nearby forests for an isolated woodlot, for example. Although the species pool for a given local community is not easy to define, it is generally considered to contain all of the species within some reasonable distance of the location. The development of local communities is thus constrained by the nature of this source pool of species.

The species pool changes on scales of space and time that are generally beyond the focus of ecological investigations, but they are important to bear in mind (Ricklefs 1987). Additions to the species pool may come from speciation within the pool area. This is particularly relevant in island archipelagos, where isolation of islands or splintering of archipelagos can foster the evolution of endemic species, enriching the species pool for the archipelago as a whole. The radiation of endemic species in the Hawaiian islands, for example, has provided the entire native (nonintroduced) species pool for landbirds of those islands (Berger 1981, Juvik and Austring 1979).

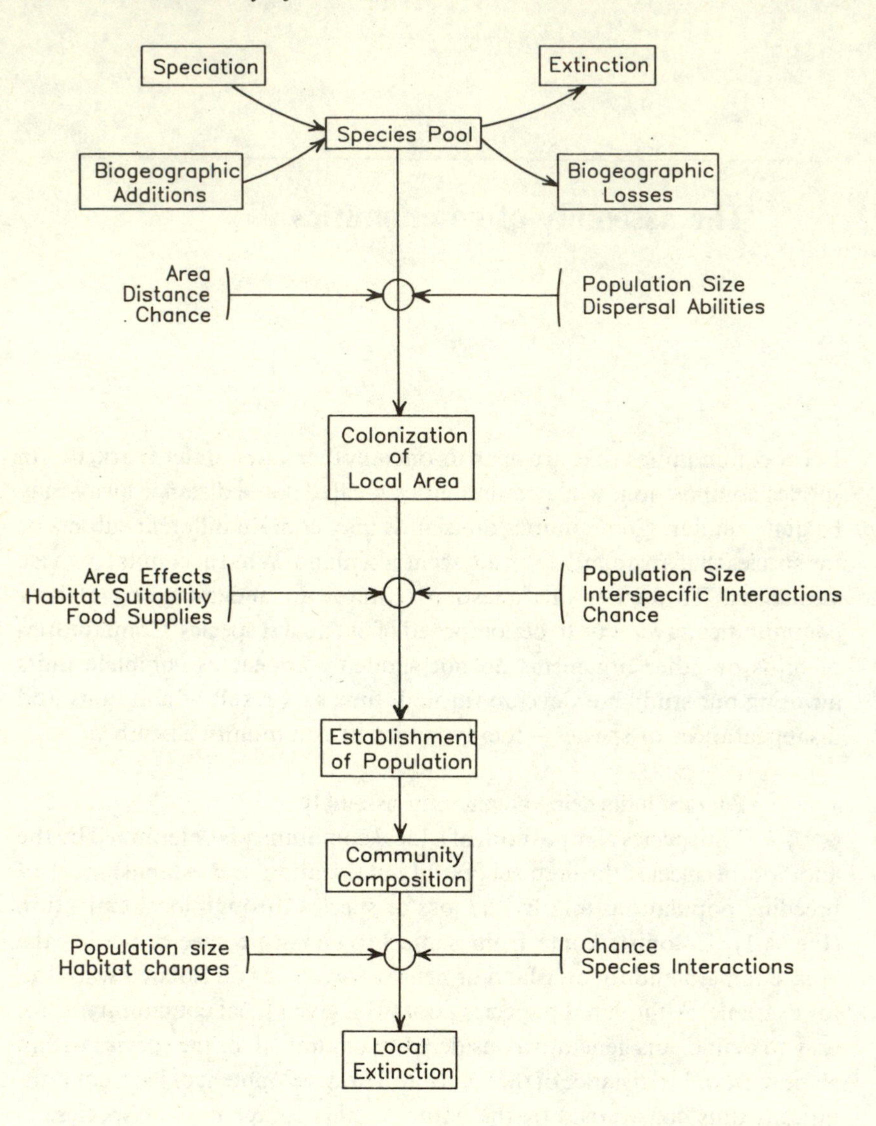

Fig. 4.1. Factors contributing to the process of community assembly.

Of a more immediate nature are additions to the species pool through recent expansions of the ranges of species. Thus, for example, the expansions of Scarlet Grosbeaks (*Carpodacus erythrinus*) in northern Europe (Stjernberg 1979, Järvinen and Ulfstrand 1980) or of Cattle Egrets (*Bubulcus ibis*) in the Americas (Bock and Lepthien 1976) or Australia (Blakers *et al.* 1984) have changed the composition of the species pool in those areas. Järvinen and

Ulfstrand (1980) calculated that the species pool of birds in Scandinavia was increased by some 88 species during the period 1850–1970, a rate of 2.8 species per decade and country.

The species pool also suffers losses. Extinction permanently removes a species from a pool, and the effects of this loss may be particularly severe in island archipelagos. Of the 36 endemic landbirds present in the Hawaiian islands during the nineteenth century, 12 are now extinct (Juvik and Austring 1979). From preliminary analyses of fossil remains, Olson and James (1982) documented an additional 39 species of endemic Hawaiian landbirds that became extinct before historic times. Similar findings have been reported for the West Indies (Pregill and Olson 1981). In addition to losses by extinction, however, the species pool may also be diminished if species withdraw from regions they formerly occupied without actually becoming extinct. In Australia, for example, Ground Parrots (*Pezoporus wallicus*) have disappeared from much of their range, and Emus (*Dromaius novaehollandiae*), which are widespread on the mainland, have disappeared from Tasmania and Kangaroo Island since European settlement (Macdonald 1973). In northern Europe, some 22 species vanished from the species pool between 1850 and 1970 as a consequence of distributional shifts, an average of 0.6 species per decade per country (Järvinen and Ulfstrand 1980). The species pool of an area is thus certainly not static.

The importance of speciation and extinction processes in determining the composition of the species pool available for formation of local or regional communities has not been widely appreciated by avian community ecologists. Brown and Maurer (1987), however, have considered how the diversification of a species pool within a geographical region may be constrained as functions of body sizes, average local abundances, sizes of geographic ranges, and the trophic roles of species, which influence both speciation and extinction dynamics. Thus, they propose that the combinations of large body size or high population density with small geographic range, for example, should not be likely to persist for very long, and there are therefore restrictions on the kinds of species that may occur in an avifauna. Brown and Maurer tested their ideas using information on nearly 400 species of terrestrial birds in North America and found general (albeit coarse and rather scattered) agreement with their expectations. This work only provides an initial set of hypotheses for more detailed examination, but it is a step toward explicit consideration of the factors influencing the species pool itself. Cracraft (1985) has also considered this issue from a different perspective.

Given the characteristics of the species pool for a region, several factors act in a more or less sequential fashion to determine which subset of species will occur in a local community (Fig. 4.1; see also Jones and Diamond 1976). First, not all of the species in a pool are equally likely to colonize a given location. The dispersal ability of a species is the primary determinant of its colonization probability. Some species are extremely sedentary or avoid movement across water or habitat discontinuities (Järvinen and Haila 1984, East and Williams 1984, Diamond 1984). A sizeable fraction of the lowland forest species and almost all of the montane avifauna in New Guinea have never been recorded even as single vagrants on islands more than a few kilometres off the New Guinea coast (Diamond 1972, personal communication). The dispersal probability of such species declines sharply with increasing distance from a source, so the distance separating an area from its source pool also figures importantly in colonization (MacArthur and Wilson 1967). Because dispersing individuals must strike the 'target' area during dispersal, the size of that area is also important; large areas are more likely to intercept dispersers than small areas (MacArthur and Wilson 1967). The population size of a species in the pool determines the rate at which dispersing individuals emerge. Rare species therefore are less likely to contribute to the assembly of a local community than are common species of equivalent dispersal ability. Finally, dispersal and colonization are probabilistic, and there is thus an element of chance in whether or not a given species in the pool will encounter a 'target' area during a given period of time. This stochastic component of colonization may be particularly important if there is a 'priority effect' in community assembly, first arrivals having a greater probability of success than later but equivalent arrivals (see below).

Second, even if colonization of an area does occur, a species cannot be considered a *bona fide* member of a local community unless it is also established as a population (Fig. 4.1). Several factors influence the likelihood that this will occur. If the area lacks appropriate habitat or food for a species or is otherwise unsuitable, the species will not become established no matter how frequently it appears as a colonist: most woodpeckers will not become established in an area lacking trees. Area may also influence the prospects for establishment. Individuals of a given species have characteristic home-range sizes and, if an area (e.g. an island or a habitat patch) is too small to contain several individuals, a viable breeding population probably will not become established. Low population size may also affect establishment if colonization is infrequent and few individuals arrive at any one time. Colonists must also deal with other species that exist in the area when they arrive. The presence of a predator may reduce the number of colonist individuals to such a low level that a persistent population does not become established. Haila and Järvinen (1983), for example, attribute the absence of Whinchats (*Saxicola rubetra*) and Meadow Pipits (*Anthus pratensis*) from an island in the Baltic Sea to nest predation by Hooded Crows (*Corvus corone*), which are unusually abundant there. In other instances, competitive interactions with ecologically similar species that are already established members of the community may prevent a species from becoming established.

Collectively, these factors act to determine which species will enter a local community during its assembly. This assembly is ongoing, in part because most communities are subjected to a continuing stream of colonists from the source pool, but also because of a third factor, the subtraction of species from the community through local extinction (Fig. 4.1). Changes in habitat within the area may render it unsuitable for some species, while other species may suffer the effects of interactions with predators or competitors that develop after their initial establishment. Rare or chance events such as hurricanes, droughts, forest fires, or tsunamis may lead to the local disappearance of a species, especially if population size is small (Wiens 1977b, Diamond 1975a).

Different views of community assembly have emphasized different aspects of this series of factors and effects (see Table 1 of Abbott 1980). Lack (1976) downplayed the influences of area, distance, and dispersal abilities on colonization probability, stressing instead the importance of ecological conditions in the target area in determining which species entered a local community. A small island, for example, might support a less diverse bird community than a larger island not because of size differences or difficulties in dispersal, but because it contained a more restricted range of habitats or food resources. Lack also considered competitive interactions to be of primary importance in adjusting the niche relationships of those species whose ecological requirements were met in a particular area, especially on small islands. MacArthur and Wilson (1967), on the other hand, stressed the importance of dispersal and extinction dynamics, especially as they related to the size of an area, but gave much less prominence to the role of habitat conditions. Diamond (1975a) considered suitable habitat to be a prerequisite to successful entry of a species into a community, but attributed greatest importance to competitive interactions among species. Abbott (1980) emphasized dispersal difficulties and habitat variation.

It is unlikely that any single factor plays an overpowering role in determining community assembly in most situations. Some indication of the multiplicity of factors contributing to the species composition of communities comes from studies of the Åland archipelago in the Baltic Sea (Haila and Järvinen 1983, Järvinen and Haila 1984). There, the species of Main Åland (970 km^2) may be considered as the pool available to the island of Ulversö (5.8 km^2, 10 km away). Of the 121 species present on Main Åland, 52 are not found on Ulversö. Many of these species, however, are rare in the source area and thus would not be expected to occur in a given year even in a small area on the mainland. Others have only recently reached the source area and entered the species pool. For others, habitat conditions on Ulversö may not be suitable. The Chiffchaff (*Phylloscopus collybita*), for example, occurs in tall spruce in this archipelago, but although a limited amount of spruce forest exists in Ulversö, the trees are apparently too short. Some absences may reflect competition: Sedge Warblers (*Acrocephalus schoenobaenus*), for example, are missing from the island, whereas Reed Warblers (*A. scirpaceus*), are abundant. Overall, the 52 species missing from Ulversö are absent for a variety of reasons (Table 4.1). Furthermore, because some of the species that are rare in the source pool might breed on Ulversö sooner or later despite their rarity, short-term and long-term views of the contributions of these factors to community assembly differ (Table 4.1). Jones and Diamond (1976) documented a similar filtering effect of multiple factors on the avifauna of the Channel Islands off California.

Assembly rules for communities

Diamond's studies of Southwest Pacific bird communities

The most extensive treatment of the assembly of bird communities is that of Diamond (1975a), who considered bird distributions on New Guinea and its satellite islands. These studies merit detailed attention, not only because they contributed to the explicit statement of several 'rules' for community assembly, but because they have generated intense controversy and the controversy itself is instructive.

Diamond addressed the notions that resident landbird species were nonrandomly distributed over the islands, that competition between species accounted for a good deal of this pattern, and that the result was the development of communities of coadjusted species that were resistant to invasion by other species. Diamond used the close fit of his observations to a species–area regression (see Chapter 5) to support his assumption of avifaunal equilibrium for most of the islands; the exceptions were landbridge islands (which contained more species than expected) and islands recently altered by volcanic explosions or tidal waves (which were depauperate). Landbridge islands may contain more species than expected because of the potential for repeated colonization from the mainland

Table 4.1. *Primary factors associated with the absence of 52 bird species found on Main Åland from the nearby island of Ulversö in the Baltic Sea, expressed as percentages.*

The short-term column calculates the percentages on the basis of absences in any given year, while the long-term column contains values adjusted to reflect the fact that some species absent because of rarity in one year might still occur on Ulversö during a longer time interval.

Reason for absence	Short-term	Long-term
Rarity in source area	75	33
History and dispersal difficulties	3	9
Habitat unsuitability	12	47
Interspecific competition	2	2
Predation	4	6
Other	4	4

Source: From Järvinen & Haila (1984).

(Gotelli, personal communication), while in some instances volcanic eruptions reduce the area of an island as well as disrupt the biota (Abele, personal communication).

Diamond began by constructing 'incidence functions', graphs in which the frequency of occurrence of species on islands of different size classes (J) is plotted against the average number of species in the island class (S) (e.g. Fig. 4.2). Diamond justified his use of S rather than island area (A) as the independent variable on the basis of the generally close regression relationship between the two (although in later papers Diamond (1982, 1984) did plot J versus A). Diamond (1975a) did not specify the criteria used to define the island size-classes (which determine the shape of incidence functions). Size-class designations differed for different species. Abbott (1980) has challenged the formulation of these incidence functions.

For most of the species these incidence functions describe nonrandom distributions among islands of different S (Fig. 4.2). Some species are absent from low- or intermediate-S islands but present on all species-rich islands ('high-S species'). Others ('supertramps') occur only on islands with few species, disappearing when values of S exceed some critical level. In between are arrayed a series of curves for species that occur on islands of intermediate as well as high species richness. Diamond arbitrarily classified these into four 'tramp' categories. Diamond also noted several correlates of the incidence-function categories. Supertramps, for example, are characterized by high dispersal ability, high reproductive potential, and unspecialized

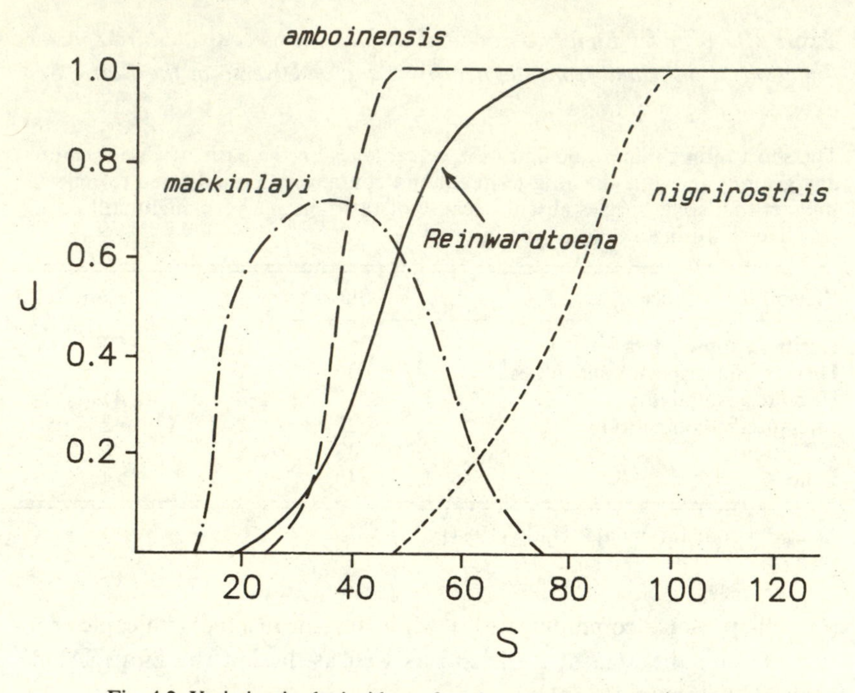

Fig. 4.2. Variation in the incidence functions of four species of cuckoo-doves on islands of the Bismarck archipelago. The functions plot the proportion of islands occupied by a species (*J*) against the mean number of species (*S*) characterizing a size-class of islands. The species *Macropygia mackinlayi* is a 'supertramp' confined to species-poor islands, whereas *M. nigrirostris* occurs only on species-rich islands. After Diamond (1975a).

habitat affinities. The high-S category, on the other hand, includes both species endemic to forests on large islands and nonendemic species occurring in scarce habitats that are generally unrepresented on smaller islands; neither has high dispersal rates. The incidence function for a species, by reflecting its distribution over islands of different sizes and species richnesses, thus combines into a single curve the effects of a great many factors.

Diamond recognized this, but he drew special attention to the relationship of the incidence functions to competitive ability. Because supertramps, with their high dispersal abilities, are missing from species-rich islands, where high-S species are abundant, Diamond (1975a: 381) inferred that 'supertramps are outcompeted and permanently excluded by guilds of K-selected, high-S specialists, each of which is better adapted to some fraction of the supertramp's potential niche'. He suggested that this competitive exclusion of poorly-adapted supertramps by high-S species might be ac-

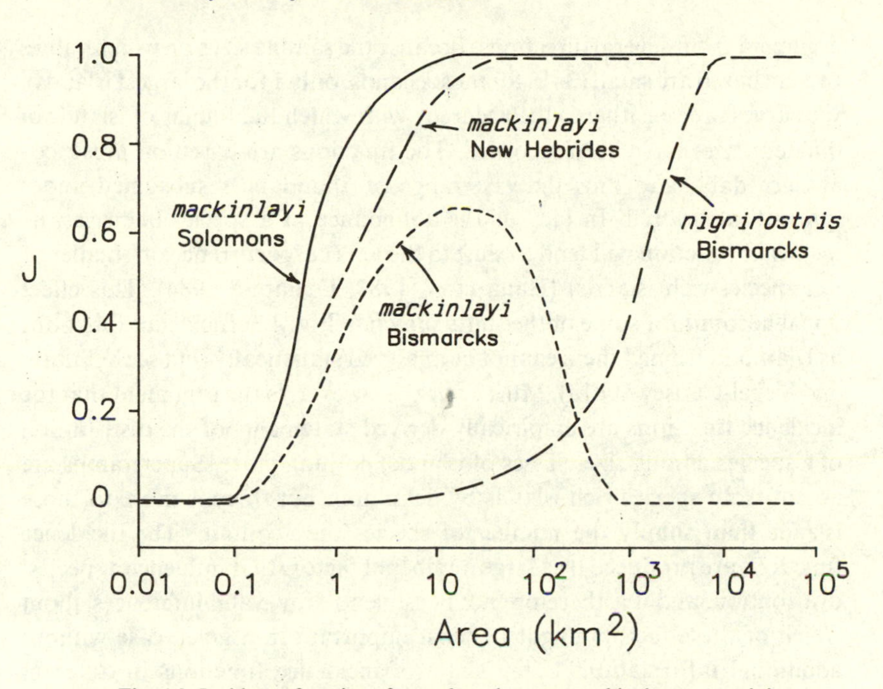

Fig. 4.3. Incidence functions for cuckoo-doves on archipelagos containing different total numbers of species. *Macropygia mackinlayi* and *M. nigrirostris* are the same size, similar in diet, and taxonomically the most closely related pair of species in the genus. *M. nigrirostris* occurs only in the Bismarcks, whereas *mackinlayi* occurs in all three archipelagos. In the Solomons and New Hebrides, from which *nigrirostris* is absent, *mackinlayi* is present on every island > 10 km^2 as well as on smaller islands. In the Bismarcks, *mackinlayi* is confined to small islands; larger islands are occupied by *nigrirostris*. From Diamond (1982).

complished through an 'overexploitation ethic', in which the high-S species, by collectively exploiting a range of food resources quite efficiently, would depress production below levels that would support populations of the less efficient supertramps. As evidence of the importance of competition in these assemblages, Diamond (1982) also noted that the shape of the incidence function of a species may shift with changes in the size of the species pool of an archipelago, as might be expected from an intensification or diminuation of competition (Fig. 4.3). Diamond concluded from his consideration of incidence functions that species were nonrandomly distributed among islands (as a random distribution would produce a linear, horizontal incidence function) and that competition played a large part in determining those patterns.

Diamond's incidence functions and their interpretation are open to argument from several directions. Because the sample sizes on which values of *J* are based are small (3–13 for most islands, only 1 for the largest islands), the curves are sensitive to the accuracy with which the faunas of islands of different sizes have been recorded. The functions are based on presence–absence data, and thus the vast range of abundances subsumed under 'present' is ignored. In fact, as the abundance of a species increases, its incidence function will tend to shift to the left (i.e. occurrence on smaller or less species-rich islands) (Haila *et al.* 1983, Diamond 1984). This effect might account for some of the shifts shown in Fig. 4.3. Incidence functions as Diamond framed them cannot be analyzed statistically (but see Whittam and Siegel-Causey 1981a). Most critical, however, is the argument that the incidence functions are empirically-derived statements of the distribution of a species among size-classes of islands, nothing more. Supertramps are absent from species-rich islands by definition, but there is more to those islands than simply the number of species they contain. The incidence functions are produced by a large number of factors that influence a species' distribution, and it is therefore not possible to draw valid inferences about which of these factors might be most important in a given case without additional information. Using shifts of incidence functions in different archipelagos as evidence of competitive effects requires the assumption that islands of a given area in the different archipelagos are equivalent in all respects except the number of species they contain, which is unlikely.

Diamond did not base his arguments about community assembly solely on incidence functions, however. Using several functionally-defined groups of species (guilds, see Chapter 6), he considered whether or not the species in a group were distributed among islands nonrandomly and, in particular, whether or not certain combinations of species could successfully resist invasion by other members of that group. One form of nonrandom distribution is dictated by the incidence functions themselves: if the incidence functions of two species in a group do not overlap, the species cannot co-occur. Similarly, if the nonzero portion of the incidence function of one species is entirely contained in the range of S values for which the incidence of a second species = 1.0, the first species will never occur in the absence of a second. Such distributional patterns could be due to any of the factors whose effects are contained in the incidence functions. In other situations, however, species with overlapping nonzero incidence functions do not co-occur, and Diamond considered these to be especially clear examples of competitive effects.

The simplest such distribution is a 'checkerboard', in which two (or more) species in a group have mutually exclusive distributions in an archipelago,

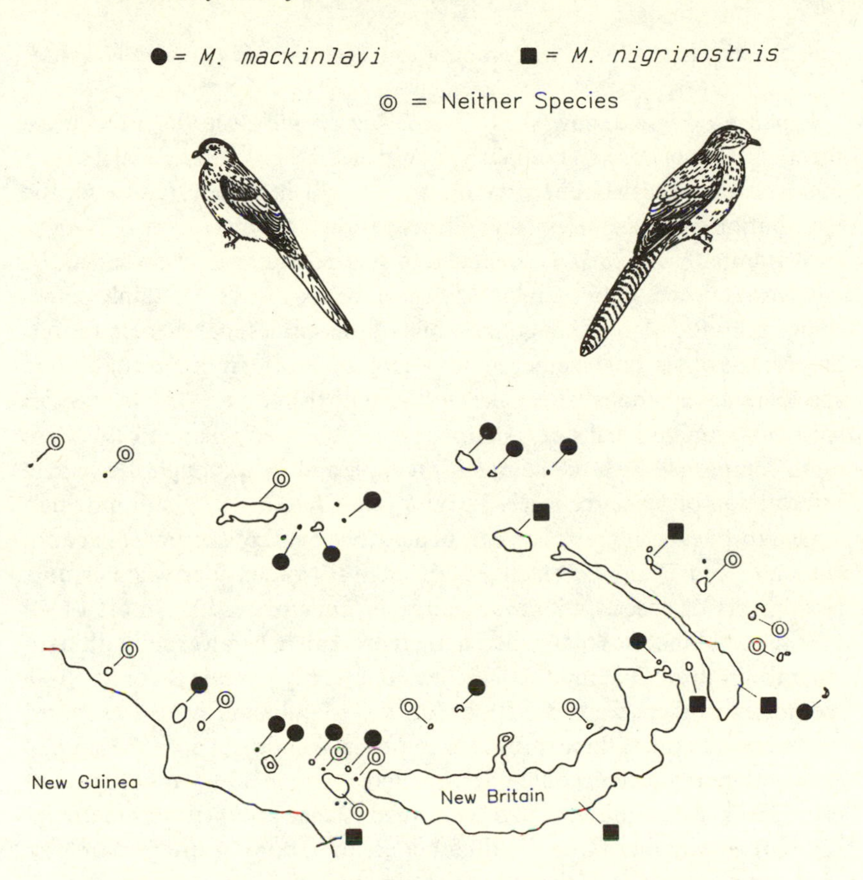

Fig. 4.4. 'Checkerboard' distribution of *Macropygia* cuckoo-doves in the Bismarck region. On islands for which the avifauna is known, either *M. mackinlayi* or *M. nigrirostris*, or neither species, occurs. After Diamond (1975a).

each island supporting only one of the species. The cuckoo-doves *Macropygia mackinlayi* and *M. nigrirostris* in the Bismarck archipelago, for example, do not occur together on any of the islands surveyed, although adjacent islands that are seemingly quite similar may contain one of the other species (Fig. 4.4). These two species are quite similar in size and ecology, and Diamond suggested that their distribution reflects mutual competitive exclusion. Exactly which species occurs on an island may be due to chance, the first species to arrive becoming established and then excluding the other. Some apparently suitable islands lack either species; these absences Diamond attributed to the presence of other cuckoo-dove species, which singly or in combination exclude *M. mackinlayi* or *M. nigrirostris*.

A glance at Fig. 4.2 shows that the two *Macropygia* species have incidence functions that overlap only slightly, however, so their checkerboard distribution is not so easily interpretable as it might first seem (although the distribution is nonetheless significantly nonrandom; see p. 87). The supertramp *M. mackinlayi* is confined to species-poor islands, whereas *M. nigrirostris* occurs only on relatively species-rich islands. Certainly something about the islands that contain one of the species makes them unsuitable for the other, but whether or not it is competition from the established species remains problematic. Checkerboard distributions involving species with quite similar incidence functions (see Gilpin and Diamond 1982) are more compatible with the competition explanation, although the critical information on resource levels and use (see Chapter 10) is still missing.

One-to-one species replacements in distributional checkerboards are not common, and Diamond extended his analysis to consider whether only certain sets of species within a group co-occur or whether most or all possible combinations of the species are represented. From examinations of several groups, Diamond concluded that only certain subsets were 'permitted'; others were not found on the islands and were considered 'forbidden'. Some of these forbidden combinations do not occur because of nonoverlapping incidence functions, whereas others are consequences of checkerboard distributions involving species that are distributionally incompatible with one another. Still other combinations of species would be expected on the basis of the incidence functions and apparent compatibilities of the species but are nonetheless missing. On the basis of incidence functions alone, for example, one might expect to find *Macropygia amboiensis* (A), *M. mackinlayi* (M), and the superspecies *Reinwardtoena* (R) occurring together on islands with medium species numbers (Fig. 4.2) (*M. nigrirostris* would not be expected to enter such combinations because of its apparent incompatibility with *M. mackinlayi*; Fig. 4.4). The combinations AM, AR, and MR are found on islands in the Bismarcks, yet the combination AMR is never recorded. Diamond reported several anecdotal observations of R occurring as vagrants on an island containing A and M without becoming established.

Diamond's assembly rules

Diamond summarized these observations of pattern in a series of 'assembly rules', as follows (1975a: 423):

1. 'If one considers all the combinations that can be formed from a

group of related species, only certain ones of these combinations exist in nature.'

2. 'Permissible combinations resist invaders that would transform them into forbidden combinations.'
3. 'A combination that is stable on a large or species-rich island may be unstable on a small or species-poor island.'
4. 'On a small or species-poor island, a combination may resist invaders that would be incorporated on a larger or more species-rich island.'
5. 'Some pairs of species never coexist, either by themselves or as part of a larger combination.'
6. 'Some pairs of species that form an unstable combination by themselves may form part of a stable larger combination.'
7. 'Conversely, some combinations that are composed entirely of stable subcombinations are themselves unstable.'

Several comments are in order. First, these 'rules' are simply statements of pattern derived inductively from Diamond's observations. Because of this, they do not 'explain' the observations. It is not logically valid to argue that some combinations of species do not occur because they would violate one or more of the rules, when the rules are no more than descriptions of such distributional patterns.

Second, the rules contain no mention of process. Nonetheless, Diamond made it quite clear that, although a variety of factors may influence the form of incidence functions, interspecific competition is primarily responsible for the failure of missing combinations to appear. Often the explanation involves the effects of diffuse competition from several ecologically similar species that act in combination to overexploit resources and thus exclude less efficient competitors. As Diamond himself noted (1975a: 348), however, diffuse competition may be used to explain anything, and in many cases its operation is likely to be so diffused that its existence becomes 'difficult to establish and impossible to refute'. Diamond's arguments for the 'exploitation ethic' are largely speculative. His conclusions regarding the importance of competition are based on distributional patterns and other observations of direct interactions, diet overlap, or niche shifts (see Chapters 10 and 11), but they are nonetheless inferences rather than empirical tests of process hypotheses.

Third, the analysis rests on the assertion that the patterns represent equilibrial conditions except where Diamond explicitly stated otherwise. The possibility that some of the distributional patterns of species on islands might not represent stable configurations is not considered seriously.

Finally, the use of terms such as 'rules', 'permitted', and 'forbidden' gives these statements an aura of generality and finality that may dissuade further research in this area. Instead, these statements represent hypotheses to be tested independently, as do Diamond's inferences regarding the role of competition. As such, these observations of island patterns and Diamond's interpretations of them represent a valuable start toward understanding community assembly, but the explicit tests of these hypotheses remain to be done. Despite these problems, Giller (1984) presented Diamond's assembly rules and their competitive interpretation with no reservations or hints of doubt. Gilpin *et al.* (1986) have conducted laboratory tests with *Drosophila* whose results generally support Diamond's empirical observations.

Criticism, controversy, and null hypotheses

Connor and Simberloff's analysis

Not all ecologists accept Diamond's views, however. In particular, Connor and Simberloff (1978, 1979, 1984a, b, 1986) have attacked virtually all aspects of Diamond's assembly rules and their interpretations. Together with Gilpin, Diamond has responded (Diamond and Gilpin 1982, Gilpin and Diamond 1982, 1984a, b). The debate has been prolonged, repetitive, and often acrimonious, but it has served to sharpen our focus on some critical issues in avian community ecology.

Connor and Simberloff (1979) concluded that Diamond's assembly rules were trivial, tautological, or statements of patterns that would be expected were species distributed on islands randomly. Because species-rich islands contain more species than species-poor islands, for example, one would expect them to contain some species combinations that are absent from the species-poor islands by chance alone. Rule 3 is thus considered to be trivial. Rule 2 rests on whether 'forbidden' combinations do not occur because they are resisted or because of chance. Connor and Simberloff refuse to accept Diamond's evidence that invasions are actually resisted by established sets of species, leaving the rule empty of stating anything more than 'permissibile' combinations occur while 'forbidden' combinations do not – a tautology. Rule 4 is considered to be nothing more than a combination of Rules 2 and 3, and thus a trivial tautology. Rule 6 refers to pairs of species forming an unstable combination by themselves. Connor and Simberloff argue that, because there are no islands containing only two species in the data sets, this proposition is untestable.

All of these conclusions may be contested. The rejection of Rules 2 and 4, for example, rests on whether there is or is not satisfactory evidence of established combinations resisting invasion. Diamond's evidence is far from compelling, but in some instances it is certainly suggestive. Rule 6 is rejected because of a misunderstanding of the scale of aplication of the rules – Connor and Simberloff apply them to entire island faunas, while Diamond restricts them to species within a group or guild (although he does not state the criteria he used to assign species to guilds; see Chapter 6).

Most of the argument, however, has revolved about Connor and Simberloff's examination of Rules 1, 5, and 7, which in one form or another make statements about the likelihood of occurrence of certain pairs or combinations of species. To Connor and Simberloff, the key question is whether the occurrence of some combination of species should be considered remarkable and suggestive of a process such as competition or merely the effect of chance associated with a random distribution of species among islands. They tested this by constructing a null model in which pairs or trios of species were placed randomly on islands within an archipelago, subject to three constraints: (1) The number of species on a given island in the model analysis must match the number of species actually observed there. (2) Each species occurs on exactly as many islands as it does in reality (these two constraints have the effect of retaining the column and row totals of the island × species matrix). (3) Each species is placed only on islands within the size range (as measured by number of species) that it is actually observed to occupy (the form of the species' incidence function is retained).

What did Connor and Simberloff conclude from this exercise? To begin with, if one analyzes the individual cases of species pairs with checkerboard distributions that Diamond presented, they all are significantly more exclusively distributed than would be expected by chance alone. For the *Macropygia* species of Fig. 4.4, for example, the probability of realizing the observed distributional pattern by chance is 0.04 (Gilpin and Diamond 1982). But Connor and Simberloff suggested that, because there are 9870 possible pairings of bird species in the Bismarcks, one would expect some pairs not to occur by chance alone. Their null-model analysis for the New Hebrides and West Indies archipelagos[1] indicated that pairs or trios of species that did not co-occur were no more frequent than expected by chance in the New Hebrides ($0.1 < P < 0.25$[2]) but were significantly more

[1] Connor and Simberloff did not have access to the original data for the Bismarcks, which Diamond did not publish. Gilpin and Diamond (1984a) have applied Connor and Simberloff's model to the Bismarck data for all species and find that there are many more exclusive species pairs than expected by chance ($P < 0.001$).

[2] From Gilpin and Diamond (1984a), who recalculated probabilities after adjusting a programming error in the original algorithm of Connor and Simberloff (1979), who calculated $P > 0.90$–0.99. Connor and Simberloff (1984b) insist that their analysis is correct, and the approach used by Gilpin and Diamond may itself contain errors (Gotelli, personal communication).

frequent in the West Indies ($P < 0.001$). Similar results were obtained when the analysis was restricted to confamilial species (following Diamond's 'group of related species' term in Rule 1). Despite finding significant departures from model predictions for the West Indies birds (and a similar result for West Indian bats), Connor and Simberloff concluded that one would expect many instances of exclusive distributions of species on the basis of chance alone. Diamond's few detailed examples were therefore considered insufficient evidence that certain combinations of species were 'permitted' or 'forbidden'. Rules 1, 5, and 7 thus dealt with patterns that could easily result from a random (but constrained) distribution of species among islands.

Criticisms of Connor and Simberloff's Approach

Diamond and Gilpin (1982, Gilpin and Diamond 1982, 1984a, b) responded to the criticisms of Connor and Simberloff, and Connor and Simberloff (1983, 1984a, b, 1986) replied in turn. The most important points of contention are these:

1. *Dilution and guilds.* Connor and Simberloff conducted their analyses by drawing species randomly from species pools containing either all of the species in the archipelago or all members of a given family of birds. As justification, they pointed out that Diamond's stated assembly rules did not specify the domain of their application except in the phrase 'group of related species' in Rule 1. Diamond and Gilpin (1982) responded by observing that Diamond (1975a) had clearly intended the rules to apply to ecologically-defined guilds of species, not entire avifaunas, whereupon Connor and Simberloff (1984a) accused Gilpin and Diamond of changing the original argument. A careful reading of Diamond (1975a), however, leaves little doubt that, although guilds were not explicitly mentioned in the formal statement of the assembly rules, he nonetheless intended them to apply at that level: the entire paper deals largely with guilds of species. Although families may be less difficult to define than guilds, using families as counterparts of such guilds is incorrect, as many families of birds contain species belonging to vastly different guilds. By failing to confine their analysis to guilds, Connor and Simberloff submerged patterns of species distributions that might suggest competitive effects in a mass of irrelevant detail.

Consider an example. If each of the 211 species in the West Indies species pool is competitively excluded by one other species, we would find 106 exclusively distributed pairs of species. But with 211 species in the total fauna, there are 22 155 possible pairs. The competitive pairs would comprise only 0.5% of these, and the importance of competition in this system would thus be diluted (Alatalo 1982a). The nature of this dilution effect is apparent in Fig. 4.5, which charts the distribution of species' associations for all possible pairs of Bismarck species using Gilpin and Diamond's (1982) null model. The potentially interesting instances of negative (and positive) species associations that do occur are overwhelmed by the many species pairings dominating the center of the distribution. Wright and Biehl (1982) and Bowers and Brown (1982) also noted how the inclusion of nonguild members can destroy patterns that are apparent within a defined guild.

Whether the total-species or the guild approach is appropriate depends on the question being asked. If one's focus is on potential competitive exclusions, there is little reason to consider species pairs between which any sort of interaction is unlikely. A large avifauna will contain a great many such pairs, and confining attention to well-defined guilds is sensible. On the other hand, asking whether or not the distribution of species combinations on islands departs from random expectations for the community as a whole is perfectly legitimate, especially as a first step in pattern analysis. The danger in this approach is that a failure to reject the null hypothesis may lead one to conclude that the species are in fact distributed randomly with respect to one another (an invalid inference) and that there is no reason to examine the data on species combinations any more closely.

2. *The identification of extremes*. In a distribution of species associations such as that shown in Fig. 4.5, there are quite a few extreme positive and negative associations. Although the approach used by Connor and Simberloff can determine the direction in which a given set of data departs from random expectations, it does not permit one to identify which specific combinations of species are responsible for that departure. Gilpin and Diamond considered this a flaw in the approach and noted that their model does permit such identification. This is important, for the rejection of a null hypothesis of random assemblage is only the first step in pattern analysis. One must then ask why the distribution is nonrandom, and an ability to identify the extremes in the distribution is essential to framing second-level hypotheses. The distribution of the Bismarck data (Fig. 4.5) departs significantly from a random expectation in the direction of an excess of positive species associations. A variety of factors might contribute to such associations: shared incidence functions, shared geographical origins, similar habitat preferences, or endemism. On the other hand, the distribution also includes several species pairs that are negatively associ-

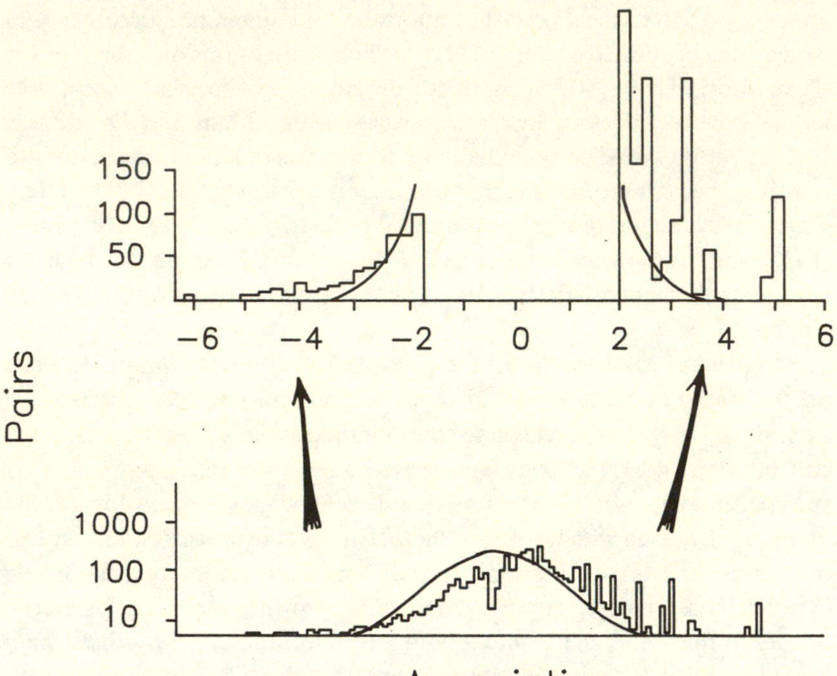

Fig. 4.5. Bottom: Patterns of associations in the 11325 pairs of bird species occurring in the Bismarck archipelago. 'Association' is the difference between the observed and expected number of islands shared by a given species pair, expressed in standard deviates of the expected number (which was derived using a random-association null model) on a logarithmic scale. Large positive or large negative values of association mean that the two species have much more concordant or much more exclusive distributions, respectively, than expected by chance. A randomly assembled community would yield the smooth curve. Top: The ordinate scale is arithmetic and expanded to show the details of distributions of extreme positive and negative species associations. Modified from Gilpin and Diamond (1984a).

ated, perhaps as a result of competitive exclusion or differences in distributional strategies, geographical origins, or habitat preferences (Gilpin and Diamond 1984a). The distributional patterns themselves do not permit one to distinguish among such alternatives, but they do suggest pathways to framing and testing more specific hypotheses.

3. *Hidden structure*. Connor and Simberloff incorporated three constraints in their null model in an attempt to retain some of the biology of the species and characteristics of the islands. This avoids the development of absurdi-

ties such as small islands containing the entire species pool or supertramps occurring only on species-rich islands. At the same time, however, it incorporates into the model whatever biological processes have contributed to the row and column totals of the species–island matrix and to the form of incidence functions. These may include competitive effects, and the model may therefore not be entirely 'null' to the factors it purports to test (Diamond and Gilpin 1982, Gilpin and Diamond 1984a). Gilpin and Diamond (1984a) admitted to their inability to deal successfully with this problem of hidden structure in their own model, and they considered it to be the most intractable difficulty in the entire null-model approach. There seems little doubt that much of what accounts for the distribution of species among islands (competition as well as other factors) is already contained in the incidence functions and row and column totals. In Connor and Simberloff's tests, they therefore considered effects that might produce nonrandomness *over and above* those already included in the model constraints. If considerable structure is hidden within the null model, the likelihood of its rejection becomes diminishingly small. On the other hand, if the assumption that the colonization probabilities of species are not equal is relaxed and the species are assigned equiprobable colonization functions, the usual result is that distribution patterns are not random (Gotelli, personal communication). Whether this assumption is realistic or not depends on the particulars of a given situation.

Another example: Galápagos finches

Following the lead of Abbott *et al.* (1977), Connor and Simberloff (1978, Simberloff 1978) also used null models of random assembly to examine the distribution of *Geospiza* finches in the Galápagos Islands. In the previous analyses, the question was how pairs of species shared islands in their distributions (the 'shared island' approach). In the finch analysis, a 'shared species' approach was used. Although this approach is not so powerful for detecting possible competitive effects as the shared island approach (Simberloff 1978, Wright and Biehl 1982, Case and Cody 1983), it suggests that interactions such as competition may be important when the number of species shared by different islands is lower than that expected by chance.

Using procedures weighted by the frequencies of occurrence of the six species over the archipelago as a whole and data revised somewhat from the earlier treatments, Simberloff (1984) concluded that the number of combinations of species observed on islands containing only one or two *Geospiza* species does not differ from that expected on the basis of independent, random, island colonization (as portrayed by the null model). On 3-, 4-, and 5-species islands, however, there are significantly fewer species combinations observed than would be expected by chance. Changing the size of the species pool not surprisingly altered these conclusions (Simberloff and Connor 1981). Grant and Schluter (1984) also conducted a null-model analysis of the *Geospiza* combinations on these islands and reached the same conclusions. Including additional information on populations that once bred on the islands but are now locally extinct altered the patterns, with the one-species islands now exhibiting significantly fewer combinations than expected. Approaching the problem from a shared islands perspective, Alatalo (1982) also concluded that the *Geospiza* finches were nonrandomly distributed. Thus, the finches clearly do not seem to occur among the islands in a manner suggesting that they colonized them independently and randomly. This documentation of pattern is suggestive of competitive effects, but other interpretations are also possible (Simberloff and Connor 1981, Case and Cody 1983, Simberloff 1984). Altering the details of the original data also may change the outcome of the analyses, indicating the critical importance of accurate censuses of species distributions in investigations of community assembly.

Some comments on the perils of testing null hypotheses

The approach advocated by Connor and Simberloff in these analyses rests heavily on testing a null hypothesis of noninteractive community assembly by generating simulated data with a model of randomness that lacks species interactions. This 'null model' approach is the focus of considerable controversy. It has been championed by Simberloff and his colleagues (e.g. Strong 1982, Strong *et al.* 1979, Connor and Simberloff 1984a, 1986) as the most effective way to tell whether a given set of observations displays a pattern that deviates from one produced by chance with respect to key variables or processes. Others (e.g. Gilpin and Diamond 1982, 1984a, Harvey *et al.* 1983) question its value, noting a series of seemingly insurmountable problems in framing rigorous and unbiased tests of null hypotheses. As May (1984: 5) has observed, 'it is just as easy – and just as foolish – to construct an inappropriate or misleading neutral model as it is glibly to deduce evidence for competition from data that are susceptible to other interpretations'.

To be valuable as a tool in research on communities, a null model must be properly constructed, free of biases, and correctly used. When applied to hypotheses of community assembly, this means that the species pool must be identified correctly, a suitable algorithm must be defined for drawing and distributing species from the pool, the statistical analysis of the observed and expected results must be sound, and the interpretation of the findings must be circumspect and logical. Harvey *et al.* (1983) and Colwell and Winkler (1984) provide throughtful discussion of some of the difficulties contained in each of these steps, but three merit special mention here.

1. How does one specify the species pool? In analyzing the bird communities of the Tres Marías Islands in the Sea of Cortez, for example, Grant (1966) considered the source pool to contain the species from an equivalent portion of the adjacent mainland (same area, range of altitude, and habitats). In his subsequent analyses of Grant's data, Simberloff (1970) included all species within 480 km of the islands (excluding Baja California) or all species resident at elevations of <900 m in the adjacent mainland states of Nayarit, Sinaloa, and Jalisco (Simberloff 1983a). Clearly, how the source pool is defined may bias an analysis toward certain results. If one wishes to know whether or not the bird community residing in a given woods is a nonrandom subset of the species pool, the answer will differ according to whether one uses a pool containing species found in woods within an area of 100 km^2 (quite likely 'no'), a pool of all woodland species recorded in the political unit containing the woods (probably 'perhaps'), or a pool of all birds recorded on the continent (certainly 'yes'). Should the pool be restricted to members of a guild or a family or contain all species in an area? Should one distinguish between the total pool for an area and the 'habitat pool', those species whose habitat preferences would enable them to occupy the area (Graves and Gotelli 1983)? Should species in the pool be weighted by their dispersal abilities and, if so, should one use incidence functions (Diamond 1975a), size of geographic range (Graves and Gotelli 1983), or some other measure to achieve this weighting? Such factors lie at the heart of the 'dilution effect' discussed above. Proper definition of the source pool for a given analysis is critical, and this requires both a clear statement of the problem and a considerable knowledge of the natural history of the species.

2. There is also the problem of hidden structure. A model that is devoid of biology avoids this problem but is worthless. We could model community assembly as the movement of gas molecules in a vacuum, but what would falsification of such a hypothesis tell us? On the other hand, if we sample from a species pool that reflects the consequences of predation, our findings will inevitably be an underestimate of that effect and rejection of the null when predation actually is important will be difficult. The severity of this problem will depend on how much a factor has influenced the data contained in the model structure. Gilpin and Diamond (1984a: 302) have stated the dilemma quite clearly: 'How can one devise a "null model" that does not

implicitly contain effects of competition [or any other factor] and that would be rejected as a result of competitive effects in the observed data base but not as a result of other effects?' One possible answer is to include several variables (e.g. habitat, geography, range, taxonomy, dispersal abilities) in the model and then test it against more than a single alternative (Gotelli personal communication).

3. Considerable care is also required in the interpretation of the results of tests of null hypotheses. If one finds no difference between field observations and the predictions of a model based on random distribution of species among islands, can one conclude that the species are randomly distributed in nature? If the observations do depart significantly from the predictions, can one conclude that species interactions determine the distributions? Certainly not, in either case. The distinction between testing pattern hypotheses and process hypotheses is critically important here. All one has done in this test is to determine the existence of a particular sort of pattern. That in itself is valuable, but any inferences of process on the basis of such a test are precarious indeed (see Table 2.1).

What is the role of such null models in community ecology? Simberloff and his colleagues have used them primarily to examine other people's data and to point out that patterns were incorrectly defined or inferences improperly drawn. This has been a worthwhile undertaking. Connor and Simberloff (1983) also have proposed that null hypotheses and models may approximate the role of a 'control' in testing an hypothesis with nonexperimental evidence by identifying what might be expected in the absence of a given factor (= treatment). This quasi-experimental role of the null model is less satisfactory, given the difficulties noted above. Null hypotheses that prompt statistical tests of *pattern* may be valuable if performed correctly. We do need to be able to determine whether a particular pattern in observations should be regarded as unusual, so that we may then inquire why. Models that are used to explore *processes* by considering the effects of all factors save the one of interest are much more difficult to construct and interpret and their role in ecology is much less certain.

Other studies of community assembly

Despite the optimism generated in some quarters (e.g. Giller 1984, Gilpin *et al.* 1986) by Diamond's 1975 paper, the search for assembly rules for communities has proven to be difficult. Sale and Dybdahl (1975, 1978) and Talbot *et al.* (1978), for example, looked with little success for permitted or forbidden species combinations in their data for coral reef fish communities. If they exist for these communities, assembly rules must be very subtle

(Sale 1984). A similar conclusion emerged from Vepsäläinen and Pisarski's (1982) studies of island ant communities and Ranta's (1982) investigations of rock-pool invertebrate communities. All of these authors suggested that a complex array of factors influences community assembly and that a rather thoroughgoing autecological approach is needed to understand why species are distributed as they are.

This is not to say that other attempts to define community assembly in simple terms have not been made. Tits of the genus *Parus* are widespread throughout Europe and occur in various multispecies combinations. Herrera (1981) asked if one could discern 'combination rules' that portrayed these patterns. He used information on the presence or absence of species in 88 local communities distributed over a wide variety of habitats in western Europe. The communities were surveyed for varying lengths of time by different observers using different procedures in areas of different size, so the data set is extremely heterogeneous. Herrera found what he considered (without tests) to be a highly nonrandom pattern of species occurrences, in which only one-third of the possible combinations of species occurred. In the combinations that did occur, differences in bill lengths among the species were greater than in the 'missing' combinations. Assemblages containing only two *Parus* species were generally comprised of the large- and small-billed species, and larger assemblages were created through the addition of species with intermediate bill sizes. Herrera did not consider the relative abundances or the habitat preferences of the species, nor did he justify his use of bill length rather than some other metric as a measure of species similarity. Thus, how and why *Parus* assemblages in Europe vary remain open questions (see also Alatalo 1982b, Alatalo *et al.* 1986).

Whittam and Siegel-Causey (1981b) specifically addressed the matter of guild definition in their investigation of Alaskan seabird communities. These assemblages are particularly well-suited for this sort of analysis: breeding-colony species compositions have been estimated with reasonable care using standardized procedures (Sowles *et al.* 1978), island area has relatively little effect on community structure, and distance from a mainland source pool seems to make little difference. Five guilds of species were defined on the basis of details of their nest-site requirements and log-linear models and multiway contingency tables were then used to generate expected values of species co-occurrences in colonies within each guild. For all of the guilds the distributions were significantly nonrandom, positive associations predominating. Whittam and Siegel-Causey discussed several sorts of mutualism that might account for such patterns, but the most direct explanation would seem to reside in the guild definitions themselves: by defining the guilds in terms of nest-site preferences, it is inescapable that the species having similar requirements will occur together quite frequently wherever these requirements are met.

Some studies have expanded on the leads of Diamond and Connor and Simberloff. Graves and Gotelli (1983), for example, used a null model to determine whether or not the avifaunas of several neotropical land-bridge islands differed from random samplings of the adjacent mainland source pools. They attempted to remedy some of the deficiencies of earlier work by defining the source pools geographically with respect to each island, estimating the dispersal powers of species from the sizes of their geographic ranges, and incorporating habitat availability into their colonization algorithm. At the family level, the island communities appeared to be a random subset of the habitat-defined mainland pools, but species having large mainland ranges were disproportionately common on the islands.

Habitat requirements of species have also been emphasized by Haila and Järvinen (1981, Haila *et al.* 1983), who proposed using 'prevalence functions' to quantify both the distribution and abundance of species over islands of differing sizes. In this approach, the population density of a species in the mainland source area provides an 'expected' value against which the average density observed on islands of a given size class may be compared. Values of 1 thus indicate equivalence of densities, whereas negative values indicate sparser populations on the islands. Haila *et al.* (1983) developed extensions of this approach that are based on the availability of habitats of different types on the islands and the population densities of a species in equivalent habitat types on the mainland. By using such prevalence functions instead of standard incidence functions, they gained a much more accurate image of the distributional pattern of species among islands and could frame more specific explanatory hypotheses. The applicability of this approach, however, is limited by the requirement for quantitative censuses and habitat measurements from a large number of islands.

An entirely different approach was developed by Pulliam (1975, 1983) to examine which species of wintering sparrows co-occurred in grassland and woodland habitats in Arizona. Pulliam censused the birds with a variable-width transect procedure and concurrently estimated seed abundances in the habitats. Using measurements of bill lengths from the literature and of seed sizes in the diet from US Fish and Wildlife Service files, he generated a theoretical bill size–diet relationship from which competition coefficients between hypothetical sparrow species of different bill sizes could be derived. By combining these competition coefficients with estimations of the carrying capacity of a site (based on seed availability and theoretical projections of seed utilization by the species) he constructed a 'coexistence matrix'. This matrix predicted the bill sizes of species combinations that could or could not coexist under given resource conditions.

Pulliam's (1975) initial application of this approach, using observations from a single year, was remarkably encouraging: the theoretical predictions were met in every case. Pulliam later (1983) extended this analysis to include additional data for the subsequent 3 yr but was unable to repeat his initial success. The model in fact did poorly in predicting both the number and the sizes of sparrows present in the habitats, leading Pulliam to ask what sort of model performance might be expected in a randomly generated community. In designing his null model he assumed that the occurrence of a species was independent of the presence of other species and of seed production but was restricted by the habitat preferences of a species (calculated from the observed frequency of occurrence in the different habitat types). Pulliam's model thus contained 'hidden structure' in this habitat constraint. Perhaps for this reason, the random model did about as well as the community-matrix model in predicting the actual community structure. Adding thresholds to account for the minimum food requirements of sparrows of different sizes improved the performance of both models, although many observations still did not match the predictions.

Some other approaches to understanding community assembly remain largely untested. Haefner (1981), for example, proposed a model of assembly based on 'generative grammars', in which various ecological attributes of species are itemized and then related in a hierarchical fashion to specifications of the environment to determine which species can be inserted into a community. Some of these species are then removed from the community according to a set of 'deletion rules', resulting in the final set of 'permitted' species that comprise the community. Although the approach has promise, it is hampered by the considerable amount of ecological information required for each species, the reliance of most of the deletion rules on some form of species interaction, and the deterministic nature of the model operations. At the opposite extreme of detail, Brown (1981) suggested approaching questions of community structuring by considering 'capacity rules' and 'allocation rules'. Capacity rules define the physical characteristics (chiefly energetics) of environments that determine their capacity to support biota, while allocation rules relate to how available energy is captured among species in terms of characteristics such as body sizes, trophic status, local abundance, or home range size. Brown admitted his uncertainty about how to proceed to develop formal statements of these rules, but he suggested that an emphasis on the energetic foundations of community dynamics offers greater potential than the traditional focus on population dynamics. Probably both perspectives are necessary.

Extensions of community assembly

Community assembly has traditionally been viewed as a largely deterministic process involving species interactions that takes place on island archipelagos. There may be some value, however, in extending the sequence shown in Fig. 4.1 to other situations.

Islands versus mainlands

Bird communities in mainland areas must undergo species assembly in a manner resembling in general terms that occurring on islands. Island communities have the advantage of being relatively closed and discrete. Diamond (1975a, 1983) has suggested that one may therefore consider the islands in an archipelago to be analogs of replicates in an 'experiment' of sorts, in which communities have undergone assembly to yield final, equilibrium configurations of species. One determines the outcome of such an 'experiment' by tabulating the presence or absence of species on islands. Whether or not one accepts this view of island comparisons (and there are compelling reasons for not doing so, such as the fact that different islands are never really 'replicates' of one another), it is apparent that bird communities in mainland situations are rarely so discrete. Habitat 'islands' are surrounded by other habitat types rather than by an expanse of inhospitable ocean. Dispersal and colonization should therefore occur much more rapidly. On this basis, Herrera (1981) proposed that the effects of competition on community assembly should be much clearer in a continental than in an insular setting. One indeed might expect that, with many of the constraints on colonization relaxed, the sequence of Fig. 4.1 would go to completion much more rapidly; other things being equal, equilibrium should be established more quickly.

Continental habitats may undergo climatic and habitat change more frequently, however. Because local systems are more open to influxes of species, they are also more prone to lose species through immigration or local extinction, especially if local populations are small. In a mainland situation it is also difficult to find clearly bounded 'replicates' in which one may document community patterns through simple distributional analyses. Instead, one must resort to more detailed studies of habitat characteristics, population densities, and the details of the species' ecology, perhaps including *bona fide* experimental manipulations. Given that distributional analyses alone provide limited insights (Case and Cody 1983, Rummel and Roughgarden 1983, Harvey *et al.* 1983, Vepsäläinen and Pisarski 1982), this approach is clearly necessary in island studies as well. There is no reason why the sequence of Fig. 4.1 should not take place in mainland situations; indeed, given the high rate of human-induced disturbance and fragmentation of habitats in such areas, ecologists are continuously provided with opportunities to study the assembly of communities directly. One does not need to go to New Guinea or the Galápagos to document community assembly.

Chance versus determinism

How important is chance in community assembly? Diamond (1975a) noted that it might play some role in the determination of alternative, stable communities through a priority effect, and Rummel and Roughgarden (1983, 1985) drew a similar conclusion from their theoretical analyses. The first of several ecologically similar species to enter a community might become so numerous by the time a competitor arrived that it would be impossible for the later arrival to establish itself. Diamond attributed the checkerboard distribution of eight species of *Lonchura* grass finches in New Guinea midmontane grasslands to such an effect, but in general chance was not given a major role in most of Diamond's arguments. On the other hand, Sale (1978, Sale and Dybdahl 1975, 1978) proposed that chance effects during colonization and establishment may have a major impact on the composition and structure of coral reef fish communities on a local scale, largely overwhelming any but the most extreme differences in competitive abilities between species in a group. Vepsäläinen and Pisarski (1982) offered similar arguments for communities of ants. It seems unlikely that birds would be more strongly governed by deterministic forces during community assembly than fish or ants, and Sale's 'lottery' hypothesis deserves closer consideration than it has received in studies of bird communities.

Interactive versus noninteractive assembly

In most studies, the importance of species interactions in community assembly has been emphasized. There is a spectrum of degrees of intensity and diffuseness of interaction among species that may come into play during assembly, however. After considering the structure of bird communities on islands in the Sea of Cortez, for example, Cody (1983a, Case and Cody 1983) concluded that the occurrence of particular species on islands depended on resource levels there (a function of island size and topography) and the ecological attributes of the species, not on which other bird species were already there. Noninteractive community assembly has been the focus of most null models, but it needs to be considered more explicitly as an alternative hypothesis in direct field studies as well.

Evolutionary considerations

Diamond's (1975a) notion that the stable combinations of species formed by community assembly are resistant to invasion suggests that the species may share mutual evolutionary niche adjustments. Rummel and Roughgarden (1983, 1985, Roughgarden 1986) included such coevolution in one of their models of community development and found that it produced quite different community patterns than did invasion and establishment of species in the absence of coevolution. Community assembly can thus be viewed on both long-term (evolutionary) and short-term (ecological) time scales.

One expression of this is the evolutionary progression that has been proposed to characterize some island species groups. In this 'taxon cycle' scenario (Wilson 1961), species are initially widespread among the islands, abundant, and occupy a wide variety of habitats. Once established on islands, the isolated populations begin to develop evolutionary adjustments to one another (niche shifts and restrictions). Their prey species may also evolve to become less vulnerable to these predators. These evolutionary changes reduce the productivity of the invader's populations and lead to greater specialization. The species become more susceptible to displacement by newly arriving species and are extirpated from some islands and restricted to isolated mountaintop habitats at low population densities on others. The species become endemic before going extinct or invading another island group to begin the cycle once more.

According to this scenario, one would predict that temporal trends of decreasing population density, increasing habitat restriction, and a shifting away from habitats used most intensively by new arrivals should accompany increasing evolutionary differentiation of island birds. Ricklefs (1970) and Ricklefs and Cox (1972, 1978) suggested that several groups of birds in the West Indies fit this pattern. Pregill and Olson (1981) noted several logical flaws in these arguments and showed that a consideration of the fossil distributions of species in the West Indies entirely changes the patterns that are observed. There is no reliable evidence of a 'taxon cycle' in island birds, although Rummel and Roughgarden's (1983, 1985) theoretical analysis does suggest a plausible mechanism for the loss of competitive ability in highly coevolved endemics. The requirements for such evolutionary changes are stringent, however, and they are unlikely to be met in mainland communities or any but the most stable and isolated island communities.

Conclusions

The factors that influence community assembly are varied and may act in complex ways (Fig. 4.1), and to consider assembly to be driven by one or a few strictly deterministic forces is an excessively narrow view. Given this complexity, it is doubtful that defining 'rules' about assembly will lead to much insight about how community composition develops. The debate over assembly rules and the role of null models has been useful in clarifying some issues (e.g. the importance of hidden structure, the dangers of tautologies), but it has drawn attention away from the real concern – the need for careful, accurate measurements of the parameters of Fig. 4.1 and the necessity of adhering to a form of hypothesis-testing in which interpretations are bounded by the data rather than extrapolated far beyond them. Snapshot distributional evidence such as that gathered from sets of islands may provide hints about patterns but they tell one little about the processes responsible for those patterns. Testing assembly hypotheses requires more than inventories of the presence or absence of species at particular times. Tests demand detailed knowledge of the habitat relationships and other ecological attributes of the species, their population densities, resource levels, the abundances of competitors and predators, and so on. It is also a mistake to think that assembly has been completed in most communities and that we therefore study the stable end products of this process. Assembly is ongoing and dynamic. In fact, the many mainland environments that are disturbed by humans provide an unexploited opportunity to study the details of assembly as it occurs.

5

Numbers of species and their abundances

The most basic patterns of communities have to do with the number of species that they contain. There are more bird species in a tropical rainforest than in a desert, and more in the hot Sonoran Desert than in the cold northern Great Basin desert. Species also differ in abundance, so there may be patterns in the commonness and rarity of species in communities as well. Such patterns have interested ecologists for more than a century, and a large mass of observation, theory, and speculation has accumulated. Hutchinson (1959) provided much of the impetus for recent activity in this area by posing the Kipling-like question, 'Why are there so many kinds of animals?' Although Hutchinson suggested that any answer to this question would probably involve a number of factors, most subsequent work focused on the limits to the similarity of coexisting species set by competition. As is so often the case, however, explanations of the presumed patterns have become rather widely accepted without detailed documentation of the actual patterns. On close examination, these turn out to be much more varied and complex than was once imagined.

In this chapter I explore some aspects in this variation in patterns, first considering simple relationships between species number and area and then species-abundance distributions. Because studies of bird communities have more often focused on patterns of species diversity that incorporate both numbers and relative abundances of species, these are considered next. Finally, I evaluate several attempts to use such patterns to document the idea that communities are 'saturated' with species.

Species–area relationships

Basic patterns and models

If one tallies the number of species (S) recorded over a number of locations of different areas (A), there is almost inevitably an increase in S with increasing A. This S/A pattern is so ubiquitous that it has been called

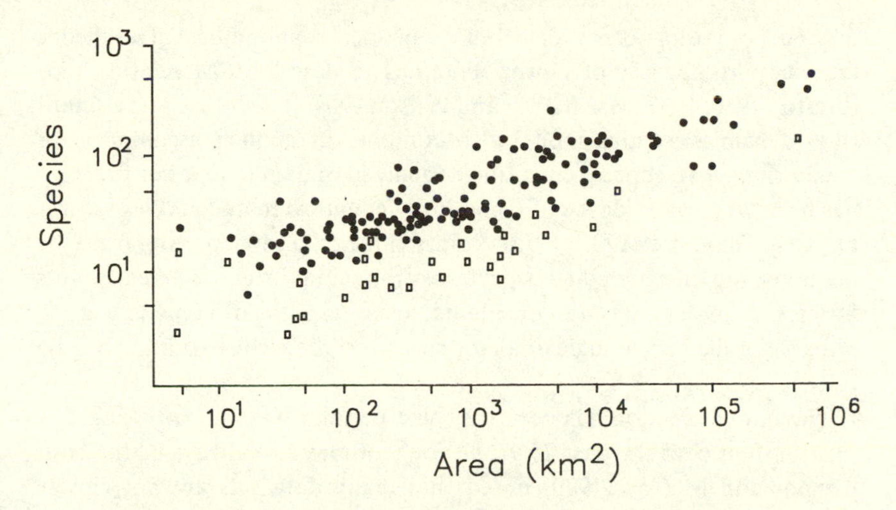

Fig. 5.1. Species-area relationships for the landbirds of islands in tropical and subtropical oceans. Open squares denote islands more than 300 km from the next largest land mass or in the Hawaiian or Galápagos archipelagos. Modified from Slud (1976) and Williamson (1981).

'one of community ecology's few genuine laws' or 'universal regularities' (Schoener 1976, 1986b). Figure 5.1 provides one example of this pattern, depicting the increase in landbird species with increasing area of 171 islands in tropical and subtropical locations. In this log–log plot, S increases linearly with A with a slope of about 0.33, indicating that a tenfold increase in A produces slightly more than a doubling in S.

To describe such patterns in a clear, simplified form, plant ecologists began in the 1920s to fit equations to S/A data. There followed several decades of argument about which model provided the best fit, with little resolution. Two models were especially popular. Initially, most attention was given to an exponential model of the form:

$$S = \log k + z \log A,$$

where k and z are fitted parameters describing intercept and slope. With the development of a richer theoretical underpinning (Preston 1962), attention later shifted to the power function:

$$S = kA^z,$$

which has usually been expressed as a double logarithmic transformation:

$$\log S = \log k + z \log A.$$

These S/A models are formally related to the distribution of species' abundances in a community (see below). In particular, the power function is derived from a lognormal distribution of species abundance. The theory that supports the use of power functions to describe *S*/*A* relationships (Preston 1960, 1962, MacArthur and Wilson 1967) is also based on equilibrium dynamics, so this model also contains an implicit assumption of equilibrium. As a consequence, an adequate fit of data to a power function has been taken as evidence of community equilibrium and species saturation (e.g. Diamond 1973, 1975a). Points in the data set departing from the fitted line are interpreted as departures from equilibrium, whereas points fitting the line closely are considered to be 'explained'. This is a good example of the sort of logical fallacy shown on the right-hand side of Fig. 2.1.

The power function has been the most popular way of expressing *S*/*A* relationships over the past 25 yr, although it may not always be the best. Connor and McCoy (1979) noted that many data sets are not closely approximated by the power function at all; in only 36% of the examples they considered did the power function provide the best fit. Sugihara (1981), however, suggested that 76% of the studies he considered fit the power function quite well. Whether the proportion of instances that are best fit by the power function is 36% or 76% is less important than the fact that not all situations are (McGuinness 1984a). No single model is 'best'. Preferences for one model over another often seem to be based more on their underlying assumptions and associated biological inferences than on the goodness of fit of the data. The expressions, however, simply represent alternative ways of examining the same data. The best fit must therefore be determined empirically by comparing these alternatives (Connor and McCoy 1979, Abbott 1980), checking, of course, to be sure that the underlying assumptions of each model are met.

The importance of z

The value of the slope parameter (z) in the power function has been of particular interest to community ecologists. If one assumes that the distribution of species abundances in the community approximates a lognormal and that larger areas contain more individuals, the values of z are expected to range between 0.15 and 0.39, with a particular form of the lognormal (the canonical) predicting $z = 0.25$ (May 1975). In fact, values of z for a wide range of island situations often do fall within the range of 0.20–0.35 (MacArthur and Wilson 1967, May 1975). Several explanations have been offered for deviations of z from the 'ideal' value of 0.25. MacArthur and Wilson (1967) reasoned that mainland values should be lower (0.12–0.17) than those found on islands (0.20–0.35) because the mainland communities are bathed in a continual influx of transients and immigrants while the island communities are not and because the probability of extinction of species increases with decreasing island size (see below). Therefore, S should be smaller for small islands than for areas of the same size on a mainland, resulting in a steeper slope of the S/A regression. If the islands are isolated, values of z might be even larger (Schoener 1976). Indeed, values of z have been used to indicate degrees of isolation of the samples (e.g. Kitchener *et al.* 1980, 1982, East 1981). In the absence of independent measures of isolation, however, such inferences rest on invalid, circular logic.

The actual values of z that have been reported for bird communities vary widely. In a survey of 75 data sets for true islands (all of which displayed significant fits to a power function), Abbott (1983) found that only 55% of the z values fell between 0.20 and 0.40, hardly an endorsement of the theoretical expectations. In isolated fragments ('islands') of forest of 0.1–7 ha in Wisconsin and New South Wales, Australia, z averaged 0.30 and 0.39, respectively (Howe 1984). The S/A slopes for larger ('mainland') tracts of forest nearby were 0.21 and 0.28 respectively. The 'island' values were indeed larger than those on the nearby 'mainland', although both were greater than generally expected. In 36 forest plots (3–100 ha) in northern Finland, z equalled 0.49 for second-growth forests 2–25 yr old, whereas in older forests (75–150 yr), $z = 0.50$ (Helle 1984). These values are higher than expected, and Helle suggested that this might be a consequence of incomplete censusing (the plots were sampled with line transects during a single visit), the lack of a close relationship between total density and area, a failure of the communities in different stands to follow the same species-abundance distribution, or a sensitivity of z to especially small areas. He tested the latter effect by excluding stands of < 25 ha from the analysis – the resulting values of z were 0.58 for the young stands and 0.60 for the older forests, opposite of expectations. For surveys in the Krunnit archipelago off Finland, Väisänen and Järvinen (1977) calculated values of z that systematically decreased from 0.35 to 0.27 over a 35-yr period following protection of the islands from human disturbance. Here S increased on all but the largest islands, but the change was especially large on the smallest islands, reducing the slope of the S/A relationship. In another analysis, Boström and Nilsson (1983) considered the S/A patterns for a series of bird communities in raised bogs arrayed along a N–S gradient in Sweden. Values of z increased from 0.31 in the northern area, through 0.46 in an intermediate location, to 0.60 in southern Sweden.

In view of such variation in z, Williamson (1981) concluded that the generalization that S/A curves have a steeper slope on islands than in mainland regions is extremely weak and plagued by many exceptions. Rafe *et al.* (1985: 333), finding values of z for communities of breeding birds in British woodlands that ranged from 0.13 to 0.37 in seemingly similar systems, more emphatically warned against 'too much emphasis being placed on quantitative interpretation of this parameter'.

May (1975), Schoener (1976) and Connor and McCoy (1979) have raised the additional point that the tendency of values of z to cluster in the range 0.15–0.40 may be a mathematical consequence of the form of the species-abundance distribution or a statistical artifact of the regression procedures used. If this is so, it suggests that attempts to attach any biological significance to values of z lying within the expected range may be in vain, although strong departures from those values might still represent interesting patterns (Harvey *et al.* 1983). Sugihara (1981) and Wright (1981) have debated this point further. The question of whether z portrays something about the biology or about the statistics of S/A relationships remains unresolved, which suggests that biological inferences should be made with considerable caution, if at all.

Nonlinearities in S/A *relationships*

In most considerations of S/A relationships, the increase in S with increasing A has been presumed to be linear on a log–log scale. Over a considerable range of A this is so, but for small areas the relationship may change. When Diamond and Mayr (1976) examined the species–area pattern for birds in the Solomon Islands, for example, they found that the slope of the S/A curve was quite steep for islands of less than 0.3 km^2 but then changed abruptly to follow a much shallower slope for larger islands (Fig. 5.2). (If the same data are fitted to an exponential model, however, the relationship is linear, with no breaks; Williamson 1981.) On islands in a Minnesota lake, z was much larger (0.46) for islands smaller than 2.5 ha than for larger islands ($z = 0.27$) (Rusterholz and Howe 1979). Here some of the smallest islands may have been too small to contain even a single territory of some of the species, and the rapid increase in S for small changes in A may therefore reflect the rapid addition of new species as thresholds of minimal territory sizes for different species are passed.

Species–area relationships may not necessarily be linear even for intermediate values of A. Cody's (1983a) analysis of the distribution of landbird species on islands in the Sea of Cortéz is especially instructive. There, S was significantly related to A by the power function, with $z = 0.27$ (Fig. 5.3, top). Area alone accounted for 79% of the variation in S. Given this fit, one might

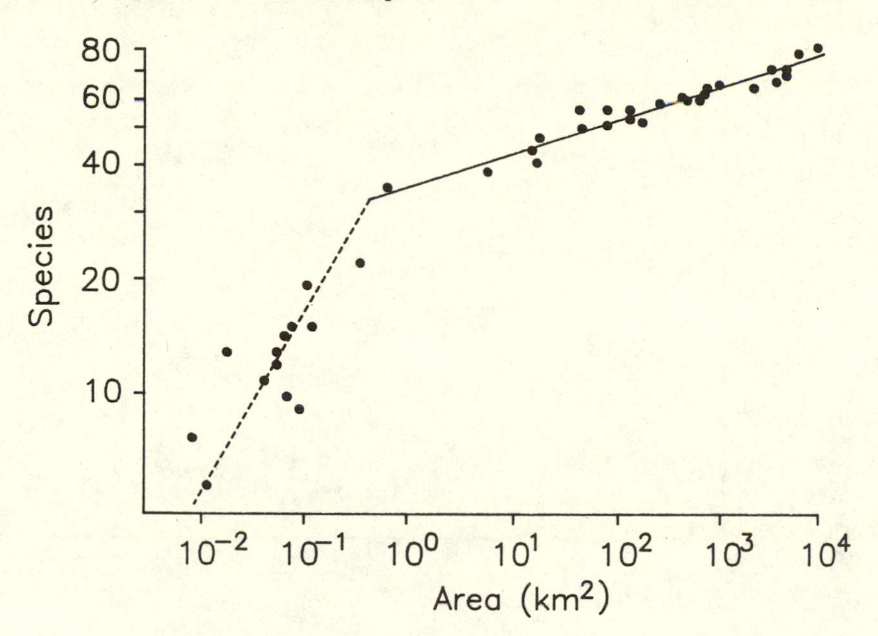

Fig. 5.2. Species–area plot for birds in the Solomon Islands. From Diamond and Mayr (1976) and Williamson (1981).

well be satisfied and proceed to draw inferences about the causes or implications of the relationship. Cody noticed that there were irregularities in the fit of data to the linear regression: S appeared to increase very little on islands between about 40 and 1000 km^2, while decreasing sharply between 6 and 1 km^2. This might be considered simply another example of the small-area effect noted above, but Cody proposed that the data instead fit a 'stadial' model, in which S increases abruptly when A passes a certain threshold and then remains relatively constant until the next threshold level of A is reached (Fig. 5.3, bottom). Cody suggested that resources may not increase smoothly with island area but instead change suddenly in diversity and availability when thresholds of A are passed. Plant species diversity appears to increase in such a stepwise fashion on these islands, perhaps in response to threshold changes in topography and drainage patterns (Cody *et al.* 1983). The empirical data from the islands fit the equation:

$$S=29.5/(1.5)^n,$$

where n = the number of islands in a given level (0–8), 29.5 = the average number of species on the largest island and three diverse mainland areas, and 1.5 designates the factor by which S is increased with each stadial level.

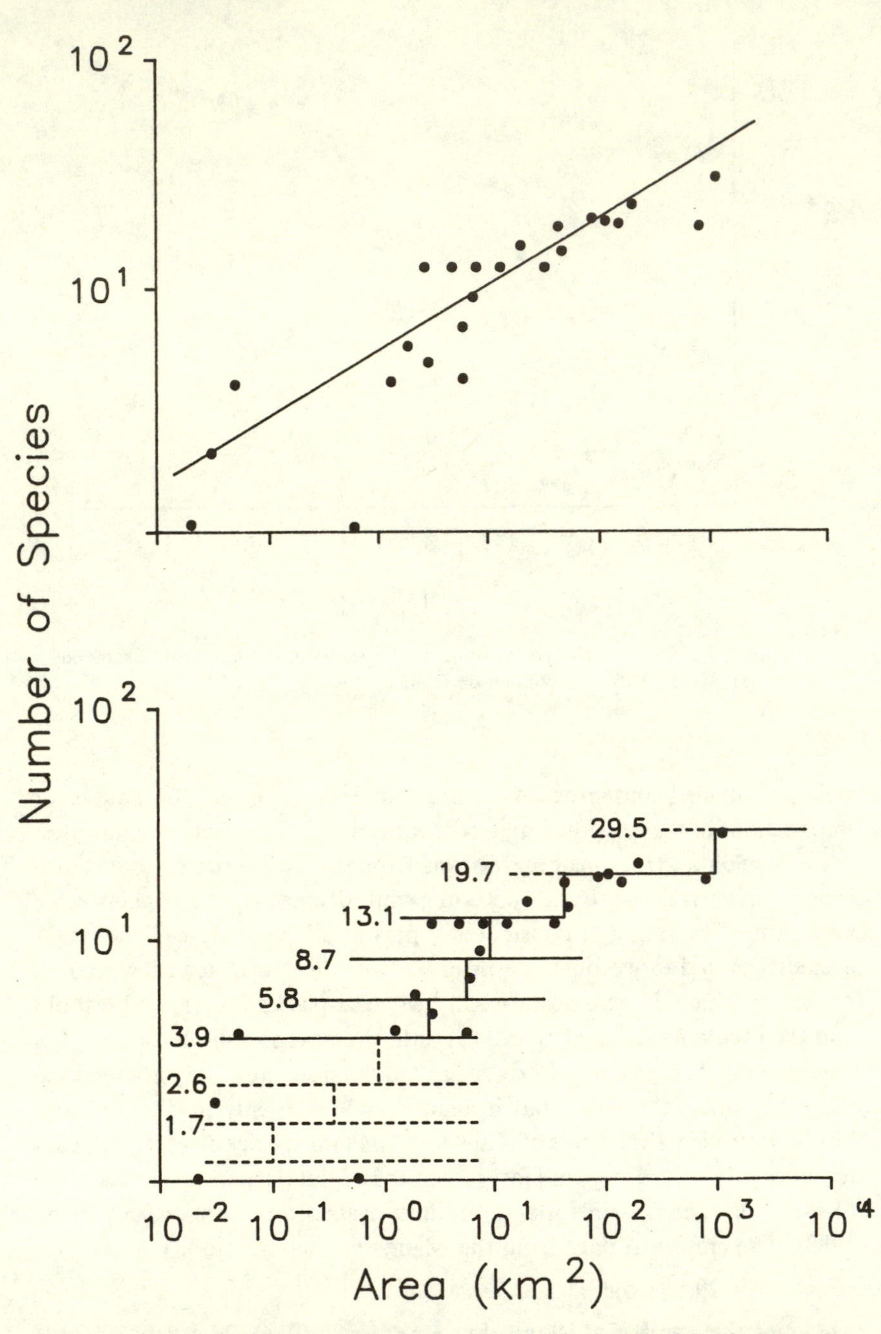

Fig. 5.3. Top: Species–area plot for 24 islands in the Sea of Cortéz. Bottom: A stadial model of species additions with island area, using the data shown in the top panel. Each stadial level represents a 1.5-fold increase in S, which is held at this plateau until a threshold of island area is reached, whence the next higher plateau is attained. From Cody (1983a).

The terms of this equation were derived empirically from the data; theoretical explorations of such threshold increases in S with A might prove rewarding.

Factors contributing to S/A *relationships*

Area and isolation. Species–area relationships are perhaps best seen on islands, where area is clearly defined and the confounding effects of transients are minimized. Much of the work that has been reported, therefore, deals with islands and addresses the question of why islands generally have fewer species than a mainland area of equivalent size (Abbott 1980). MacArthur and Wilson (1967) approached this question by developing a theory of island biogeography that emphasized the roles of area and, less importantly, isolation in determining S. The apparent agreement of many S/A curves with the predictions of the theory and its appealing structure have made it extremely popular.

A full treatment of this equilibrium island biogeography theory is beyond the scope of this chapter (see Abbott 1980, Williamson 1981, Brown and Gibson 1983). The basic structure of the theory, however, is deceptively simple. MacArthur and Wilson proposed that the number of species on an island is determined by the relationship between the rates of immigration from a larger species pool, which adds species, and extinction, which deletes them. These, of course, are basic elements of the assembly of any community (Fig. 4.1). MacArthur and Wilson related immigration and extinction rates to features of the islands themselves. Thus, the rate at which new species are added to an island is considered to be a declining function of S, because in a rich community the likelihood that an immigrant species either will already be present on the island or will be excluded by competition with established species is high (Fig. 5.4). Because of the difficulties of long-distance dispersal, successful immigration is more likely for islands near to the source pool than far away. In fact, distance is the only variable that is considered to have a major effect on immigration; area is relatively unimportant.

Area is important, however, in determining the shape of the extinction curve. Extinctions should be more likely on species-rich islands than on those containing a few species, simply because there will be more rare species present. The probability of extinction of any single species, however, is expected to increase as area decreases, as population sizes should be smaller and therefore more sensitive to chance catastrophes. Given the forms of these curves (Fig. 5.4), it is a logical necessity that an equilibrium in species numbers will occur where immigration exactly balances extinctions

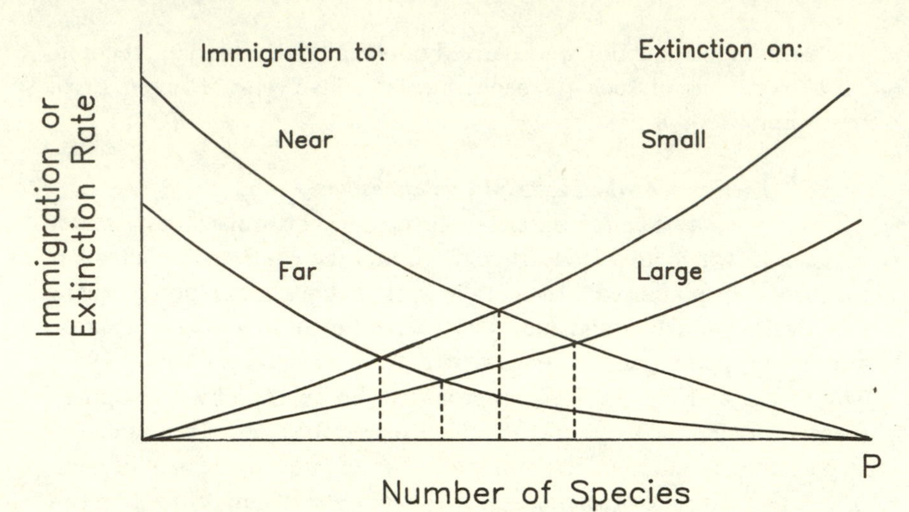

Fig. 5.4. The patterns of immigration and extinction developed in the MacArthur–Wilson theory of island biogeography. Equilibria occur at the intersections of the immigration and extinction curves; P is the total species pool. The patterns differ depending on island distance from the source pool and on island area.

(Williamson 1981). The point at which equilibrium is reached, however, will vary depending on distance and area – a small, distant island will contain fewer species than a large, near island (Fig. 5.4).

Some elements of the theory have a degree of empirical support. Diamond and Jones (1980), for example, presented data on species turnovers for the Californian Channel Islands that suggest a relationship between extinction probability and population size, and Moulton and Pimm (1986) found a significant relationship between extinction rates of introduced species and S on the Hawaiian Islands. Gilpin and Diamond (1976) showed that extinction rates of bird species in the Solomon Islands vary as the reciprocal of area and that the effect of area on immigration rates is slight.

Gilpin and Diamond extended their analysis to develop a model of S/A relationships that included distance (D) as well. When the simple S/A model was applied to the 37 central islands in the archipelago (among which D was small), 98% of the variation in S was accounted for (Diamond and Mayr 1976). When a model including D was applied, 98% of the variation in S among all 52 islands surveyed (including 'remote' islands) was explained. Gilpin and Diamond concluded that, in general, A influences S much more strongly through its effects on extinction rates than on immigration. Abbott (1980), on the other hand, argued that the interaction between distance and the dispersal propensities of species may be a critical determinant of S on islands. By incorporating restrictions on the ability of species to colonize islands of varying area into a model of S/A relationships, he obtained values of z similar to those reported in the literature as being consistent with the MacArthur–Wilson theory (Abbott 1983). Using other information on habitat relationships and competitive interactions in the model produced similar values of z.

So area and other factors may have a greater effect on immigration rates than MacArthur and Wilson considered in their theory. Moreover, extinction may not be completely independent of distance. Brown and Kodric-Brown (1977) noted that islands located close to a source pool may receive immigrants so frequently that many populations close to extinction are 'rescued' by continuing influxes of individuals. The value of S for such islands would be greater than predicted on the basis of area alone, due to the reduced extinction. Kobayashi (1983) and Järvinen and Ranta (1987) noted that the S/A relationship may be distorted *within* an archipelago because of interisland immigration. Once again, the theoretical relationship between S and A seems not so simple as was first thought.

How does all of this relate to the reality of species numbers on islands? A particularly instructive example is provided by G.R. Williams' (1981) studies of resident, native land and freshwater birds on New Zealand islands. The general relationship between S and A for all 34 islands he considered fits a power function, with $z = 0.16$ (Fig. 5.5). There is considerable scatter, however, and several exceptions stand out. Whale Island (the most heavily modified of the offshore islands) and White Island (an active volcano) lie below the regression line. The islands of the Chatham group, which are about 650 km from the mainland, all lie above the regression line, whereas all other islands > 100 km from the three 'mainland' islands lie well below it. The regression for the Chatham islands has a much shallower slope ($z = 0.13$) than does the (nonsignificant) regression for the other outlying islands ($z = 0.37$). The reasons for the unexpected richness of islands in the Chatham group are not immediately apparent, although the cluster lies directly in the path of winds blowing from the mainland and the larger islands support a considerable diversity of habitats. Isolation clearly seems to depress the species lists for the other outlying islands. With few exceptions, the 17 offshore islands (see Fig. 5.5) lie within 30 km of the mainland and are separated by water gaps generally of 15 km or less. For those islands alone, the S/A regression gives a value of z (0.28) very close to that expected from equilibrium theory.

Williams examined the reasons for the absences of species from islands in

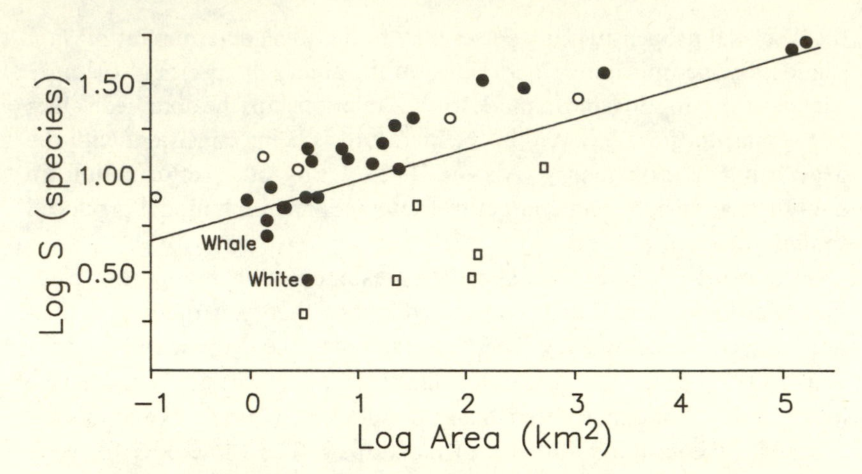

Fig. 5.5. The relationship between numbers of bird species and island area for islands in the New Zealand region. Solid circles denote offshore islands, open circles indicate islands in the Chatham group, and open squares represent outlying islands. From G.R. Williams (1981).

considerable detail. Some of the species present on the mainland do not cross water barriers of any kind, either because they are flightless or because they are 'psychologically incapable' of doing so. Others appear to cross such barriers sufficiently to occur on some of the offshore islands, but not beyond. Still others seem able to disperse widely but are absent from some islands because specific habitat requirements are missing. In some instances, habitats on the islands are not readily equivalent to their mainland counterparts. This may affect especially the native forest species, whose need for large areas of habitat increases with the degree of endemism of the species (East and Williams 1984). Finally, the New Zealand source pool itself is depauperate, a consequence of its own relative isolation. The New Zealand avifauna contains only 60 resident native species (19 families), whereas Tasmania, a land-bridge island that is only 25% of New Zealand's area, has 97 species (30 families). Some 34 species of land and freshwater birds have been introduced by humans and are now established on the New Zealand mainland. Interestingly, many of these have colonized the offshore and outlying islands, and most of these islands now support breeding populations of up to 10 species of exotics. The number is apparently unrelated to island area. Williams argues from such evidence that these New Zealand islands are not saturated with an equilibrial number of species, and Abbott and Grant (1976) reached the same conclusion from similar data (but see Diamond and Veitch 1981).

Habitat features. Differences in habitats among islands can also distort expected *S/A* relationships. In the original MacArthur–Wilson model, habitat complexity was simply indexed by area: large islands were assumed to support a greater diversity of habitats than small islands. It is doubtful that habitat complexity increases monotonically with area, however. Islands off southwestern Australia that are exposed to the full force of gales support only stunted heath vegetation, for example, while islands in the same area in more sheltered locations develop forests (Abbott, personal communication). Moreover, as area increases, the proportion of 'interior' relative to 'edge' habitat changes (e.g. Fig. 3.2). Because there are often more species in interior than in edge habitats (Helle and Helle 1982, Ambuel and Temple 1983), these changes influence the form of the *S/A* relationship. It may be, as Kangas (1987) has suggested, that *A* is a good predictor of *S* only when the analysis is confined to a within-habitat, alpha-diversity scale. At the broader scales indexed by beta or gamma diversity (see below), habitat differences may reduce the fit of observations to a simple *S/A* curve (Boecklen and Gotelli 1984, Boecklen and Simberloff 1986).

Relatively few studies have included direct measurement of habitat features, but among those that have, a large number report that *S* is more closely related to habitat characteristics than to island area alone (Lack 1969a, 1976, Yeaton 1974, Williams 1964, Abbott 1980, Haila 1981, Järvinen and Haila 1984; see Table 4.1). Reed (1984) found a strong correlation between *S* and *A* for islands of Scilly in Great Britain when he used logarithmic transformations, but when the analysis was conducted using untransformed variables, *S* correlated most strongly with the number of habitats on the islands, a relationship that was confirmed with partial correlation tests. (Note that the patterns detected using different analytical procedures were quite different.) In the Hawaiian Islands, Scott *et al.* (1986) found a relationship between the number of native bird species present and the area of blocks of rainforest habitat on the islands, but the fit was much closer when elevation was included as a variable in the analysis. There, *S* increased at higher elevations, although lowland areas typically contain more bird species, as in the West Indies (Kepler and Kepler 1970, Lack 1976), the Galápagos Islands (Harris 1973), the Solomon Islands (Greenslade 1968), New Guinea (Diamond 1972), and in temperate (Miller 1951, Able and Noon 1976, Sabo 1980, Sabo and Holmes 1983) and tropical (Moreau 1966, Terborgh 1971, 1977, Haffer 1974, Pearson and Ralph 1978) continental areas. The departure of the Hawaiian Islands from the 'typical' pattern may be associated with widespread habitat destruction in the lowlands and the presence there of avian malaria (van Riper *et al.* 1986). In any case, habitat variations associated with elevation clearly contribute to variations in S in all of these locations.

One way to examine the contributions of habitat and other factors to changes in S among islands is through multiple regression analysis. Abbott *et al.* (1977) applied this procedure to an analysis of the numbers of landbird species on the Galápagos Islands. There, the number of plant species on an island was most highly correlated with S (an observation whose meaning is clouded somewhat by Connor and Simberloff's (1978) observation that the number of plant species on these islands is most highly correlated with the number of collecting trips that have been made to the island by botanists). Power (1972) found that the number of native plant species accounted for 67% of the variation in bird species on a series of islands off the Californian and Baja California coasts, with island isolation also contributing. The number of plant species, in turn, was largely explained by island area, secondarily by latitude. When Power (1976) restricted his analysis to the California Channel Islands, maximum island area and plant species numbers were the best predictors of S. Johnson (1975) used the same procedures to examine patterns of variation in bird species numbers on 20 mountaintop 'islands' and 11 sample areas within adjacent 'continents' in the western USA. An index of habitat diversity was highly correlated with increasing S (Fig. 5.6) and in a stepwise multiple regression it explained 91% of the variation in S. Habitat diversity itself was under complex control, with the area of forest woodland, latitude, elevation, distance from other islands, and other unknown factors all important (Fig. 5.7).

This last example illustrates that aspects of habitat complexity may be more important determinants of S than area or isolation alone, but it also demonstrates an inherent difficulty of multiple regression techniques. Because variables such as area, habitat diversity, isolation, elevation, and the like are all usually intercorrelated to a degree, their use as 'independent' variables in an analysis is problematic (Vuilleumier and Simberloff 1980). The patterns that emerge may be statistically sound (provided the assumptions of the tests are met), but biological interpretation of those patterns is difficult because of the intercorrelations among variables.

Species differences. These considerations of habitat factors indicate that islands with the same values of A are not necessarily ecologically equivalent. In a similar manner, bird species differ ecologically, and thus not all units of S are equal. These differences can affect S/A patterns. Terborgh (1973), for example, calculated S/A regressions separately for different families of

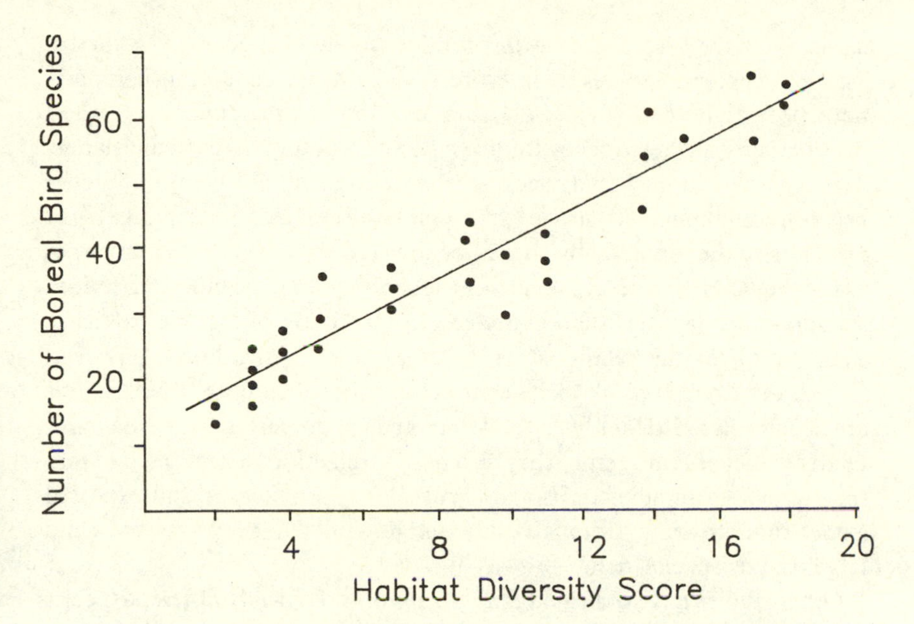

Fig. 5.6. The relationship between the number of species of boreal birds occurring in montane 'islands' in western United States and a habitat diversity score (see Fig. 5.7.). From Johnson (1975).

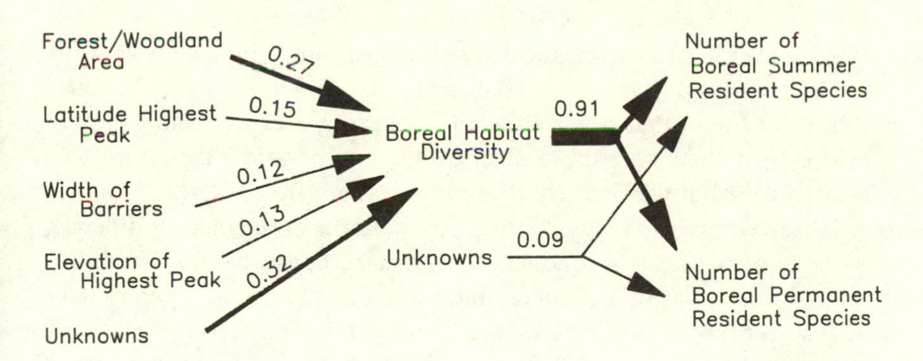

Fig. 5.7. Path diagram showing the relationships among variables affecting the number of boreal bird species in montane 'islands' in western United States, as revealed by stepwise multiple regression analysis. Modified from Johnson (1975).

landbirds in the West Indies. Many families followed different *S*/*A* curves. The differences in slopes were interpreted as evidence of differing colonization potentials and dispersal abilities: low slopes indicated good dispersability and a tendency for all species in a family to reach all islands. Terborgh also argued that, because competition should be more intense between confamilial bird species than less closely related forms, sympatry of members of the same family should be low and the number of familes per island high. He presented evidence suggesting that islands did indeed support more families than expected. On the basis of a more extensive examination of the relationships of these West Indies birds, however, Gotelli and Abele (1982) concluded that the slopes of family *S*/*A* regressions are closely related to family size and that much of the variation in *z* may thus be an artifact of family size. They also used rarefaction to demonstrate that the number of families on islands generally approximates that expected from a random draw, although small islands tended to be species-poor and large islands species-rich.

Other studies have indicated that groups of ecologically different species may follow different *S*/*A* patterns. Blake (1983; see also Blake 1986) found that numbers of foliage–insectivore species increased more rapidly with increasing *A* in forest 'islands' than did other groups. Most of these species are long-distance migrants, and this pattern may thus reflect the general predilection of such species for forest interior habitats (Whitcomb *et al.* 1981), which increase dramatically with increasing *A*. In a study of bird-species distributions on over 500 Bahamian islands, Schoener and Schoener (1983a, b) found that occurrences of resident species were more closely related to *A* than were those of migrants, which varied haphazardly in relation to *A* but were closely related to distance from a larger island.

These studies clearly indicate that taxonomically or ecologically different species groups differ from one another in their responses to area. The moral is that calculations of *S*/*A* curves and *z* values will be sensitive to exactly what sorts of species are included in the analysis. Different combinations of species are likely to yield different patterns. Inferring processes from these patterns may be especially risky.

Passive sampling. In the MacArthur–Wilson equilibrium model, islands are viewed as accumulating species up to some point by 'sampling' from the species pool via the deterministic processes of immigration and extinction. Alternatively, one may consider island faunas to be built up by a random, noninteractive 'passive' sampling of species from the pool (Connor and McCoy 1979, Coleman 1981, Coleman *et al.* 1982, Haila 1983). Given an unequal species-abundance distribution, if species are distributed at random among islands, larger areas will inevitably contain more species simply because they contain larger samples. Connor and McCoy (1979) suggest that the passive sampling hypothesis should be considered as the appropriate null hypothesis because it invokes no biological processes at all.

Constructing and testing models of passive sampling is not easy. A simple random sampling of species from a species pool ignores important features of their biology, notably dispersal abilities (Grant and Abbott 1980). This could produce absurdities such as flightless emus dispersing great distances over water to colonize small islands. On the other hand, incorporating information on dispersal abilities or habitat preferences into the model immediately raises the spectre of hidden structure discussed in Chapter 3.

Other, more analytical models of passive sampling tend to be complex and require considerable information. Coleman (1981), for example, developed a 'random placement' model, in which the probability of an individual residing on an island is proportional to the area of the island and independent of the presence or absence of other individuals or species. Coleman's model requires information on the abundances of species on the islands, not just their presence or absence (but see Simberloff 1976, McGuinness 1984b for alternatives based on presence–absence data). It also assumes that the slope of the S/A regression decreases with increasing A, which may not be unreasonable. Coleman *et al.* (1982) applied this model in the analysis of resident bird distribution on a series of small islands in Pymatuning Lake, Pennsylvania and concluded that variations in S with A were what one would expect were the birds distributed at random. Haila (1983) developed a 'sampling metaphor' model that is similar in some respects to the random placement model, except that it incorporates aspects of the 'sampling efficiency' of an island, at least part of which is due to its habitat composition. Whereas Coleman's model requires that the abundances of species on islands be known, Haila's approach is to derive expected abundances from prevalence functions (see Chapter 4) based on the regional abundances of species and their habitat requirements. Haila also found that, within the ecological constraints of his model, the distribution of S among islands of differing A matched expectations based on random sampling closely. Bilcke (1982) reached much the same conclusion from a different analysis, although he also noted that species number was correlated with variations in vegetation structure as well.

What should one conclude from this? The number of species clearly varies with area, although it does so in several ways. Simply fitting data to a power function (or any other equation) does not allow one to distinguish between the effects of area, isolation, habitat complexity, passive sampling, or other factors that may change with area (e.g. disturbance, McGuinness 1984a, b). A positive *S*/*A* relationship provides only weak and ambiguous support for the MacArthur–Wilson model because the same relationship follows from other models as well. There are multiple pathways of influence on *S*/*A* relationships (e.g. Fig. 5.7), and to understand these may require the formulation of specific hypotheses and detailed tests of each.

The importance of censusing

Determinations of *S*/*A* relationships require only lists of species found in areas of differing sizes and therefore do not demand the detailed estimation of abundances that often makes censusing so difficult (Chapter 3). This should not lead to a cavalier attitude toward censusing, however. Because small changes in the number of species estimated to occupy an area can alter the shape of an *S*/*A* relationship or its closeness of fit to one or another model, an accurate determination of *S* is essential. When Schoener and Schoener (1983a) surveyed Bahamian islands, for example, they did so in an admittedly incomplete fashion. The errors in estimation of *S* were probably greater for migrant than for resident species and it is therefore possible that some of the pattern they reported reflected sampling inadequacies. To obtain a complete census of small areas containing low vegetation may be relatively easy, but enumeration of all of the species present becomes increasingly difficult as area or habitat complexity increases. There is thus a possibility that census adequacy may be biased in an area-dependent fashion, producing an artifactual change in z. The way to control for this bias is by standardizing the survey efficiency in areas of different sizes (Abbott 1983), but this is rarely done.

Patterns of species abundances

Consideration only of the number of species in a series of communities may reveal interesting general patterns, but it masks important differences between species. Species differ in their abundances in local areas, and these differences may also exhibit interesting patterns. Røv (1975), for example, tallied the abundances of passerine species breeding in four forest types on a gradient from meadow to subalpine birch forest in Norway. When the species are ranked in order of decreasing abundance and their abundances expressed on a logarithmic scale, smooth curves result (Fig. 5.8). Røv attributed the changes in the shape of the curve along the gradient to increased competition. Although this explanation rests on an unwarranted inference of process from pattern (see Chapter 2), the patterns clearly do differ between the forest types.

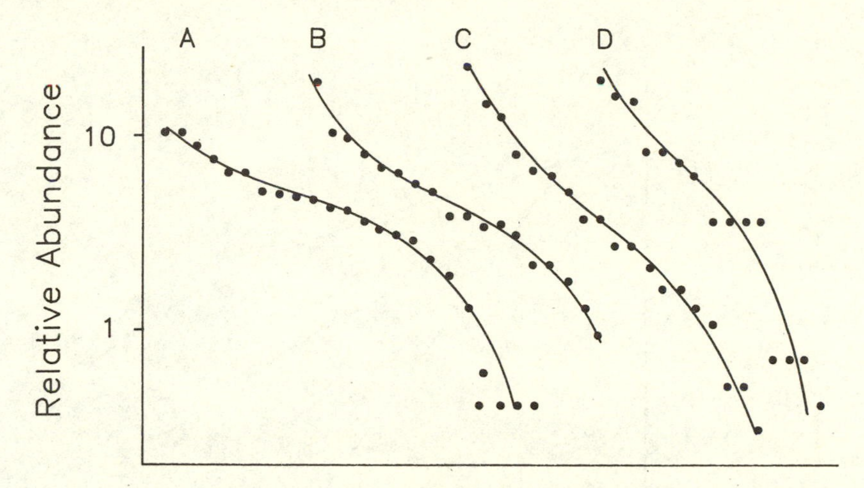

Fig. 5.8. Relative abundance distribution curves for a sequence of habitats from meadow (A) to forest (D). Each point represents the relative abundance of a bird species on a log scale plotted against the species' abundance rank in the community. Curves fitted by eye. After Røv (1975).

Species abundance models

As soon as ecologists recognized that species differed in their abundances in communities, they began to fit the data to various statistical distributions, perhaps hoping that a greater understanding of the communities might emerge. Several models were proposed; each was accompanied by a biological mechanism, and a fit of data to a given model often led directly to the conclusion that the mechanism associated with that model had in fact produced the pattern. The models are treated in detail by Williams (1964), Pielou (1977), Whittaker (1965), and May (1975).

Three models have been especially popular. If the numerical dominance of a few species in a community is strong, abundances will be quite uneven. Plotted on a log scale, abundance will fall off sharply and linearly with decreasing species rank, approximating a *geometric* series or logseries (Fig. 5.9). This is especially likely in communities containing few species. This model has been interpreted in terms of 'niche pre-emption', in which the dominant species occupies a certain proportion of the community niche space, the next species a certain fraction of the remainder, and so on. This competitive dominance is the basis for Røv's (1975) inference from the patterns of Fig. 5.8.

Alternatively, a set of species might divide the niche space randomly among themselves, so that they occupy nonoverlapping niches. If abun-

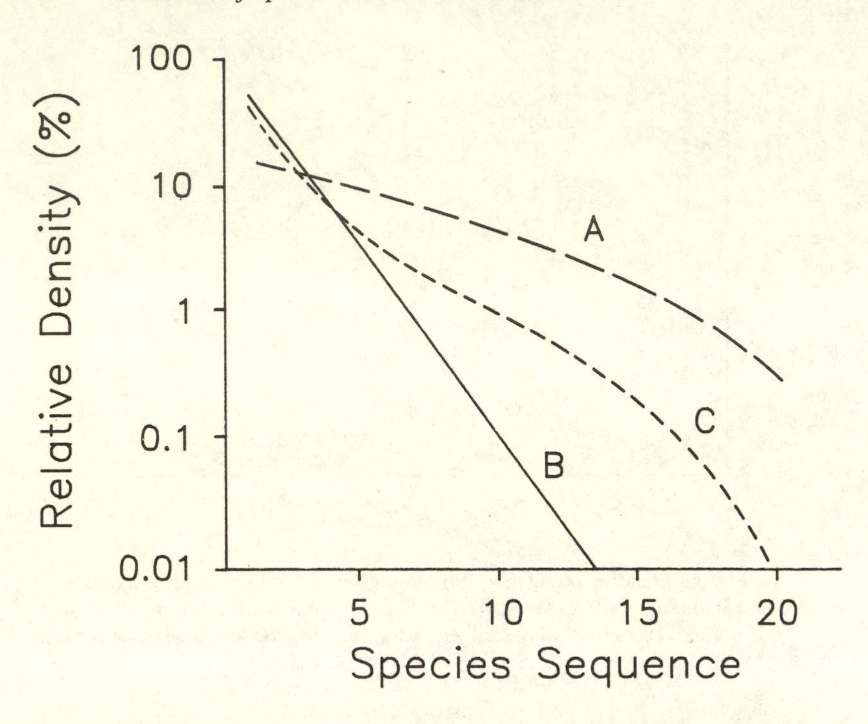

Fig. 5.9. The major species-abundance distributions. A = broken-stick, B = geometric series, C = lognormal. After Whittaker (1965).

dance is proportional to niche size, the distribution of species abundances will be relatively even, with little numerical dominance by one or a few species (Fig. 5.9). This fits the '*broken stick*' model proposed by MacArthur (1957). MacArthur (1966) subsequently disavowed the model on both mathematical and biological grounds, but the model nonetheless continues to receive attention (De Vita 1979, 1982, Pielou 1981).

The distribution of bird abundances appears most often to coincide with a *lognormal* distribution (Fig. 5.9), especially if the community contains many species. The original biological rationale underlying this model was that the distribution of niche sizes among species should be normal if the niche occupancy of different species is determined by a large number of independent factors; this results in communities with rather few very common or rare species and many species of moderate abundance (Preston 1948). Mathematically, the distribution follows from the Central Limit Theorem: if the relative abundances of species are governed by the interplay of many independent factors, the factors compound multiplicatively and the lognormal distribution is produced. When data fit the pattern it is difficult to determine whether the result is of mathematical or biological interest. May (1975) has concluded that the lognormal distribution is no more than a statistical consequence of the Central Limit Theorem and that the 'empirical rules' derived from the canonical form of the lognormal (Preston 1948, Hutchinson 1953) are simply mathematical properties of the more general lognormal distribution. Sugihara (1981) challenged May's conclusion, noting that natural communities adhered more closely to the predictions of the canonical lognormal model than would be expected from mathematical arguments alone. May (1984) conceded that his earlier conclusion may have been 'too glib', but the issue of whether the lognormal distribution and empirical generalizations derived from it reflect biology or mathematics remains unresolved.

Preston's (1948, 1960) derivation of the lognormal model was based on the assumption that the community is at equilibrium. Ugland and Gray (1982) found that species-abundance distributions in fact fit the lognormal only if the community is in equilibrium; data from disturbed, nonequilibrial communities do not fit this distribution. This suggests that some of the scatter of data sets about a lognormal distribution may reflect varying degrees of departure from equilibrium. It does not warrant the conclusion that a close fit of observations to the model certifies that the community *is* in equilibrium, however.

Analysis by rarefaction

Nowhere is the need to standardize data to equivalent levels of census intensity more apparent than in considerations of the relationships between species number, numbers of individuals, and area. In the absence of appropriate adjustment, as by rarefaction (see Chapter 3), apparent patterns may be quantitatively or qualitatively wrong. Consider an example (James and Wamer 1982: 162): In a census of breeding birds taken on 11.6 ha in a floodplain forest in Maryland, 35 species and 107 individual territories were recorded, in comparison with 13 species and 30 territories recorded in a census of 19.2 ha of aspen woodland in British Columbia. Some of the species were included only on the basis of fractions of territories within the plot; if these are excluded, the values of S are 27 and 10, respectively. If one assumes that individuals are regularly distributed in space (a troublesome assumption of rarefaction procedures), the densities for the two areas can be standardized to a common area (10 ha): 93 and 16 individuals, respectively. By rarefaction, the numbers of species in randomly drawn samples of 93 and 16 individuals in these communities are 25.7 and 7.3. Thus, although the unadjusted censuses suggest that the aspen

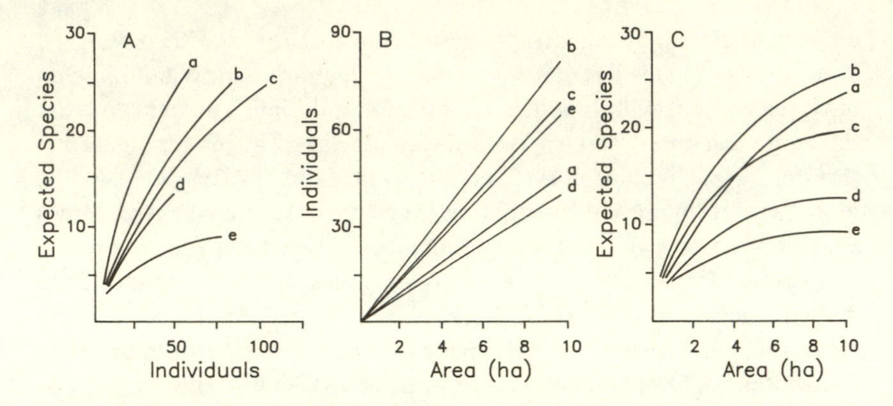

Fig. 5.10. Differences among five bird communities (a–e) as shown by (A) rarefaction curves for the expected number of species/individuals; (B) individuals/area; and (C) the expected species/area relationship. The habitats are (a) maple-pine-oak second growth; (b) tulip tree-maple-oak; (c) cottonwood floodplain forest; (d) mature jackpine with birch; and (e) wax myrtle forest. From James and Wamer (1982).

woodland contains 37% of the species number in the floodplain forest, this proportion drops to 28% when standardized to plots of equal size.

An impression of the interplay of numbers of species, individuals, and area for rarefied data can be gained from an examination of James and Wamer's (1982) comparison of several habitat types (Fig. 5.10). The rarefaction curves for the expected number of species in samples of a given number of individuals (Fig. 5.10A) are steepest in the second-growth and mature deciduous forest stands and lowest in a single-species wax myrtle forest. The increase in individuals with increasing area (*B*), on the other hand, is greatest in mature deciduous forest and lowest in the coniferous and second-growth forest. The species–area relationship (*C*) combines the previous two functions. The expected number of species is high in both second-growth and mature deciduous forests, but for different reasons. The second-growth habitat contains fewer individuals per unit area but accumulates species more rapidly with increasing individuals. For two of the habitats (a and c in Fig. 5.10C) the species–area curves cross at 5.5 ha; estimates of relative species richness in these communities would thus be reversed by a consideration of plots above and below this size.

Species diversity

Most investigations of patterns of variation in species numbers or relative abundances have been conducted under the rubric of species diversity. Although diversity may be measured most directly as number of species (S), more often it has been expressed as an index that incorporates the interplay of species richness and the relative abundances of species (evenness) into a single value for a given community. Most of the indices that have been proposed have severe analytical or statistical drawbacks, and the application of different diversity measures to the same set of data may produce quite different patterns (Kempton and Taylor 1976, Pielou 1975, Hurlbert 1971, Whittaker 1972, Hill 1973, Routledge 1979, 1980, Shmida and Wilson 1985). The most widely used diversity index, the Shannon–Weaver information measure (H') (MacArthur and MacArthur 1961), is particularly sensitive to the contributions of rare species, and this effect varies with sample size. The size of the area sampled also affects diversity calculations through its influences on both species numbers and abundances (Fig. 5.10). Such limitations have often been disregarded in the search for patterns of diversity, but the use of indices such as H' really tends to obscure rather than clarify such patterns. Although some ecologists have abandoned indices such as H' in favor of less biased and direct measures of species numbers and relative-abundance distributions (Routledge 1980, James and Wamer 1982), most of the literature on bird species diversity is based on unrarefied estimates of H'. This makes any overall evaluation or comparison of published diversity patterns rather pointless. Still, there are some interesting features of diversity variations that merit attention regardless of how diversity is indexed.

Scales of diversity

Diversity may be considered over a wide spectrum of spatial scales, from variations across continents, where diversity patterns may be influenced by speciation and biogeography (Rosenzweig 1975), to variations from point to point within a small field, where patterns may be consequences of individual habitat selection and territorial overlaps between species (Wiens 1969) (Fig. 5.11). Several categories of diversity have been defined on this gradient (Whittaker 1960, 1977, MacArthur 1965, Cody 1975, Shmida and Wilson 1985); Whittaker's classification is given in Table 5.1. Most attention has focused on patterns of alpha, beta, and gamma diversity. It is important to remember that these scales of diversity are rather arbitrary and that they bear no formal quantitative relationship to area *per se*. The spatial scale on which beta diversity is measured, for example, depends on the structure and size of the habitat mosaic in a landscape and upon the movement and dispersal distances of the species one considers. Different factors contribute to the determination of diversity at the different scales, and this may help to explain why general models of

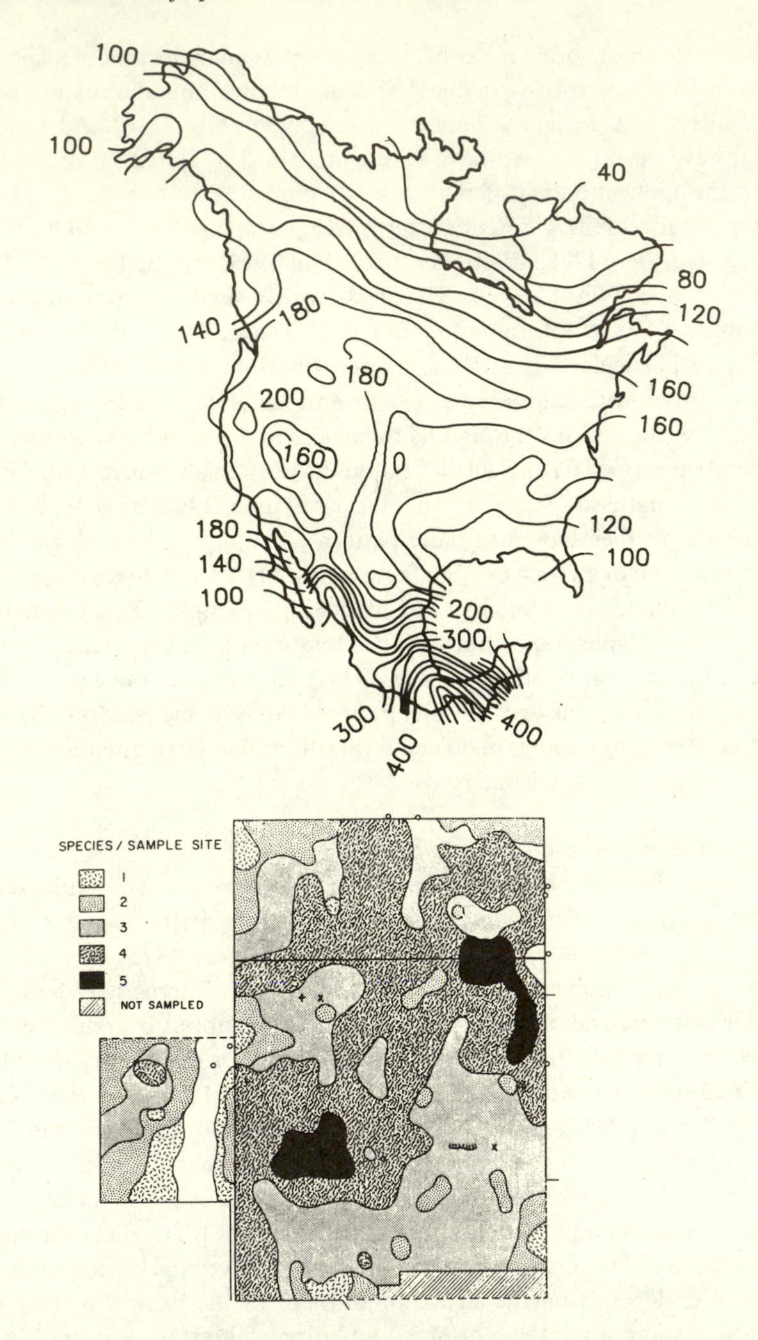

Fig. 5.11. Variations in regional bird species diversity (above, number of species; after Cook 1969) and local point diversity within a 40-ha grassland in Wisconsin (below, from Wiens 1969).

Table 5.1. *Levels and types of species diversity.*

Inventory diversities	Differentiation diversities
1. For a small or microhabitat sample within a community regarded as homogeneous, internal alpha or subsample diversity, *point* diversity.	2. As change between parts of an intracommunity pattern, internal beta or *pattern* diversity.
3. For a sample representing a community regarded as homogeneous (despite its internal pattern), *alpha* or within-habitat diversity.	4. As change along an environmental gradient or among the different communities of a landscape, *beta* or between-habitat diversity.
5. For a landscape or set of samples including more than one kind of community, landscape or *gamma* diversity (in the sense of Whittaker, 1960).	6. As change along climatic gradients or between geographic areas, geographic differentiation or *delta* diversity.
7. For a broader geographic area including differing landscapes, *regional* (epsilon) diversity.	

Source: From Whittaker (1977).

S/*A* relationships or species-abundance distributions often do not fit observations gathered over a broad array of scales very well (Shmida and Wilson 1985, Kangas 1987). Underwood (1986a) has discussed the difficulties of defining the spatial scales for measuring alpha, beta, and gamma diversity.

Broad-scale regional diversity variations have been mapped for bird communities in a number of areas (North America, Cook 1969, MacArthur and Wilson 1967; British woodlands, Fuller 1982; tropical and subtropical Africa, Crowe and Crowe 1982; Argentina, Rabinovich and Rapoport 1975; Australia, Schall and Pianka 1978). Because such treatments are based on variations in the numbers of species occurring in large map-grid blocks, they obscure much detail, but they nonetheless reveal interesting general patterns. The common feature of all these maps is a general failure to document smooth, monotonic changes in bird diversity along linear transects placed in any direction through the area. In Australia, for example, diversities of insectivorous birds are high at midlatitudes on the east coast, decreasing both northward and southward and toward the arid interior, but diversities rise again on the northwest coast (Schall and Pianka 1978). Among Argentinian passerines, there is a clear latitudinal change in diversity, but there are many distortions of this pattern produced by changes in habitat or biogeographic regions or by reductions in diversity in mountainous areas (Rabinovich and Rapoport 1975). In North America there is also a distinct increase in diversity from the arctic to the tropics, but diversities increase as well in the western mountains and in northern forests (Fig. 5.11). The pattern changes somewhat if one considers alpha-diversity variations over the same region. Short (1979) documented high diversities of nesting landbirds in the northeastern forests and along the Mississippi River drainage but lower diversity in the western mountains. Thus, areas that support large numbers of species at a regional scale may or may not also contain many species within local habitats.

Latitudinal changes in diversity have attracted particular attention. Tropical areas often have substantially greater alpha- and beta-diversities than do temperate areas of the same size (MacArthur 1969), although the details of this pattern vary between habitats and between continents. Howell (1971), for example, found diversities in rainforests in Nicaragua to be substantially greater than those in temperate forests, but diversities in tropical and temperate pine savannahs were equivalent. Howell attributed this to the isolation and recent development of the Nicaraguan pine-savannah habitat. Karr (1971) compared alpha diversities in several habitat types in Illinois and Panama. Although the tropical grasslands contained more irregular, transient species than the temperate grassland sites did, the number of resident species and overall species diversity (H') were quite similar (see also Wiens 1974a). Scrub and forest habitats in Panama, however, supported nearly twice as many species as were present in these habitat types in Illinois. Many of the additional species were quite rare, and evenness was therefore somewhat greater in the temperate areas, with the result that overall diversity was only slightly greater in the Panama locations. When the Panama locations were compared with similar habitats in the African tropics (Liberia), other differences emerged. Forest bird communities were generally more diverse in Panama, whereas grassland and savannah habitats in Liberia supported more species (Karr 1976). Karr attributed these differences to differences in the overall geographical extent of the habitats in the two continents and to the greater degree of Pleistocene fragmentation of habitats in the neotropics.

Even before the details of latitudinal diversity patterns (and the many exceptions to them) were defined, ecologists were busy generating explanations for the apparently greater tropical diversity (see Pianka 1966, Slobodkin and Sanders 1969 for reviews). It was suggested that the enhanced diversity was due to the greater age and stability (Sanders 1968) or greater structural complexity (MacArthur *et al.* 1966) of habitats, greater opportunities for and/or rates of speciation (MacArthur 1969, Haffer

1974), the presence of resource categories not present or poorly represented in the temperate zone (Karr 1971, Orians 1969, Schoener 1971a), higher predation rates (MacArthur 1969), or narrower niches (Klopfer and MacArthur 1961). Few of these hypotheses have been carefully framed or rigorously tested, perhaps because it has become apparent that all of these factors may contribute to specific latitudinal diversity patterns to varying degrees (MacArthur 1969).

Terborgh's (1980a) solution to this problem was pragmatic: he compared a lush temperate forest bird community in South Carolina (40 species) with an Amazonian tropical forest (207 species) and attempted to categorize the 'extra' species in the latter. Of these, 56 (34%) belong to ecological groups or guilds (see Chapter 6), such as obligate ant-followers or dead-leaf gleaners, that are not represented in the temperate communities because opportunities for those modes of life do not exist there. Some of the groups that are represented in both areas contain far more species spanning a broader spectrum of resource conditions in the Amazon; 17% of the 'extra' species can be categorized as reflecting a larger tropical 'guild niche'. Finally, there is often a tighter packing of species within the niche space for a given guild in the tropics (narrower specialization), and this accounts for the presence of 82 (49%) of the additional species. Terborgh suggested that this increased species packing might reflect a greater rate of speciation in the tropics, perhaps coupled with lower extinction rates. Elsewhere in Amazonia, unique habitat types may further enhance diversity. Remsen and Parker (1983), for example, noted that perhaps 15% of the avifauna in the Amazon Basin has evolved in habitats created by the dramatic seasonal fluctuations in water levels of the river systems.

Within broad regions, diversity may depart markedly from this temperate–tropical latitudinal trend. Järvinen and Sammalisto (1976) found that bird diversity increased with *increasing* latitude in mire habitats in Finland, and Boström and Nilsson (1983) documented a slight trend toward more northerly raised peat bogs in Sweden containing more species (chiefly waders). Tramer (1974) noted that, although alpha diversity of wintering North American birds followed a relatively smooth increase toward the tropics, diversity patterns in summer indicated a 'plateau' between 45° and 25°N in eastern North America in which diversity did not change appreciably. In fact, the diversity of breeding passerines increases from south to north in this region in both deciduous and coniferous forests (Rabenold 1978, 1979; see Fig. 5.11). Rabenold speculated that this pattern might be a consequence of thresholds in seasonality of food supplies. North of about 45°, seasonal oscillations may be so strong that few species are able to breed quickly and then either overwinter or undertake long migrations. South of about 25°, winter food supplies may become dependable enough to support increasing numbers of year-around resident species. Between these thresholds, the magnitude of the summer burst in production increases toward the north, fostering a parallel trend in species packing. Unfortunately, the information on food-resource levels and their seasonal dynamics necessary to test this interesting idea rigorously will be difficult to obtain.

A somewhat different approach to assessing patterns of alpha- and beta-diversity variations was developed by Cody (1975). Using measures of vegetation height and half-height (the height dividing the vertical foliage profile into two equal areas), Cody derived a synthetic habitat gradient by principal components analysis. Sites were then positioned on this gradient and curves depicting the rates at which new species are added or existing species are lost as one moves along the gradient were described. The difference between these curves is the number of species present at any point along the habitat gradient, its alpha diversity. For a series of several habitat gradients in mediterranean environments in California, Chile, Sardinia, and South Africa, alpha diversity increases with increasing habitat structure in a generally similar fashion, although in Sardinia and one of the African areas diversity decreases in more mesic woodlands (Fig. 5.12). By deriving a curve halfway between the species-gain and species-loss curves, one can describe the rate at which species are accumulated from point to point along the habitat gradient; the derivative of this curve defines the rate of species turnover attributable to change on the habitat gradient, beta-diversity. When these curves for the five mediterranean areas are compared, striking differences are apparent (Fig. 5.12). The beta-diversity curve for Chile is quite low, with a maximum of about 16 species turnovers per unit of H. Beta-diversities are highest in South Africa, where they peak at lower values of H (structurally simpler habitats). Using the same procedures in a comparison along a shrub-forest habitat gradient, Karr (1976) found that both alpha- and beta-diversities differed markedly between areas in Illinois, Panama, and Liberia. He attributed the difference between this and Cody's findings to the more mesic and seasonally stable nature of the habitat gradient he studied.

The five mediterranean areas Cody studied differ in the areal extent of habitat types at points along the habitat gradient. Local sites thus may have different reservoirs of potential colonists to draw from in accumulating alpha- or beta-diversities. Cody (1975, 1983b) performed some analytical gyrations to calculate the amount of 'source' habitat available in each area and discovered that variations in alpha-diversity values largely disappear

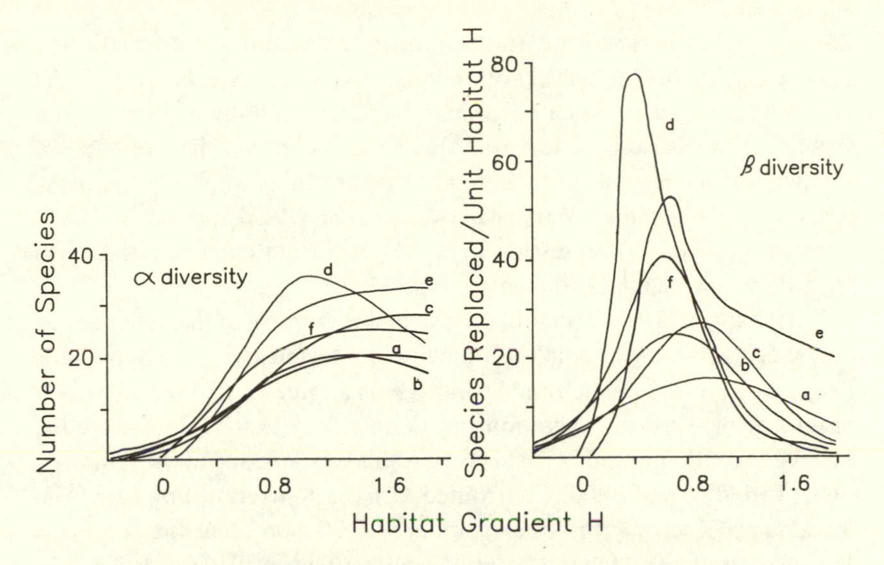

Fig. 5.12. Plots of alpha-diversity and beta-diversity over a gradient of mediterranean-climate habitats in six areas on four continents; (a) Chile, (b) Sardinia, (c) California, (d) Cedarberg, South Africa, (e) Outeniqua Mountains, South Africa, (f) Cape Town area, South Africa. After Cody (1983b).

when this factor is considered. To a considerable degree, differences in beta-diversity may also be related to differences in regional habitat availability. Karr (1976) suggested that such factors might also have contributed to the between-continent differences he observed.

Habitat influences on species diversity

Foliage height diversity. Cody's approach uses a certain amount of analytical wizardry to determine the relationship between habitat and diversity patterns. More often, these relationships are explored through direct correlation analyses. When MacArthur introduced information-theory measures of diversity to community ecology, he did so in the context of relating bird species alpha-diversity (BSD) to habitat diversity (MacArthur and MacArthur 1961). Habitat structure was expressed by a measure of the evenness of the distribution of foliage among vertical height strata, foliage height diversity (FHD). MacArthur and MacArthur (1961) found that increases in BSD over a range of habitat types were closely correlated with increases in FHD (Fig. 5.13). The initial derivation of this relationship was entirely empirical, although MacArthur later (1965) developed a theoretical foundation for the relationship, based on resource use and limiting similarity among coexisting competitors. Recher (1969) gathered data on

BSD and FHD for five woodland and heath communities in Australia and discovered that they fell nicely on the regresssion line calculated from the North American data, despite the clear differences in both bird species and vegetation. Recher concluded from this observation that the avifaunas of the two regions responded to habitat structure by subdividing the niche space in similar fashions. When MacArthur *et al.* (1966) applied the North American FHD–BSD regression to census data from Panama and Puerto Rico, however, the data fit poorly (Fig. 5.13).

What happened next provides an instructive example of the use of *ad hoc* adjustments to bring anomalous data into conformity with the hypothesis (Fig. 2.1). MacArthur and MacArthur (1961) originally used several subdivisions of the vertical vegetation profile to calculate FHD, selecting the layers 0–0.6, 0.6–7.6, and > 7.6 m because this subdivision provided the best fit between BSD and FHD. Confronted with the nonconforming data from Panama and Puerto Rico, MacArthur *et al.* (1966) proceeded to subdivide the foliage profile for Panama locations into four layers (0–0.6, 0.6–3.1, 3.1–15.2, > 15.2 m) and that for Puerto Rico into two (0–0.6, > 0.6 m). This produced a remarkably good fit to the North American regression. MacArthur *et al.* (1966: 322) then concluded that the birds in Panama 'appear to act as if their habitat were subdivided into finer layers than the temperate, and Puerto Rican birds act as if their habitat were subdivided into coarser layers than the temperate'. A questionable demonstration of pattern led immediately to an unsubstantiated inference about process.

Other investigators have used different subdivisions of three layers (Austin 1970) or different numbers of layers (4: Røv 1975, Karr 1968, Moss 1978; 4–5: Terborgh 1977; 5: Beedy 1981; 6: Siegfried and Crowe 1983; 7–8: Lovejoy 1975) to calculate FHD. As a way of detecting pattern, of course, there is nothing inherently wrong with using the subdivision that maximizes the fit of data to the BSD–FHD regression, although the procedure does smack of data-dredging (Chapter 3). It would be better if there were some *a priori* biological justification for the selection of a particular subdivision scheme. Inferences about ecological relationships drawn from the observation that different subdivisions work best in different situations, however, are unwarranted.

Not all studies have substantiated the simple BSD–FHD relationship proposed by MacArthur and his colleagues. Willson (1974) and Erdelen (1984) found a reasonable relationship between BSD and FHD when they considered a spectrum of habitats ranging from open grasslands to mature forests, but within any subsection of that spectrum the relationship did not hold. Other workers investigating habitats as varied as grasslands and

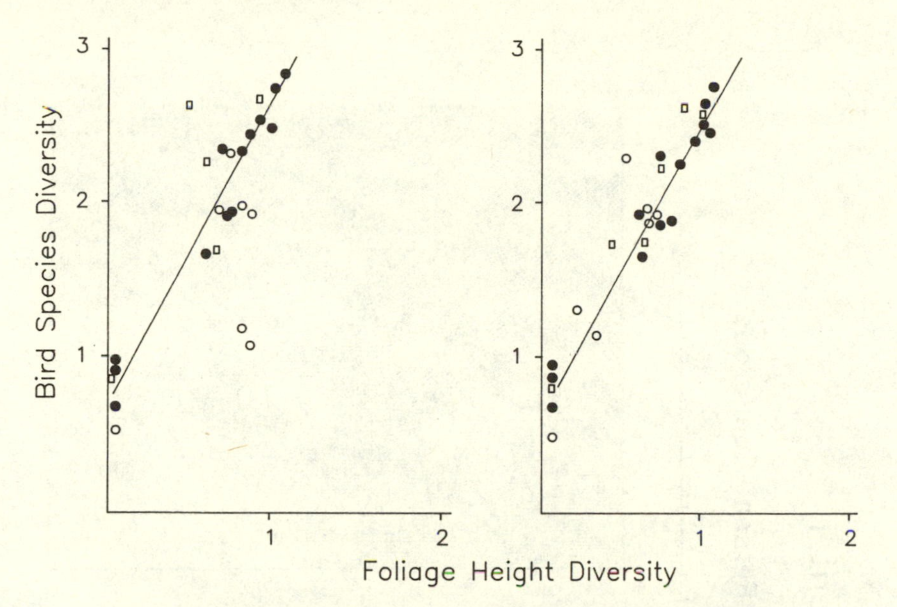

Fig. 5.13. Bird-species diversity plotted against foliage height diversity for censuses from Puerto Rico (open circles), Panama Canal Zone (open squares) and temperate United States (solid circles). Left: regression determined for the temperate locations only, using three vegetation layers to index foliage height diversity. Right: foliage height diversity for Puerto Rico calculated from two vegetation layers, that for the Canal Zone determined from four layers. Note that adjustment of the number of layers considered in the calculation of foliage height diversity improves the 'fit' of the Puerto Rico and Canal Zone data to the North American regression. After MacArthur *et al.* (1966).

shrublands (Karr 1968, Karr and Roth 1971, Wiens 1974b, Roth 1976), deserts (Tomoff 1974), riparian woodlands (Rice *et al.* 1983a), temperate or Bahamian pine forests (Balda 1975, Szaro and Balda 1979, Emlen 1977b), tropical pine savannahs (Howell 1971), South African fynbos (Siegfried and Crowe 1983), or Swedish forests (Nilsson 1979a) have failed to find a significant relationship between BSD and FHD. This is not to say that habitat structure is unimportant in these situations; indeed, most of these investigators demonstrated a close relationship between BSD and some other aspect of vegetation structure. Noting that horizontal patchiness might be more important than vertical vegetation stratification in low-stature habitats such as grasslands or shrublands, Wiens (1974b) and Roth (1976) developed measures of horizontal heterogeneity. Roth found a close relationship between his measure and BSD over a range of habitats from shrubby fields through forests, but I was unable to demonstrate any relationship over a series of grassland and shrubsteppe locations. Roth may

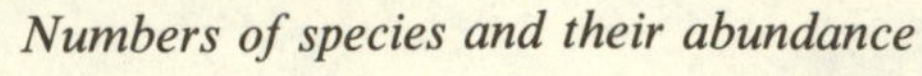

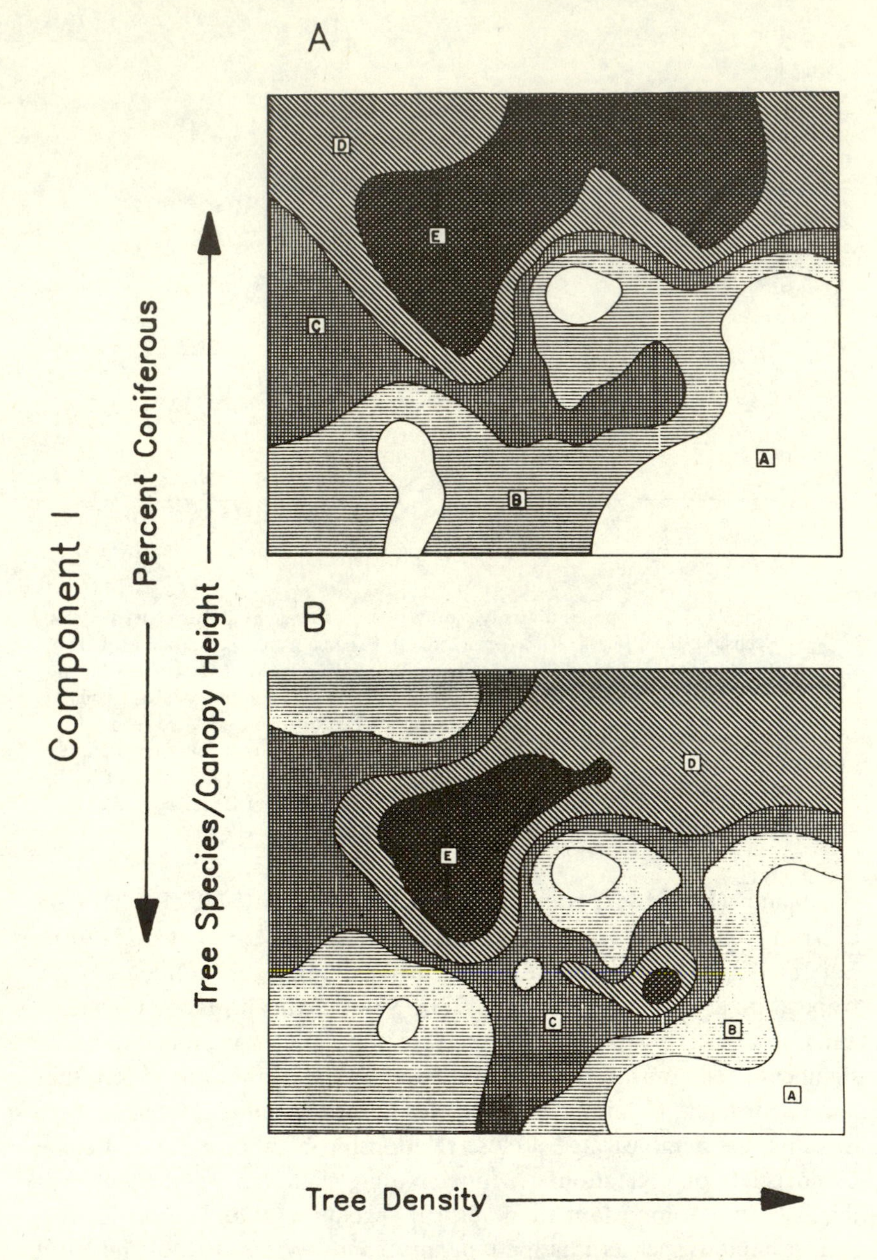

Fig. 5.14. Contour-map diagrams for (A) the number of territorial pairs of birds and (B) the number of bird species expected on plots of 10 ha, plotted in a bivariate space determined by principal component axes. PC I = increasing tree species and canopy height and decreasing conifer cover; PC III expresses

be correct in suggesting that the gradient I considered may have been too restricted in vegetation coverage and height to produce clear heterogeneity–BSD correlations, but we are still left with the conclusion that variations in diversity may not be closely associated with simple (or even complex) measures of habitat variation in all situations.

Things aren't so simple. Two studies indicate something of the subtlety of bird diversity–habitat relationships. Terborgh (1977) examined BSD variations along an elevational gradient from lowland rainforest through montane rainforest and cloud forest to high-altitude elfin forest in the Peruvian Andes. He found a significant relationship between FHD (measured with either four or five subdivisions) and BSD. Instead of stopping at this point and fashioning some explanation of the pattern, Terborgh went on to consider how the diversity of three major trophic subgroups of the avifauna varied. Although the number of insectivores decreased by over fivefold from the bottom to the top of the gradient, frugivores decreased by a factor of only 2.3, and nectarivores remained unchanged. Clearly, diversity patterns are determined by many influences beyond simply the vertical structure of the habitat. To Terborgh, this 'demonstrated the folly of taking a good correlation too seriously in the absence of compelling *a priori* logic'.

Following an entirely different approach, James and Wamer (1982) explored the relationships between bird diversity and habitat changes over a spectrum of forest types in North America. Principal components analysis ordered the 56 stands with respect to dimensions representing tree species richness and canopy cover and height, variation in conifer importance, variation in canopy height, and tree density. Both the number of species in a standardized area of census and the total bird density varied markedly with respect to these dimensions, but in ways that could not possibly be discerned from a simple BSD–FHD correlation (Fig. 5.14). The number of species was maximal in mature deciduous forests, but not in the forests that had the highest tree species richness, canopy height, or tree density. Bird densities followed a somewhat similar pattern, except that maximal densities occurred at maximal values of tree species richness and canopy height. Both species numbers and densities were lowest in coniferous forests with high tree density, low canopy, and few tree species. It is worth noting that the coniferous stands included in this study were primarily from the Pacific

Fig. 5.14. (*cont.*)
primarily increasing tree density. In (A), region A = 4–22, B = 22–40, C = 40–58, D = 58–76, and E = 76–94 pairs; in (B), A = 4–8.4, B = 8.4–12.8, C = 12.8–17.2, D = 17.2–21.6, and E = 21.6–26.0 bird species. From James and Wamer (1982).

Northwest, where bird species numbers are considerably lower than in coniferous forests in northeastern North America (Wiens 1975). Inclusion of such stands presumably would have produced different patterns.

This study is noteworthy in several respects. James and Wamer used rarefied census data standardized by area, they considered diversity in terms of numbers of species and densities per unit area rather than by a synthetic index such as H', they used multivariate procedures to order a large number of habitat measures rather than arbitrarily selecting a single measure they deemed important, and they explicitly avoided drawing conclusions about anything other than the patterns their analyses demonstrated. Their study was also unusual in that plant *species* were considered as variables to which the birds might be responding. MacArthur and MacArthur (1961) had included a measure of plant species diversity (PSD) in their initial consideration of bird diversity–habitat patterns. It was in fact a reasonably good predictor of BSD, but not as good as FHD. Because PSD was correlated with FHD, it had little additional explanatory power in a partial regression following the BSD–FHD correlation. Subsequent workers largely ignored PSD as a habitat feature of any importance, often citing MacArthur and MacArthur as justification. Several recent studies, however, have demonstrated clear correlational patterns between the plant species composition of habitats and bird community patterns in both low-stature shrubsteppe and grassland (Wiens and Rotenberry 1981a, Rotenberry 1985), and mixed hardwoods forests (Holmes and Robinson 1981, Robinson and Holmes 1984). Clearly, floristics can no longer be ignored.

Temporal changes in diversity

Succession. Ecological succession involves changes in both the floristic and structural composition of habitats, so it is reasonable to expect bird species diversity to change systematically during succession. The conventional wisdom of succession theory (e.g. Odum 1969) is that diversity increases during succession, in accordance with increasing production, standing crop biomass, and structural complexity. Mature 'climax' ecosystems may experience some reduction in overall diversity, however, as they often tend toward dominance by a small number of tree species and may contain reduced vertical layering. In a postulated successional series from meadow to spruce forests in western USA, Smith and MacMahon (1981) found that the 'trends' Odum (1969) predicted for production and energetic parameters of ecosystems generally were expressed in bird assemblages. Trends relating to species diversity and various life-history attributes, however, were not. Moreover, most patterns varied considerably between years. Smith and MacMahon concluded that the successional relationships were more complicated than Odum's simple model would predict. Odum's 'trends' are derived from views of north-temperate forests (largely deciduous), and their applicability to temporal sequences of community development in other situations may be limited.

Although bird diversity generally increases during vegetational succession (e.g. Kendeigh 1948, Odum 1950, Johnston and Odum 1956, Karr 1968, Shugart and James 1973), this increase is not always monotonic. Smith and MacMahon (1981), for example, found that bird diversity was greatest in the preclimax fir forest. In a sequence of deciduous forest stands of different ages following clear-cutting in Poland, bird diversity increased slowly at first, then rather rapidly between 10 and 20 yr. This was followed by a decline in diversity at 30–40 yr, but diversity then peaked in the mature stands (100–150 yr) (Głowacinski and Weiner 1983). Bird densities followed a generally similar pattern, although the increase in 10–20-yr stands was more dramatic. May (1982) recorded a similar pattern of initial rapid increase, then decrease, then increase to peak diversity and density in a sequence of deciduous forest succession in eastern North America. Głowacinski and Järvinen (1975) determined that the rate of avian successional change in the Polish oak-hornbeam series decreased monotonically with age, although in Finnish coniferous forests the rates peaked after 5–25 yr. Once again, general patterns seem elusive.

These studies, like most, analyzed successional change by comparisons among stands of different ages. Lanyon (1981) combined this comparative approach with continuous monitoring of individual plots over a 20-yr period to chart the patterns of avian succession on fallow farmland in New York. There the number of breeding species increased rapidly during the first 15 yr of succession from bare soil to open shrubland. Species number then declined slightly and held relatively steady over the next 25 yr (open shrubland to shrubby woodland). A more mature oak woodland (60 yr after cutting), however, contained more species than any of the stages of old-field succession. Bird density also increased during the first phase of plant succession but continued to increase after diversity levelled off at 15 yr following cultivation. Peak densities were reached in shrubby woodlands 30–40 yr of age, and density was reduced in later stages of these stands and in the more mature oak woodland.

Lanyon's study clearly indicates that changes in diversity and density may not be parallel during succession. This raises the possibility that the changes in diversity are simply consequences of passive sampling involving a larger number of individuals in older stands. Helle (1984) used rarefaction procedures to control for this factor in a comparison of forest stands of different ages in Finland. There, older stands (> 75 yr) supported roughly twice the number of species as did younger stands (< 25 yr), but they also supported greater densities. Following rarefaction, the older stands still supported 20% more species than did younger stands in samples of the same number of individuals. In this case, part, but not all, of the increase in S with age is a consequence of increasing densities.

Seasonal changes. Most terrestrial environments undergo seasonal changes in habitat structure and food abundance, and these changes are likely to influence diversity. Rabenold's (1979) suggestion that a reversed latitudinal diversity gradient in eastern temperate North America is associated with thresholds of seasonality has already been mentioned. For many temperate areas, the observations of Holmes and Sturges (1975) may be typical: in a hardwoods forest, they found greater diversity in summer and reduced evenness in winter (due to the influx of large flocks of granivorous finches). Tramer (1969) argued that bird communities should generally have lower evenness in winter than in summer, because the unpredictable and harsh environment should reduce evenness in winter and territoriality should increase it in summer. He extended this argument to suggest that diversity generally should change through variations in evenness in rigorous, unpredictable environments but through changes in species richness in more benign, predictable environments.

These ideas have received varying support. Kricher (1972) provided data consistent with the evenness predictions, although Rotenberry *et al.* (1979), Austin and Tomoff (1978), and Alatalo and Alatalo (1980) found no distinct seasonal changes in evenness. Alatalo and Alatalo (1980) suggested that, if samples are relatively small, clumping by dominant species that form flocks during winter may lower evenness because rare species are excluded from the sample. In samples from a larger area of habitat, such seasonal changes in evenness disappear. Tramer's more general hypothesis relating evenness and richness variations to environmental rigor and predictability was tested by Rotenberry (1978), who derived a gradient of climatic variability and rigor for a large area in northwestern United States. Diversity changes from the mild, moist, uniform climate of coastal areas to the severe, arid, variable inland region were produced largely through changes in evenness rather than species richness. Alatalo and Alatalo (1980) also observed that a tendency for evenness to decrease from central to northern Europe parallels a trend toward increasing climatic unpredictability and harshness toward the north (Järvinen 1979). Nudds (1983), however, failed to find support for Tramer's hypothesis in an analysis of diversity variations in waterfowl assemblages in Canada. To be rigorously tested, Tramer's hypothesis requires a clear, quantitative definition of environmental predictability and rigor, which is by no means easy to produce. Moreover, the mechanisms by which evenness varies have to do with the dynamics of population stability or instability of species in the community and the degree to which their population fluctuations are synchronized (Alatalo and Alatalo 1980).

Seasonal changes in diversity are likely to be quite different in subtropical or tropical areas. In a comparison of bird communities in Wisconsin (41°N) and new South Wales (31°S), for example, Howe (1984) recorded much greater seasonal flux in species numbers in the former. There, diversity peaked in April–May and again in September–October with the passage of large numbers of migrants; numbers of breeding species (*ca.* 30 per 16-ha plot) were substantially lower, but were greater than the numbers of wintering species (*ca.* 11 species). In New South Wales, on the other hand, migratory movements are less pronounced and more complex and are complicated by nomadism related to droughts in the interior (Rowley 1975, Recher *et al.* 1983). There, no peaks associated with migratory movements were evident. Species numbers were greater during the prolonged September–January breeding season (*ca.* 40 species per 16-ha plot) than in winter (*ca.* 30 species). In subtropical and tropical areas that are not so isolated as Australia, however, large seasonal changes in diversity may be produced at least in some habitats by the arrival of migrants from the Palearctic or Nearctic regions (Keast and Morton 1980).

Many tropical environments are sharply seasonal, of course, but the changes are associated with rainfall rather than temperature variations. These changes affect habitat structure and food supplies, and one would expect the bird species to respond. Karr (1976) found that seasonal variations in the diversity of resident bird species in several habitats in Panama decreased with increasing vegetation complexity. Karr attributed this to the increased buffering of the physical environment by the more complex vegetation, and he suggested that the same relationship might hold in temperate areas as well. He also considered how diversity varied among different subgroups or guilds of species and hypothesized that, in general, one might anticipate increasing variability in diversity from omnivores to frugivores to insectivores. Among species groups defined by foraging positions, variability in diversity might be expected to increase from bark foragers, to ground feeders, to species foraging in low, medium, or high foliage strata, to those feeding in open spaces within or above the vegetation. These are intriguing hypotheses, but they remain largely untested.

Are bird communities saturated?

Equilibrium saturation

The notion that animal communities might contain as many species as ecologically possible goes back at least as far as Elton (1950), who observed that 'the number of different kinds of animals that can live together in an area of uniform type rapidly reaches a saturation point'. Birds are more strongly limited in their opportunities for niche diversification than are plants or many invertebrates, which can respond to small differences in chemistry or ratios of critical nutrients, and this may place lower limits both on opportunities for speciation and packing into local habitats (alpha-diversity) for birds (Whittaker 1977, Tilman 1982). As a consequence, the observation that species diversity reaches generally similar values in structurally similar but widely separated habitats has been taken not only as evidence of saturation, but as implying that 'the kinds of species occupying the habitat are those most appropriate, and hence are better adapted than an equal number of a rather different set of species, as well as better than a different number of species' (Cody 1966: 375). MacArthur (1965) argued that the close fit of BSD to a linear regression with FHD clearly indicates that the bird communities are saturated, and Recher (1969) concluded from the close fit of data from Australian temperate forest and scrub habitats to the North American BSD–FHD regression that equilibrium and saturation levels in the two regions are similar and appear to be independent of the differing histories and ancestries of the avifaunas. Some of the apparent consistency of such comparisons may reflect artifacts introduced by the behavior of the H' diversity index used (Whittaker 1977), but this has not discouraged rather widespread acceptance of the reality of the pattern of saturation in bird communities.

One study of community saturation merits detailed inspection, because it reveals some important aspects of pattern detection and bears directly on the points developed in Chapter 2. Terborgh and Faaborg (1980a) examined the variations in the numbers of species found in small, uniform tracts of sclerophyll scrub and of lower montane rainforest on 12 islands in the West Indies. The islands differed in area and thus in the sizes of the species pool available to provide colonists to the local habitat patches on the different islands. Terborgh and Faaborg graphed the number of species recorded in census surveys and mist-nest samples in the sample areas in the two habitats against the size of the island's species pool. After an initial rapid increase, values of S stabilized and remained unchanged with further increases in the size of the species pool (Fig. 5.15A). Thus, although numbers of species on the islands continued to increase with increasing area, the number recorded within habitats did not. Terborgh and Faaborg drew an analogy between the curves of Fig. 5.15A and saturation curves in physiology and concluded that the curves provided clear evidence of community saturation. Having accepted this pattern, they then proceeded to explore hypotheses that incorporated mechanisms related to the use of niche space, habitat specialization, and competition to explain this pattern. The finding that species richness at saturation is higher in the sclerophyll than in the rainforest habitats, for example, was attributed to closer species packing due to greater invasion pressures. They concluded that the communities were 'noninvadable' and interpreted the pattern of Fig. 5.15A as indicative of 'a coevolved upper limit of tolerance to interspecific competition' (1980a: 190). This conclusion is not supported by the work of Pregill and Olson (1981), who noted that most of the Lesser Antillean area was covered with schlerophyll shrub during Pleistocene glacial periods. The greater contemporary species diversity in such habitats may therefore simply be a consequence of the development of a large, scrub-adapted avifauna during this period.

The inferential propositions put forth by Terborgh and Faaborg rest on the reality of the saturation pattern, but how real is it? To begin with, although both axes of Fig. 5.15A are measured in the same units (numbers of species), they are scaled differently. When they are scaled uniformly (Fig. 5.15B), the visual impression of saturation is diminished considerably. This does not mean that the pattern is not the same (none of the numbers has been changed), only that it is less apparent. Terborgh and Faarborg conducted no statistical analyses to determine whether their data fit an appropriate nonlinear model but only approximated the curves of Fig. 5.15A by eye. Such analyses seem necessary, given that the different scalings change the placement of the points and therefore the form of their relationships. Note also that all of the saturation curves of Fig. 5.15A are firmly anchored at the origin, even though there is no biological justification for this (one could record 0 species when the pool is, say, 5 or 10). This 0, 0 anchor point is not an actual data value and should not be included in statistical tests. Linear and negative exponential regression tests using the nonzero values for each data set statistically confirm a saturation pattern for three of the four data sets. The census data agree almost as well with a linear model, however, suggesting that increases in within-habitat S with larger species pools may be gradual and continuous.

Terborgh and Faaborg (1980a) did not specify their netting and censusing procedures in any detail. The number of individuals captured per

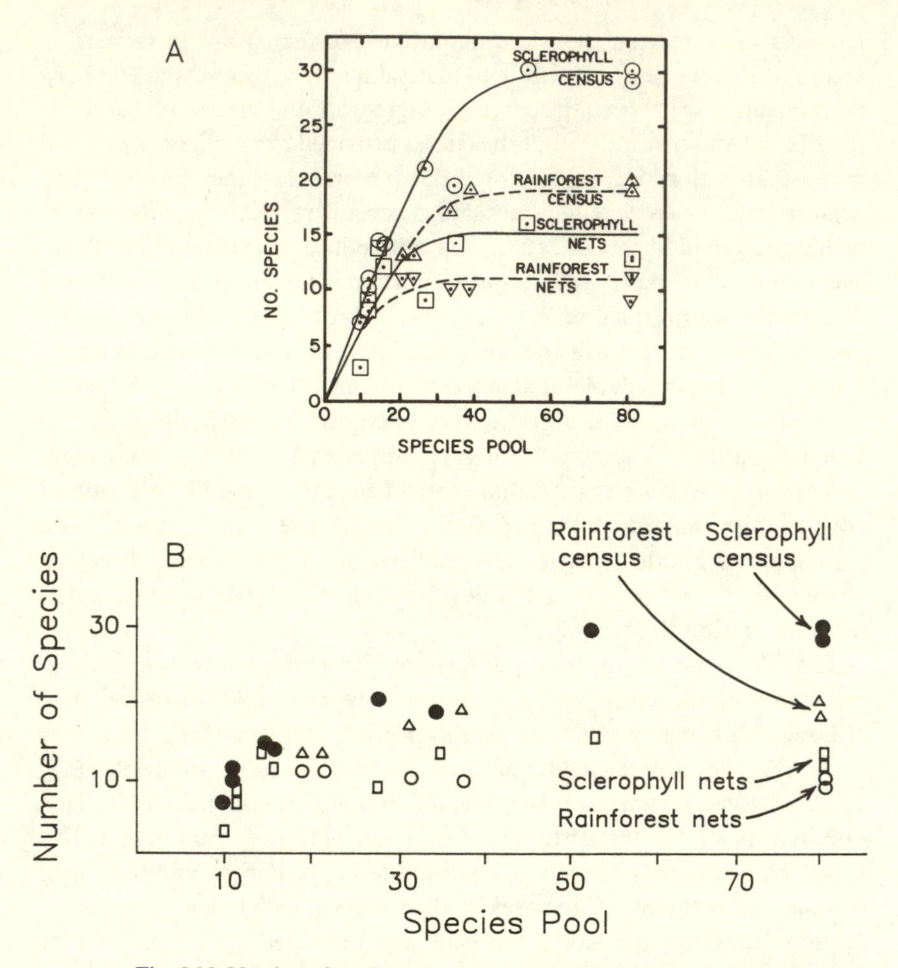

Fig. 5.15. Number of species recorded in observational censuses and in netting surveys within rainforest and sclerophyll shrub habitats in relation to the size of the species pool present on 12 West Indies islands. Top: graph from Terborgh and Faaborg (1980a); bottom: same data replotted with symmetrically scaled axes.

net sample varied widely, but Terborgh and Faaborg argued that this was unlikely to introduce a systematic bias to the results. Smaller samples, however, will record fewer species than larger samples. Censuses did not follow established procedures but consisted of records of the species seen during the period of fieldwork in a given area. To use the net or census values one must assume that all of the species present in a habitat are counted or that the proportion missed is constant or varies randomly with respect to island area. Larger species pools, however, are likely to contain more rare species that may easily be missed in surveys. A smaller proportion of the total avifauna may therefore be recorded in within-habitat surveys on large than on small islands.

There are also ambiguities in the determination of the census values and the sizes of species pools for the different islands. According to Terborgh and Faaborg (1980a), Desirade, for example, has a species pool of 17 (extracted from Bond 1971). Terborgh and Faaborg (1980b), however, give a value of 15 for the species pool on this island, while Terborgh *et al.* (1978) used a value of 19, in agreement with Lack (1976). The values used by Terborgh and Faaborg (1980a) for other islands also included recently extinct species. It is rather difficult to imagine how such species can be considered potential colonists of a local habitat. For several reasons, then, Terborgh and Faaborg's documentation of a pattern of species saturation in West Indies bird communities is equivocal.

Claims of community saturation generally rest on documentations that the number of species in a system is relatively constant, that species number can be predicted from habitat measures, that similar habitats in different regions contain similar numbers of species, or that guild sizes are similar in similar habitats. From his observation that grassland and shrubland breeding bird communities in North and South America usually contain 3–8 species, Cody (1966) concluded that the communities are saturated. My own observations from other areas in North America (Wiens 1974b) and Australia (unpublished) agree with Cody's findings. I have suggested, however, that this apparent constancy in species numbers may not reflect a maximal packing of the available niche space but rather the effects of frequent but unpredictable environmental variations (Wiens 1974a, 1977b). Flack (1976) applied similar arguments in explaining why bird species diversity in aspen forests in western North America appears to be less than would be expected on the basis of habitat structure and productivity. The number of species recorded in the undergrowth strata of humid seasonal forest in Panama also appears to be relatively constant from year to year, suggesting that those systems might be saturated and at equilibrium. This impression belies the considerable flux in species composition and activity levels that is found between sites in the same habitat and between times at the same plot (Karr and Freemark 1983). In all of these situations, species number is seemingly in equilibrium, but this simple observation of pattern alone cannot be used to conclude that the community is ecologically saturated or that the community represents a coevolutionary optimum (Cody 1966). Whittaker (1977) regarded the concept of community saturation as 'scientific quicksand' and suggested that the term be used guardedly, if at all.

Supersaturation of faunas

If one accepts the assumption that a fit to a species–area curve such as the power function is indicative of equilibrium, then it follows that points lying much above that curve depict 'supersaturated' faunas that contain more species than they should (Diamond 1972, 1975a). In Fig. 5.5, for example, many of the offshore islands (especially the Chatham group) appear to fit this description. Diamond (1975a) attributed the 'excess' species on several islands in the New Guinea region to Pleistocene connections either with New Guinea or with larger satellite islands, or to larger island areas caused by the lowering of sea levels during that period. These factors allowed the islands to accumulate larger numbers of species than their present area would permit. Through time, Diamond argues, the 'supersaturated' avifaunas 'relax' back to the predicted, equilibrium number of species. This may occur relatively rapidly for small islands, but for large islands it may take a very long time, so we still find the faunas to be 'supersaturated'. Terborgh (1974) offered similar arguments for the avifaunas of some islands in the West Indies. In both Diamond's and Terborgh's studies, the assertions that the islands are indeed supersaturated and that relaxation times are long are based on no evidence other than the observed departure of the points from the calculated S/A regression, accompanied by untested theories about Pleistocene effects. Simpson (1974) followed a similar argument in a more rigorous manner to determine whether or not the flora of the Galápagos Islands represents an equilibrium. Her propositions, however, have been challenged by Connor and Simberloff (1978).

Cody (1975) used a different approach, based on the relationships between alpha-diversity curves and contemporary regional habitat availability, to document his suggestion that, while mediterranean habitats in Chile and South Africa and low-stature shrublands in California have reached equilibria, woodland and forests in California are 'supersaturated'. He attributed this to carryover from the more extensive distribution of such habitats during the Pleistocene. Karr (1976) followed the same approach to explain the observation that grassland and savannah habitats in Africa contain more species than similar habitats in Central America, while the reverse is true in a comparison of Neotropical and similar African forests.

In theory, one should be able to use S/A curves to predict the number of extinctions that would accompany the reduction in size of an area. If one is dealing with a portion of the S/A curve that is relatively flat (i.e. large areas), however, substantial reductions in A may be expected to produce only slight reductions in S – the effects of changing A on extinctions are different in different portions of an S/A curve. Scatter about the S/A curve may also diminish the accuracy of predictions of the number of extinctions to be expected with a reduction in area.

Some direct information on island relaxation times is afforded by the history of Barro Colorado Island (BCI), which was abruptly isolated from the Panamanian mainland during the construction of the Panama Canal in 1914 (Willis 1974). By comparing the avifauna of BCI with that of a large tract of lowland forest on the adjacent mainland, Karr (1982a) documented that as many as 50–60 species of forest birds are missing from BCI. Faunal relaxation has been rapid. Karr (1982a, b) suggested that several factors accounted for the extinctions on BCI. The habitat mosaic on BCI is restricted, and this has undoubtedly contributed to some extinctions. Recolonization rates from the adjacent mainland may also be low, as many of the undergrowth bird species are reluctant to cross even small water gaps. Rarity alone is not a good predictor of extinction probability, but high variability in population sizes does predispose a species to extinction, especially in the context of the restricted habitat mosaic. Species dependent upon patchily distributed or variable resources (some insectivores as well as frugivores and nectarivores) are especially likely to have high population variability. Karr's studies on BCI are also relevant to the effects of fragmentation of large continuous areas of habitat into smaller, isolated parcels; these effects are considered in detail in Volume 2.

Conclusions

As MacArthur (1965) observed, 'patterns of species diversity exist'. The problem is to determine exactly what they are and then to decide what to make of them. As the examples described in this chapter have shown, the basic documentation of patterns is not easy but may be easily abused. There has been a tendency to believe that matching a set of data to some mathematical function is equivalent to understanding the pattern. Carried to an extreme, such mindless function-fitting can only justify views such as that of Poole (1974), who described species-abundance distributions as 'answers to which questions have not yet been found'.

If properly conducted, of course, documentations of species–area relationships, species-abundance distributions, or diversity gradients can be quite valuable. These patterns are the foundation for comparisons – between islands and mainland, between stages of forest succession, between temperate and tropical regions, between Australia and North America, between large and small forest tracts, and so on. They provide a way to recognize situations that depart from a general relationship and that may warrant additional study, and they offer the potential to explore in greater detail how specific parameters of the models or features of the community may influence the observed pattern. Such efforts are likely to be most rewarding when the comparisons are made in the context of some *a priori* biological hypothesis.

There are two major difficulties in the way a good deal of this work has been done, however. First, many ecologists have been satisfied to obtain a reasonable fit of data to some model and then to stop, accepting some inferential explanation as now proven without either considering alternative explanations or pursuing further, deeper tests of the pattern or explanations. Alternative explanations almost always exist, and they must be considered. Several of the patterns I have discussed above may be derived just as easily from mathematical or statistical propositions as from biological mechanisms. A fit of species–area data to a power function, for example, may be 'explained' by hypotheses of area effects alone, habitat diversity, random sampling, or disturbance. To distinguish among these hypotheses requires more detailed and specific tests. To accept a biological inference without considering nonbiological alternatives may cause efforts to be diverted into an ultimately unproductive area of study.

Second, simple synthetic indices or functions have generally been used to express relationships wherever possible. The H' diversity index is a good example. Not only is the statistical behavior of this index unreliable, but it combines into a single measure the effects of numbers of species, relative abundance distributions, density, and the area sampled. In the process, important details are obscured, assuring that the relative effects of the contributing parameters cannot be determined (James and Rathbun 1981). For example, Järvinen and Väisänen (1976), Karr (1976), and Rice *et al.* (1983b) have all reported situations in which the species diversity of a community changes little through time, even though the species *composition* of the community undergoes tremendous changes. By using such general terms, we may detect a pattern but have little idea of what is really going on. Development of second-level hypotheses is thwarted. The search for patterns should be conducted using direct measures with clear biological meaning, not abstract, dimensionless indices.

All of this is not to say that searches for general patterns are a waste of time. They are not. Properly done, they are an important starting point toward understanding how communities are put together and what factors influence that assembly. But detection of patterns in species–area relationships, species–abundance distributions, diversity variations, and the like is not an end. The activity is futile if it does not lead to the development of more refined, detailed, and testable hypotheses of pattern and process.

6

Niche theory and guilds

The concepts and studies described in the previous chapter involve features of entire communities in which the particular identity or attributes of the species present are largely unimportant. If the details of the species composition of communities are considered, more specific questions can be asked. How are resources shared among the species in a community? Are there limits to the degree of ecological similarity of species? How do the ranges of environmental conditions used by species change with variations in the environment or in the number of species in the community? Do sets of ecologically similar species form repeatable groupings in communities? Are increases in the species richness of communities attained by adding species that are ecologically similar to those already present, by incorporating new sets of ecologically different species, or by both?

To address such questions requires knowledge of the parameters of the ecological niches of species. Many of the investigations of bird communities conducted during the past three decades have dealt in one way or another with niches, and a large body of concepts and hypotheses collectively known as 'niche theory' has developed. Much of the impetus for this development came from the studies of MacArthur and his associates, so the emphasis of this theory has not surprisingly been on resource-defined niches and their relation to competition. In particular, examinations of niche overlap among species and the limitations to that overlap have been predominant. One nearly universal finding of such studies is that the distribution of niche overlap among the species in a community is neither even nor entirely random. Instead, overlap usually tends to be high among sets of a few species, which are ecologically differentiated from other sets of overlapping species. The notion that communities may be structured into clusters of species of similar ecology – ecological guilds – has been useful in its own right. Because they occupy a position somewhere between individual species and the entire community, guilds provide a way to explore community features that are not revealed by a consideration of species numbers and abundances alone without overwhelming one with a mass of species-specific detail that might obscure interesting community patterns. This may be especially important if one wishes to compare communities with quite different species compositions.

In this chapter I develop some of the basic elements of niche theory. I will not spend much time critically evaluating the theory but will present it as it has been expressed within the framework of the MacArthurian community paradigm. This provides the background for many of the studies that will be discussed in the remaining chapters of this section, where both the problems with the theory and its contributions to our understanding of communities will become clear. I also explore some aspects of the guild concept – its definition, how one determines guild boundaries and membership, and what sorts of patterns emerge from guild analyses.

Niche theory

Niche concepts

'Niche' must surely be one of the most variably defined terms in ecology. Most definitions, however, relate closely to one of two major concepts of the niche. On the one hand, Grinnell (1917, 1924, 1928) considered the niche of a species to be the set or range of environmental features that enable individuals to survive and reproduce. His focus was primarily on factors determining the distribution and abundance of species. Competition between species might be one such factor, but it was not the primary object of niche analysis. Contrasting with this is the view championed by Elton (1927), who described the niche of a species as its functional role in the community, especially its position in trophic interactions. Elton's emphasis on the community context of the niche was reinforced in 1957, when Hutchinson cast the niche in terms of mathematical set theory. He suggested that an environment may be viewed as a large number of dimensions, each representing some resource or factor of importance, on which different species exhibit frequency distributions of performance, response, or resource utilization. Collectively, the dimensions define an n-dimensional space, and the frequency distributions for a species define an n-dimensional hypervolume within this space – its niche. Hutchinson emphasized the role of interspecific competition in determining the niche actually occupied by a species: the 'fundamental' niche that a species can occupy in the absence of competitors is reduced to a 'realized' niche in the presence of competitors. As a consequence, niche size, shape, location, and overlap with other species will shift in response to changes in competitive pressures. The focus is on coexistence of exclusion of species rather than on variations in their abundances.

Grinnell's view of the niche was thus primarily autecological. In its emphasis on the multiple determinants of the distribution and abundance of single species, it paralleled the later individualistic views of Gleason (1926) and the population focus advocated by Andrewartha and Birch (1954, 1984). Hutchinson's niche concept stressed the position of a species in a community relative to those of other species. It led to different questions and prompted different sorts of studies than those fostered by the Grinnellian concept (James *et al.* 1984; Table 6.1). Hutchinson's view of niches strongly influenced MacArthur's thinking about communities (although MacArthur also related some of his distributional work to Grinnell's concepts; see MacArthur 1972: 149–53) and provided the framework for a good deal of the empirical and theoretical work in community ecology during the following three decades.

To draw a firm distinction between the Grinnellian and Hutchinsonian concepts of 'niche' and to advocate one as being better than the other (e.g. James *et al.* 1984) seems no longer necessary nor helpful. Although most recent applications of Hutchinson's niche concept to community investigations have emphasized resource dimensions and competitive interactions, there is nothing in Hutchinson's original formulation that necessarily makes it so restrictive. As Grinnell suggested, a wide range of factors (many of them not entailing biotic interactions and operating with different strengths in different areas) may influence the distribution and abundance of species. These factors may provide an appropriate focus for single-species investigations, to be sure, but they may also be considered as dimensions of Hutchinsonian niches in multi-species studies.

Niche overlap

A primary emphasis of niche theory has been upon the close relationship between niche overlap among species in a community and interspecific competition. According to the Gauseian competitive exclusion principle, ecologically similar species using the same set of limited resources cannot coexist at equilibrium. It follows that, by defining the niches of such species, one can determine their similarity (= overlap) and thence should be able to assess their potential for competition. Under conditions of resource limitation, a large amount of overlap between species on one or more niche dimensions should therefore lead to a contraction in niche space by one or more of the species in order to lessen competition (Fig. 6.1A). If resources vary in abundance through time, one might expect a reduction in overlap

Table 6.1. *A comparison of attributes of the Hutchinsonian and Grinnellian niche concepts.*

Hutchinsonian niche model	Grinnellian niche model
Objective To understand communities in terms of resource division among coexisting species; to study the assembly of communities and their evolution.	*Objective* To understand population regulation in terms of the resources that limit the distribution and abundance of a single species throughout its geographic range; this understanding must precede analysis of communities.
Focus The fundamental niche is an *n*-dimensional version of the Grinnellian niche. The observed (realized) niche is a synecological construct; emphasis on (direct or indirect, present or past) interactions among co-occurring species in guilds and communities; competitive exclusion; coexistence mechanisms.	*Focus* Primarily autecological; emphasis on the relationship between a species and attributes of the environment that might impinge on its life history; relegates interspecific interactions to a lesser role in affecting distribution and abundance than the Hutchinsonian niche.
Communities are organized units because membership and densities of component species are affected by co-occurring species. The role of interspecific interactions is important in species' distributions, population regulation, and evolution. These interactions may be competition for resources, predator–prey effects, or other interspecific interactions.	Communities are sets of species that occur in one's study plot; guilds are sets of species that use resources in a similar way and sometimes co-occur.
Relationship to evolution Invokes natural selection in the evolution of communities and lineages, and the associated constructs of limiting similarity and character displacement.	*Relationship to evolution* Invokes natural selection for intraspecific adaptations of species to different parts of their geographic ranges. Gene flow constrains local genetic adaptation; evolution in relation to membership of local communities is relatively less significant than evolution in relation to the geographic distribution of resources.
Methods *Mathematical* – Theoretical models based on Lotka–Volterra equations; set theory; niche overlap is sometimes equated with interaction coefficients.	*Methods* Direct gradient analysis, axes are defined by resources not necessarily interpreted as shared with other species and initially of unknown relevance to the species in question.
Empirical – 1. Quantify resources (such as food, habitat and physical	1. Consider species-specific environmental requirements in terms

Table 6.1. (*cont.*)

Hutchinsonian niche model	Grinnellian niche model
environment) on a study plot. Locate the responses of two or more species along shared-resource axes.	of life history, physiology, behavior, and habitat in many places.
2. Characterize species-specific feeding methods; express within-habitat differences along axes for the set of species.	2. Determine the geographic distribution and abundance of the species; the average values of resources and their variance. Determine what resources are common to all points in the geographic range and those not in evidence beyond the range.
3. Measure niche breadth and overlap along axes as utilization functions.	3. Find resource axes that are good predictors of the density of the species. These will probably represent sets of covarying resources. But it may be possible to decipher which members of the sets are important in different parts of the geographic range.
4. Quantify morphological differences; infer ecological differences and proceed as above.	4. Examine the density of other species in the resource space of the species in question.

Source: From James *et al.* (1984).

during periods of relative resource scarcity, when competition intensifies, and an increase in overlap during resource abundance, when competition is relaxed (Wiens and Rotenberry 1979, Schoener 1982). In either case, high overlap and the attendant competition are held to produce various patterns of niche divergence or partitioning. Several avenues of niche divergence may be followed (Fig. 6.2); each of these is considered in detail in subsequent chapters.

Niche theory also predicts that, because the resource use of species may be scaled on a variety of dimensions, species that exhibit high overlap on one dimension should differ on other dimensions (MacArthur 1972, Schoener 1974a). Indeed, MacArthur (1970) and others (Diamond 1975a, Cody 1981, Giller 1984) have argued that the patterns of such 'niche complementarity' among species in a community with optimal species packing should produce an aggregate resource utilization function that closely approximates actual overall resource availability, thus minimizing wastage.

The notion of niche complementarity has two corollaries. First, it pre-

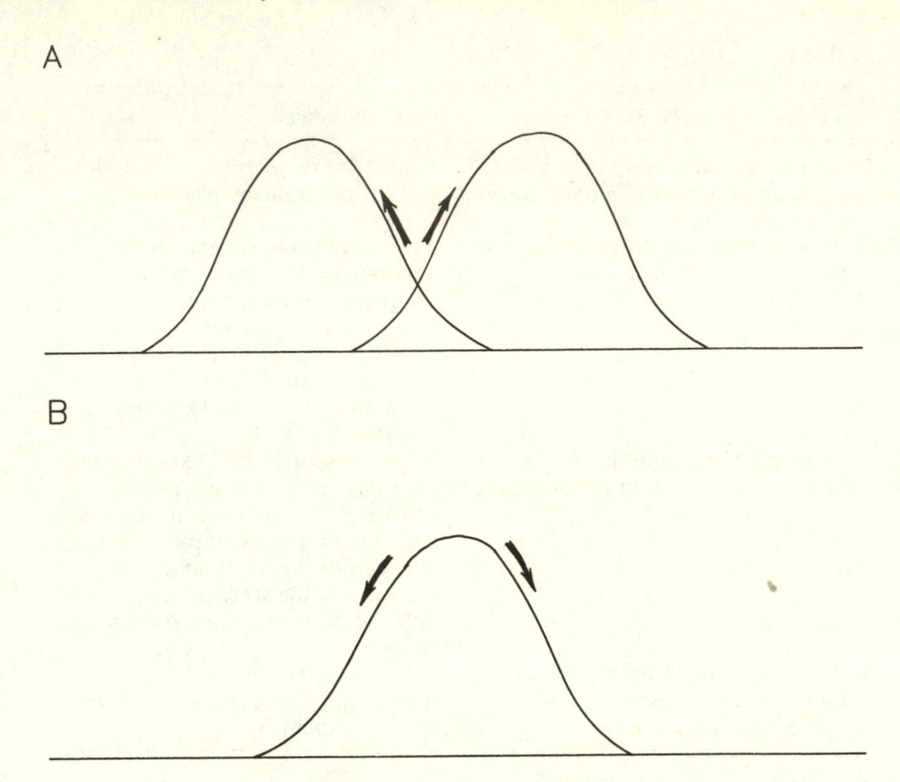

Fig. 6.1. Diagrammatic representation of the responses of populations to increasing interspecific (A) and intraspecific (B) competition. The curves represent utlization functions on niche axes; as densities (and thence competitive pressures) increase, the distribution of the population over the potential niche space changes as shown by the arrows. After Svärdson (1949).

dicts that resource use in communities containing many species should separate on more niche dimensions than in simpler communities. Second, because niche dimensions are often difficult to define, high overlap among species on several dimensions may be a consequence of our failure to measure or categorize the dimensions properly rather than an indication of the absence of niche separation among the species. Of course, because more diverse communities are usually associated with more diverse environments, the pattern of increased niche dimensionality in more species-rich communities may reflect specialization of the species on different aspects of the environment, quite independently of any species interactions. Moreover, high levels of overlap may be maintained on several *bona fide* niche dimensions if resources are not limiting (Fig. 6.2; Wiens and Rotenberry 1979), so a failure to find the expected separation among species does not necessarily imply that niche dimensions have been defined incorrectly.

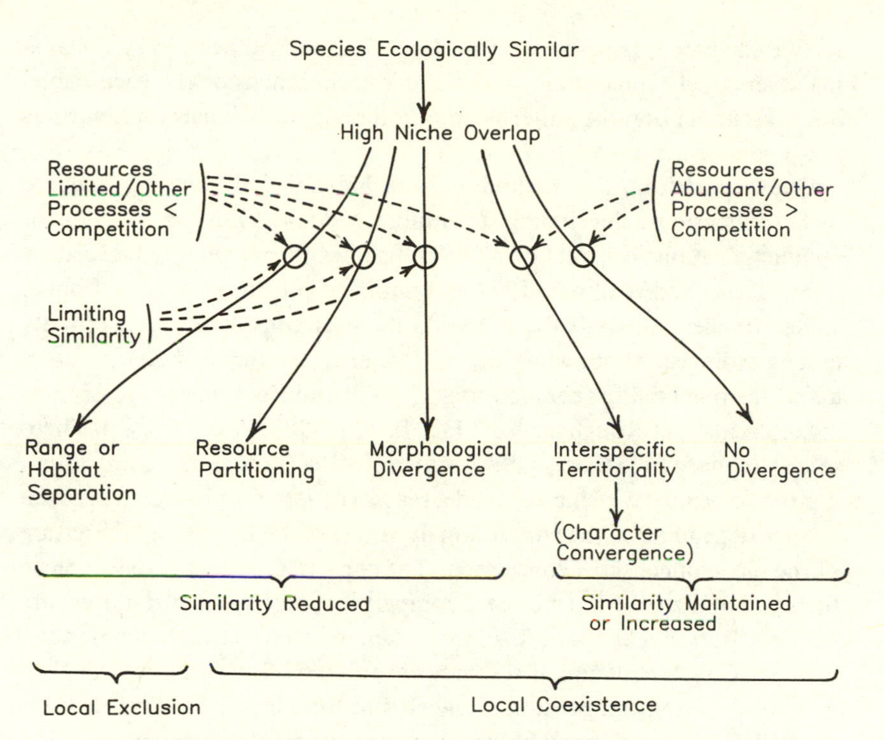

Fig. 6.2. Flow chart showing the avenues of response to high niche overlap among co-occurring species anticipated from niche theory. Alternative interpretations of the resulting patterns can be offered, as detailed in later chapters.

To consider only single niche dimensions to the exclusion of others, however, certainly may be misleading. Pianka (1981), Cody (1974a), and others have argued that niche dimensions should be combined into an overall, multidimensional measure of niche overlap. If the niche dimensions are truly independent of one another, this multidimensional overlap can be calculated as the product of the overlaps on single dimensions. If the dimensions are totally interdependent, however, there is really only a single dimension, and the average of the functions is the most appropriate multidimensional measure (May 1975, Pianka 1981). In fact, most situations fall somewhere between these extremes, and neither calculation then provides an accurate measure of multidimensional niche overlap (May 1975, Hanski 1978, Alatalo 1982c). Multivariate statistical procedures such as principal components analysis or discriminant functions analysis provide a way of dealing with a multiplicity of niche dimensions that avoids the decision of whether niche dimensionality is multiplicative or additive, but these procedures present numerous problems when applied to niche analysis (see Chapter 3; James *et al.* 1984, James and McCulloch 1985). Collapsing several niche dimensions into a single multidimensional measure may also obscure important patterns among the separate dimensions, such as niche complementarity.

Quite apart from the methodological difficulties of combing multiple niche dimensions, developing a suitable measure of niche overlap, or conducting statistical tests of niche difficulties (Schoener 1970, Hurlbert 1978, Harner and Whitmore 1977, Abrams 1980, Linton *et al.* 1981, Loman 1986, Mueller and Altenberg 1985), there is considerable uncertainty among ecologists about what high or low niche overlap actually indicates about the potential for competition (Colwell and Futuyma 1971, Abrams 1980, Pacala and Roughgarden 1982, Lawlor 1980, Wiens 1977b). Theoretical analyses by Abrams (1986) and Holt (1987) show that competition may be associated with increased, decreased, or zero niche overlap between species, depending on the population dynamics of the species and the nature of the niche dimensions considered. The expectation from classical niche theory, that high overlap = intense competition, requires that resources are limiting. Resource limitation is more often inferred than measured, however (see Chapter 10) and, if resources are abundant, overlap may be high without fostering competition (Colwell and Futuyma 1971, Pianka 1981, Wiens 1977b). Overlap might also decrease during resource scarcity if the different species specialize on resources that they use most efficiently, quite independently of any competitive effects. Pulliam (1986) provided an especially instructive example of this problem. He found that at times when seed supplies are high in southern Arizona grasslands, wintering sparrow species occupy a wide, overlapping range of habitats, but the species specialize on particular seed types that are most profitable to them. During periods of relative food scarcity, on the other hand, each species occupies a narrow habitat range but consumes a wide variety of seeds. Calculated average dietary overlap may thus underestimate the potential for food competition within habitats when food is scarce. In contrast, average habitat overlap values may overestimate the potential for competition, as each species occupies a different habitat during shortages.

Limiting similarity

According to niche theory, ecologically similar species coexist by virtue of niche differences. At the same time, each species has certain minimal requirements on the niche dimensions, so there are limits to the degree to which its niche can be compressed by competition without the species being forced out of the community. There should therefore be some

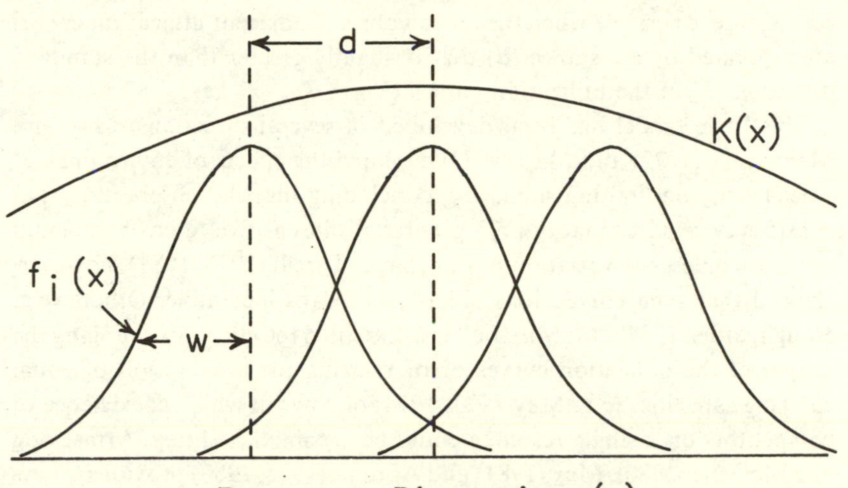

Fig. 6.3. Limiting similarity theory. The curve $K(x)$ represents a resource carrying capacity function (e.g. food quantity, K, as a function of food size, x) for several species with normally distributed resource-utilization functions $f_i(x)$ characterized by a breadth (w) and between-species separation (d). After May (1981).

level of niche differentiation that is just sufficient to separate species and yet ensure that all resources will be utilized; otherwise, the community may be susceptible to invasion.

This thinking was formalized in the theory of limiting similarity developed by MacArthur and Levins (1967), May and MacArthur (1972), Roughgarden (1974a), and others. If one assumes that the species use resources that can be linearly ordered along some single dimension, the utilization functions of the species can be portrayed as frequency distributions on this dimension, indicating the proportion of the total resource use (diet, for example) contributed by resources at each position on the spectrum (prey size, for example). In their original model (Fig. 6.3), MacArthur and Levins considered three species occupying different positions on such a dimension, with utilization functions that were normal curves of identical shape (i.e. symmetrical about the mean with equal standard deviations). The species were assumed to have equal carrying capacities on the resource base. By inserting the species' utilization functions into a derivative of the Lotka–Volterra competition model, one could relate the degree of overlap between the functions of adjacent species to competition and solve for coexistence. It turns out that under these rather restrictive conditions, coexistence is possible when the mean values of adjacent utilization curves are separated by a distance (d) that is slightly greater than the standard deviation (w) of the utilization curves (Fig. 6.3).

This basic model has been developed in several directions. May and MacArthur (1972) and May (1973) explored the effects of environmental stochasticity on limiting similarity, concluding that the observation that coexistence requires that $d/w > 1$ generally is insensitive to environmental variation unless the variation is quite large. Turelli (1978, 1981), however, showed that such conclusions are by no means inevitable. Others (e.g. Roughgarden 1974b, McMurtrie 1976) explored the effects of changing the shapes of the utilization curves or of relaxing the assumption of equal carrying capacities (e.g. May 1973, 1974) or ways in which coexistence of competitors on a single resource could be accomplished (e.g. Armstrong and McGehee 1980). May (1981) and Abrams (1975, 1983) provide reviews of much of this work.

Models derived from coevolutionary arguments (e.g. Roughgarden 1976, Case 1982) offer a somewhat different perspective. These models indicate that the optimum similarity between competitors may be substantially less than the maximum predicted by limiting similarity arguments. If this is the case, d/w values should be considerably greater than 1 and the communities should be susceptible to invasion. As a consequence, we may be likely to observe such coevolved communities only in situations in which invasion is otherwise restricted, such as on remote islands (Case 1982).

Limiting similarity theory has generated considerable interest, but it has not weathered well (Abrams 1983, Brown 1981, Case 1982). The model derives an equilibrium solution to the problem of ecological similarity and thus may not apply to assemblages that are not close to equilibrium. The approach is based on the uncertain assumption that overlap in resource use translates directly into competitive effects. There is an inherent fuzziness in how the single resource dimension is to be defined, how 'similarity' is to be gauged, how carrying capacities are to be determined, or how utilization functions vary in shape (much less how these parameters can be measured in the field!). As a consequence, values of limiting similarity in nature are likely to depend on so many variables that they will be different for almost every pair of species. For both theoretical and operational reasons, then, the generalization that coexistence requires $d/w > 1$ must be rejected (Abrams 1983).

Niche breadth

The range of a particular niche dimension or the volume of multidimensional niche space occupied by a species is its niche breadth.

Niche breadth is, of course, related to niche overlap, in that species with broad niches are more likely to overlap in a given situation than are narrow-niched species. Like overlap, niche breadth can be expected to change as resource levels change. According to most optimal foraging theory, it is increasingly expensive for an individual to pass by many prey items as resources become scarcer (Emlen 1968, MacArthur and Pianka 1966, Schoener 1971b, Charnov 1976). Niches should therefore expand as resource availability decreases. The same arguments can be applied to nonfood resources. If resources are abundant, on the other hand, such theory predicts niche narrowing, as individuals can afford to specialize on the resources that to them are 'best'. If the penalties for deviating from the theoretical optimum under such conditions are slight, however, there may be considerable variation and opportunism in resource use both within and among individuals in a population (Wiens and Rotenberry 1979, Wiens 1984b). Under these conditions, decreasing resource abundance might be accompanied by decreased niche breadth, at least to a point (see Chapter 10). These optimal foraging arguments apply to species in homogeneous environments, however. If the environment is patchy, an optimal forager may reduce the range of patches visited but continue to take the same or an increased variety of food types within the patches as resource availability declines (MacArthur and Pianka 1966, MacArthur and Wilson 1967). This generally matches the patterns that Pulliam (1986) observed in the Arizona sparrow assemblages.

These arguments derived from single-species optimal foraging models suggest the patterns to be anticipated in multispecies assemblages. Because different species have different resource utilization functions, increased interspecific competition associated with increased niche overlap should foster niche narrowing, as the species specialize on the resources they use most efficiently (Svärdson 1949, MacArthur 1972; Fig. 6.1A). This prediction is based on the unrealistic case of equivalent resource utilization functions among species, however. Different patterns of overlap in resource use theoretically may lead to either a contraction or an expansion of niches with increased interspecific competition (Abbott *et al.* 1977). In particular, the MacArthur–Pianka (1966) patch-use theory predicts that species will respond to increased competition by narrowing habitat niche breadth while maintaining or increasing diet breadth. If the primary form of competition is intraspecific rather than interspecific, the reduction in resource availability to individuals should be greatest in that portion of the niche space occupied by that species. This will result in a broadening of that species' niche (Fig. 6.1B).

Guilds and their definition

The distribution of the species in a community on niche dimensions is not as even as modeled in theory (e.g. Fig. 6.3), nor is it random or haphazard. Instead, groups of species cluster together in niche space, separated from other sets of species. Even with an assemblage as narrowly defined as shorebirds feeding in an intertidal mudflat, such clustering is apparent (Fig. 6.4). This grouping of species by ecological similarities is an important structural feature of communities. The existence of such 'guilds' indicates how consideration of communities solely in terms of species numbers of diversities may obscure important details and oversimplify community patterns.

Root's definition

Although several ecologists (e.g. Grinnell 1917, Salt 1953) had previously considered the structure of communities in terms of the ecological 'roles' played by different species, Root (1967) formalized the guild concept and showed how it could be applied in ecological studies. Root focused his study on the ecology of a single species, the Blue-gray Gnatcatcher (*Polioptila caerulea*), in California oak woodlands, but he was also interested in describing how the foraging ecology of this species related to that of other species in the community. Recognizing that the conventional way of assessing potential ecological similarity by close taxonomic relationship was inadequate in this situation (there were no congeneric or confamilial species present), Root described a 'foliage gleaning guild' containing five species that overlapped in their foraging repertoires, movements, use of substrates, and diets.

Root (1967: 335) defined a guild as 'a group of species that exploit the same class of environmental resources in a similar way'. The term thus 'groups together species, without regard to taxonomic position, that overlap significantly in their niche requirements'. Moreover, the concept 'focuses attention on all sympatric species involved in a competitive interaction, regardless of their taxonomic relationship'. The key elements of this definition are thus (1) the species are syntopic; (2) the similarity among the species is determined by their use of resources (niche requirements) rather than their taxonomy; and (3) competition is expected to be especially important within guilds. The general view of guild structuring of communities that emerges from these statements is that shared adaptations to a particular mode of life act to cluster species together in niche space into guilds, while competition separates the species ecologically within the region of niche space occupied by a guild.

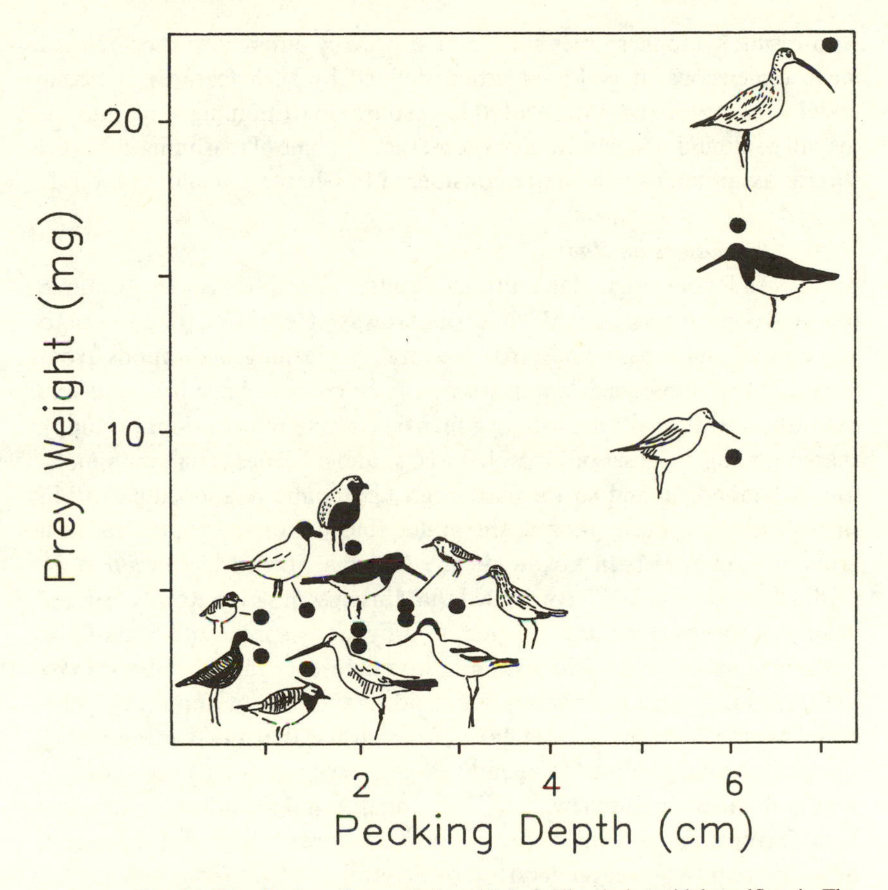

Fig. 6.4. Distribution of several shorebirds feeding in intertidal mudflats in The Netherlands in a space defined by the depth to which they probe for prey and the mean ash-free dry weight of prey (*Nereis*) obtained. One cluster of species feeding on small prey obtained from shallow depths is evident; the remaining species form a less cohesive cluster. After Zwarts (1980).

It follows from this view that an examination of resource partitioning or of any other patterns presumed to arise from competitive interactions will be most profitable if it is focused *within* guilds, as Root (1967) suggested. If Root is correct in proposing that guild membership may bear little relationship to taxonomic relationships among species, then conducting tests of these patterns based on groupings such as families (e.g. Connor and Simberloff 1979, Strong *et al.* 1979) is probably invalid. The close linkage between niche theory and guilds also explains why so many guild analyses have emphasized feeding relationships among species. The argument goes like this: If food is an important limiting resource, communities should be structured on the basis of how food is partitioned among coexisting species.

As a result, syntopic species should be especially different in food-related niche dimensions. If guilds are then defined by such features, a better understanding of the nature of this resource partitioning, and thus of overall community structuring, may emerge. This line of reasoning contains several assumptions, which are considered in Chapter 10 and Volume 2.

Variations on Root's Theme

In Root's original definition of guilds, he emphasized the resource base and consumers that used it in a 'similar way'. He applied the concept to a group of small passerine birds feeding by gleaning arthropods from foliage. Most subsequent applications of the concept have followed this example, operationally describing guilds by features of behavior or ecology shared among some set of birds, lizards, spiders, beetles, small mammals, fish, salamanders, and so on. Although taxonomic relationships within such guilds have been ignored, the guilds (indeed, the communities) at a broader level have been taxonomically circumscribed. MacMahon *et al.* (1981) and Jaksić (1981) have argued that this misconstrues Root's original definition. A given resource is often used by many species other than just birds or lizards or spiders, for example, and to restrict guild membership by arbitrary taxonomic boundaries may lead to a neglect of important influences among distantly related taxa. They argue that guild membership should be broadly defined to include all of the species using a resource in a similar manner. In this view, 'similar manner' is judged not by comparing such features as foraging tactics (which of course differ between, say, flycatchers and orb-web spiders) but by determining the *effects* of consumer use on the resource itself. As MacMahon *et al.* (1981: 301) observed, 'it does not matter whether an organism removes a tree leaf for nesting material, for food, or as a substrate to grow fungi which in turn are eaten; the leaf is gone and the leaf users belong to the same guild.' This seems quite different from Root's original view of 'similar way'.

Recognizing that it is rarely possible to include all of the species using a resource in a study, Jaksić (1981) drew a distinction between 'community guilds', which include all of the species that are known to use a particular resource in a similar manner, and 'assemblage guilds', which are defined within a taxocene such as birds or lizards or spiders. Most guild studies have been and will continue to be of the latter variety, although some attempts have been made to study community guilds (e.g. Mares and Rosenzweig 1978, Fenton and Fleming 1976, Brown *et al.* 1986). Of course if guilds are defined by resources rather than by consumer attributes, as MacMahon *et al.* and Jaksić propose, then situations may arise in which one species has a large influence on the resource even if that resource is relatively unimportant to the species. For example, 90% of the consumption of resource *X* might be by species *A*, even though *X* comprises only 5% of *A*'s diet. If guilds were delineated following Root's original definition, species *A* would not be likely to be clustered with other species for which *X* is a major resource or niche dimension, but by the MacMahon–Jaksić definition species *A* would clearly be an important member of the guild founded on resource *X*. Its activities would be likely to influence that of the other species consuming *X* (especially if *X* constituted a significant portion of their diets).

Viewed in this light, the guild concept of MacMahon *et al.* and Jaksić is almost a mirror image of the niche concept (R. Bradley, personal communication). The niche concept describes how a species utilizes several resources and how different species overlap in resource use. Root's guilds represent clusterings of such overlapping species. The MacMahon–Jaksić guild concept, on the other hand, examines how a resource is used or affected by several consumers and how different resources share consumers.

One other variation on Root's theme deserves mention. In applying the guild concept to his study of salamanders, Hairston (1984) restricted guild membership to species that were *known* to interact. He thus explicitly required that the resource dimension(s) in question be limiting. He concluded that we would thus 'need to know the exact ecological relationships to a degree that is unlikely to be attained for this or for most other guilds, or . . . this and many other claimed examples of guilds are products of our imaginations' (1984: 23). The primary value of the guild concept is to focus attention on sets of species sharing positions in niche space or influencing resource dynamics in similar ways, so that the consequences of these similarities may be determined. Were we to possess the information Hairston requires, application of the guild concept would be redundant and unnecessary. Hairston's requirements are unrealisitically demanding, and his despair thus comes as no surprise.

Guild classification

How does one subdivide a community into guilds and assign species to them? To a large extent, this depends on the objectives of a given investigation. If one's focus of attention is within a single guild, then a suitable definition of the boundaries of that guild may be all that is required. If one wishes to compare that guild composition of, say, bird communities in mediterannean woodland habitats on different continents, a suitable classification should contain a number of categories small enough so that each is not hopelessly general and large enough so that each does not contain only one or two species. Whether the categories are best defined in terms of diets, foraging behavior, nest-site locations, residency status, or some combination of these or other criteria may also depend on whether one is testing hypotheses related to, for example, foraging niche dimensions or the effects of seasonality.

Ecologists have followed two rather different approaches to categorizing guilds and species in communities (Jaksić 1981). Most have made *a priori* guild designations by using a small number of criteria (usually determined subjectively) to define guilds and their membership before the data have been gathered or analyzed. Others have performed more objective *a posteriori* analyses of the observations themselves to determine guild boundaries and membership, often employing multivariate statistical techniques and, of necessity, basing their analyses on quantitative measures of a large number of variables.

A priori *classifications*

A standard approach to designating the guilds in a community is to define a small number of very general ecological categories (e.g. nectarivore, frugivore, insectivore, omnivore) more or less intuitively, and then to assign the species present in the community to these categories on the basis of one's own observations, published information, or previous guild classifications of other authors. Sometimes guild assignments are based upon detailed, quantitative information, as in Root's original study, but more often the determinations are quite subjective. Cody (1983c), for example, assigned species in South African montane woodland communities to guilds largely on the basis of his own observations, but he defined some of the guilds on the basis of foraging habits and diets (foliage insectivores) and others on the basis of taxonomy alone (e.g. Turdidae). He excluded some species from the foliage insectivore guild because of differences in body size, but in the 'slow-searching omnivore' guild he included species differing in size by more than a factor of 2. These procedures may not influence the within-guild patterns he discerned, but they do affect between-guild comparisons. Diamond's (1975a) guild assignments were similarly subjective and of mixed criteria, and his failure to describe his protocol for determining guild boundaries and membership contributed to Connor and Simberloff's (1979) disenchantment with his analyses.

A wide array of criteria and levels of subdivisions has been used in making *a priori* guild assignments (Table 6.2). Most classifications have been based on features of diet and foraging behavior or location, some on

Table 6.2. *A sampling of guild categorizations used in studies of avian community structure.*

Because the studies differed in objectives and in the fineness with which guilds were defined, differences in guild numbers between studies indicate nothing about the relative structures of the communities.

Study	Community	Number of guilds	Classification criteria[a]
Howell 1971	Tropical pine savannah	5	R
		5	D
Karr 1971	Grassland → forest	4	R
		5	L
		5	D
Emlen 1972	Grassland → woodland	6	D,L
Lein 1972	World faunal regions	7	D,L
Emlen 1972	Urban desert	6	D,L
		7	N
Karr 1976	Tropical shrub-forest	7	D
		9	B
		7	L
		4	R
Emlen 1977b	Bahama pine forest	17	D,B,L
		5	V
		10	N
		9	P
Terborgh 1977	Tropical forest-woods	3	D
Szaro & Balda 1979	Ponderosa pine forest	4	B
Karr 1980	Tropical forest undergrowth	10	D,B,L
Terborgh 1980a	Tropical forest	10	D,B,
Folse 1982	Serengeti grassland	4	D
May 1982	Deciduous forest	15	D,L
Blake 1983	Forest fragments	7	D,L
Bradley & Bradley 1983	California lowlands	6	D,B,L
		3	L
		4	R
Case *et al.* 1983	West Indies	4	D,B
Manuwal 1983	Coniferous forest	7	D,B,L
Recher *et al.* 1983	Australian woodland	12	D,B,L,S
Keast 1985	Australian woodlands	20	D,B,L
Recher & Holmes 1985	Australian eucalypt forests	see Fig. 6.5	D,B,L,S,H
Wong 1986	Malaysian forests	10	D,B,L

Notes:

[a] D = diet, B = foraging behavior, L = foraging location, R = residency, N = nest site, S = body size, V = singing location, P = resting location, H = habitat.

nest–site locations and residency status, and some even on sites selected for resting. In several studies separate guild analyses based on different classifications have been conducted. Despite the subjectivity of these guild assignments, they generally are not biased. They thus provide a useful, if rather coarse, way of dissecting community structure beyond simple species listings. Often the detailed quantitative information required to conduct more objective *a posteriori* analyses of guild structuring is simply not available, and careful, thoughtful, qualitative guild classifications may be of considerable value in revealing patterns of community structure.

A posteriori ***classifications***

Although most *a posteriori* guild classifications employ statistical procedures to define guild memberships, not all do. In a remarkably detailed analysis of guild structuring in Bahama pine-woodland bird communities. Emlen (1977b) defined 18 different foraging guilds on the basis of 10 categories of habitat position, 10 food-type categories, and 7 kinds of foraging movements. To determine the 'guild role' for a particular species, he multiplied the density or biomass value of the species in a census by the species' fractional representation in a guild (e.g. 70% foliage insectivore, 30% bark gleaner). This produced a quantitatively based and reasonably objective assignment of species to guilds that was weighted by both their abundance and their ecological diversity. The guild categories, however, were still subjectively defined. In a similar manner, Recher and Holmes (1985) used observational data on eucalypt forest birds in Australia to define 9 foraging-tactic categories, 4 foraging-height zones, 8 substrate types, and 7 prey-attack behaviors. This information was then used to array the 41 species into hierarchically nested guild clusters (Fig. 6.5). In this arrangement the guilds are based on different levels of detail – the two parrot species are grouped solely by diet, whereas the guild containing the Golden Whistler (*Pachycephala pectoralis*), Black-faced Flycatcher (*Monarcha melanopsis*), and Rose Robin (*Petroica rosea*) is defined by diet, foraging height, foraging substrate, behavior, and habitat.

Various statistical procedures (e.g. cluster analysis, PCA, canonical correlation, DFA) have been used to determine both the boundaries and the membership of guilds (Crome 1978, Holmes *et al.* 1979a, Folse 1981, Landres and MacMahon 1983, Adams 1985; see Capen 1981). The approach followed by Holmes *et al.* (1979a) provides a good example. They studied a community of 22 species of insectivorous birds breeding in a northern hardwoods forest. The foraging behavior of each species was determined from opportunistic spot-observations of individuals. Holmes

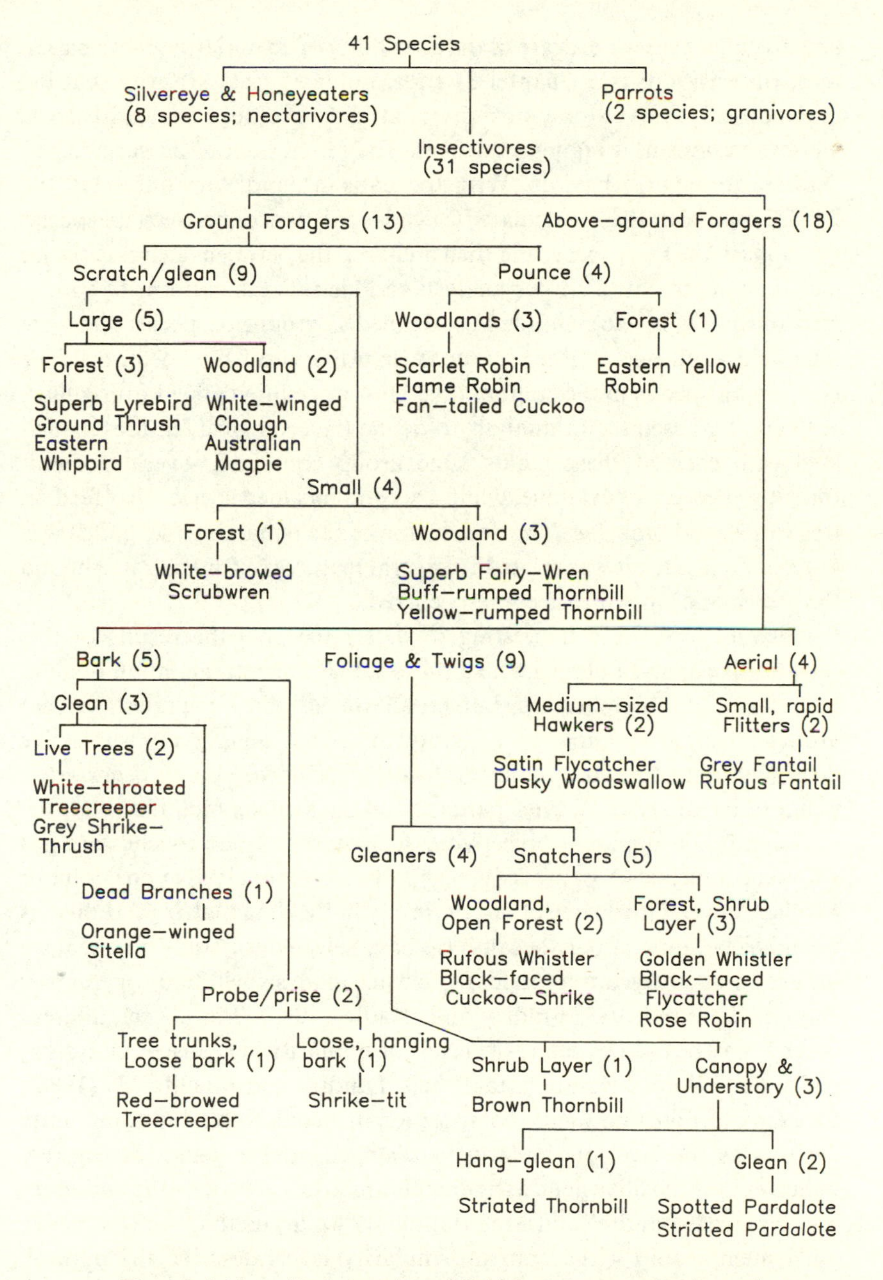

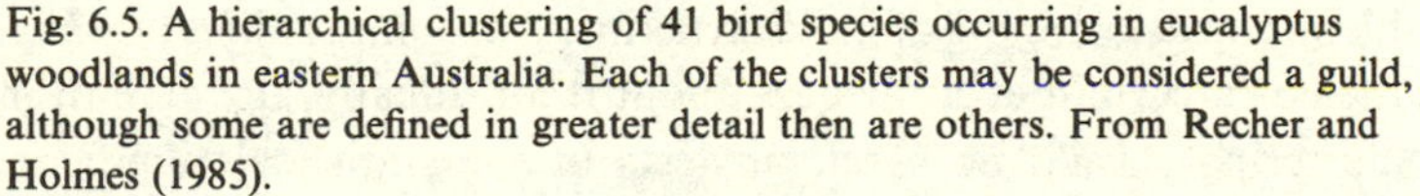

Fig. 6.5. A hierarchical clustering of 41 bird species occurring in eucalyptus woodlands in eastern Australia. Each of the clusters may be considered a guild, although some are defined in greater detail then are others. From Recher and Holmes (1985).

and his colleagues were aware of the possibility of an initial visibility bias in these observations (see Chapter 3) and conducted tests to verify that the initial observations of foraging individuals did not differ significantly from all observations in foraging sequences. They used a set of 27 variables to describe foraging behaviors. With the data in hand, they subjected the 27 × 22 matrix to PCA, calculated Euclidean distances between all species positions in the PCA space, and then analyzed the between-species distance matrix by hierarchical cluster analysis to illustrate the patterns of species relationships (Fig. 6.6). Guilds were defined as groups of species that were separated from one another by more than the mean Euclidean distance between all species in the community. This procedure defined four guilds. Holmes *et al.* used factor analysis to define the ecological features associated with each of these guilds. One group contained several ground-foraging species, for example, while a second included species that feed on tree trunks and branches. The description of the remaining two guilds was more complex, for it was related to several features of foraging height and the tree species in which feeding occurred.

It would have been interesting to determine how the results of this quantitative analysis agreed with a more subjective determination of guild boundaries and membership, but Holmes *et al.* did not undertake that exercise. Indeed, I know of no study in which an *a priori* subjective determination of guild structuring has been followed by an *a posteriori* quantitative analysis of guild patterns for the same community.

Despite their apparent objectivity, such statistical approaches are not without problems. Most procedures involve cluster analysis at one point or another. Given any degree of discontinuity in the data matrix (and there is bound to be some, if only because we have selected for study a system we believe to be arrayed into guilds), a cluster analysis will find (or impose) clusters (Landres 1983, Bradley and Bradley 1985). Moreover, different criteria may be used to define the levels of similarity in a cluster analysis at which guild boundaries are established. Landres and MacMahon (1980), for example, noted how guild composition in an oak woodland community changes as the level of similarity is changed, and a glance at Fig. 6.6 indicates how sensitive guild assignments are to exactly where the boundary line is placed. Landres and MacMahon (1980) argue that the changes in guild membership with changing similarity levels describe the natural, hierarchical grouping of guild structuring in communities (see Fig. 6.5), but there is a certain arbitrariness about it all. Ideally, the definition of guild clustering should be based on tests of the statistical significance of the

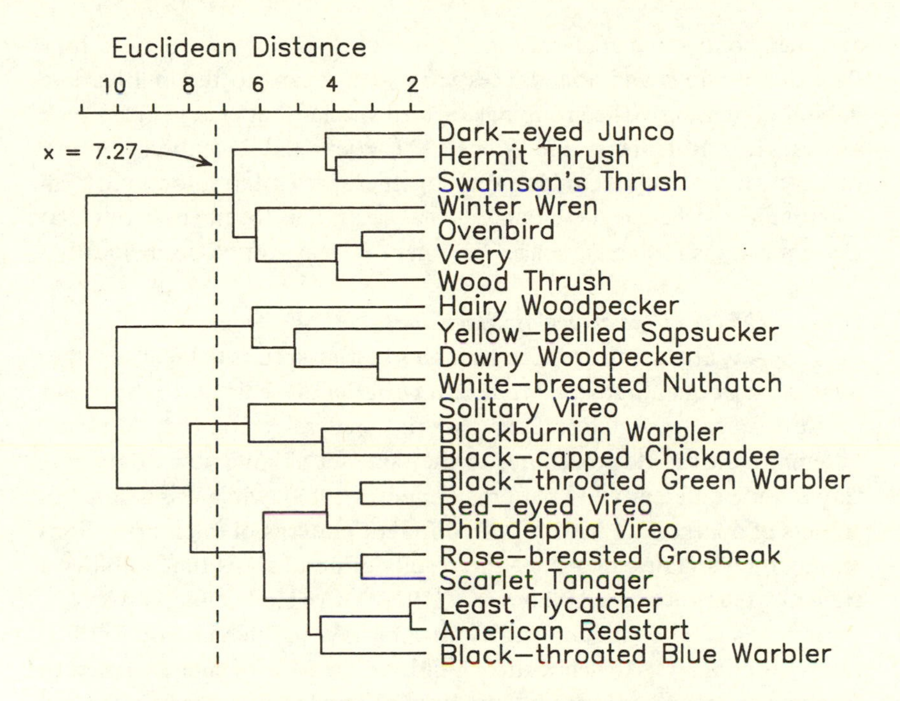

Fig. 6.6. Dendrogram of Euclidean distances between species in a hardwoods forest in northeastern USA, based on foraging behavior. The dashed line indicates the mean Euclidean distance between all combinations of species pairs. From Holmes *et al.* (1979a). (Scientific names for species not previously mentioned: Swainson's Thrush, *Catharus ustulatus*; Winter Wren, *Troglodytes troglodytes*; Ovenbird, *Seiurus aurocapillus*; Veery, *Catharus fuscescens*; Wood Thrush, *Hylocichla mustelina*; Hairy Woodpecker, *Picoides villosus*; Yellow-bellied Sapsucker, *Sphyrapicus varius*; Downy Woodpecker, *Picoides pubescens*; Solitary Vireo, *Vireo solitarius*; Blackburnian Warbler, *Dendroica fusca*; Black-capped Chickadee, *Parus atricapillus*; Black-throated Green Warbler, *Dendroica virens*; Red-eyed Vireo, *Vireo olivaceus*; Philadelphia Vireo, *Vireo philadelphicus*; Rose-breasted Grosbeak, *Pheucticus ludovicianus*; Scarlet Tanager, *Piranga olivacea*; Least Flycatcher, *Empidonax minimus*; American Redstart, *Setophaga ruticilla*; Black-throated Blue Warbler, *Dendroica caerulescens*.)

clusters, but this is by no means easy (Strauss 1982, Bradley and Bradley 1985).

Another analytical and interpretational problem arises when multivariate techniques are applied to analyze bird-habitat relationships over a series of locations. The analysis generally takes the form of recording several measurements of habitat features at the locations, using PCA or some similar technique to order the locations in multivariate habitat space, and then conducting analyses that cluster bird species according to how their distributions and abundances vary with regard to the multivariate habitat dimensions. The groupings of bird species that emerge from such analyses (e.g. Rotenberry and Wiens 1980, Wiens and Rotenberry 1981a), however, are *not* guilds. Guilds group together species that co-occur in local communities, whereas these habitat analyses group together species that share positions in habitat space; they may or may not co-occur locally.

Effects of the choice of dimensions

Any attempt to describe guilds and assign species to them, whether qualitative or quantitative, *a priori* or *a posteriori*, is sensitive to the choice of dimensions or variables used and the analyses employed. Applying different multivariate procedures to the same set of community data may reveal quite different guild patterns (Landres 1983), while using different subsets of a large data matrix, with different patterns of intercorrelations among the variables, may alter the results of an analysis that employs a standardized procedure (Holmes *et al.* 1979a). In a study of guild structuring in a waterfowl community on a eutrophic lake in Finland, Pöysä (1983) used cluster analysis to define three guilds on the basis of characteristics of feeding habitat and, in a separate analysis, five guilds on the basis of feeding methods. When the data from the two analyses were combined, these patterns of guild structuring broke down, and all that could be discerned were two large, general guilds: diving ducks and grebes, and dabbling ducks and coots. In more subjective analyses, the patterns determined on the basis of finely subdivided categories (e.g. May 1982, see Table 6.2) are likely to be different from those derived from more general categories. The patterns that emerge from a guild analysis are thus very much dependent upon a series of initial decisions by the investigator – what criteria to use; which attributes of species measure at what level of detail; which species to include; which clustering algorithm or similarity level to use, and so on. There is thus something of an art to guild analysis.

Some guild patterns

The guild concept has been used in a large number of studies of birds since Root formalized the notion and the term. Some examples have already been noted, and additional applications of the idea will be described in subsequent chapters. At this point, I will describe only briefly a few selected examples of the use of the concept in pattern analysis. The studies fall naturally into two groups: single-guild studies in which intensive within-guild analysis is the focus, and among-guild studies, in which the guild

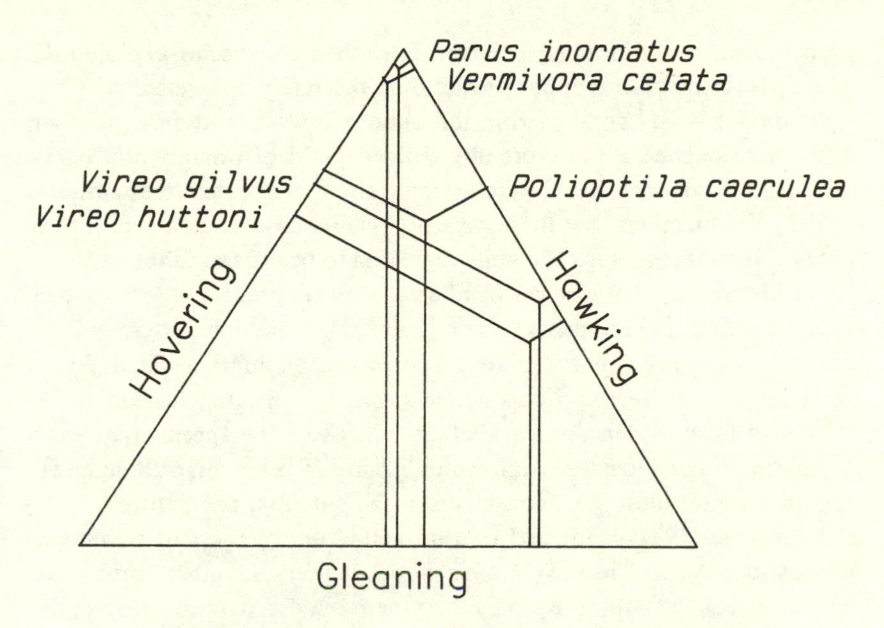

Fig. 6.7. The foraging tactics of members of the foliage-gleaning guild in a California oak woodland. The length of the lines perpendicular to an axis is proportional to the relative amount of foraging time devoted to that tactic. After Root (1967).

structuring of communities is first categorized and then examined for variation in different sorts of comparisons.

Within-guild patterns

The focus of Root's (1967) initial application of the guild concept was upon a single guild of foliage-gleaning insectivore species inhabiting oak woodland in California. He gathered behavioral observations of the five species by using a modified spot-observation procedure (he recognized the possibility of an initial visibility bias and corrected for it by confining his observations to oak woodlands, where the open nature of the vegetation made visibility quite good). The species within the guild differed in the details of their use of different substrates for foraging, their use of different movements in the foraging repertoire (Fig. 6.7), the sizes of prey they captured, and in aspects of their morphology. In all of these features, however, the guild members overlapped considerably. This indicates both the accuracy of the guild-membership determinations and the subtlety of ecological differentiation within the guild. Root concluded that partitioning of food was not based upon a single factor but instead on the effects of all of these elements of foraging behavior. Even then, the species' exploitation strategies seemed more opportunistic than discretely segregated.

In another study conducted in the same oak woodland area, Wagner (1981) investigated a more broadly defined guild of foliage- and bark-gleaning insectivores. There were seasonal changes in behavioral patterns of the guild members, but these were not as great as between-year differences at the same season. The guild appeared to be organized into a 'core' group of three resident species, together with two migrants that occurred in fall and winter and another spring migrant. The foraging patterns of the residents were not influenced by the presence of migrants at different seasons, and the seasonal changes in foraging appeared instead to be responses to changing food availability. Most of the species that were similar in foraging locations differed in bill size or behavior, although two species did not; these were considered to be potential competitors.

James *et al.* (1984) conducted a within-guild study by focusing on a single species, the Wood Thrush (*Hylocichla mustelina*) and other thrush-like species associated with it. By extending their analysis over the geographic range of the Wood Thrush, they were able to assess how its habitat occupancy patterns differed in areas with and without other guild members. In a habitat-based PCA space defined by axes representing forest maturity and the presence of small trees, places containing only Wood Thrushes had 'average' habitat for the species (central in the overall PCA space occupied by the species). Places in which Wood Thrushes co-occurred with other guild members, on the other hand, were located toward the periphery of the PCA space (Fig. 6.8). This an intriguing guild pattern that should be examined in other within-guild studies to determine whether it is general, a quirk of Wood Thrush ecology, or simply an artifact.

One additional example is instructive. Feinsinger (1976) examined in detail the behavior and relationships of species within a guild of nectarivorous hummingbirds in a tropical location. The guild was structured rather tightly: it contained a territorial species that dominated all of the other species at resource clumps, a nonterritorial traplining species, and 12 other species that were either important in nearby communities, specialists on particular resources, or highly migratory opportunists. The recognition of this guild provided the framework for a detailed study of resource–partitioning patterns within the guild. Feinsinger found that the species diverged from one another in features of the floral resources they used, the strata in which they foraged, and the time of day (which was related to nectar availability). Seasonal changes in within-guild patterns were related to shifts in the resource base. Overall, each species in the guild responded to a particular portion of the resources, with the consequence

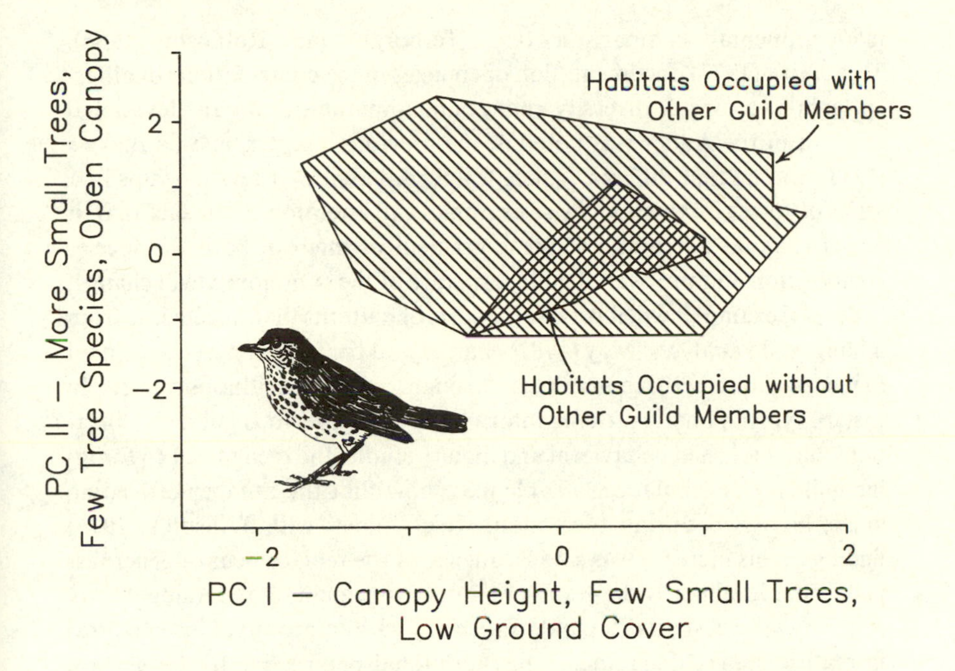

Fig. 6.8. The habitats of Wood Thrushes and other members of the 'thrush guild' in the geographic range of the Wood Thrush. The habitat space is defined by principal components derived from measures on 24 plots in which Wood Thrushes were present. The enclosed spaces include survey locations where Wood Thrushes occurred with and without other guild members. After James *et al.* (1984).

that the guild as a whole tracked the shifts in the entire resource base. Feinsinger and Colwell (1978) extended this analysis, describing six 'community roles' within the nectarivore guild (see Volume 2). Each of these roles (e.g. 'low-reward trapliner', 'territorialist') was associated with a particular suite of morphological and behavioral features and tended to exploit certain categories of floral resources. Thus, once they had defined a broad, general guild (nectarivore), Feinsinger and Colwell were able to use detailed information on the behavior, morphology, and resource use to subdivide the guild into smaller groups. Because each of these groups contained several species, they could legitimately be labelled as guilds in finer-scale analyses.

Among-guild patterns

In quite a few studies, investigators have categorized the guilds present in a community and then compared this guild structure with that of other communities, generally in relation to some environmental gradient or intercontinental comparison (e.g. Terborgh and Robinson 1986). Terborgh's (1977) documentation of changes in the contributions of different guilds to overall diversity changes in communities on an elevational gradient in the Andes was noted in the previous chapter, as was Blake's (1983) finding that different guilds follow different *S*/*A* relationships in a series of forest fragments. These examples call attention to the fact that, if guilds respond differently to some gradient of interest, both the species composition and the overall guild structure of the community will change.

Several examples will illustrate the sorts of patterns that can emerge from among-guild analyses. May (1982) charted guild patterns on a successional sequence from herbaceous fields through mature deciduous forests in eastern North America. Using information from previous guild classifications and published behavioral and dietary studies, he recognized 15 foraging guilds. His tabulations (Table 6.3) show that the number of feeding guilds increased during succession, which agrees with Willson's (1974) findings. This increase was a consequence of the replacement of generalist guilds (predominantly the granivore–insectivore guild) that dominate early stages by several specialist guilds (especially various groups of insectivores) in mature stages. This adds some additional perspective to the general observation that species diversity as a whole increases during succession (Chapter 5).

May also found (*contra* Willson 1974) that the number of species per guild increased during succession, at least for some guilds (Table 6.3). This pattern parallels the increase in guild richness with increasing total community species richness or increasing area found in several studies (e.g. Johnson 1975, Blake 1983, Faaborg 1985). Comparing community guild structure among several West Indies islands, Faaborg (1985) found that the rate of increase in guild richness with increasing total *S* differed considerably among guilds; frugivore richness, for example, increased much more rapidly than that of nectarivores (Fig. 6.9). Faaborg extended his analysis from the 12 islands that he studied to a larger set of 26 West Indies islands and found much the same patterns. Moreover, with the exception of the nectarivores, the regressions of guild richness on total *S* from the West Indies predicted the patterns for several neotropical landbridge islands fairly well. To Faaborg, this agreement of patterns suggested 'some overriding controls in avian community organization' (1985: 642). The consistency of the guild pattern is intriguing, and it merits further study.

In other studies, differences or similarities in the 'guild signatures' of communities (the distribution of species and individuals among guilds; Karr 1980) have been examined in comparisons among similar habitats in

Table 6.3. *Changes in guild composition and importance over a series of successional stages of eastern deciduous forests.* N = *10 for each stage.*

Attribute	Successional stage[a]			
	A	B	C	D
Number of guilds	2.3	6.8	6.6	8.8
Species/guild	2.3	2.3	1.9	2.7
Percent total density				
Omnivore	0	5.3	4.1	5.0
Granivore/insectivore	80.5	44.8	16.1	4.3
Frugivore/insectivore	0	6.4	0.6	0.9
Ground insectivore	8.5	2.7	26.3	13.9
Undergrowth insectivore	3.1	13.7	5.2	5.9
High foliage insectivore	0	19.5	35.7	38.5
Twig/branch insectivore	0	0.5	3.4	5.9
Trunk surface insectivore	0	0	2.6	7.6
Sallying flycatcher	0	3.4	3.6	15.4
Number of species				
Omnivore	0	1.2	0.8	1.6
Granivore/insectivore	3.7	4.3	1.6	1.9
Frugivore/insectivore	0	1.4	0.2	0.4
Ground insectivore	0.9	0.7	2.1	3.0
Undergrowth insectivore	0.5	2.1	0.7	1.9
High foliage insectivore	0	2.9	4.4	6.2
Twig/branch insectivore	0	0.1	0.7	1.7
Trunk surface insectivore	0	0	0.7	3.3
Sallying flycatcher	0	1.3	0.8	2.6

Notes:
[a] A = herbaceous, B = herb-shrub-sampling, C = young forest, D = older forest.
Source: After May (1982).

different areas. For example, Landres and MacMahon (1983) compared the structure of arboreal bird communities in oak woodlands in Sonora, Mexico, with those in California. Guilds and their memberships were defined by multivariate and clustering procedures. Based on calculations of the relative contributions of the guilds to the community using species' densities, foliage gleaners and bark gleaners together constituted 49% of the community at the Sonoran site (Table 6.4). Bark gleaners were absent from the California location, but foliage gleaners accounted for 50% of the individuals. The contributions of the other guilds, and the overall contributions of gleaners, probers, and salliers, were remarkably similar in the two locations. Birds in the various guilds differ in size, however. Landres and MacMahon considered this effect by using consuming biomass (calculated for each species as density times body mass raised to a fractional

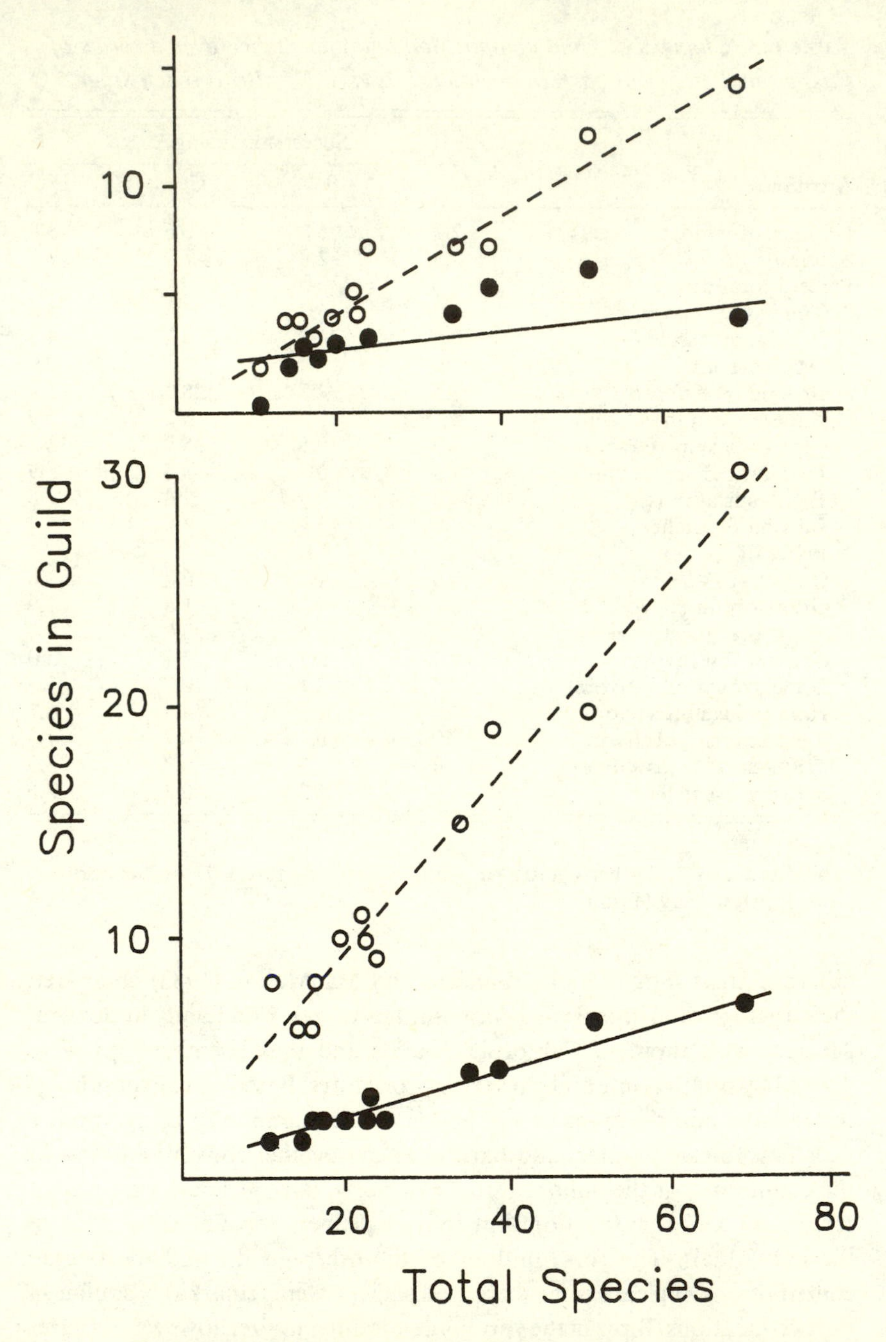

Fig. 6.9. The relationship between the number of species within each of several guilds and the total number of species present for 12 islands in the West Indies. Top: open circles = gleaners, solid circles = nectarivores; bottom: open circles = frugivores, solid circles = flycatchers. After Faaborg (1985).

Table 6.4. *Percentage contributions of five foraging guilds to total arboreal bird community density and consuming biomass in oak woodlands in Sonora, Mexico and central coastal California.*

	Density		Biomass	
Guild	Mexico	California	Mexico	California
Foliage gleaners	26	50	13	35
Bark gleaners	23	0	17	0
Bark probers	14	18	32	18
Air salliers	30	22	31	37
Ground salliers	7	10	7	10

Source: After Landres & MacMahon (1983).

power to account for energy consumption; see Chapter 13) to calculate guild contributions to the communities. The patterns changed (Table 6.4). Both systems still contained similar levels of gleaners as a group, but prober biomass was roughly 80% greater at the Mexican site than at the Californian woodland, whereas sallier biomass was 30% greater in California. Thiollay (1986) also reported large differences in the relative importance of various guilds in French Guiana rainforests when guild contributions were calculated using biomass instead of density. These observations indicate that there are systematic differences in the size distributions of species among guilds.

Broader geographic comparisons have been conducted by Karr (1980) and Remsen (1985), among others. Remsen compared high-elevation humid forests in the Bolivian Andes with mature forests in New Hampshire. The Bolivian communities contained more species and more guilds than those in North America, mostly due to the addition in Bolivia of 'tropical' resource types such as fruit, nectar, epiphytic vegetation, and bamboo thickets (Fig. 6.10). Karr derived guild signatures for tropical forest undergrowth habitats in Central America, Africa, and Malaysia. These patterns were relatively similar among samples from the same geographic region, but there were major differences among the regions. The neotropical samples contained the greatest number of guilds, supporting groups such as ground-feeding frugivores or professional ant-swarm followers that were absent in the other regions. The African habitats supported a reduced diversity of frugivores and bark gleaners in comparison to the neotropics, whereas undergrowth frugivores were absent and terrestrial insectivore diversity was low in Malaysia. Karr attributed the regional differences in guild

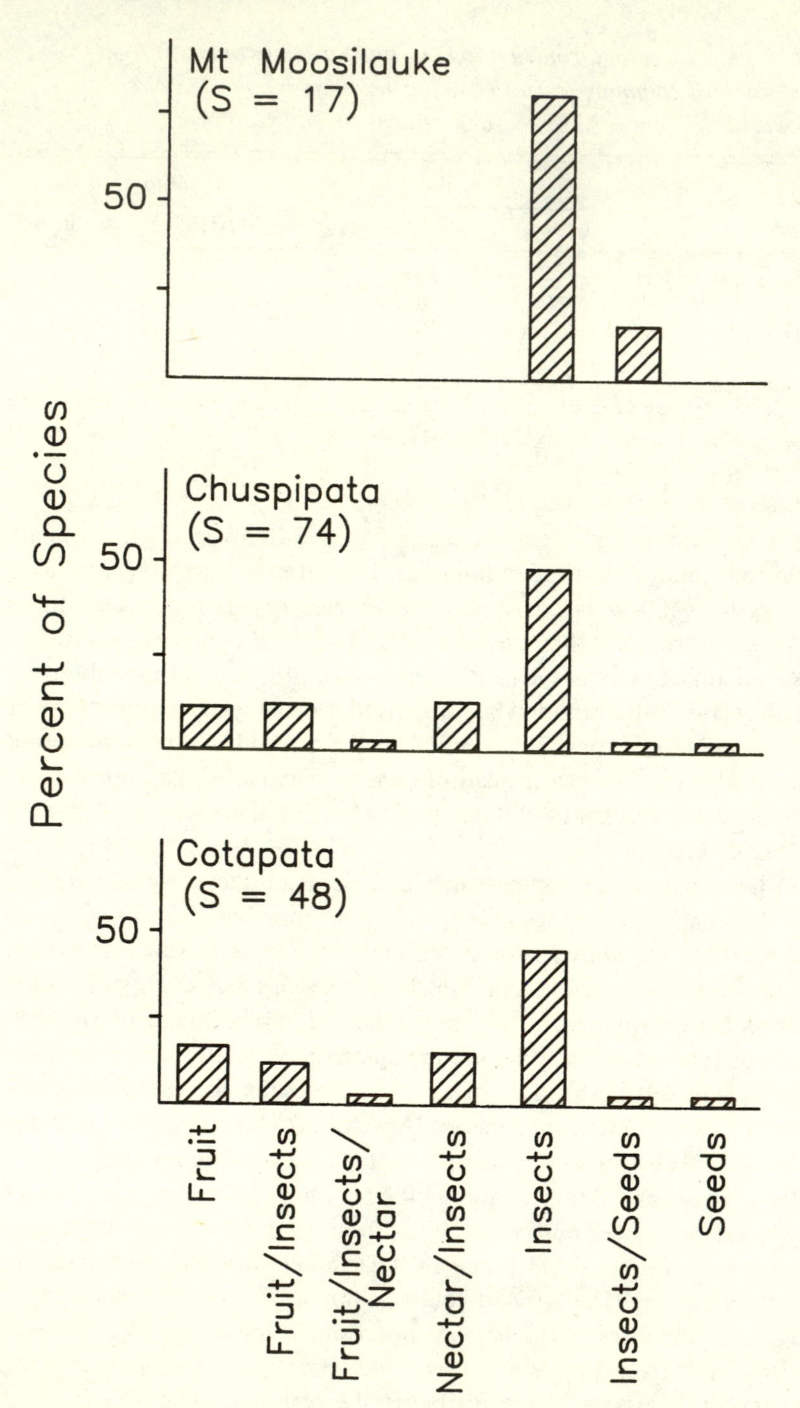
Mt Moosilauke
(S = 17)
50
Percent of Species
Chuspipata
(S = 74)
50
Cotapata
(S = 48)
50
Fruit
Fruit/Insects
Fruit/Insects/
Nectar
Nectar/Insects
Insects
Insects/Seeds
Seeds

signatures to differences in past history (e.g. Pleistocene habitat changes) and the effects of differing patterns of speciation and coevolution.

Karr's Malaysian data were derived from a single location. Wong (1986) also observed that frugivore diversity was low in that region. Wong found lower species richness and abundances of individual species in a regenerating dipterocarp forest than in a virgin forest, where food resources were more plentiful. Despite these differences, the two forests did not differ in the relative importance of various guilds, suggesting that similar resources were present in similar proportions in both areas. Wong noted, however, that resources such as flowers and fruit are generally scarce in Malaysian forests in comparison with tropical forests elsewhere, so resource differences may also contribute to the regional differences that Karr observed.

Random tests of guild patterns

Patterns such as those described above are clear enough, but it is legitimate to ask whether or not they depart from those that might be expected from a random distribution of species in the niche space. Such exercises are not simple (see Joern and Lawlor 1980), and few attempts have been made to treat bird guild patterns in this way. Bradley and Bradley (1983) used randomization tests to inquire whether or not variations in the abundances of wintering bird in lowland southern California were concordant among species belonging to the same guild. They used several sets of criteria to define guilds, grouping species by diet and foraging microhabitat, foraging location only, migratory/wintering status, and their taxonomic affiliation (family) (Table 6.2). On the basis of the randomization tests, two of the groups (guilds defined by total foraging ecology and by residency status) exhibited significant within-guild distributional concordance – species within guilds covaried in their wintering distribution and abundance. The other two categorizations failed to produce significant patterns. Bradley and Bradley used these findings to hypothesize that at least some of the concordance might reflect ecological similarities among boreal and montane species that move into the area during winters in which their food supplies become limited.

The among-guild studies I just discussed indicate that differences in resources may contribute to the patterns observed. This raises the prospect

Fig. 6.10. Percentages of total species (S) present at study sites in montane forests in New Hampshire (Mt Moosilauke) and in the Bolivian Andes (Chuspipata, Cotapata) belonging to several trophic guilds. Modified from Sabo (1980) and Remsen (1985).

that the departure of species from a strictly random distribution in niche space within a community might be a consequence of a nonrandom distribution of resources (Jaksić 1981, Pianka 1981). If there are gaps in the resource spectrum on which a community is built, they will be reflected in the absence of species that might be adapted to exploiting such missing resources, leaving the remaining species clustered in niche space by default. This suggestion does not bear one way or the other on the hypothesis that competition might be more intense within than between guilds, although it does imply that the observed guild structuring of a community may be due as much to characteristics of the resources as to interactions among the consumers.

If this notion is to be taken seriously, we must determine that discontinuities in the resource base can produce guild-like patternings of communities. Bradley and Bradley (1985) developed a null model algorithm that assembles species in communities independently and noninteractively while still preserving the patchiness or discontinuities of resource states in the environment. In their model analyses based on a homogeneous resource distribution, the species were distributed evenly and no guild structure was recognizable, as expected. When a patchy resource distribution was used, however, species clustered into groups that showed a remarkable similarity to the guild structuring observed in real communities. The 'resource-gap' hypothesis thus merits serious consideration in testing explanations of observed community guild patterns.

Conclusions: the utility of niche theory and the guild concept

Perhaps not surprisingly, the predictions of niche theory turn out to be less clearcut than was once thought. Nonetheless, this body of theory has been the foundation of a good deal of community ecology, and many of the studies described in the following chapters were couched in the predictions of this theory. The theory addressed questions that were and continue to be interesting, and it directed our attention to aspects of communities that otherwise might not have been noticed. What happens to niche overlap or niche breadth in communities with changes in resources levels, population sizes, or community composition, however, depends on more variables than are included in the simplified theory, and the relationships of these patterns to competition are equivocal.

These problems diminish the current value of niche theory, but they do not alter the fact that the theory generated a tremendous amount of interesting research that advanced our understanding of community patterns, if not of niches *per se*. Rather than abandoning niche theory as now useless, however, we need to develop a fresh approach to 'niche theory' that is based on the sorts of patterns and variations in those patterns described in the following chapters. This will require that factors other than competition be considered as determinants of the niche occupancy of species and a realization that nonresource niche dimensions may be very important. This approach should incorporate Grinnellian as well as Hutchinsonian views instead of relying on further manipulations of Lotka–Volterra competition equations to produce theoretical insights.

Because of its close association with resource dimensions and niche overlap, the guild concept shares some of the problems that restrict the usefulness of traditional niche theory. Still, by drawing attention to the discontinuous distributions of species in community niche space, the guild concept has led us to ask a variety of new and different questions about communities. Broad-scale comparisons of the guild signatures of communities in different regions suggest some distinctive patterns, but the number of rigorous comparative studies is too small to determine whether or not such patterns might have any generality.

Not everyone wishes to undertake broad-scale comparative studies of entire communities, of course. For these individuals, the recognition of guilds in communities can provide a focus for investigations of subsets of communities containing only one or a few species. If one wishes to test hypotheses of interactions among species, for example, a general guild analysis of a community may be extremely valuable in directing attention toward a particular guild and in indicating the boundaries and defining the membership of that guild. In the absence of a guild analysis, it becomes more likely that such a study may be misdirected toward sets of species among which interactions are unlikely (their niches are coarsely partitioned) or that key participants in the web of interactions may be inadvertently omitted.

This is not to say that all is rosy and guild analyses are *the* way to do community ecology. Several problems and constraints must be considered. First, if one takes the admonitions of Jaksić (1981) and MacMahon *et al.* (1981) seriously, defining and classifying guilds in nature may be far more difficult than the current literature would lead one to believe. Defining a resource base for guilds is not itself easy, as we shall see (Chapter 10). Even if it can be done, designing an observational or experimental study that simultaneously considers organisms as different as, say, birds, spiders, and mantids is difficult. Perhaps we must be content, after all, to conduct studies of 'assemblage guilds'.

Second, there is a danger that, by compartmentalizing a community into guilds and then studying the guilds in detail, we may ignore critical webs of interactions between guild members and other elements of the community that were not included in the guild on the basis of a particular set of classification criteria. Traditionally, the definition of guilds has been closely allied with niche theory and has shared with it the presumption that competition is the primary determinant of the patterns observed. Aspects of resource use have been emphasized, especially as they relate to the potential for competitive interactions among species. This may be appropriate if one's objective is to study such interactions, but the guild concept may be more useful in ecology if its application is not restricted to such situations. If our interests are in predators and their effects on community patterns, for example, defining guilds by aspects of resource use related to potential competition might yield few insights. It might be more appropriate to emphasize instead such features as the use of cover or the relative values of species as primary or secondary prey to predators in defining guild categories and memberships. Unless caution is exercised, we are likely to define guilds in a particular way and then proceed as if the community background in which they are embedded is irrelevant, as if it does not change among the communities we compare. This, of course, would be shortsighted nonsense. Schaffer (1981) has considered this problem in detail.

A third concern reiterates a point I made in Chapter 2. If we demonstrate that the species in a particular community can be objectively clustered into guilds to which we can attach ecological meaning, we have documented the existence of a pattern (within the constraints of our methodology), nothing more. I suggested previously that this may be quite useful, especially in leading to more detailed investigations or more formal testing of hypotheses. It does not justify any conclusions about the causes of that pattern. We may suggest that it is related to a clustering of species in the niche space to minimize diffuse competition (Pianka 1980), that it is a consequence of gaps or patchiness in the resource distributions or niche dimensions (Terborgh 1980a, Jaksić 1981, Wiens 1983, Bradley and Bradley 1985), or that it is related to some other cause. Additional information may lead us to lean more toward one explanation than another, but that 'explanation' remains only a hypothesis. Each requires separate testing. It is important to exercise restraint in drawing explanatory inferences from guild patterns.

Finally, one must remember that simply because a concept such as the ecological guild is useful does not mean that it is necessarily biologically valid (Landres 1983). There is an arbitrariness to guild classification and the determination of guild membership, which is especially evident in subjective *a priori* classifications. This raises the prospect that the guild 'patterns' that emerge from studies based on such classifications are consequences of imposing an arbitrary arrangement on a community that is actually structured ecologically in some other way altogether (or is not structured at all). Using multivariate statistical procedures does not grant immunity from this problem. Once again, sound biological judgment, based on an intimate knowledge of the species and the environments, must underlie our attempts to understand communities through guild analysis. Properly applied, however, the guild concept represents an important approach to dealing with the complexity of natural communities.

7

Ecomorphological patterns of communities

It was part of Darwin's (1859) thinking that there is a close relationship between the structure of a species and its ecology and that morphological differences between species therefore indicate ecological differences between them as well. This idea was given impetus in this century by the work of Huxley (1942) and, especially, Lack (1944, 1947). Lack in fact devoted an entire book (1971) to documenting the ecological differences among species that fostered their coexistence, and morphological features figured conspicuously in his arguments. There is a certain intuitive appeal to this notion. Most introductory biology texts, for example, contain a figure showing how bird bills of different shapes are associated with different feeding habits. Other features of bird morphology also have clear ecological correlates: the tarsus (tarsometatarsus) of birds that forage on tree trunks is generally shorter than that of ground-foraging species (Richardson 1942, Dilger 1956) and the wings of birds that maneuver through closed habitats are more rounded than those of birds occupying open habitats (Savile 1957, James 1982). Given such correlates, one might expect to find clear patterns of species differences in bill sizes, wing or leg structure, body size, or other morphological features among members of guilds or entire communities.

Studies of the ecomorphological patterns of communities have become quite popular for several reasons, but primarily because obtaining measurements of morphological features from museum specimens or published accounts is far easier than gathering the information on food habits, habitat use, or behavior directly. It has also been suggested that, because the effects of selection on morphological features occurring during different seasons over an individual's lifetime carry different weight in their influence on fitness, by analyzing morphological patterns 'the investigator need not guess the season during which competition exerts its strongest influence on community organization' (Ricklefs and Cox 1977). Moreover, morphology can be measured independently of the habitat or taxonomy of the bird species, facilitating broad comparisons (Ricklefs and Travis 1980, Miles *et al.* 1987).

Countering these advantages, however, are several weaknesses. The morphology of an individual is unlikely to change rapidly in response to changing environmental conditions (Ricklefs *et al.* 1981) nor may it reflect the consequences of opportunistic behavior, so it may not be a good indicator of short-term, proximate ecological conditions. In addition, features of the morphology of an organism are not subject to selection in isolation but exhibit allometric interrelationships and intercorrelations. A given ecological circumstance therefore may influence several features of morphology in concert, so specific morphological structures are likely to reflect a compromise solution to several sources of selection. Perhaps because of an implicit faith in the optimizing power of natural selection, however, such difficulties have generally been ignored, and investigations have been carried out under the presumption that the match between specific features of morphology and of ecology is quite close.

Evaluating the premise: does morphology reflect ecology?

General correlative evidence

The value of an ecomorphological approach to community analysis rests on the closeness of the relationship of morphology to ecology, so it is appropriate to consider some evidence bearing on that relationship. There are, for example, general relationships between body size and prey size that hold within broadly defined trophic categories such as carnivore or herbivore for terrestrial vertebrates as a whole (Vézina 1985). At a more relevant scale, groups of bird species within families often are separable morphologically, and these patterns reflect ecological differences between the groups. Among tropical-forest pigeons and doves, for example, ground-feeding species are smaller and have shorter wings and longer tarsi than canopy feeders, which also have stouter bills (Karr and James 1975). This pattern holds whether one considers birds in Panama or in Liberia. Species of meliphagid honeyeaters that feed chiefly on insects gleaned from leaves and bark or captured in the air and on flowers of a few plant species tend to be short-billed, whereas species that feed primarily on flowers of a wide variety of plants and that also feed on insects hawked from the air often have longer bills (Ford and Paton 1977). Among European sylvine warblers, species that occur in reed-beds have a hind-limb structure suitable for clinging to vegetation. They are strongly differentiated in leg and foot morphology from the species occupying wooded habitats, which, in turn, are morphologically distinct from the ground-dwelling species (Fig. 7.1;

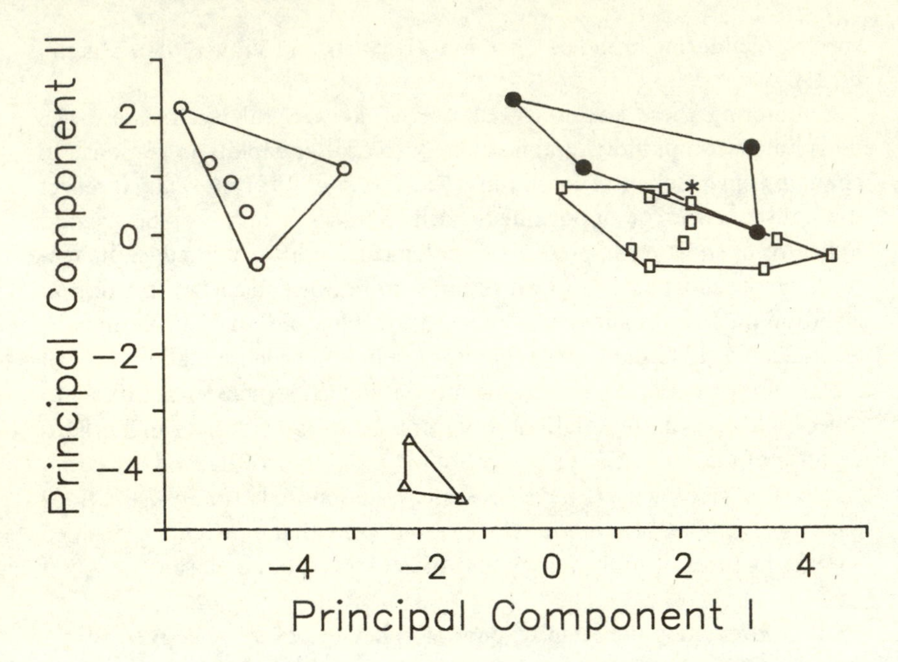

Fig. 7.1. Arrangement of 25 sylviid warbler species in a multivariate morphological space defined by a PCA of hind-limb characters. Open circles = *Acrocephalus* species, solid circles = *Phylloscopus*, open squares = *Sylvia*, open triangles = *Locustella*, * = *Hippolais*. After Winkler and Leisler (1985).

Leisler and Winkler 1985). In this case, the clusters of species with similar hindlimb structure are composed almost entirely of congeners, indicating that there are evolutionary as well as ecological restrictions on the morphological patterns.

The linkage between morphology and ecology seems especially clear when one makes comparisons across a range of related species sharing a common and somewhat demanding form of resource exploitation. Insectivorous tyrannid flycatchers, for example, have undergone a substantial radiation in the neotropics that has produced several distinctive variations on the general fly-catching theme. These foraging guilds also differ morphologically (Fitzpatrick 1985; Fig. 7.2). Species that capture prey with upward-directed strikes have wide bills, short wings, long tarsi, and reduced tails. Both perch gleaners and ground foragers, on the other hand, have narrow bills, and the ground foragers have longer legs as well. Aerial hawkers have triangular bills, extremely long wings, and short legs. The differences between guilds are not absolute in either behavior or morphology, however (Fig. 7.2). Thus, for example, the ratio of wing length to tarsus

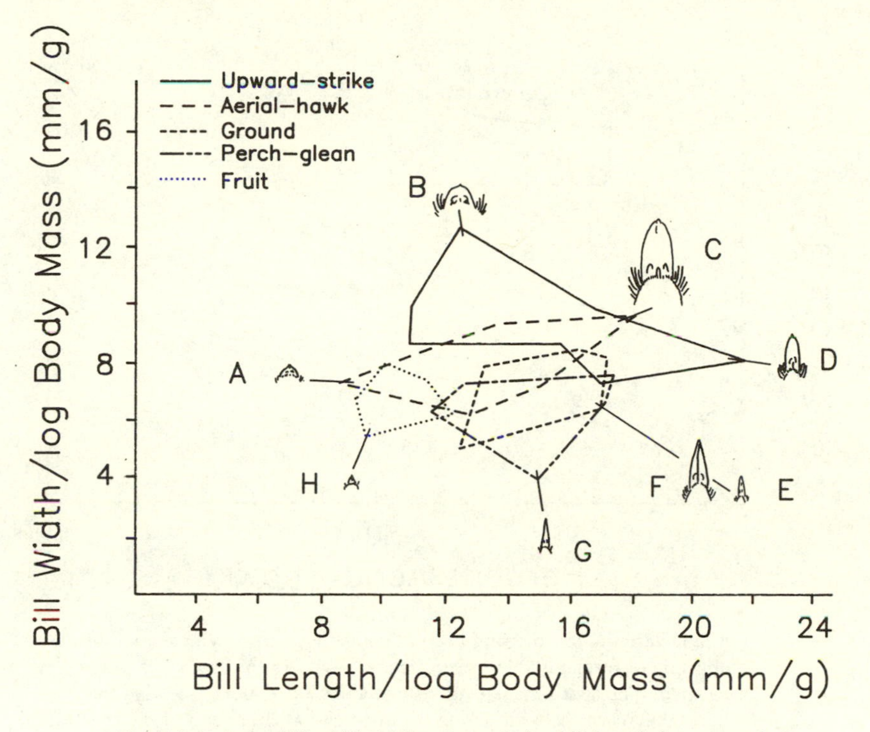

Fig. 7.2. Relative bill width (at base) vs relative bill length for neotropical tyrannid flycatcher guilds. Extreme bill shapes, drawn to scale, are of *Colonia colonus* (A), *Platyrinchus platyrhynchos* (B), *Megarynchus pitangua* (C), *Todirostrum cinereum* (D), *Anairetes paulus* (E), *Agriornis montana* (F), *Tachuris rubrigastra* (G), and *Tyrannulus elatus* (H). After Fitzpatrick (1985).

length varies more or less continuously among species with changes in the proportion of time they devote to aerial hawking, although the morphological patterns at the extremes are distinctive (Fig. 7.3). Similar (but not identical) relationships between morphology and foraging behavior have been documented for a smaller set of neotropical tyrannid species by Leisler and Winkler (1985).

Insectivores such as flycatchers feed on prey that have various antipredator adaptations, and the morphology of the insectivores is correspondingly specialized for prey capture and handling (Lederer 1975, 1984). Fruits, on the other hand, often have characteristics that serve to attract frugivorous birds, which disperse the seeds. Fruit is thus more readily accessible than are insects, and one might expect the morphology of frugivores to be more closely related to prey manipulation than to prey capture. Variations in the sizes of fruit consumed by different species should

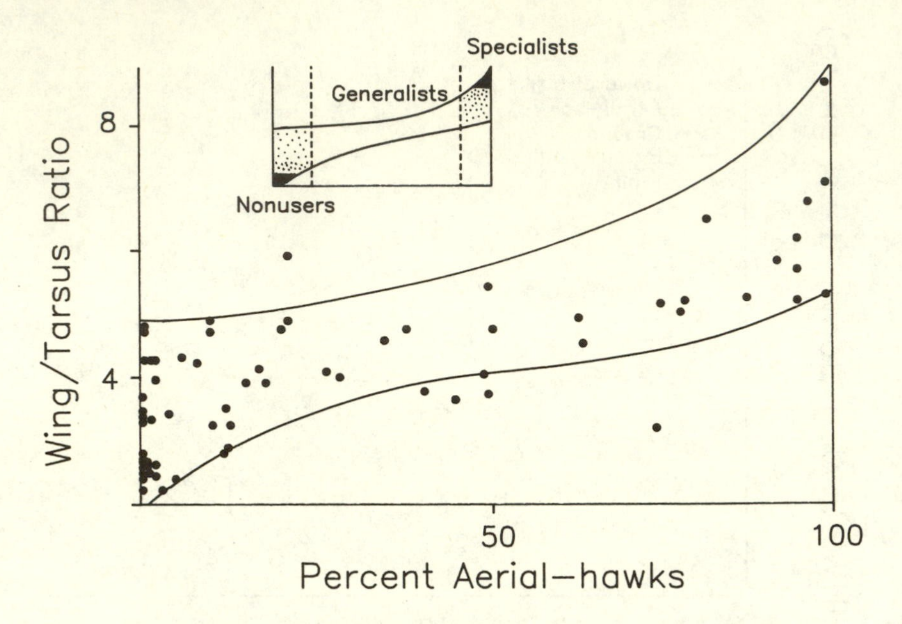

Fig. 7.3. Variation in the wing-length:tarsus-length ratio as a function of the proportion of foraging time spent in aerial hawks for 67 species of neotropical flycatchers. The inset schematically summarizes the pattern, showing the extreme differences between the ratios of species at either end of the spectrum compared to the ratios of generalist species. After Fitzpatrick (1985).

therefore be closely associated with morphological differences among the species. Such relationships do seem apparent, at least to a degree. In an analysis spanning a broad taxonomic spectrum, Ricklefs (1977) used discriminant function analysis (DFA) to determine the morphological patterns among 82 species feeding on four different types of fruits in Central America. The birds were indeed differentiated on the basis of features such as the ratio of wing length to body length, the ratio of tarsus length to total length, or the ratio of tarsus and body length to toe length, but no relationships to bill characteristics emerged. In a more detailed study of frugivores in montane forests of Costa Rica, on the other hand, Wheelwright (1985) found that species with large gape sizes consumed a greater variety of fruit species than did small-gape species and that the mean size of fruits in the diet correlated significantly with gape width. The size of the largest fruits eaten by a species was clearly a function of gape width as well, but the size of the smallest fruits taken was unrelated to gape size (Fig. 7.4). The maximum size of fruit that can be consumed is therefore constrained by morphology but, because it is easy to 'capture' even very small fruits, having

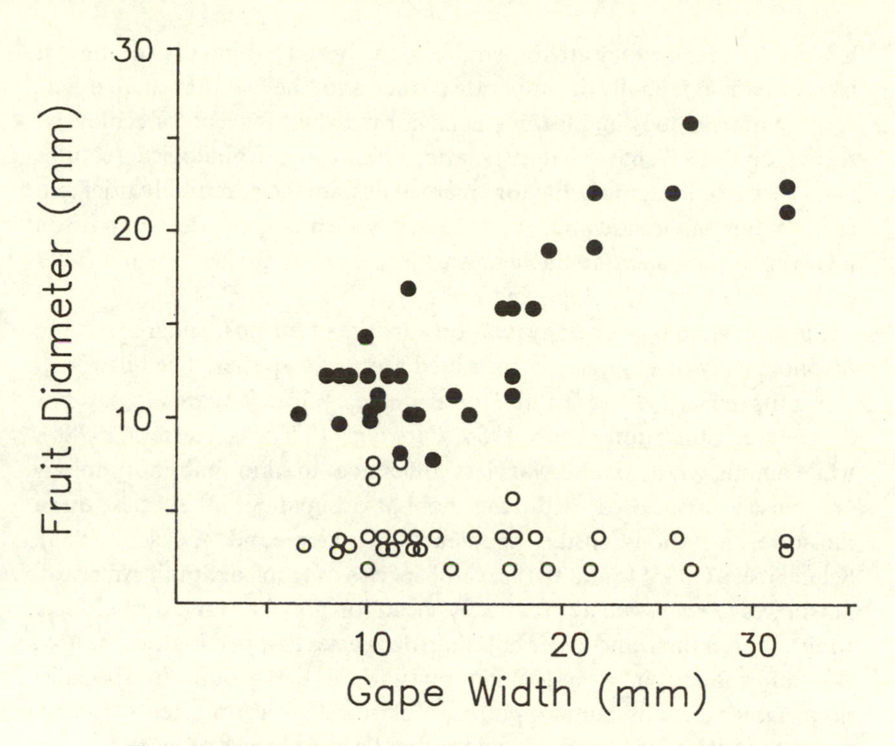

Fig. 7.4. Maximum (solid circles) and minimum (open circles) diameters of fruit species included in the diets of 32 neotropical frugivores in relation to the gape width of the bird species. After Wheelwright (1985).

a large gape does not preclude their consumption. As a consequence of the ready availability of small fruits, there was substantial overlap in fruit sizes taken by these frugivores.

If the morphology of a species is closely related to its ecology, then the amount of ecological overlap among species should match the degree of morphological similarity among the species. Ricklefs and Cox (1977) examined such relationships by comparing the morphological similarities among 11 passerine species occupying second-growth forest on St Kitts in the West Indies with overlaps in their foraging behavior. The morphological analysis was based on measures from 2–3 museum specimens of each sex; the behavioral data (which included as few as 8 observations on some species) were gathered by spot-observations made during the mornings of 7 days in August. Ricklefs and Cox found that overlap in feeding methods was strongly correlated with the wing/tarsus ratio, whereas overlap in feeding location was correlated with both the wing/tarsus ratio and similarities in bill morphology. They found no relationship between morphology and habitat occupancy; they attributed this to the broad habitat distribution of bird species on faunally depauperate islands and 'the fact that small differences in morphology and feeding behavior may be sufficient for ecological segregation' (1977: 65). That associations between morphological features and aspects of foraging behavior emerged despite the admittedly brief and crude nature of the study may not be terribly surprising, as the comparison involved species such as finches, warblers, and flycatchers, which differ greatly in both morphology and ecology.

Close morphology–ecology relationships are often most apparent within taxonomically or ecologically restricted groups of species. The bill size of *Parus* tits in Europe and Britain, for example, is closely related to patterns of resource utilization (Snow 1954, Partridge 1976a, b, Herrera 1978b), while among *Acrocephalus* warblers differences in hind-limb morphology are closely associated with the habitat-occupancy of species on a moisture/vegetation-density gradient (Winkler and Leisler 1985). Schulenberg (1983) found that several species of tropical South American antshrikes (*Thamnomanes*) that sally-glean for prey have relatively longer wings, shorter tarsi, and wider bills than do species that perch-glean; he used this information in a systematic revision of the group. In specialist nectarivores such as hummingbirds, a close relationship often exists between the bill length of species and the corolla-tube length of the flowers on which they forage (Wolf *et al.* 1976, Kodric-Brown *et al.* 1984, Feinsinger and Colwell 1978), although this relationship may erode for species that also consume large quantities of insects (Hespenheide 1975). Among shorebirds, variations in bill lengths among species may be related in a general fashion to prey sizes (Holmes and Pitelka 1968, but see Lifjeld 1984) or to features of the foraging microhabitat (Baker 1979), while tarsal length may be related to water depth in the foraging habitat (Baker 1979) and body size correlated with mean prey size (Baker 1977, Lifjeld 1984).

Contrary evidence

These patterns, based largely on the demonstration of a correlation between some feature of morphology and aspects of diet or foraging behavior, bolster our confidence in the premise underlying ecomorphological studies of communities. The results of other studies, however, are not so encouraging. In the guild of foliage-gleaning insectivores studied by Root (1967), for example, bill-length differences between the two *Vireo* species corresponded with differences in the mean sizes of prey consumed, but there was no correlation between bill size and prey size for the guild as a whole and the guild members overlapped almost completely in the size ranges of prey consumed. Willson (1971) likewise found no correlation between bill size and preferred seed size in laboratory tests with eight finch species. In a field study of wintering finches in North Carolina, Pulliam and Enders (1971) documented a faint tendency for larger-billed species to prefer larger seeds, but each species took a considerable variety of seed sizes. Consequently, overlap in their use of seeds was almost complete within habitats, despite the twofold difference in bill lengths. Thompson and Lawton (1983) tested seed preferences of nine granivore species during winter in Britain and also found no relationship between either bill size or body size and the sizes of seeds preferred.

In these examples, the relationship between a morphological measure and its presumed ecological correlates is absent or weak, perhaps because the form of the morphological attribute is influenced by other selective forces or because the expectation of a relationship is in error. In other situations, ecomorphological patterns that are apparent from one perspective vanish when the focus of the analysis is changed. There is a weak but general relationship between body size and mean prey size over a broad range of temperate and tropical seabird species. When one considers the variation present in local assemblages, however, the ecological significance of this relationship diminishes. On one island, for example, three tern species whose body mass varied from 93 to 98 g consumed prey whose mean size varied over an order of magnitude (0.2–3.5 g) (A.W. Diamond 1984). In other situations, patterns that are present at a restricted scale disappear at a broader level. Bill size is related to prey size within several guilds of insectivores occupying wet sclerophyll forest in Western Australia (Wooller 1984), but Wooller and Calver (1981) found no relationship between morphological similarity among species and their overlap in diets or foraging behavior for an assemblage that included several guilds of insectivores. Similarly, Jaksić and Braker (1983) found a significant relationship between prey size and body size within guilds of diurnal raptors from several locations, but this pattern disappeared in between-guild comparisons. In this case, the body size–prey size relationship was obscured because raptor species of the same size captured prey of quite different sizes in different areas.

If the relationship between morphology and ecology is really tight, one might expect individuals with different bill dimensions within a population of a species to consume different sizes or types of prey (Grant *et al.* 1976, B.R. Grant 1985). John Rotenberry and I looked for such relationships in several species of North American grassland and shrubsteppe birds for which we had measures of morphology and diet from the same individuals

(Wiens and Rotenberry 1980). The patterns were inconsistent, and for most comparisons no significant relationships were evident. When we considered relationships over all of our samples of species populations, however, bill length was significantly related to prey length (Fig. 7.5), as was body mass. This pattern must be viewed with some caution, however, because it is determined entirely by the differences between one group of several large species and another group of several small species; within either group, prey size varies independently of bill or body size.

The points plotted in Fig. 7.5 represent collections made at several widely separated locations, and the more meaningful ecological comparison is among the species co-occurring at a specific place. Such comparisons sometimes reveal significant differences in prey sizes between species that correspond to their morphological differences, but in other situations they do not (Fig. 7.6; Wiens and Rotenberry 1979). Rotenberry (1980a) examined the diets of several species of quite different morphology at the shrubsteppe location in Fig. 7.6 throughout the year and found that the composition and size distribution of prey in the diet changed seasonally in parallel ways for the different species. Thus, at no time were features of the diet closely related to the morphological differences among the species.

Analyses of functional morphology

This sampling of studies indicates that, although there is some support for the notion that morphology and ecology are closely linked in birds, there are many exceptions. All of these findings are based on rather broad comparisons and correlations, however. Greater insight into the nature of the linkage might emerge by considering how morphology is related to ecology in a functional rather than a correlative sense. The force exerted by a bird's bill, for example, is most closely related to its depth, slightly less so to its width, and considerably less to its length (Lederer 1975). This functional relationship is expressed in the variation in bill depths among species feeding on seeds of different size and hardness (Newton 1967, Abbott *et al.* 1977), and it suggests that bill depth may be more closely related to ecology than is bill length. James (1982) reviews other examples of the detailed functional connection between morphology and ecology, but one set of studies merits special mention.

In northern Europe, Goldcrests (*Regulus regulus*) and Firecrests (*R. ignicapillus*) are ecologically quite similar and may co-occur with overlapping territories. Leisler and Thaler (1982) used large aviaries in which prey availability was standardized to observe the foraging behavior of these species. Goldcrests clearly preferred to forage in spruce, whereas Firecrests

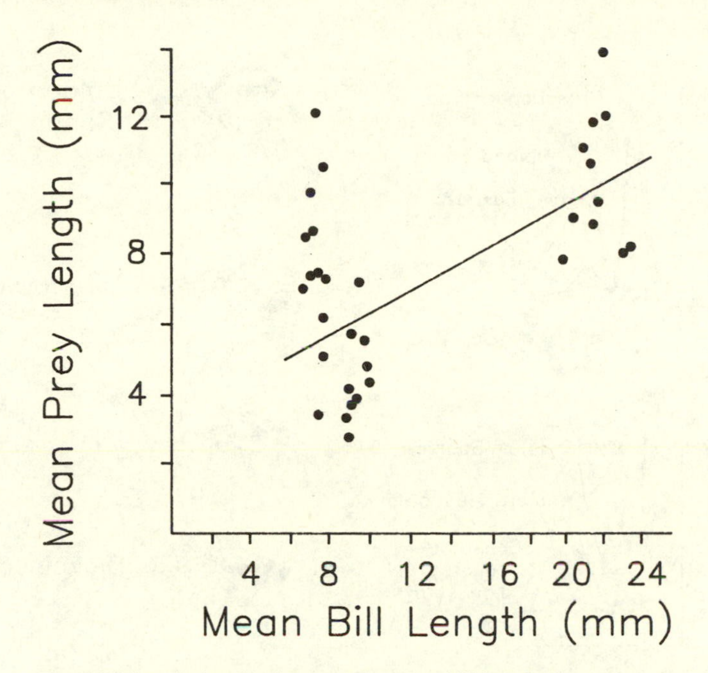

Fig. 7.5. The relationship between the mean bill length of populations (species/year/location samples) of shrubsteppe and grassland birds and the mean length of prey items measured for those populations. After Wiens and Rotenberry (1980).

foraged equally in spruce and beech. Goldcrests tended to cling vertically and hang while foraging, whereas Firecrests moved more rapidly, and they more often foraged standing. These differences in behavior are associated with several differences in the foot and bill morphology of the species; collectively, these attributes suggest that Goldcrests are specialized for foraging in conifers.

Norberg's (1981) studies have provided some additional perspective on Goldcrest morphology and ecology. Although the birds are insectivorous, they are resident throughout the year in many northern areas. In order to balance their energy demands during winter, they must capture an arthropod every few seconds throughout the day, so selection on foraging efficiency must be intense. In the autumn, when prey items are abundant, the birds forage primarily by hovering. As food becomes scarce in winter, however, the birds shift to less energy demanding (but less efficient) forms of locomotion. These may yield a lower gross rate of energy capture per unit time but, because of the reduction in power costs, the *net* energy gain is greater. As a consequence of this seasonal shift in foraging patterns, the

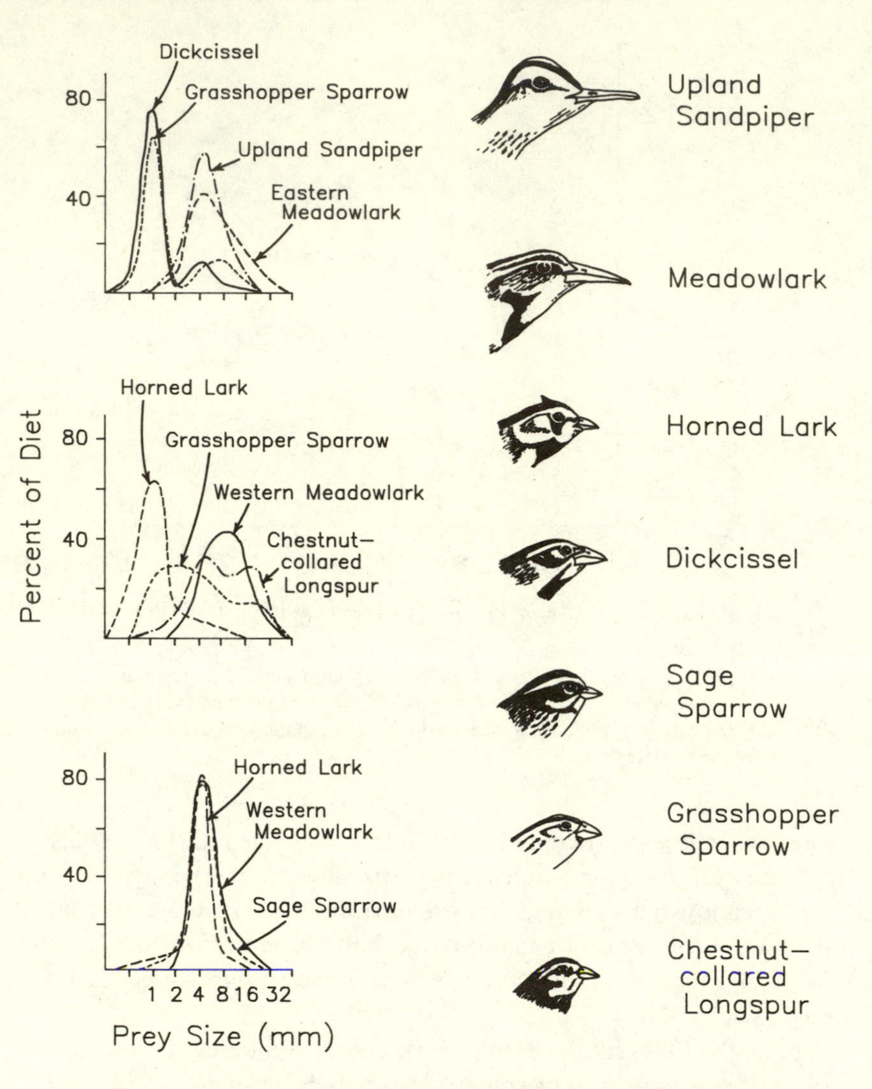

Fig. 7.6. Prey-size distributions in the diets of breeding birds at tallgrass prairie (top), mixed-grass prairie (middle), and shrubsteppe (bottom) locations in North America. Profiles of the heads of the species are drawn to scale at the right. Modified from Wiens and Rotenberry (1979). (Scientific names of species not mentioned previously: Upland Sandpiper, *Bartramia longicauda*; Eastern Meadowlark, *Sturnella magna*; Western Meadowlark, *S. neglecta*; Horned Lark, *Eremophila alpestris*; Dickcissel, *Spiza americana*: Sage Sparrow, *Amphispiza belli*; Grasshopper Sparrow, *Ammodramus savannarum*; Chestnut-collared Longspur, *Calcarius ornatus*.)

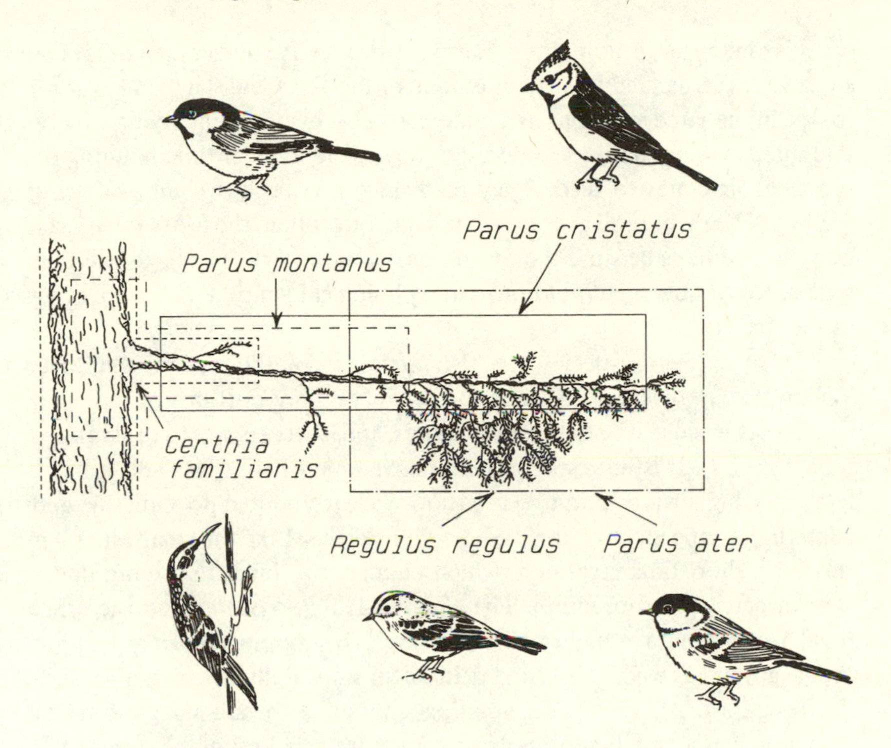

Fig. 7.7. Schematic diagram of the foraging-site selection by three tit species, the Goldcrest, and the Treecreeper in spruce trees in Scandinavia. After Haftorn (1956) and Norberg (1979).

wing morphology of Goldcrests is probably not ideally suited to either form of locomotion but reflects a compromise. Nonetheless, there are significant differences in the structure of Goldcrest and Firecrest wings that are associated with their different rates of movement and habitat preferences (Leisler and Thaler 1982).

Norberg (1979) also studied the functional morphology of several other species of the guild of wintering conifer-associated insectivores that includes the Goldcrest. The species occupy somewhat different foraging positions in the trees (Fig. 7.7), and associated with this are differences in their movement patterns and their morphology. Goldcrests forage primarily on the foliage of the outer reaches of branches. They have a short arm in relation to the total length of the wing and very low wing loading and body weight, features associated with slow flight and hovering and with maneuverability. Low wing loading enables a bird to produce sufficient lift in slow flight and hovering to avoid the use of high wing-beat frequencies, which would result in large moments of inertia and excessive power demands. The short arm further reduces the moment of inertia. Coal Tits (*Parus ater*) forage in the same locations as Goldcrests, also by hovering, flying slowly, and maneuvering; unlike Goldcrests, they spend a lot of time hanging on branches or cones to feed. They also have low body weight and wing loading. Their wing and arm are somewhat longer than those of Goldcrests, however, perhaps because the tits are partly migratory. The birds are thus well suited to slow flight but not so morphologically modified for hovering as Goldcrests are.

Two other *Parus* species are also members of this assemblage. The Willow Tit (*P. montanus*) forages primarily on bare portions of the main branches, on which it both hops and hangs, and it often clings to the trunk as well (Fig. 7.7). It is not especially maneuverable when it flies and it has a relatively high wing loading. It is more clearly suited to clinging and climbing, as reflected by a long tail and long claws. Like the Coal Tit, it has relatively short tarsometatarsi, which characterize birds that hang under slender perches and/or climb. The Crested Tit (*P. cristatus*), on the other hand, uses the main branches, where it forages by hopping, and it often hops on the ground as well. It is suited neither for slow flight and maneuverability, because of its high wing loading and short, broad wings, nor for climbing, because of its short tail and toes. It is the largest of the species and thus may have a greater ability to withstand severely cold winter nights. Finally, the Treecreeper (*Certhia familiaris*) feeds mainly on the trunk, in a head-up position. For a bird climbing vertically on bark, long toes and tail reduce the horizontal pull on the claws of the upward toes (which hold on to the bark) and hence reduce energy expenditure during climbing. These traits, along with shortened tibiotarsi, characterize Treecreepers. Low wing loading, relatively long wings, and short arms render them well suited to slow flight, maneuverability, and hovering.

In this guild of species, then, there are clear relationships between morphology and ecology at a functional level. The relationships are complex, however, and they are complicated by phylogenetic constraints (the three *Parus* species, for example, are quite similar in wing skeleton, but differ in wing-feather morphology). Such patterns would not be apparent from a general, correlative comparison of the species.

Complications: morphological variation

The premise that morphology and ecology are closely linked often includes the assumptions that a particular relationship holds over a wide array of species and that morphological variation within species is unimportant. Hespenheide's (1975) regressions of prey size against body size for several bird groups confirm the premise but indicate that the first assumption is invalid, as prey size increases much more rapidly with body size among vireos than among swallows.

Such differences may arise because allometric patterns of size and shape variation differ between the groups. These allometric patterns differ not only for different morphological features but for the same feature between groups of species. Among insectivorous honeyeaters, for example, bill length increases with the cube root of body mass at a faster rate than in sympatric insectivorous non-honeyeater species. Among nectarivorous honeyeaters in the same habitats, however, there is no clear relationship between these variables (Wooller 1984).

Ecologists have also preferred to use linear size measures of morphology (e.g. bill or tarsus length) rather than shape variables in their analyses (Mosimann and James 1979, Mosimann and Malley 1979). Shape is thereby assumed to be a passive correlate of selection on size. There is no *a priori* reason to consider size to be more important than shape (Lande 1979, Simberloff 1983b). Indeed, James and McCulloch (1985) have proposed that shape variables may have more striking ecological correlates than size variables alone, and they attribute our failure (Wiens and Rotenberry 1980) to find a clear ecomorphological patterns among grassland and shrubsteppe birds to our neglect of shape variables. We consider it unlikely that we would find a match to resource-use patterns such as those shown in Fig. 7.6 using shape variables either, but the point is well-taken. Analyses based solely on shape, however, may also be misleading (Bookstein *et al.* 1985). Size and shape should be considered together, as Mosimann and James (1979) advocated.

It is also not always clear which variables (be they size or shape) should be considered in ecomorphological studies. Bill length has been used more than any other variable, perhaps because it is easily measured, perhaps because it has been assumed to be most closely governed by selective forces related to competition for food (Cody 1974a). It could also be argued that, because foraging habitat and behavior are more easily partitioned than food type and size, competitive effects should be expressed in terms of foraging morphology (wing, tarsus) rather than bill morphology (Hespenheide 1975). Because larger jaws close faster (Beecher 1962, Bock 1964, Lederer 1975), bill size may actually bear a closer relationship to prey mobility than to prey size, and body mass may more often be a better indicator of prey size. In some terns, gape width and esophagus length are more closely related to prey size than is bill length (Hulsman 1981), and in other species there may even be a negative correlation between bill length and prey size because shorter bills are usually stronger (Lederer 1984).

The difficulty of determining which feature of morphology to measure is apparent from Carothers' (1982) study of three Hawaiian honeycreepers. The species differ in general diet, foraging tactics, and feeding rates on flowers of the same species. They also differ in bill lengths. The species seemingly best suited to feed on the short-corolla flower on the basis of bill morphology, however, actually feeds at a slower rate than the species with a long, curved bill. Feeding efficiency, in fact, varies opposite to the pattern expected from bill morphology. It turns out that the feeding patterns are more closely related to the structure of the tongue and associated differences between the species in their ability to extract nectar than to bill morphology, although differences in foraging behavior and in the overall spectrum of flower types exploited by each species are also important. Conclusions derived on the basis of bill-size comparisons in this group would reveal rather little of their ecology. Any patterns of morphological relationships among members of a community or guild pertain only to the features measured, of course, and different features may well produce quite different patterns. This point, although obvious, seems not to be widely appreciated.

Morphology varies within a species, and this variation can influence both our ability to detect ecomorphological patterns and the accuracy of such patterns. If variation is substantial, measures of morphological states for a species derived in small samples from museum collections may be inaccurate. If there is geographic variation in morphology (e.g. James 1970, Wooller *et al.* 1985, Murphy 1985), the use of specimens from other parts of the range or of literature values accumulated from scattered locations (e.g. Ridgway 1901–1918) may be inappropriate. In our collections of breeding birds from grassland and shrubsteppe locations in North America, most of the species varied significantly in most features that we measured (Wiens and Rotenberry 1980). There was considerable between-sex variation, only a moderate amount of variation due to locality, but a surprising amount of between-year variation, especially in bill features (Fig. 7.8). The features of morphology covaried but did so in different ways for the different species.

How spatial variation can contribute to morphological patterns on a more restricted scale is apparent from two studies in which features of birds in different habitat types in a local area were measured. Ulfstrand *et al.* (1981) found that both sexes of Great Tits (*Parus major*) occupying coniferous forests in southern Sweden had significantly shorter bills than birds in nearby deciduous woodlands; males also had shorter wings in the conifer-

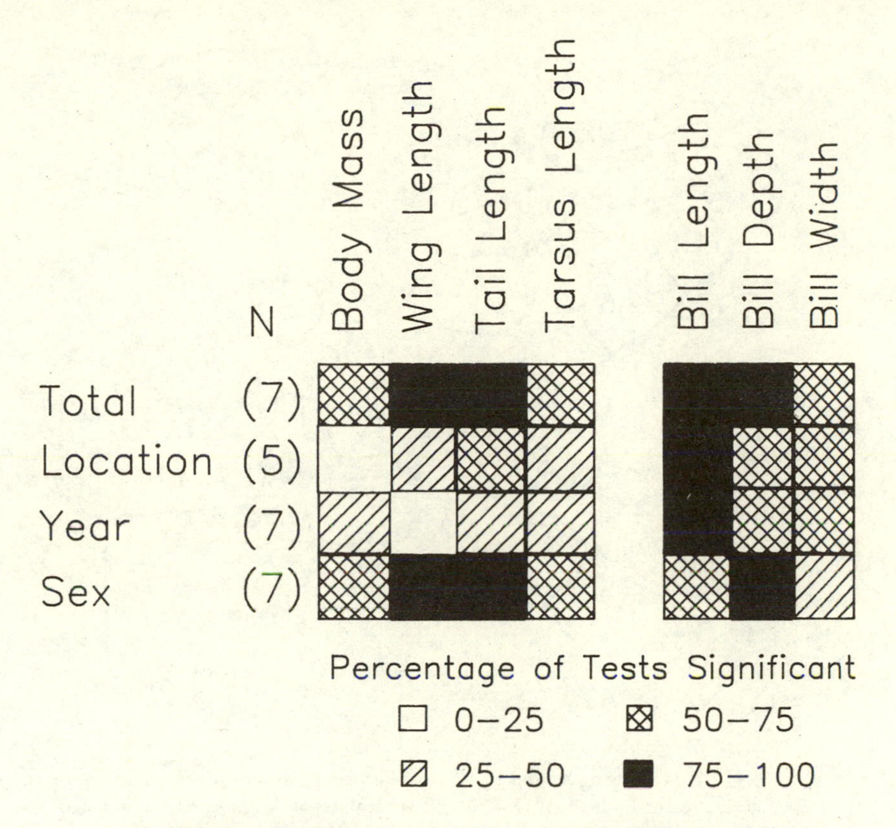

Fig. 7.8. The proportions of statistical tests showing significant variation ($P < 0.05$) attributable to location, year, and sex for morphological features of grassland and shrubsteppe bird species. After Wiens and Rotenberry (1980).

ous habitat. They suggested that this pattern was probably a reflection of localized social dominance, which forced smaller, subdominant birds into less suitable habitat. Studying Willow Warblers (*Phylloscopus trochilus*) in Finland, Tiainen (1982) showed that males holding territories in a mixed deciduous woodland had significantly longer wings that those in adjacent spruce forest (Fig. 7.9). In the deciduous habitat, males returning to the same territory occupied in the previous year or to one immediately adjacent to it tended to be larger (wing length) than those returning to more distant territories.

In these examples, the patterns of variation in morphology are influenced by individual differences in habitat occupancy associated with social dominance and site tenacity. More generally, the morphological variation in a population is an expression of both environmental and genetic influences.

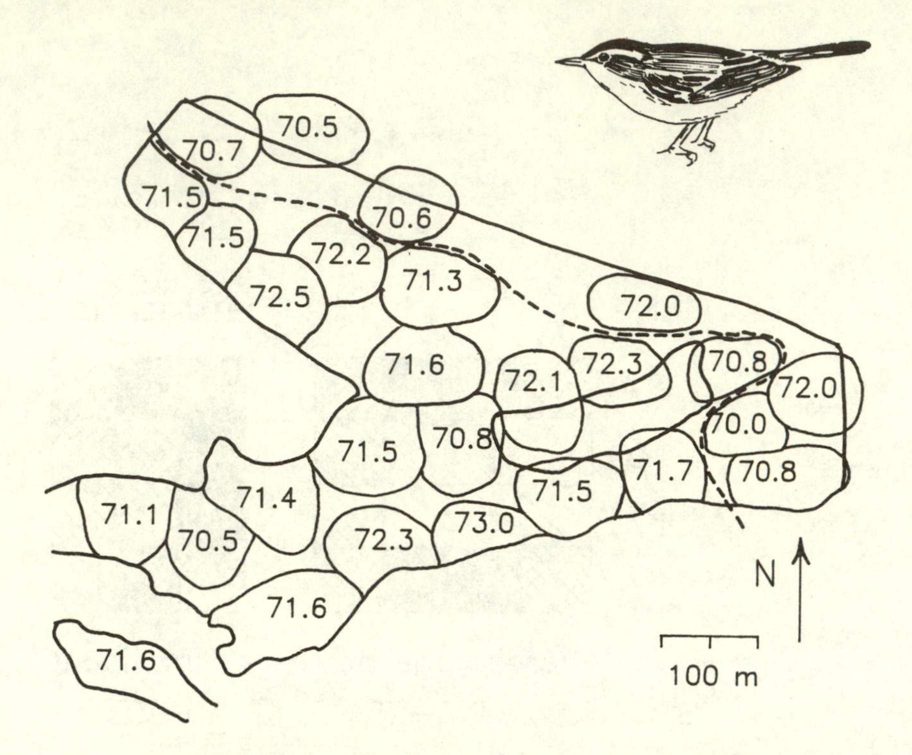

Fig. 7.9. Mean wing lengths of Willow Warbler males in various territories of a study area in Finland, 1972–1977. The overall mean in the spruce-dominated area east and north of the broken line (70.7 mm) was significantly smaller than that in the remaining deciduous-dominated part (71.6 mm). From Tiainen (1982).

To obtain accurate estimates of the heritability of a trait in a free-living bird population requires that one obtain measurements of adults and of their known offspring when they in turn reach adulthood. The few calculations of heritabilities that have been made have yielded varying results (Hailman 1986). In Song Sparrows (*Melospiza melodia*), heritability values for bill dimensions and body size are rather low (Smith and Zach 1979). On the other hand, in Great Tits, the heritability of traits such as tarsus length (Garnett 1981) or body weight (van Noordwijk *et al.* 1980) is relatively high, and heritabilities of bill features are high in populations of Galápagos finches (*Geospiza*) (Grant 1983a, Price *et al.* 1984a). In at least some situations, then, the amount of morphological variation that has an underlying genetic basis may be large.

That environment can also exert a powerful influence on morphology is apparent from the elegant experiments of James (1983). By transferring

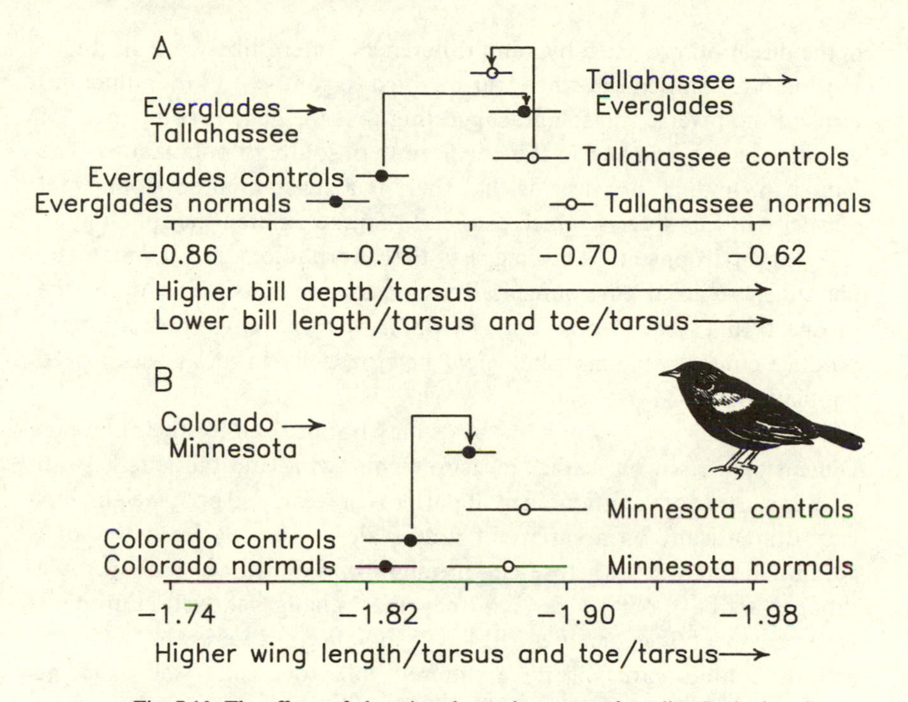

Fig. 7.10. The effects of changing the environment of nestling Red-winged Blackbirds on phenotypic development. The axes in these diagrams are significant dimensions of variation discriminating normal nestlings in (A) two Florida locations and (B) Colorado and Minnesota sites. Means and 95% confidence intervals are shown for scores on these axes for groups of unmanipulated control nestlings and for groups of nestlings hatched from eggs transplanted between localities, as well as for normal groups. In each case, the shape of the transplanted birds shifted toward that of normal nestlings in the foster population. After James (1983).

eggs of Red-winged Blackbirds (*Agelaius phoeniceus*) between nests in southern and northern Florida and from nests in Colorado to nests in Minnesota and then monitoring the development of chicks in control and foster nests, James sought to determine whether or not the considerable regional differences in morphology in this species represent local genetic adaptations. Normal nestlings in the different areas can be discriminated by shape characteristics, and comparisons of the experimental nestlings with chicks from their origin versus their foster location demonstrate clear and dramatic shifts (Fig. 7.10). In the reciprocal transplant across Florida, the ratio of bill length to tarsus length shifted the most, in the direction expected on the basis of differences among adults. In the Colorado to Minnesota transplant, the ratio of wing length to tarsus length shifted the most, again in the direction predicted by adult differences. Interestingly, the northern Florida birds shifted less when transplanted to southern Florida than did birds in the reverse transplant, suggesting that the genetic component of variation may be greater in the northern than southern populations. The important finding, however, is that there is a substantial developmental plasticity in this species, which can lead to quite different morphology in different environments. This suggests that morphology might indeed be fine-tuned to local environmental circumstances, but it also raises the prospect that year-to-year changes in a variable environment might produce considerable morphological heterogeneity in an age-structured population.

Even the morphology of adult birds may not be fixed for their lifetime. Dimensions based on feather measurements (wing and tail lengths) can change because of variations in molt patterns or wear, and body weight may vary dramatically on a variety of time scales, especially in species that accumulate fat reserves before lengthy migrations. Bill dimensions are often thought to be more stable, but they may change as well. European Oystercatcher (*Haematopus ostralegus*) populations wintering in The Netherlands contain birds of three different bill forms, as well as some intermediates. The distribution of bill types varies as a function of age and sex, but each bill type is suited for a specific way of capturing and handling prey. Swennen *et al.* (1983) experimentally demonstrated that, if individual oystercatchers of a particular bill type are forced to change their feeding habits, the shape of the bill gradually changes to match the new feeding regime. Bill form thus influences prey choice and feeding method, but these in turn affect bill form.

Morphology and species-abundance patterns

The species in a community differ in abundance as well as in morphology (Chapter 5), although measures of abundance and of morphology are usually considered separately in community analyses. What emerges when they are considered together? Are species of different sizes equally likely to be abundant or rare, or are large species generally less abundant (as one might expect from body size–home range or trophic pyramid relationships)?

Brown and Maurer (1987) addressed this question by plotting the average abundances of nearly 400 species of North American landbirds recorded on Breeding Bird Surveys as a function of their body mass. The data were distributed within a fairly well-defined polygon (Fig. 7.11), and Brown and Maurer offered hypotheses to explain each of the boundaries.

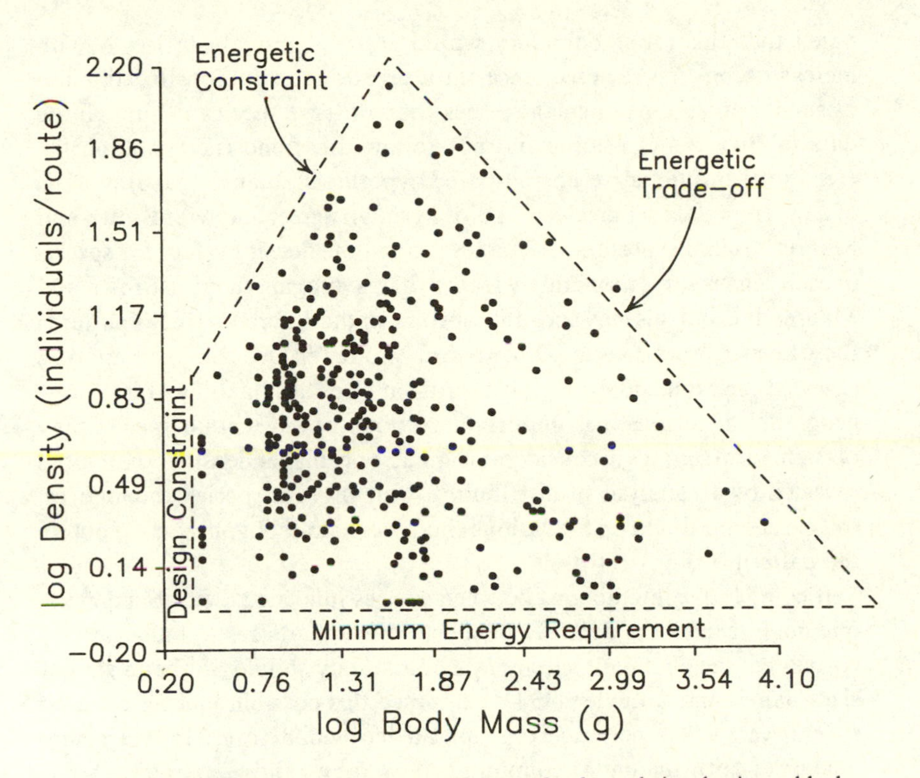

Fig. 7.11. The relationship between average local population density and body mass for North American terrestrial bird species. The points fall within the polygon indicated by the dashed line; for each boundary, hypothetical constraining factors are indicated. Modified from Brown and Maurer (1987).

The tendency for abundance (log *N*) to decrease with increasing body size (log mass, *M*), for example, was explained by proposing that there is a constant maximum rate of energy intake for a species within a taxon that is a result of a trade-off between having high densities of small individuals or lower densities of larger individuals. Below a size threshold of roughly 50–100 g, however, log *N* decreased with *decreasing* log *M*. Brown and Maurer attributed this to a different sort of energetic constraint, arguing that, because small organisms have increasingly high energy demands per unit of mass, they may be restricted to more concentrated energy sources below a certain size; as a consequence, the availability of *useable* resources declines with decreasing log *M*. To explain the relatively constant lower limit of log *N* over a wide range of values of log *M*, they suggested that there is a lower limit to the ability of species to extract energy from the environment, independent of body size. In fact, Brown and Maurer had initially anticipated that this lower boundary would show a decrease in log N with increasing log M as a consequence of the greater susceptibility to extinction of small populations of small species than of large species. Although the data of Brown and Maurer did not follow this trend (necessitating the erection of an alternative, energy-based hypothesis), Juanes' (1986) analysis of a different data set produced a plot in rough agreement with Brown and Maurer's initial expectations. Juanes plotted the density values for species in each census survey separately rather than averaging them as Brown and Maurer did, and this may account for some of the differences in the patterns these investigators detected. Brown and Maurer argued that their analysis revealed the broad adaptive constraints on birds as a taxon within a large geographical region. It is important to note, however, that species may occur in local settings in considerably greater or smaller densities than those revealed by an analysis of distribution-wide *average* species abundances; this casts some doubt on the various energy-constraint hypotheses, if not on the pattern of Fig. 7.11 itself.

If competitive interactions between species influence their abundances, one might expect the densities of species of similar sizes to be negatively correlated; the abundant species in a community should not be of similar sizes. James and Boecklen (1984) examined this possibility using censuses taken over a 7-yr period in an upland hardwood forest in Maryland, analyzing both the entire community and four guilds separately. Some species pairs did indeed vary inversely (although positive associations between species' densities were more frequent), but these species were not unusually similar to one another in morphology. There were also no indications that abundant species were more widely separated from other species in a multivariate morphological space than were rare species. In this community, the abundances of species relative to one another varied quite independently of their degree of morphological similarity. On the other hand, Griffiths (1986) concluded that the distribution of abundances of various size classes of birds in several communities was often polymodal, species of certain, similar sizes being much more likely to be abundant than species of other sizes. Such patterns might reflect evolutionary convergence toward a similar body size by several members of a particular guild (e.g. granivores, insectivores) as a consequence of size-related constraints on resource exploitation, or they might be determined by discontinuities or polymodality in the abundance of resources of different sizes. Because such analyses have been conducted for few avian communities, the generality of the pattern (much less its causes) is uncertain.

Niche variation and morphology

Species that are abundant may occupy a wider range of environmental conditions than do rare species (Fig. 6.1), and it has been suggested that they may be more variable in morphology as well (Van Valen 1965, McNaughton and Wolf 1970, Rothstein 1973). Van Valen (1965), for example, found that populations of six species of passerines on the Canary Islands are seemingly more variable in bill lengths than are mainland populations, presumably because of reduced competition and niche expansion on the islands. Grant (1979a) also found that populations of the Chaffinch on the Azores are more variable than those on the Canary Islands; he attributed the pattern to the elevational range occupied, noting that the more variable populations occurred over broader elevational ranges, both within the Azores and between the Azores and Canary Islands. In other studies of island birds, however, no pattern of increased morphological variation on the islands has been discerned (e.g. Crowell 1962, Grant 1967, Keast 1976, Abbott 1977), and Beever (1979) has questioned the validity of the patterns on which Van Valen based his original development of the theory.

If the relationship between morphological and ecological variation is real, it should also be present in mainland communities. Indeed, Rothstein (1973) found a significant positive correlation between morphological variability and abundance in a selected sample of species, but Willson *et al.* (1975) considered the same relationship over a larger series of species from central Illinois and found no consistent pattern between variation in bill lengths and abundance. They also tested the relationship between variability and various measures of the species' ecology and found few nonrandom patterns. Willson (1969) also compared variation in bill sizes in populations of birds from lowland tropical and from temperate habitats, reasoning that niches should be larger in the less diverse temperate communities. There was no evidence of any consistent increase in variability in the temperate samples. In our analysis of grassland and shrubsteppe birds, we found no relationship between the variation of several morphological features within populations and either the abundance of the species or the diversity of their diets (Wiens and Rotenberry 1980).

The evidence supporting the 'niche variation hypothesis' is thus scant and inconclusive (Grant and Price 1981). This is not really too surprising, for it is more likely that a population would occupy a broad niche through behavioral flexibility than by virtue of morphological variety. In particular, the niche width of a population may reflect both differences between individuals (the 'between-phenotype' component of niche width; Roughgarden 1972, 1974b) and the range of ecological situations exploited by single individuals (the 'within-phenotype' component). Morphological measurements relate only to the between-phenotype component of niche variation and may fail to portray the within-phenotype features that relate most strongly to the actual ecological versatility of a species.

Character ratios

Most ecomorphological analyses of communities have been based on simple tallies of the presence or absence of species rather than on abundances. Not surprisingly, the focus of many of these investigations has been on patterns of morphological differences among coexisting species. The idea that these species might be regularly spaced on size gradients was originally put forth by Huxley (1942; see Carothers 1986). Following this lead, Hutchinson (1959) calculated the ratios of the bill sizes or body lengths of several sets of coexisting, ecologically similar species, and found an average ratio (larger/smaller) of approximately 1.3. He concluded that this value might indicate the amount of difference necessary to permit the coexistence of such species. The idea rapidly gained favor among ecologists, and the ratio value of 1.3 for linear dimensions (2.0 for body mass) came to be regarded as somewhat of a constant. Maiorana (1978) proposed that such 'ecological constants' provide 'a firm base from which to investigate the nature of competition interactions in natural communities'. Some features of the structure of this hypothesis and its tests are examined in Chapter 2 (Fig. 2.2).

The logic underlying the expectation of a constant size ratio among coexisting competitors is straightforward: larger organisms require more food but feed upon larger prey items that are scarcer than smaller prey. They will therefore use a broader range of prey sizes than will smaller organisms. Given a limit to overlap in prey-size utilization between species, the broader utilization functions of progressively larger species on a size sequence will result in a wider spacing (and therefore approximately constant ratios) between them (MacArthur 1972). Resource monopolization by the larger species (Brown and Maurer 1986) may also lead to such spacing.

Some evidence

To support this argument, MacArthur (1972) used the example of the North American *Accipiter* species studied by Storer (1966). Males of the three species weigh on average 99 g (*A. striatus*), 295 g (*A. cooperi*), and 818 g (*A. gentilis*) – body-mass ratios of 2.98 and 2.77 (females of each species are considerably larger than males, but the basic between-species pattern holds for them also). Using data gathered from various locations in all seasons from a number of years, Storer determined that these body-size relationships corresponded with differences in both the mean and the range of prey sizes taken by each species (Fig. 7.12A). This has become a textbook example of both the constancy of size ratios among ecologically similar species and of the close relationship between predator size and prey size.

The latter relationship, however, may be more complex than Storer or MacArthur realized. Reynolds and Meslow (1984) conducted a careful study of the food habits of these *Accipiter* species in coastal northwestern Oregon, where only *striatus* and *cooperi* occur, and in the eastern portion of that state, where all three species are present. At this more intensive scale of analysis, the general relationship between predator size and prey size niche-width presisted. With the exception of *striatus*, however, the mean prey sizes for the species were quite different from those reported by Storer (Fig. 7.12B). There were also regional variations in the diets of the raptors, which to some degree were associated with differences in the prey sizes available to them. In the northwest, bird prey items were on average half as large as those in the east, whereas mammal prey averaged twice as large. The mean prey size of *striatus* (which fed almost entirely on birds) was significantly smaller in the northwest (12.8 g) than in the east (28.4 g). The average prey sizes for *cooperi*, however, did not differ between regions (134.7 g vs 136.3 g); this species fed on a mixture of bird and mammal prey and the diet of birds in the east included more mammals and fewer birds than in the west, cancelling out the effects of regional differences in the sizes of prey available. The general absence of very small prey may account for the truncation of the *striatus* prey-size curve at the low end and the similarity to the value reported by Storer. The calculated limiting-similarity (d/w) values for the Oregon raptors were: *striatus-cooperi*, 2.2 (northwest), 1.8 (east); *cooperi-gentilis*, 0.8. Interestingly, there was no evidence of a shift in the prey-size utilization of *cooperi* between eastern Oregon, where the larger *gentilis* also occurred, and the northwest, where it was absent.

Several studies have provided circumstantial evidence that high morphological similarity between ecologically similar species may lead to the local extinction or distributional exclusion of one of them, thus promoting the interspecific spacing predicted by the hypothesis. In the New Guinea avifauna, sympatric pairs of congeneric species with weight ratios of < 1.67 do not coexist spatially unless they differ in diet or foraging, but most pairs

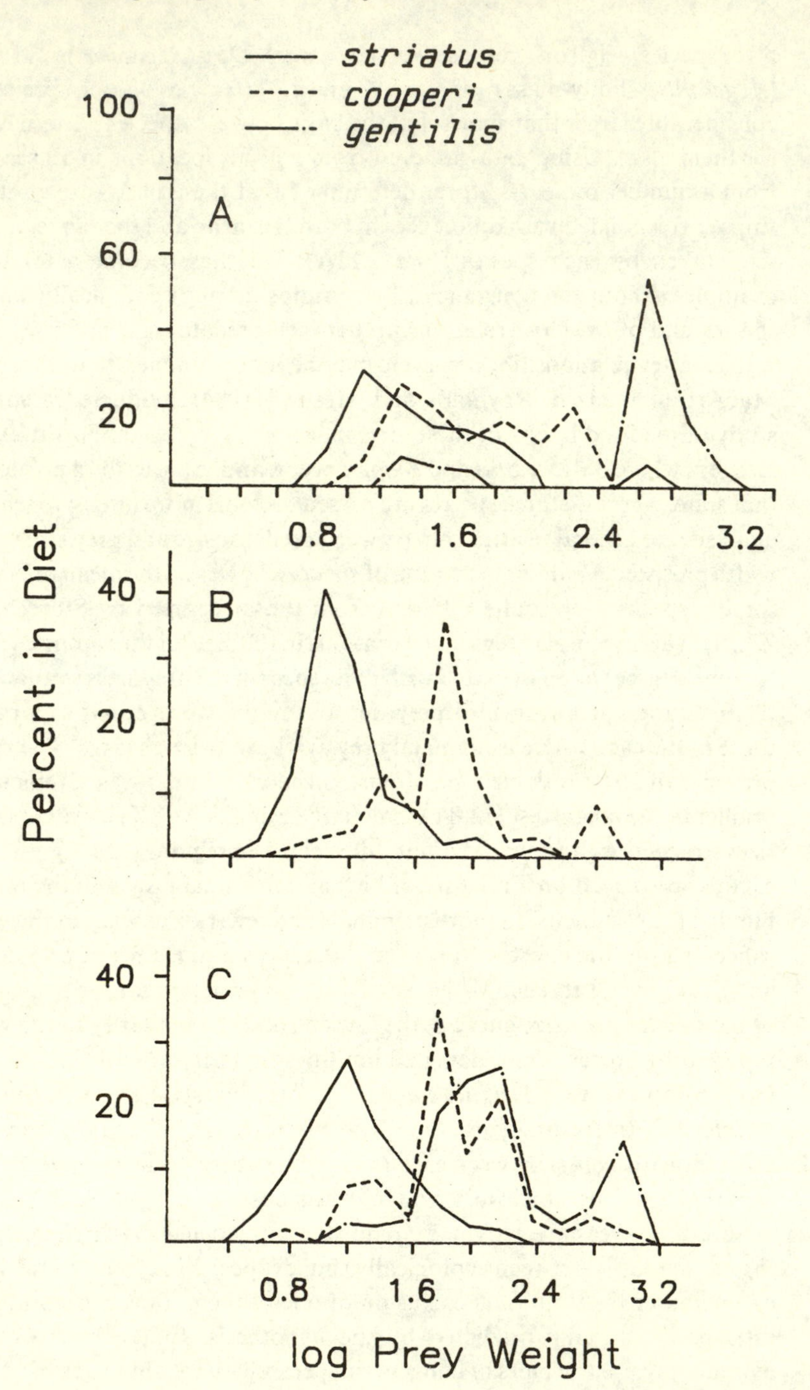
striatus
cooperi
gentilis
100
60
20
A
0.8
1.6
2.4
3.2
Percent in Diet
40
20
B
40
20
C
0.8
1.6
2.4
3.2
log Prey Weight

with size ratios > 1.79 overlap extensively in space and often share diets and foraging behaviors as well (Diamond 1986b). Among 46 pairs of allopatric New Guinea congeners, on the other hand, only three have a weight ratio > 1.75. Similarly, Brandl and Utschick (1985) found that the degree of morphological dissimilarity among wading birds in the Okavango Delta of Botswana was positively related to their distributional overlap – similar-sized species tended not to co-occur. In the Hawaiian Islands, Moulton and Pimm's (Moulton 1985, Moulton and Pimm 1986, 1987) analysis of the fates of introduced species suggested that mean difference in bill lengths between members of congeneric pairs in which one species became extinct was significantly less than that between members of pairs that persisted.

All of these findings are consistent with the suggestion that some minimal morphological separation is necessary to permit the coexistence of congeners, although they do not justify the conclusion that interspecific competition is actually the primary cause of the extinctions or of the patterns of sympatry and allopatry. Simberloff and Boecklen (personal communication), however, criticized Moulton and Pimm's analysis of the Hawaiian Island introductions on several counts (e.g. inappropriate statistical tests, incomplete documentations of introductions and/or extinctions, incorrect definition of species pools) and failed to confirm the finding that congeneric pairs that persisted differed in bill lengths more than those in which one species became extinct. They concluded that the patterns of success and failure of introductions are those one would expect on the basis of the colonization abilities of the birds and the habitats of the islands.

Other evidence bearing on the size-ratio hypothesis is even less conclusive. Consider, for example, the values of size ratios for several morphological traits among several woodpecker species that are sympatric in Malayan lowland dipterocarp forests (Table 7.1; Short 1978). The species range in body mass from 9 g to 430 g, but seven of the species weigh between 50 and 100 g; correspondingly, some weight ratios are large but most are quite small. Although all of the species have typical woodpecker bills, there is considerable variation on this theme. Species sequences on bill-size axes differ from those based on body size or tarsus length, but again the ratios vary and many are small. All of these species may occur in the same forest, but they are not all ecologically similar. Four of the species (2, 6, 11, and 12

Fig. 7.12. Size–frequency distributions of prey in the diets of *Accipiter* hawk species. A, samples taken from several North American locations (Storer 1966, MacArthur 1972); B, samples from northwestern Oregon, and C, samples from eastern Oregon (Reynolds and Meslow 1984).

Table 7.1. *Morphological size sequences and character ratios for an assemblage of 13 species of sympatric Malaysian woodpeckers for several morphological features.*

Values for males were used in dimorphic species.

Species	Body weight (g)	Ratio[a]	Wing length		Tarsus length		Bill length		Bill width		Bill depth	
			Species sequence	Ratio	Species sequence	Ratio	Species sequence	Ratio	Species sequence	Ratio	Species sequence	Ratio
1. *Sasia abnormis*	9	3.56	1	1.63	1	1.25	1	1.39	1	1.31	1	1.32
2. *Hemicircus concretus*	32	1.44	2	1.16	2	1.17	3	1.10	2	1.18	2	1.05
3. *Meiglyptes tristis*	46	1.15	3	1.05	3	1.06	2	1.10	3	1.10	3	1.09
4. *M. tukki*	53	1.25	4	1.10	5	1.07	5	1.02	5	1.06	5	1.06
5. *Celeus brachyurus*	66	1.26	5	1.07	4	1.08	4	1.12	4	1.03	7	1.02
6. *Blythipicus rubiginosus*	83	1.01	6	1.06	8	1.00	8	1.17	7	1.06	4	1.10
7. *Picus puniceus*	84	1.06	8	1.01	10	1.00	7	1.09	6	1.10	8	1.01
8. *P. miniaceus*	89	1.08	7	1.04	6	1.01	10	1.02	9	1.04	9	1.03
9. *Dinopium rafflesii*	96	1.02	10	1.06	7	1.05	9	1.05	8	1.01	6	1.06
10. *Picus mentalis*	98	1.58	9	1.11	9	1.22	6	1.27	10	1.05	10	1.19
11. *Reinwardtipicus validus*	155	1.45	11	1.43	11	1.20	11	1.27	11	1.35	11	1.12
12. *Dryocopus javensis*	225	1.91	12	1.05	12	1.07	12	1.07	13	1.09	12	1.17
13. *Mulleripicus pulverulentus*	430		13		13		13		12		13	

Notes:
[a] Larger ÷ smaller.
Source: Calculated from data in Short (1978).

in Table 7.1) engage in extended excavating and drilling, and these species are reasonably dissimilar in body and bill size. Another group of four species (numbers 3, 4, 5, and 8) are more generalized, foraging by gleaning ants and termites from the tree surface or from shallow chambers that are reached with a few light taps. These species are quite similar in morphology. The remaining species vary in their foraging habitats or habitat preferences. In eastern North America, the bill lengths of woodpeckers that feed on intermediate-sized prey (which tend to be relatively abundant) are nearly identical, whereas ratios between species feeding on less abundant very small or very large prey are greater (Woods 1984). The pattern in both woodpecker assemblages thus shows little overall adherence to the Hutchinsonian expectations, although some corroborating patterns emerge when subsets of species are considered separately.

Even when character ratios are calculated within guilds, however, the patterns are by no means always consistent with theory. Character ratios of bill size or wing length in the guild of foliage-gleaners studied by Root (1967) were generally less than 1.15. Among species within guilds of granivores (Thompson and Lawton 1983), diurnal raptors (Jaksić and Braker 1983), and grassland and shrubsteppe breeding birds (Wiens and Rotenberry 1980, Wiens 1984a), body- and bill-size ratios are variable and many are quite small; neither is at all closely related to dietary overlap among the species. In fact, among some of the grassland and shrubsteppe birds we studied, sequences based on morphology and on prey sizes are reversed – larger birds consume smaller prey, on average, then the next smaller bird species (Wiens 1984a).

In a broad comparative analysis, Schoener (1965) restricted his calculations of bill-length ratios to sympatric, congeneric bird species. Large ratios were found among some food specialists using foods of relatively low abundance or (to put it another way) among species whose body sizes are large relative to overall food abundance (e.g. hawks, owls, toucans). Most small-billed species feeding on more abundant foods had rather small character ratios. When Roth (1981) and Tonkyn and Cole (1986) plotted the overall distribution of 314 of the size-ratio values calculated by Schoener, however, both the mean and the modal values were less than the Hutchinsonian value of 1.3. There were a great many small values, and the overall distribution of values approximated that generated by a random or independent allocation of size differences among species. Moreover, ratio values decreased with increasing guild size. In communities structured according to the dictates of competition theory, one would expect this to occur only to the point of limiting similarity among guild members. Certainly, if constant size ratios do occur as some sort of 'ecological law' among certain kinds of guilds, these guilds are poorly represented in Schoener's data (Tonkyn and Cole 1986), as well as in many other studies.

Null model analyses

In analyzing size-ratio patterns, one is faced with the difficulty of determining how one recognizes a given ratio as being unusual and therefore possibly indicative of competition or some other process. Simberloff and Boecklen (1981) used a null-model approach to tackle this problem, reanalyzing Schoener's values as well as those from a number of other investigations of birds in which the values were claimed to exceed some necessary minimum and/or to be relatively constant. Very few of the individual ratio values were significantly greater than those expected from a null model based on noninteractive community assembly. May (1978), for example, stated that the 11 bird communities studied by Cody (1974a) conformed to a 1.3 ratio; by Simberloff and Boecklen's analysis, however, the null hypothesis of random and independent arrangement of species between the smallest and largest sizes observed cannot be falsified for 10 of the communities. Of the 130 ratios in Cody's data set, 120 have values of less than 1.3. Simberloff and Boecklen found only one instance (a guild of fruit- and seed-eating species from an island in the West Indies reported by Terborgh *et al.* (1978)) in which the ratios among all species in the assemblage were significantly larger than expected by their null model. Of 21 studies that claimed constancy in size ratios, only 4 actually met the criteria of constancy applied by Simberloff and Boecklen. One of these was the assemblage of New Guinea fruit pigeons described by Diamond (1975a). Simberloff and Boecklen discounted the importance of this example, suggesting that it was selected from a large number of possible assemblages of New Guinea birds and thus proves nothing about the generality of the pattern of constant or large character ratios in that avifauna. This is true, but Diamond claimed no such generality and the example is of interest in its own right.

In Schoener's (1965) analysis, large character ratios were often associated with assemblages occurring on islands, especially small ones. Grant (1968) saw this pattern to be a consequence of restrictions in the variety and abundance of resources on islands, which lead to intensified competition and stronger selection for morphological divergence between species than occurs in mainland areas. Indeed, greater character differences have been documented between some species pairs on islands in comparison with adjacent mainland locations (e.g. Grant 1968, Keast 1968, 1970, Higuchi

1980), but both the interpretation and the reality of these patterns are open to question (Simberloff 1983b). Other morphological patterns on islands are interpreted in terms of a release from competition and responses to resource abundance (see Chapter 11), so it seems that competition in one form or another can be used to explain almost any conceivable island pattern. In the absence of direct measurements of resource levels, such 'explanations' are only speculations. The studies of Galápagos finches to be discussed later in this chapter show how such speculations can be strengthened.

If competition is a mechanism of general importance in island avifaunas, this pattern of increased size ratios among island species might be expected to be a community-wide pattern, characterizing virtually all sets of ecologically similar species on islands. Strong *et al.* (1979) used a null-model approach to examine such patterns in several island groups. They assembled null communities of birds for the Tres Marias and California Channel islands by randomly selecting confamilial species from a mainland source pool so that the family composition and number of species in the null and observed island communities were the same. Ratios of bill and wing lengths were then calculated between adjacent species in the size rankings within each family, and the procedure was repeated for 100 iterations to generate a distribution of expected size-ratio values. In comparisons of the observed communities with the null assemblages, 17 of 30 bill-length ratios and 15 of the wing-length ratios for the Tres Marias were greater than the expected values. For the California Channel Islands, 12 of 20 bill ratios and 9 wing ratios were larger than expected. These proportions do not differ significantly from those expected by chance, and on this basis Strong *et al.* were unable to reject their null hypothesis that the community-wide pattern of size ratios in these island assemblages did not differ from that expected from a random and independent draw of species from the mainland source pools. Although they did not discount the possible effects of competition on the morphologies of these avifaunas, they concluded that a wide array of deterministic and stochastic factors probably influence morphology in different ways on different islands, destroying any community-wide patterns. Clearly, there were some families within which character ratios *were* greater than expected by chance, but, by seeking community-wide patterns, Strong *et al.* explicitly ignored such details.

These null-model analyses have generated considerable criticism. Hendrickson (1981) noted several errors in the data treatment and statistical tests of Strong *et al.*, reanalyzed the data using a somewhat different model, and concluded that there was evidence of community-wide increases in wing-length ratios for the Tres Marias birds. The other conclusions of Strong *et al.* regarding the Tres Marias and California Channel islands remained largely intact, prompting Strong and Simberloff (1981) to conclude that their approach and findings were vindicated; they explained away the wing-length pattern by suggesting that it could be related only tenuously to competitive interactions.

Other criticisms have centered on the structure of the null model (Grant and Abbott 1980, Harvey *et al.* 1983, Case *et al.* 1983, Colwell and Winkler 1984, Schoener 1984, Diamond 1986b, Tonkyn and Cole 1986). Strong *et al.* grouped species by family, but, as the species belonging to a single family may represent considerable ecological diversity, this categorization may dilute actual morphological patterns with values from ecologically unrelated species. Simberloff (1984) attempted to address this difficulty in a reanalysis of the Tres Marias data by using genera; for the few instances that could be considered, he found that island size ratios were not remarkably large. Strong *et al.* also constructed their null communities by a process simulating equally probable colonization from the mainland pool. This ignores important differences among species in dispersal and colonizing abilities and may produce a systematic bias against rejection of the null hypothesis (Grant and Abbott 1980, Colwell and Winkler 1984). Many of the problems with the null-model approach mentioned in Chapter 3 are apparent in these examples.

The various treatments have also differed in the size and composition of the mainland source pool used, although Simberloff (1984) suggested that variations in pool size had little effect on the patterns produced. Even so, it seems likely that so long as the source pool contains actual species, their morphology may reflect adjustments to past competitive interactions whether they be from island or mainland situations. The species pool will thus contain 'hidden structure'. Tonkyn and Cole (1986) considered this problem in some detail and concluded that the patterns predicted in any null-model analysis are strongly dependent on the form of the underlying size distribution of the species in the pool. Because this is generally not known it must be assumed. This makes it difficult to construct a realistic null model. In the absence of some set of null expectations, however, inferences based on the observed patterns alone may be of little value.

Two other attempts to analyze character differences among coexisting species by the use of null models are noteworthy. Case *et al.* (1983) analyzed data gathered by Faaborg (1982) from 15 locations on 12 West Indies islands for evidence of size differences between species within subjectively-defined foraging guilds. Null communities were generated from a pool containing all species recorded from the locations (use of an alternative mainland species pool to reduce the magnitude of hidden structure in the pool did not alter the outcome of the tests). The results (Table 7.2) represent a striking rejection of the null hypothesis – the island size distributions are significantly different from a random assortment of species and populations. The differences are also in a direction consistent with a competition argument, although Case *et al.* cautioned that this consistency itself does not prove that hypothesis. The size differences among species are also more constant than expected by chance, although ratio values show little tendency to approximate 1.3.

Schoener (1984) conducted an entirely different sort of analysis: the focus was on a restricted group (bird-eating hawks, primarily *Accipiter* species), but the species pool contained all of the bird-eating hawk species in the world. Schoener generated null distributions of wing-size ratios (following Schoener's (1965) suggestion that wing length is a more reliable indicator of feeding differences among raptors than are bill dimensions; see also Andersson and Norberg 1981) by randomly drawing pairs, trios, etc. of species from this pool. The size ratios among pairs, trios, etc. of 'sympatric' hawks (as gauged by co-occurrence in areas of unspecified size and very general habitat type) were then compared with these null distributions. Among pairs of *Accipiter* species, the observed ratios were significantly greater than expected; size ratios of less than 1.15 were notably lacking, even though 35% of the expected ratio values were that small. Schoener suggested that this absence of low size ratios might reflect a limiting similarity to the sizes of co-occurring *Accipiter* pairs. The patterns for trios were similar, although not generally statistically significant. Other tests also suggested that assemblages of bird-eating hawks often differ in size more than would be expected were they a random sample of the world's pool of such species, although the ratios are by no means constant nor do they cluster about a value of 1.3. Schoener's analysis is relatively coarse – he averaged the sizes of males and females in the many *Accipiter* species that are strongly dimorphic, used very general criteria to gauge sympatry, and did not constrain the species pool biogeographically. Because the null distribution was derived from the observed distribution rather than independently of it, the null distribution may also have contained some of the effects of the competitive processes Schoener was testing for (Tonkyn and Cole 1986). Schoener was well aware of this potential problem but could not offer a better approach. The size-ratio patterns are nonetheless of interest

Table 7.2. *The number of islands with minimum size differences between species above, equal to, and below the median for null communities with the same species number drawn in that guild/habitat.*

	Wet habitats			Dry habitats			Guild totals		
Guild	Above	Equal	Below	Above	Equal	Below	Above	Equal	Below
Frugivores	4	3	0	11	0	0	15	3	0[a]
Gleaning insectivores	4	3	0	7	2	0	11	5	0[a]
Flycatchers	5	1	0	8	0	0	13	1	0[a]
Nectarivores	5	1	0	7	1	1	12	2	1[a]
								Grand totals[b]	
Habitat totals	18	8	0[a]	33	3	1[a]	51	11	1

Notes:
[a] $P < .01$; [b] $P < 10^{-6}$.
Source: From Case *et al.* (1983).

and should prompt more detailed investigations to determine how well the patterns hold at local or regional scales and what factors might be responsible.

Gaps in size arrays

Size ratios in assemblages often stray considerably from constancy. In some instances, however, a series of species with relatively small ratios may be separated from the next larger species in a sequence by an unusually large ratio. In the grassland and shrubsteppe bird communities that Rotenberry and I studied, for example, there are several small species, a few larger ones, and no intermediates (Fig. 7.5). A typical sequence of bill-length ratios for such a community is 1.03–1.28–2.26 (Wiens and Rotenberry 1981b). There is a gap in the array, into which another species should be able to fit. For reasons that are not apparent (but which may be related to the polymodal size-abundance relationships noted by Griffiths (1986)), the regional biogeographic pool of ground/shrub-foraging species seems to lack any candidate to fill this gap. Oksanen *et al.* (1979) also noted large gaps in several guilds. A guild of dabbling waterfowl in ponds of Finnish boreal peatlands, for example, contains the Teal (*Anas crecca*, 300 g), Pintail (*Anas acuta*, 737 g), the Mallard (*Anas platyrhynchos*, 1100 g), and the Whooper Swan (*Cygnus cygnus*, 9000 g), a ratio sequence of 2.46–1.49–8.18. Given regular spacing, three species could fit between the mallard and the swan. Emboldened by some modeling, Oksanen *et al.* proposed that the gap is a consequence of aggressive resource defense by the largest species, which in open habitats is able to exclude (biogeographically!) the next smaller two or three species from the guild. In closed habitats such as marshes, such aggressive pre-emption of resources is not possible and the size sequences display no such gaps.

In the absence of firmer evidence, it is difficult to accept the reasoning of Oksanen *et al.* Some (Nudds *et al.* 1981, Simberloff and Boecklen 1981) have found it difficult to accept the pattern itself. Simberloff and Boecklen were unable to differentiate the size sequences from those generated by a random, independent assembly of species in the guilds (perhaps because sample sizes were quite small). Nudds *et al.* objected to the attempt by Oksanen *et al.* to extend their suggestion to guilds of North American dabbling waterfowl and argued that the swan is not a legitimate member of the assemblage. Among both dabbling and diving waterfowl guilds in North America, Nudds *et al.* found that body-weight ratios deviate rather little from a mean value of 1.2. There is no evidence of any 'gap', and Nudds *et al.* concluded on this basis that these arrays evolved in accordance with assembly rules that dictate maximal packing of species, possibly because of competition over limited resources. Pöysä and Sorjonen (personal communication), on the other hand, have argued that the swan is a *bona fide* member of the dabbling guild in Scandinavia but that neck length may be a better measure of feeding behavior than body weight. In the Teal–Mallard–Pintail–Swan guild, neck-length ratios are 1.67–1.02–2.57 (note that the positions of the Mallard and Pintail are reversed on this sequence). The corresponding ratios of mean feeding depths, however, are 5.05–0.91–2.17. Thus, although the morphological 'gap' is much less for neck length than for body weight, the ecological gap occurs between the smallest species and the species with the most similar neck length. The ecological relevance of differences in neck length or body weight is unclear.

Several factors may contribute to unevenness or 'gaps' in arrays. Aggressive dominance, as argued by Oksanen *et al.*, is one. The expectation of even spacing, however, is based on the assumption of a continuous resource spectrum. If there are discontinuities in the availability of resources, size ratios may be unusually large, especially in assemblages in which morphology is tightly linked to resource differences. On the other hand, Fretwell (1978) has argued that the species in species-poor guilds will be of similar, intermediate sizes when resources occur in distinct size categories. When the resources vary continuously, however, such a guild will be dominated by species of quite different sizes, creating apparent gaps. Because assemblages of *Parus* species apparently follow the latter pattern, Herrera (1981) suggested that the species were competing along a continuous resource gradient (this is a splendid example of the sort of flawed logic discussed in Chapter 2). The expectation of even spacing of species on size gradients is also based on an assumption of equilibrium, so 'gaps' might arise in nonequilibrium arrays if the local disappearance of a species is not immediately followed by invasion of a similar-sized species. In view of all these complications, unevenness in size arrays should not surprise us.

Operational problems

Part of the difficulty in deriving generalizations about size-ratio patterns results from a lack of agreement about which ratio values are interesting. Hutchinson's value of 1.3 has become a standard, but it is not a theoretical necessity, and other minimum ratio values can be derived from theory (Roth 1981, Rummel and Roughgarden 1985). In fact, a wide array of minimum ratios has been used – 1.05, 1.10, 1.14, 1.2, 1.4, 1.5, 2.0 (Simberloff 1983b, Wiens 1982) – so it is difficult to know what values would *not* be considered as conforming to expectations. Values quite close to 1.0 occur in some data sets (Simberloff and Boecklen 1981, Roth 1981, Wiens and Rotenberry 1981b; see Table 7.1). Rather than being seen as falsifications of a particular prediction, however, such observations are often either ignored or explained by recourse to partitioning on other niche dimensions (see Fig. 2.2). Null models may be used to generate expected minimum ratios against which observed values may be compared, but exactly what emerges from such analyses is dependent on the model algorithm used and how the source pool is specified. Part of the problem is that we know very little about how morphological differences scale to ecological differences among species. A rather small morphological difference might translate into considerable ecological separation in some situations, whereas a large difference in morphology might collapse into inconsequential ecological differences in others. This diminishes the likelihood that any particular ratio value will have much generality and may help to explain why such an array of values occurs in nature.

It is also obvious that statements about size ratios are dependent on just which morphological variables are selected for measurement, as the discussion of size gaps in dabbling waterfowl arrays or the tabulation of woodpecker character ratios (Table 7.1) clearly shows. There is not a great deal of agreement about which features are most relevant, and the use of almost any measure may be rationalized, as we have seen. The decision is not trivial, for ratios calculated for different characters for a set of species not only produce different spacing patterns but may alter the sequence of the species (Fig. 7.13). Characters should be selected on the basis of established correlative or functional associations with ecological features of importance, but this is rarely done. The spacing of species on gradients defined by such characters should also be analyzed in a way that includes measures of variations of the traits (e.g. Fig. 7.8). Slatkin (1980) has shown that differences in the relative abundances of species may have important effects on our ability to discern the mechanisms responsible for size differences among species, so calculations should be adjusted for densities as well. The pattern of body-mass spacing of vultures and other bird species feeding on carrion in Kenya (Anderson and Horwitz 1979), for example, changes dramatically when the relative abundances of the species are considered (Fig. 7.14).

The results of examinations of size patterns in assemblages are also sensitive to which species are included in the analysis. Moulton and Pimm (1986) found no evidence that the introduced Hawaiian bird species not suffering extinction are any more widely separated in morphological space than would be expected by chance alone when all introduced species were considered together, even though they found such patterns when they

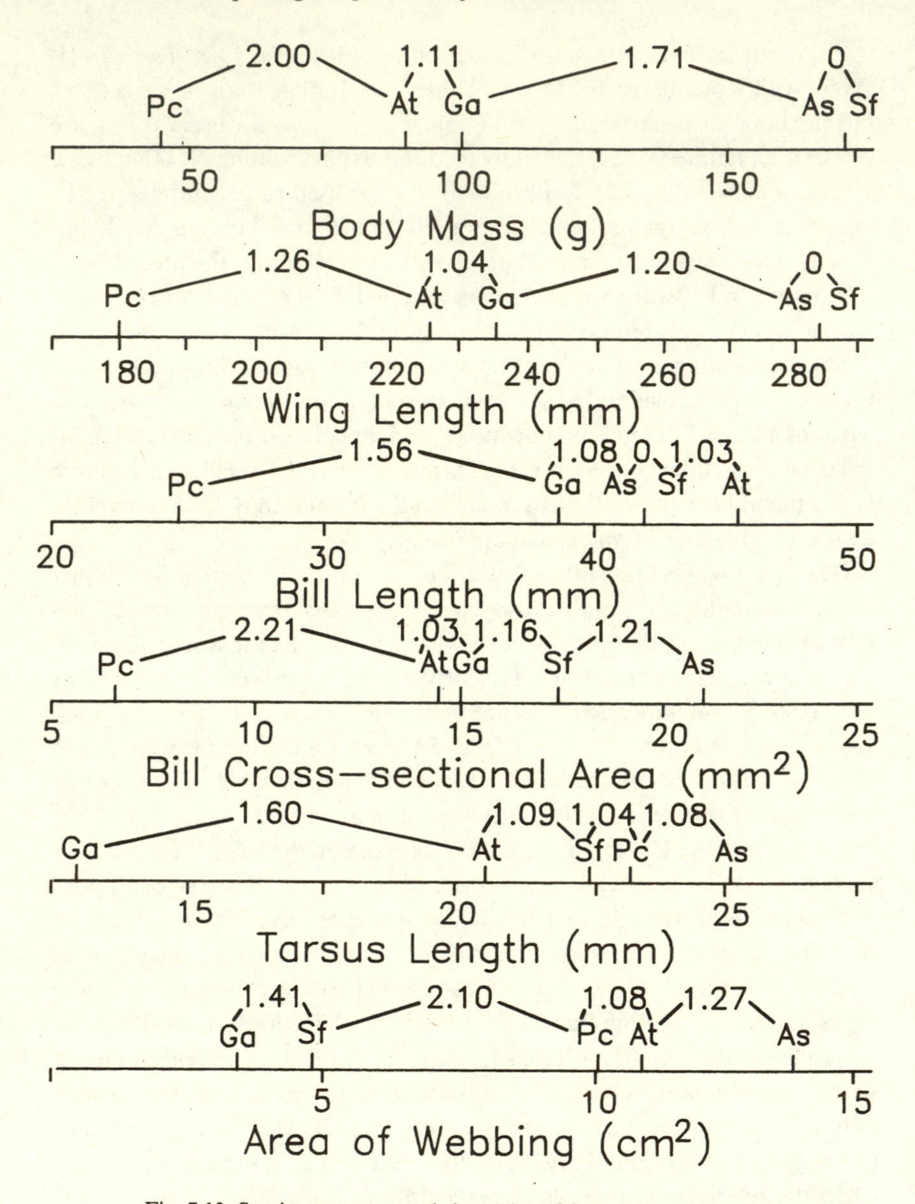

Fig, 7.13. Species sequences and size ratios of five species of tropical terns with respect to several morphological features. Values are from Ashmole (1968). As = *Anous stolidus*, At = *A. tenuirostris*, Ga = *Gygis alba*, Pc = *Procelsterna cerulea*, and Sf = *Sterna fuscata*. Bill cross-sectional area is measured at the proximal end of the gonys; area of webbing represents the approximate area of the lower surfaces of the two feet when the toes are fully spread out. From Wiens (1982).

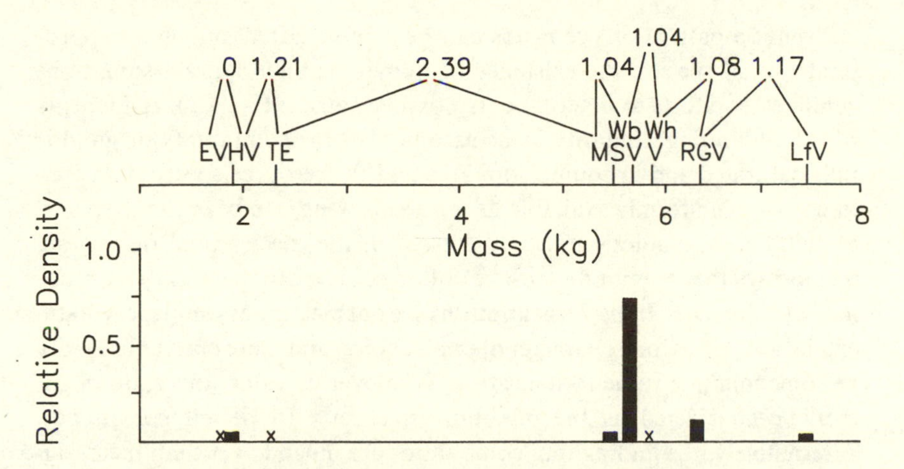

Fig. 7.14. Top: species sequence and size ratios for eight species of carrion-feeding birds in Kenya. Bottom: relative densities of these species (x = <0.01). The species are: EV = Egyptian Vulture (*Neophron percnopterus*), HV = Hooded Vulture (*Necrosyrtes monachus*), TE = Tawny Eagle (*Aquila rapax*), MS = Marabou Stork (*Leptoptilos crumeniferus*), WbV = White-backed Vulture (*Pseudogyps africanus*), WhV = White-headed Vulture (*Trigonoceps occipitalis*), RGV = Rüppell's Griffon Vulture (*Gyps rueppellii*), and LfV = Lappet-faced Vulture (*Torgos tracheliotus*). From data in Anderson and Horwitz (1979).

focused instead on congeneric species pairs. Much of the controversy over the null-model analyses of size patterns also relates to whether pairs of congeners, guilds, families, entire communities, or some other grouping is the appropriate unit for analysis. The problem, of course, is that some real patterns of size relationships among species may not be seen if 'relevant' data are scrambled together with 'irrelevant' data (Diamond 1986b). Thus, the inclusion of species that occupy different habitats or differ ecologically in other major ways may obscure patterns characterizing ecologically similar species within habitats and may not be a fair test of the size-ratio hypothesis. Moreover, some types of resources may promote size differentiation among consumers whereas others may not: insects, for example, can be captured and consumed in many ways that are unrelated to consumer size, but fruit consumption may be closely linked to frugivore gape size (Fig. 7.4). Thus, obligate herbivorous birds in New Guinea lowlands appear to segregate by body mass, but carnivores and omnivores do not (Bell 1984). There are differences among guilds in the likelihood of ecologically meaningful size differences. Judging which data or species are 'relevant' or 'irrelevant' to morphological analyses of course rests on a sound knowledge of the natural history of the species.

Even if a pattern of size ratios can be demonstrated among an appropriate set of species, the existence of numerous underlying assumptions complicates efforts to determine its possible causes (Fig. 2.2). Arguments based on limiting similarity ideas assume that the community or guild is fully saturated and in equilibrium and that the resource spectrum is continuous and uniformly available. If species packing is to be optimal, species of suitable morphology must be present in the species pool or existing members of the community must be able to coevolve the necessary morphological differences. If these assumptions are not met, the assemblage will still exhibit size ratios for a character of one's choice, and there must of necessity be some minimal value to the ratios. Whatever its value, this ratio might then be considered as the minimum necessary to permit coexistence. Alternatively, the finding that some ratios are much lower than the designated minimal value permitting coexistence is often explained by arguing that some other, unmeasured difference between the species *is* large, permitting the limits to size similarity to be transgressed. Or suppose that one were to find all members of a rigorously defined foraging guild to be separated by bill-length ratios of exactly 1.3. This would seem to be a textbook example of a Hutchinsonian ratio, and the conclusion that it reflects a spacing produced by competitive interactions would be tempting. It might even be correct. But all that has been documented at this point is a pattern, and, although the pattern may suggest a process hypothesis, it does not test it. That must be done separately, and it is difficult to conceive of such tests for size-ratio patterns. Finally, the possibility exists that size-ratio patterns, while real, are artifacts of the statistical or mathematical properties of systems rather than direct consequences of biological processes (Horn and May 1977, Eadie *et al.* 1987).

Character displacement

Character displacement is a specific variation on the theme of morphological differentiation between coexisting ecologically similar species. Brown and Wilson (1956) coined the term to describe the situation in which two species that are quite similar where they are allopatric diverge evolutionarily where they occur sympatrically. Grant (1972) redefined the term to apply to the *process* of evolutionary change in the morphology of a species due to its co-occurrence with one or more species that are ecologically and/or reproductively similar to it. Grant's definition thus includes both divergence and convergence in morphology between the species as consequences of character displacement. Here, following Arthur (1982), I will treat divergence (=displacement) and convergence separately.

For the purposes of this discussion, then, character displacement refers to a change in the morphology of one or more species that reduces their morphological similarity among them where they coexist. It is most readily recognized by a pattern of greater morphological divergence in sympatry than in allopatry. For example, two warbler species (*Dendroica pinus* and *D. dominica*) occur sympatrically in southeastern United States but are allopatric in the northeast and southern Florida (*pinus*) and in the midwest (*dominica*) (Ficken *et al.* 1968). Bill lengths in allopatric populations of the species are almost identical. In the zone of sympatry, bill lengths of *pinus* do not change, but *dominica* bills are much longer; the bill-length distributions of the two species do not overlap. Ficken *et al.* associated the longer bill of *dominica* with an increased frequency of feeding from pine cones.

Another example concerns the populations of chaffinches that occur on the Canary Islands (Grant 1979a). Only one species, *Fringilla coelebs*, is found on the adjacent European and African mainlands, as well as on the Azores. Island populations have diverged considerably in morphology from their mainland counterparts. Another species, *F. teydea*, occurs on the Canaries. Where the two species co-occur, on Tenerife and Gran Canaría, they occupy separate habitats, *coelebs* the broad-leaved and tree-heath forests at intermediate elevations and *teydea* the pine forests at high elevations. Only *coelebs* occurs on La Palma, Hierro, and Gomera, where (as on the mainland) it occurs in pine forests. Where the two species are sympatric, *coelebs* does not feed on pine seeds, in contrast to its behavior elsewhere in the Canaries. The species differ morphologically where sympatric as well, *coelebs* having a longer, thinner bill and *teydea* having a deeper bill, which presumably enables it to feed more efficiently on pine seeds.

This is the sort of pattern that is consistent with character displacement. When Grant compared bill dimensions of *coelebs* on the two islands where *teydea* is present with those on the three islands where it is absent, however, he found no consistent differences. He nonetheless concluded that character displacement was involved at least in part in producing the present differences between *coelebs* and *teydea*, basing his conclusion on two observations. First, selection for a relatively long, thin bill may not have been reversed on islands lacking *teydea* because the bill shape is adequate for cracking pine nuts. Second, populations of *coelebs* occurring on the Azores are morphologically similar to *teydea* on the Canaries, suggesting that *coelebs* may have diverged from that pattern after colonizing the Canaries, where *teydea* was already established. This argument supposes that the morphology of *coelebs* where it now occurs in the Canaries in the absence of *teydea* reflects evolutionary responses to earlier contact with *teydea*.

Grant (1972) carefully reviewed much of the evidence for character displacement and found nearly all of it unconvincing. The basic problem, as in the Canary Island chaffinches, lies in determining that the character states of a species where it co-occurs with another similar species are any different from those that would be predicted from patterns of character variation in the species elsewhere. Many species vary clinally in morphology (e.g. James 1970), so differences between species where they occur sympatrically might simply be an extension of independent clinal responses to environmental gradients rather than a response to the presence of the other species (Fig. 7.15). Grant suggests that this possibility can be tested by determining separate regressions for the sympatric and allopatric samples of a species; if the slopes and intercepts do not differ significantly, there is little to support the argument for character displacement. It is possible, of course, that the character differences between the species arose from competition in sympatry and later spread by gene flow into the allopatric areas, producing the even clines. On the other hand, an abrupt change in clinal variation might occur as a consequence of the interplay between gene flow and selection in the absence of another species or of an abrupt environmental change (Endler 1977). Such abrupt character changes therefore may not indicate character displacement unless they coincide with the region of sympatry (Grant 1972, Simberloff 1983b).

Grant (1972, 1975) applied the clinal variation argument to a reexamination of a frequently cited example of character displacement, the rock nuthatches *Sitta neumayer* and *S. tephronota*. The species are ecologically quite similar and occur sympatrically in Iran and adjacent areas in Iraq and the USSR (Fig. 7.16). Brown and Wilson (1956) noted that the species differ markedly in bill size and the prominence of a black eye-stripe where they are sympatric, whereas allopatric populations of *neumayer* in Yugoslavia and Greece are almost identical to *tephronota* from the eastern portion of its allopatric range. In fact the species do differ markedly in bill length in the zone of sympatry, but this difference appears to be largely a consequence of clinal variation in *tephronota* over its entire range (Fig. 7.16). Grant found inconclusive indications of character displacement of body sizes of the species in sympatry and suggested that some of the difference in bill sizes might reflect allometric consequences of selection on body size. The foraging behavior of each species does not differ in sympatry and allopatry, although in sympatry the two species exploit seeds of different sizes. The character displacement in eye-stripe boldness is apparently a real pattern, which Grant's field experiments indicated is associated with mate recognition. Grant concluded that the adaptive changes that characterize the species in allopatry are sufficient to permit their ecological

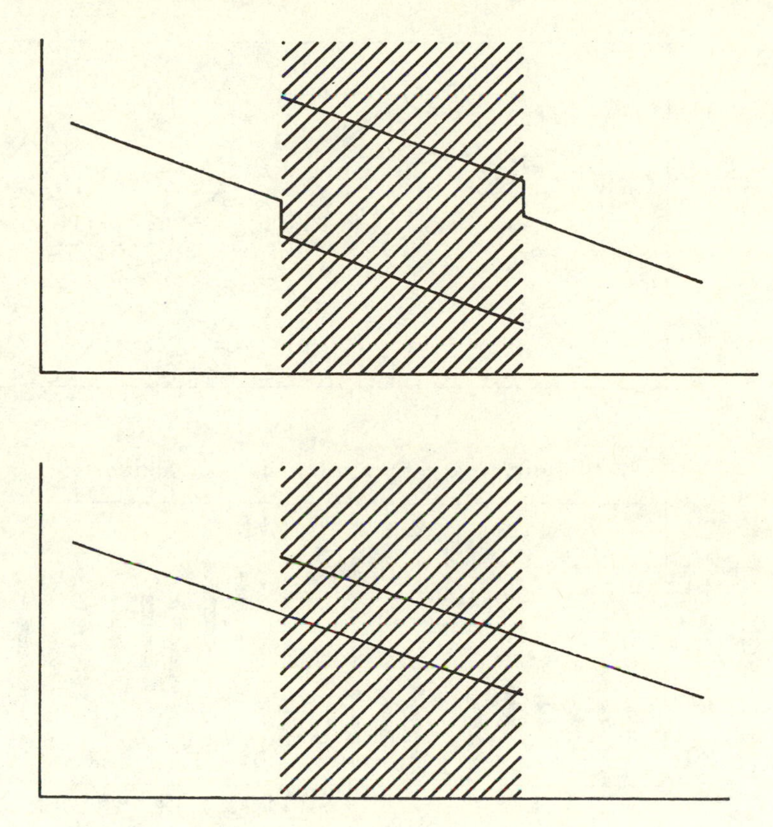

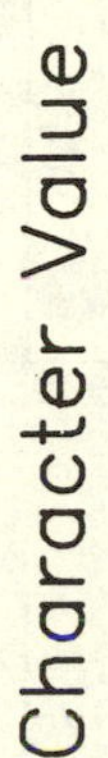

Fig. 7.15. Hypothetical example of geographical variation in morphological character states of two species that do (above) or do not (below) exhibit character displacement in sympatry (hatched area). The sharpness of the step in the cline is exaggerated; a less sharp step would be expected in nature as a consequence of gene flow across the sympatry–allopatry border. Modified and corrected from Grant (1972).

coexistence in sympatry, although displacement to reduce reproductive confusion has occurred.

The example of the *Dendroica* warblers given above can be explained in much the same manner (Grant 1972). Bill size in *dominica* varies clinally from south to north, and populations in coastal pine and mixed forests also tend to have larger bills than those in inland deciduous forests. Both of these patterns predict that *dominica* with the largest bills will occur in the area in which Ficken *et al.* (1968) found character displacement, so the pattern may

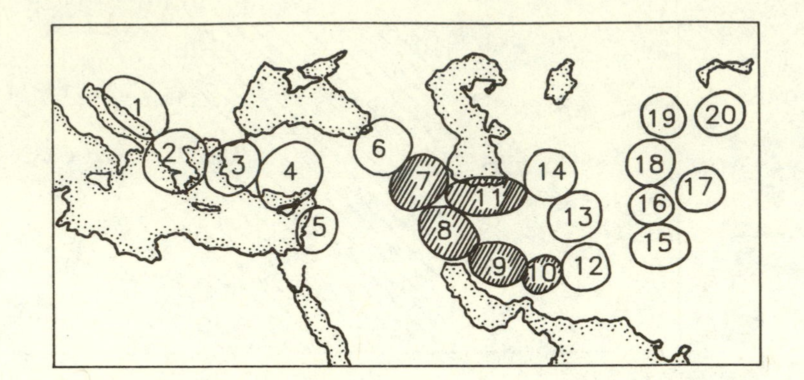

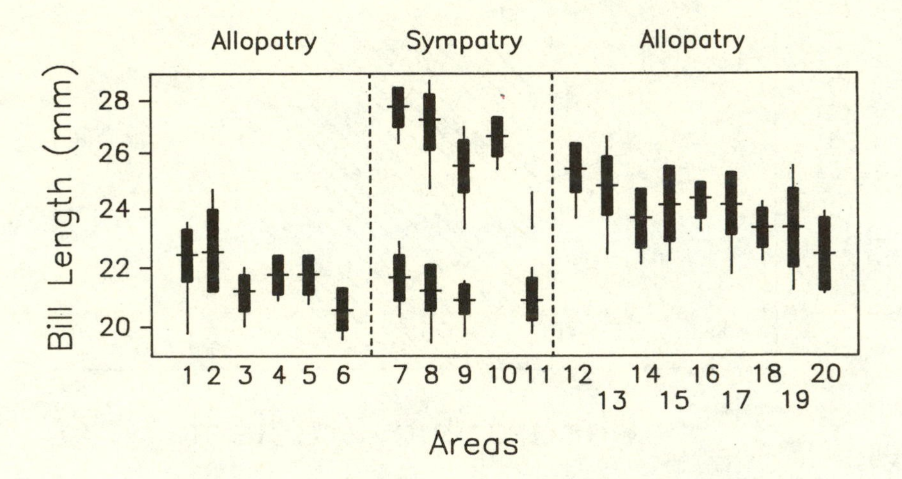

Fig. 7.16. Above: Areas from which populations of *Sitta neumayer* and *S. tephronota* were sampled in Grant's (1975) study. The area of sympatry is hatched. Below: Bill-length values of *neumayer* (lower, areas 1–11) and *tephronota* (upper, areas 7–20) for adult males. The horizontal line indicates the mean, the solid bar one standard deviation on either side of the mean, and the vertical line the range. Modified from Grant (1975).

simply be a consequence of variation in *dominica* that is related to geography and habitat but is unaffected by the presence or absence of *pinus*. Ficken *et al.* were aware of this difficulty, but their attempt to test its affects was inconclusive. As in the rock nuthatches, the differences in bill sizes between the species can be explained without recourse to character displacement

The situation among species of grebes, however, is not so easily explained. On a world-wide basis, grebes that occur in the absence of other grebe species exhibit a strong tendency to converge toward a single 'all-purpose' bill shape, but where two or more species occur sympatrically the bills diverge from this shape, often in opposite directions (Fjeldså 1983). Several pairs or triplets of species that occur in local sympatry show bill-length patterns suggestive of character displacement, but the most convincing case involves four species of Andean grebes (Fjelsdå 1983). The White-tufted Grebe (*Rollandia rolland*) and the Silvery Grebe (*Podiceps occipitalis*) are both widespread and often co-occur; their bills are similar in length but differ in depth, and dietary overlap between the species is generally low. In the area of Lake Titicaca, a third, endemic species (the Titicaca Grebe, *R. microptera*) occurs and Silvery Grebes are absent. There is a slight reduction in the bill length of White-tufted Grebes only in this area. Farther north, at Lake Junín, another large, endemic grebe (the Junín Grebe, *P. taczanowskii*) is present together with the White-tufted and Silvery grebes. There, *R. rolland* is generally stouter and has a longer, more powerful bill than elsewhere. Silvery Grebes, on the other hand, are remarkably short-billed, and this morphological change is restricted to the localized area of sympatry with the Junín Grebe. Fjeldså's analyses of the stomach contents of these species showed that even slight differences in bill lengths among the species in the areas of sympatry at Titicaca and Junín were associated with clear changes in the degree of interspecific diet overlap.

This example and that of the Galápagos finches to be discussed later in this chapter illustrate that character displacement may indeed occur in bird communities. Given the attractiveness of the idea and its frequent appearance in textbooks, however, it is remarkable that there are not many more relatively unambiguous examples. To some degree this reflects the difficulty of documenting character displacement: one must establish that the differences between species in sympatry cannot be predicted from their states in allopatry (which should be based on several transects from allopatric to sympatric areas), are related to overlap in resource use, and are evolved characteristics of the populations (which requires identifying the original and derived populations and the precontact character states) (Grant 1972, Arthur 1982). Beyond this, character displacement may operate only in rather restricted circumstances. If one species is established in an area and the invading species is nearly identical in ecology (and morphology), competitive exclusion of one or the other species may be almost inevitable (Fig. 7.17). If the invader is quite different from the resident, no modifications may be required to permit coexistence. Only in a narrow range of intermediate values of difference between invader and resident is the outcome likely to be in doubt; here, character displacement may affect that outcome by reducing the ecological overlap between the species (Grant

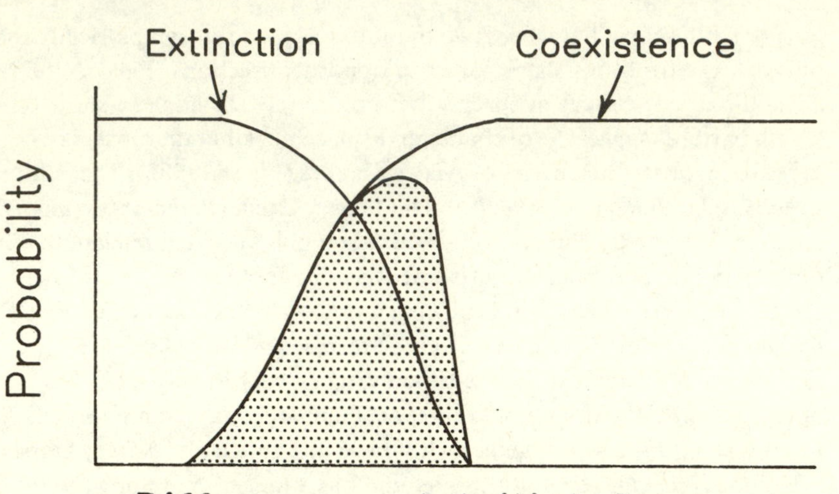

Fig. 7.17. A hypothetical model of the probability of character displacement as a function of the difference between two species when they establish contact. If the species are very similar, the probability of extinction of one or the other is high, whereas if the species are quite different, coexistence is likely. The stippled area indicates the region in which character displacement leading to coexistence is likely to occur. After Grant (1975).

1975). For this to happen, however, small differences in morphology must translate into important differences in ecology. Fjeldså (1983) suggests that this may be likely only in relatively productive, stable habitats. All of these arguments, of course, presume that the community is in a competitively determined equilibrium, which may not be the case.

Character convergence

When a resource-limited community is invaded by another ecologically similar species, there are conditions in which some of the species might converge morphologically or behaviorally. Suppose, for example, that a community containing a large species that uses the top end of the prey-size spectrum is invaded by a slightly smaller species. Evolutionary divergence by the large species is prohibited by the absence of any larger prey, but the overlap between the species is not so great that either species is excluded. In this situation, the larger species may converge morphologically toward the next smaller species (Schoener 1970). Such patterns are found in some arrays of *Anolis* lizards on Caribbean islands (Roughgarden 1986), but I know of no examples among birds.

Another route to character convergence is behaviorally mediated. Cody (1969, 1973a, 1974a) proposed that when two species of very similar morphology and ecology establish local sympatry, each must forage over large areas to obtain sufficient food. This is an inefficient arrangement, and selection may favor the reduction of spatial overlap through the exclusion of individuals of the other species as well as conspecifics from territories. Such interspecific territorial defense will be facilitated by convergence in the signals used by each species in territorial defense. Moynihan (1968) suggested that a similar convergence in signalling behavior or morphology might occur among species that form interspecific feeding flocks.

Cody supported his proposition with several examples, including two species of *Thryothorus* wrens that appear to sing convergent songs and avoid interspecific overlap where they co-occur in Mexico and a group of three species of towhees that are sympatric in some areas of central Mexico. The towhee situation was described by Cody and Brown (1970), who studied the species in one location in Oaxaca over a 9-day period. There, two species, *Pipilo ocai* and *Atlapetes brunneinucha*, are extremely similar in appearance, while a third species, *P. erythropthalmus*, is distinct from the other two but sings a song that Cody and Brown had difficulty distinguishing from that of *ocai*. On the basis of their observations of the movements of unmarked pairs of towhees, Cody and Brown concluded that the species that converged in either appearance or song overlapped spatially very little, whereas the species that were not convergent held overlapping territories. No details of the ecological similarity of the species were reported.

This pattern is nicely consistent with the character convergence hypothesis, but it may not be an accurate characterization of the relationships between the species. Murray and Hardy (1981) spent 6 weeks mapping the positions of marked birds and recording their songs in the same location and found that the species occupied completely overlapping territories and sang distinctive songs. There was little evidence that the birds responded to the songs of the other species. *Pipilo ocai* was substantially larger than the other two species, and all of the species differed in foraging behavior. In short, there was no evidence of character convergence, and ecological differences between the species were considerable.

The example of song convergence between the sympatric *Thryothorus* wrens also breaks down on close examination (Brown and Lemon 1979), as do most other instances of apparent convergence in signalling cues (Brown 1977). Interspecific territorialism between Eastern and Western meadowlarks (*Sturnella magna* and *S. neglecta*) in areas of sympatry is apparently accomplished in the absence of song convergence, although there is weak evidence of convergence in plumage pattern and coloration. The *Empidonax* flycatchers studied by Johnson (1966) also defend territories interspecifically in their small zone of sympatry, but the appearance and behavior of these species do not differ from those of either species in allopatry, probably as a consequence of their close relationship. The spatial exclusion is associated with occupancy of different habitat patches in a local mosaic, in which the birds forage in quite different places and ways. Among other species in which interspecific territorial exclusion may be a manifestation of competition (e.g. the blackbirds studied by Orians and Collier (1963) and Miller (1968)), the behavior is accompanied by no apparent convergences. Murray (1971) offered entirely different arguments to explain the occurrence of interspecific territoriality among birds. The case for Cody's hypothesis is weak.

These examples relate to character convergence that is presumably associated with the presence in the same area of another ecologically similar species – a subset of Grant's (1972) 'character displacement'. Convergence might also be expected to occur if species evolutionarily respond to the same set of resources or environmental conditions in similar ways. Thus, the unusual degree of morphological similarity among temperate North American hummingbirds may be related to their use of a suite of specialized flowers that themselves have converged toward similar morphologies and nectar production rates (Brown and Kodric-Brown 1979). Most examples of apparent convergence to similar environmental circumstances involve comparisons of communities in widely separated regions; these are treated later (Chapter 12), as is the somewhat related pattern of character change in the absence of a presumed competitor (Chapter 11).

Multivariate approaches

The emphasis in much of this chapter is on patterns within guilds. The one attempt to discern community-wide patterns of morphology that we have considered (Strong *et al.* 1979) was flawed and inconclusive, partly because it sought to document one specific pattern among all species in the community. Less restrictive explorations of morphological patterning at a community level have relied upon multivariate statistical procedures to synthesize complex variation in many morphological attributes into a small number of interpretable dimensions. Like other ecomorphological investigations, these studies rest on the premise that morphological relationships closely mirror ecological patterns, but the morphological patterns may be of interest even if their relationship to ecology is uncertain.

Karr and James (1975) initiated this approach to the study of ecomorphological patterns in communities by using canonical correlation analysis to document the relationships between a matrix of information on food habits and foraging behavior and a morphology matrix for assemblages of birds occurring in a variety of habitats in Liberia, Panama, and Illinois. Six significant canonical variates emerged, reflecting patterns such as a tendency for birds with long, thin bills and small bodies to forage by hover-gleaning and for ground-feeding species to have relatively long legs and narrow bills.

These relationships are not unexpected, and it might be argued that multivariate procedures are no more than a fancy way of stating the obvious. By determining how species are positioned in a multivariate space, however, it may be possible to determine whether or not their relationships are in any way remarkable. It was in this spirit that Ricklefs and Travis (1980) used Cody's (1974a) information on 11 bird communities of open habitats to determine how coexisting species differ in overall morphology. They applied Principal Components Analysis to a covariance matrix of the logarithms of eight morphological measures (obtained from the literature or museum specimens). The components that emerged reflected variation in features such as overall size (including some allometric shape features), the relative length of legs, bill dimensions, and the ratio of bill depth to wing length. Species were positioned in the space defined by these dimensions, and Euclidean distances were calculated for species pairs. For each community, the density of species packing in this space can be expressed by the mean nearest-neighbor distance (NND). These values were compared with NND's determined for random communities of the same size generated (a) by drawing species at random from the 'pool' of 83 species for all 11 communities or (b) by randomly determining the position of 'synthetic' species in the PCA space and calculating their Euclidean distances.

For 8 of the 11 observed communities the mean NND's fall below the values expected on the basis of noninteractive, random community assembly; that is, the species are more closely packed in morphological space than expected (Fig. 7.18). This distribution, however, does not differ significantly from that of chance departures from the expected values, so there is no evidence of an overall pattern in the morphological structuring of these communities that is especially remarkable. The larger communities, however, occupy a larger volume in the morphological space than do smaller communities, and the additional species frequently occur in extreme positions on the minor PCA components. In other words, all of the communities seem to have a set of species that occupies a common morphological space defined by the first 3–4 components, and additional species are added to the

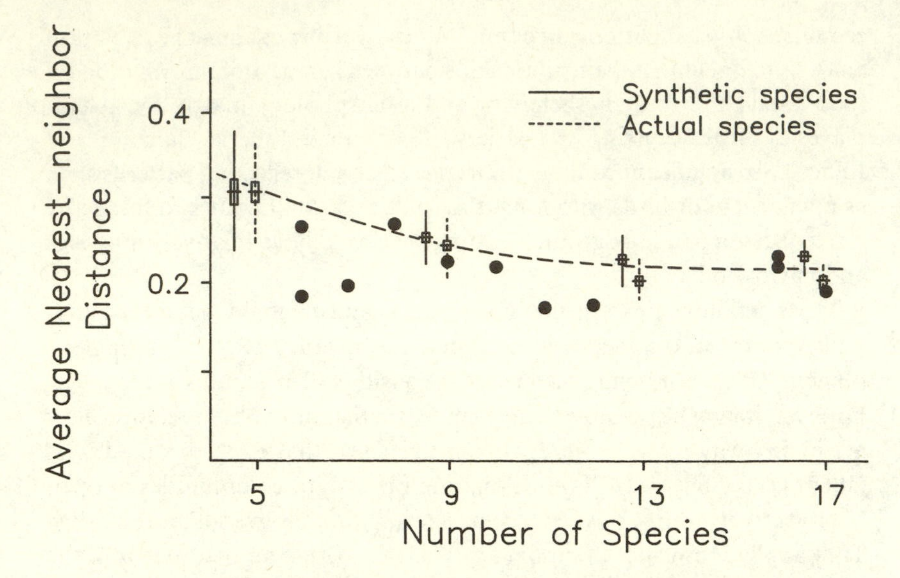

Fig. 7.18. The relationship between the number of species in 11 scrub-habitat communities studied by Cody (1974a) and the average nearest-neighbor distances among the species in multivariate morphological space. Solid circles indicate the observed community values. Random communities were generated in two ways, using (a) a pool composed of the 83 species actually observed, and (b) a pool containing species whose morphology was synthesized by combining separate measures from the observed species. For the random communities, the vertical lines show the standard deviations and the boxes the standard errors ($N = 20$). After Ricklefs and Travis (1980).

community on peripheral morphological dimensions rather than by increased packing in the 'core' of the morphological space. Karr and James (1975) likewise found that much of the increase in diversity in tropical versus temperate communities is associated with an expansion of the morphological space occupied by the communities, and the same pattern emerges among West Indies bird communities of different sizes (Travis and Ricklefs 1983). Findley and Wilson (1982) discussed this pattern with reference to bat community structuring.

Much the same procedures have been used to discern patterns in several other communities. Miles and Ricklefs (1984), for example, used canonical correlation analysis to determine whether or not the relationships among 19 hardwood-forest bird species in morphological space corresponded with their relative positions in a space defined by features of foraging behavior. Overall, 60% of the variation in the morphological features explained 55% of the variation in behavior. There are thus some relationships (e.g. between substrate use and tarsus and middle toe lengths), but a good deal of the morphological variation seems to be unrelated to foraging behavior. In several Arizona habitats, Miles (MS) found no relationship between mean NND in morphological space and the number of species in a community. Most habitat transitions on the elevational gradient he studied were not associated with major changes in the distribution of species in the morphological space, although there was a significant shift in community morphology at the oak–pine transition. In communities of quite different taxonomic composition in five neotropical locations, the volume of the overall community morphological space is quite similar, suggesting that morphological variation in these assemblages follows many of the same themes in different areas (Travis and Ricklefs 1983). Multivariate procedures were also used by Niemi (1985) to examine patterns of ecomorphological convergence between continents; this example will be considered in Chapter 12.

Because large numbers of variables are considered simultaneously when multivariate procedures are used, it may be possible to discover ecomorphological patterns that would not be apparent from single-factor analyses. There is little agreement, however, on the details of exactly how such multivariate investigations should be conducted (e.g. Ricklefs and Travis 1980, James 1982, Miles and Ricklefs 1984, James and Boecklen 1984, Bookstein *et al.* 1985, James and McCulloch 1985). Concerned that this situation might lead to a preoccupation with the machinery of multivariate analysis, Leisler and Winkler (1985) have emphasized instead the need to study a broad range of communities to begin to determine the generality of the patterns that have been found. Regardless of how this approach to ecomorphological analysis develops, the studies described here indicate that the relationships between morphological variation and features of ecology or behavior at a community level are complex and not invariably tight. This should provide a caution against making inferences uncritically from simple single-factor analyses, especially if the morphology–ecology linkage has not been substantiated.

A case study: the ecomorphology of Galápagos ground finches

The six species of ground finches of the genus *Geospiza* on the Galápagos Islands comprise perhaps the most intensively studied guild of birds in the world. Lack's (1947) monographic treatment of the group established a way of thinking about both evolution and ecology whose impact continues to be felt (Ratcliffe and Boag 1983). Much of Lack's emphasis was on the morphology of the birds. He noted, for example, that the *Geospiza* species coexisting on an island are regularly spaced along a bill-size axis with little overlap. Individuals of one species, *fortis*, have a larger bill size on islands that are shared with the smaller *fuliginosa* than do individuals on islands where it occurs alone (Grant (1972); Boag and Grant (1984a) considered this an example of competitive release, and it will be considered separately in Chapter 11). Lack interpreted such size patterns as evidence of competition for food, produced either through competitive exclusion or evolutionary character displacement. More recently, Grant and his colleagues (Grant 1986a) have spent over a decade studying the ecology and behavior of this group in detail, also emphasizing morphology. These studies provide a foundation for considering many of the points made previously in this chapter with regard to a single group of species.

Bill morphology, diet, and selection

The premise that morphology bears a close relationship to ecology can be substantiated for many members of the group. Bill features, especially depth, determine the size and hardness of seeds that can be cracked, and variations both within species (Bowman 1961, Grant *et al.* 1976, Grant 1981) and between species (Abbott *et al.* 1977, Grant 1981, Schluter 1982a, Schluter and Grant 1984) are associated with differences in feeding efficiency and diet. Larger species (*magnirostris*, *conirostris*) are about as efficient as smaller species (*difficilis*, *fuliginosa*) in feeding on small seeds but are much more efficient with larger seeds. Their diets are more diverse than those of smaller species and include both small and large seeds. Whereas small species can subsist entirely on a diet of small seeds, however, the greater metabolic demands of the larger species require that they consume larger seeds as well. One species, *scandens*, specializes on fruits and seeds of *Opuntia* when they are available (Abbott *et al.* 1977, Grant and Grant 1981).

Populations differ in morphological variability (Grant *et al.* 1985), which suggests that these differences might be associated with differences in niche breadth or abundance. Populations of *fortis* are more variable in morphology on Santa Cruz than on Daphne, and within these populations different phenotypes distribute themselves into different habitat patches, select different types of seeds, and exploit them with different efficiencies (Grant *et al.* 1976). Moreover, there is a tendency for larger species to be more variable than smaller species in both morphology and diet (Grant *et al.* 1985). More abundant populations are no more variable in morphology, however, than are sparse populations (Abbott *et al.* 1977).

Variations in bill dimensions within populations of several species are largely heritable. Heritabilities of bill length, depth, and width estimated by offspring–midparent regression all exceed 0.50 (Boag and Grant 1978,

Boag 1983, Grant 1983a, Price *et al.* 1984a). There is little evidence of direct environmental effects on these aspects of phenotypic variation (Boag 1983, Grant 1983a).

Given the above findings, it is not surprising that variations in food supplies may be accompanied by selection on bill morphology and phenotypic changes in populations. During a period of drought on the island of Daphne in 1977, seed abundance was severely reduced, and small seeds were especially scarce (Fig. 7.19). Mortality in the population of *fortis* on the island was great, and birds (especially males) with large body and bill sizes were selectively favored, apparently because they were able to crack the large, hard seeds that remained available through the drought (Fig. 7.19) (Boag and Grant 1981, 1984a, b). Similar but less severe episodes of selection occurred on Daphne in 1980 and 1982. During all three periods of high mortality, bill depth and body size were both under direct selection to increase, but bill width was directly selected to decrease (Price *et al.* 1984b). Bill shape as well as size is therefore subject to selection. During the 1977 drought, populations of *scandens* on Daphne were subject to stabilizing selection, as variances in bill dimensions were reduced (Grant and Price 1981, Price *et al.* 1984b, Boag and Grant 1984a). Overall, these findings suggest that, of the morphological features of the finches, bill depth may be especially important in feeding ecology. Studies of *fortis* on Daphne and of *fuliginosa* and *conirostris* on Espanola (Price 1984a, b; Downhower 1976) indicate that sexual selection may also contribute to larger body size in males, while natural selection may favor smaller females, which breed earlier, especially in dry years. On Genovesa, an association between male bill dimensions, song types, and mating patterns that changes between dry and wet years has been documented for *conirostris* (Grant and Grant 1983, B. Grant 1985). There is thus a reproductive component to selection on morphology as well.

Morphological patterns among coexisting species

Different numbers and combinations of *Geospiza* species occur on different islands, and analyses conducted by parties of contrasting initial persuasions (Abbott *et al.* 1977, Grant 1981, 1983b, Grant and Abbott 1980, Alatalo 1982a, Simberloff and Connor 1981, Simberloff 1983a, 1984, Case and Sidell 1983) all point to the conclusion that a large part of this pattern departs from an expectation based on random assembly of species from the archipelago pool (see Chapter 4). To what extent are patterns in morphology associated with the distinctive distribution of species among islands?

This question has been approached in several ways. One emphasizes the

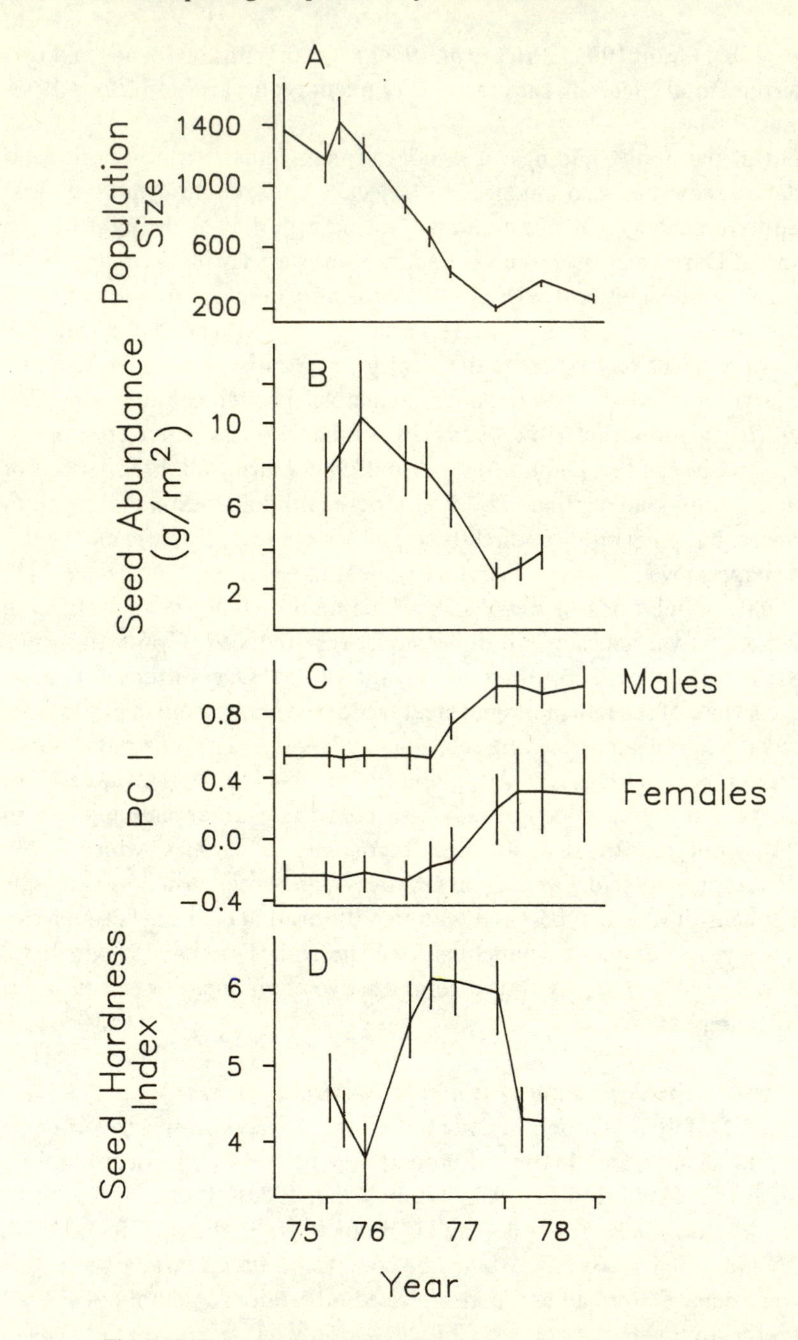

Fig. 7.19. Temporal changes in finch numbers, seed abundance, morphology, and average seed size on Daphne Major, Galápagos Islands, during a drought in 1977. A, population estimates (mean ± 95% confidence intervals); B,

combinations of species that do *not* occur together anywhere in the archipelago. There are two such combinations, *conirostris–fortis* and *conirostris–scandens*. Grant (1983b) and Grant and Schluter (1984) analyzed the morphological relationships between these and other species pairs and concluded that the missing species combinations are morphologically more similar than expected by chance and are more similar than are pairs of species that do occur together. Simberloff (1984) challenged this conclusion, finding that the bill dimensions for these species pairs were not statistically more similar than for random, independent pairings of species. Analytical differences between these studies may be responsible for the differences in results, and Grant and Schluter's arguments are the more convincing.

Another approach focuses on differences among species that are sympatric on islands. Strong *et al.* (1979) used a null model to test the hypothesis that coexisting finches are unusually different in bill structure (Abbott *et al.* 1977, Lack 1947). Species and then island races of those species were randomly drawn from a pool containing all of the species and races found in the archipelago to create an assemblage of the same size as that observed. Character ratios between adjacent species on morphological axes were then determined, and the distributions of such ratios in 100 iterations of the null model were compared with the character ratios actually observed. For minimum bill-depth ratios, 9 of 15 tests yielded observed ratios larger than those expected on the basis of random community assembly; tests for other morphological ratios produced similar results. Ratios on some islands thus *are* unusually large (Simberloff 1983b), but on the community-wide level the proportion of large values does not exceed that expected by chance. Strong *et al.* also found some evidence of character convergence, in that there is a tendency for the finch species occurring on an island all to have longer or shorter bills than would be expected from a random draw of the races of those species in the entire archipelago. They suggested that this might reflect a response to differences in habitats and food supplies between the islands (Bowman 1961).

A corrected reanalysis of these data produced much the same results, although a significant community-wide pattern for bill-length ratios to be

Fig. 7.19. (*cont.*)
estimates of the abundance of seeds eaten by the finches (mean ± standard error); C, mean (± standard error) of scores on the first component of a PCA based on finch morphology, for birds alive in each sample period; D, estimates of an average seed-hardness index (mean ± standard error) for seeds available in each sample period. Modified from Boag and Grant (1981).

greater than expected was evident (Hendrickson 1981; Strong and Simberloff (1981) disputed this finding). A somewhat different null model approach indicated that co-occurring species are more dissimilar in bill depths than would be expected by chance (Alatalo 1982a). Case and Sidell (1983) used yet another null model to support their contention that the species assembled to form communities on the islands represented nonrandom subsets of the overall pool but that there was little evidence of character divergence among species coexisting on a given island. Grant (1983b; Grant and Schluter 1984) subjected several morphological measures of the birds to principal components analysis and then calculated the Mahalanobis distances in the PCA space between all pairs of size neighbors on an island. The observed distances were then compared with the mean distance between all possible pairs of species on islands containing different numbers of species. For all four classes of islands considered, the observed mean distances are greater than expected, significantly so for two- and four-species islands. Moreover, the average distance between sympatric species pairs of *Geospiza* is greater than expected for 11 of the 13 observed species pairs (Fig. 7.20). The pairs that do not differ from expected (*difficilis–fuliginosa* and *difficilis–fortis*) are the two morphologically most similar pairs (although Strong and Simberloff have obtained a similar result using less complete data for *magnirostris–fortis* (Simberloff 1984)). The tendency for bill-depth ratios to decrease in size with increasing species number on an island (Abbott *et al.* 1977) is not supported by the Mahalanobis distance analysis (Simberloff 1984).

All of these analyses are sensitive to the hidden-structure bias, in that they are based on comparisons of observed patterns with those generated by various random reassortments of the same species and populations. This places a severe restriction on any tests of 'null' hypotheses for the Galápagos finches. To circumvent this difficulty, Simberloff (1984) proposed that the tests are not intended to address the role of competition in determining the features of the species or races in the pool, but instead to inquire whether or not the patterns among *coexisting* races and species offer evidence of competition above and beyond that already contained in the pool. This is a different claim than was originally made (Strong *et al.* 1979).

Schluter and Grant (1984) avoided this problem by approaching the question of morphological patterns among coexisting finches in an entirely different way. Using information on the relationship between bill depth and the hardness and size of seeds preferred by the primarily granivorous finches (excluding *scandens* and, for some islands, *difficilis* and *conirostris*), they determined the range of seeds that might be consumed by a hypotheti-

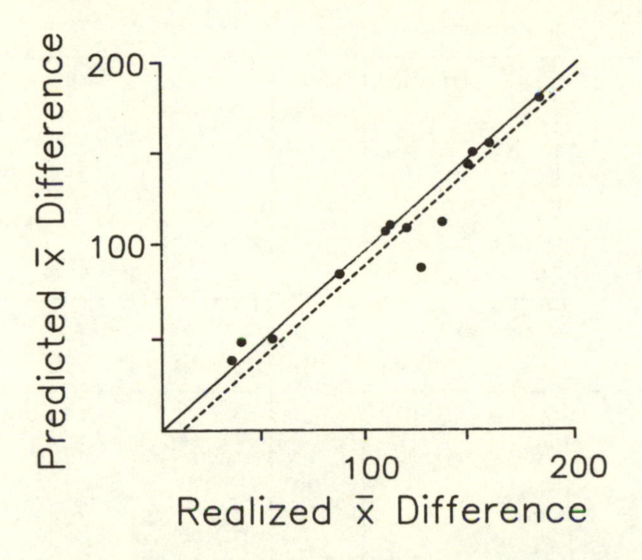

Fig. 7.20. Multivariate bill-size differences between coexisting *Geospiza* species on various Galápagos islands versus those predicted from a random combination of species. The predicted average differences between coexisting species are the differences between species means for all island populations, whether or not they actually coexist. Realized differences are averages of the actual differences between coexisting populations of those species. If the predictions were correct, all points would lie on the solid line; the dashed line is a least-squares best fit to the actual points. Most bill-size differences are greater than predicted. *Geospiza conirostris* and *G. scadens* are predicted to occur with a multivariate difference of 67 units, and for *G. conirostris* and *G. fortis* the predicted difference is 87 units; neither of these species pairs, however, actually occurs on an island. From Grant (1983b).

cal solitary finch of any size. Then, using field estimates of seed abundances, they calculated the combined abundances of all available foods within the food-size range of such a species for each island. Finally, the observed relationships between finch abundance and seed biomass were used to determine the number of individuals of the hypothetical species that could be supported by that quantity of food. This procedure was followed for all feasible hypothetical bill morphologies to derive a curve giving the expected population density of a solitary finch species as a function of its morphology (Fig. 7.21). These curves were distinctly non-normal: density is a polymodal function of bill depth that varies considerably between islands. This polymodality follows directly from gaps and irregularities in the frequency distribution of seed size and hardness values. The similarities to the patterns described by Griffiths (1986) are clear.

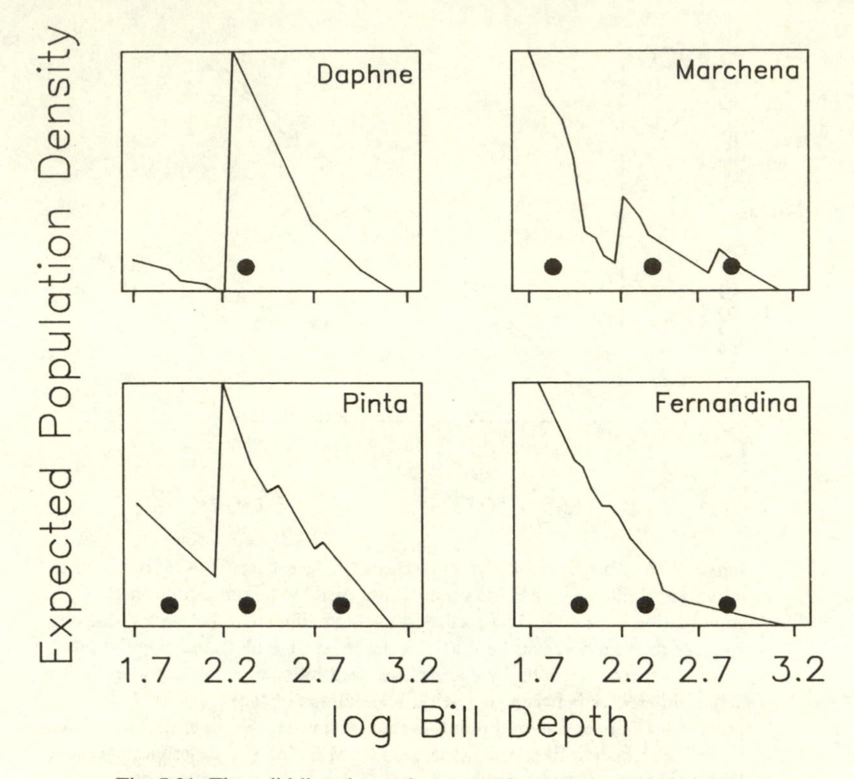

Fig. 7.21. The solid line shows the expected population density of a solitary granivorous finch species on Galápagos islands as a function of bill depth, derived from a model based on the abundance of seeds on the islands and the utilization of seeds of different sizes by finches with bills of different sizes. The heights of the curves are scaled and are not comparable between islands. The mean log bill depths of adult males in actual populations occurring on the islands are indicated (solid circles) for each curve. After Schluter and Grant (1984).

Given these curves, one may then construct hypothetical communities of finches by using models that differ in the extent to which the evolution or assembly of species is determined by food-supply effects on densities and/or by competition. The morphological patterns in the actual communities can then be compared with those of the hypothetical communities. In fact, the occurrence of finches on islands frequently (though not always) coincides with peaks in the density curves expected on the basis of food supplies (Fig. 7.21), but two species never occupy the same or immediately adjacent peaks. The observed species turn out to be more widely spaced, morphologically, than would be expected from models that lack interactive components. Models that incorporate competitive exclusion or coevolutionary displace-

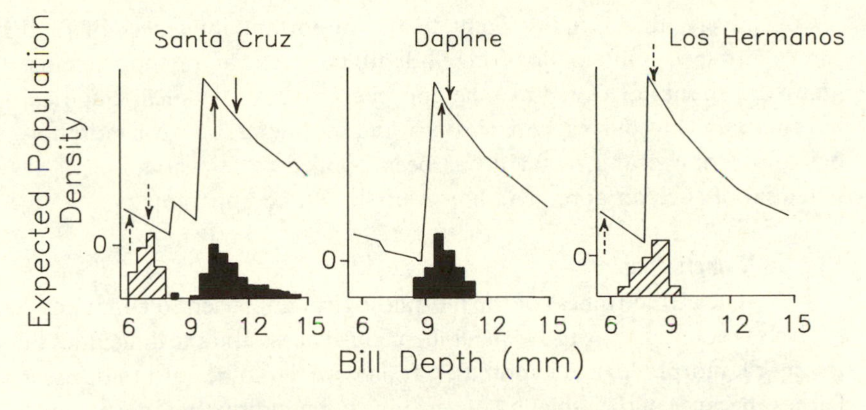

Fig. 7.22. Apparent character displacement in *Geospiza* species. The histograms depict bill depths for adult male *fuliginosa* (hatched) and *fortis* (solid). The curves indicate the population densities of a hypothetical solitary finch species expected as a function of bill depth, derived from a model based on the abundances of seeds of different sizes on the islands and the efficiency of utilization of those seeds by finches with different bill sizes. The arrows indicate the predicted (upward arrows) and observed (downward arrows) sizes for *fuliginosa* (dashed arrow) and *fortis* (solid), From Schluter *et al.* (1985).

ment of bill depths, on the other hand, fit the observed patterns reasonably well.

This approach may also be applied to a consideration of character displacement among the finches. *Geospiza fortis* and *G. fuliginosa* occur sympatrically on Santa Cruz. Their bill sizes are different here, closely matching the two sizes predicted on the basis of the food abundance–population density model (Schluter *et al.* 1985) (Fig. 7.22). Only *fortis* occurs on Daphne, and its bill size matches that predicted by the model. This size, however, is only slightly different from that of *fortis* on Santa Cruz (Fig. 7.22). An allopatric population of *fuliginosa* occupies Los Hermanos, and there its bill size has shifted from that corresponding with a low-abundance peak matching the bill size of Santa Cruz *fuliginosa* to a high-abundance peak more closely resembling the bill size of *fortis* where the species are sympatric (Fig. 7.22). During the 1977 drought on Daphne (where *fuliginosa* is absent), some *fortis* individuals were able to survive by exploiting the small seeds used elsewhere by *fuliginosa*, and their continued presence in the population shifted the mean morphology of the offspring subsequently produced toward an intermediate bill shape (Boag 1983). In this example, both the pattern of character displacement and some of the processes contributing to it are especially clear.

Collectively, these studies point to the important influences of food supply and species interactions on bill-depth patterns among the generalist granivore species of *Geospiza*. The approach requires detailed, empirical information on resources, bird densities, and the linkage between morphology and ecology, and so far it has been applied to few situations. The potential of such an approach, however, should be apparent.

Conclusions

The attractiveness of the morphological approach to community analysis is countered by its frequent inconclusiveness and the difficulties of testing ecomorphological hypotheses. It has proven to be hard to devise a fair test because of the difficulty of verifying the premise (that morphology really is related to ecology in some specific way) and of determining what is an unusual pattern. Testing patterns against the predictions of null models, while admirable in principle, has generated more controversy than insight.

A large part of the problem is related to the basic premise. Often, morphology does not seem to be very closely related to the ecological parameters of interest. Birds that are similar morphologically because of shared phylogeny may differ ecologically for behavioral reasons, while species that are morphologically and taxonomically quite different may be very similar in ecology (Hespenheide 1975). The same structure may bear a different relationship to ecology in different groups, or may be linked to other features of size or shape that are of greater ecological importance. If the premise is not valid, one really has no business expecting morphology to follow the community patterns predicted from ecological theories. Our failure to find size-ratio patterns among grassland and shrubsteppe birds that agree with theoretical expectations, for example, should come as no surprise, as the morphological measures we considered are unrelated to the ecological variables of presumed importance (Wiens and Rotenberry 1979, 1980). In other situations, however, the match between morphology and ecology seems to be quite close.

What factors influence the likelihood that morphology and ecology will be closely linked? At a local scale, matching of morphology to environmental conditions may be promoted by site tenacity, habitat selection, social dominance, selection for optimal phenotypes, and environmental effects on development, but it may also be disrupted by dispersal, emigration, differential mortality among migrants elsewhere, or environmental variability (James *et al.* 1984). If resources vary in abundance, morphological features may be closely tuned to resource characteristics in times of stress or scarcity (Boag and Grant 1981, 1984b, Grant 1986) but the fit may erode during periods of resource superabundance (Baker 1979, Wiens 1977b, 1986). It is also during times of scarcity that the dimensionality of important resources may be reduced, increasing the probability that ecology and morphology can be linked by simple analyses involving a few variables and providing a clearer indication of which morphological features might be most important. Tight coupling of morphology and ecology is also more likely in species that are resource specialists, such as some nectarivores, obligate insectivores, or strict granivores. If dietary specialization is associated with full-year residency and sedentariness, the relationship should be even closer, individuals may have no escape in space from temporal variations in resource levels. Not many species or situations fit these requirements, so it is perhaps not too surprising that finding clear ecomorphological patterns has been difficult. One group and situation that does possess many of these characteristics is the array of *Geospiza* finches on the Galápagos Islands. There, clear ecomorphological patterns have emerged.

Another difficulty is related to the level of inclusiveness at which patterns are sought. By focusing on a small, closely related guild (such as *Geospiza*), complications produced by different patterns of covariation of size and shape, differences in foraging strategies, and effects of phylogenetic and taxonomic diversity are minimized (James 1982). At the other extreme, community-wide analyses that are conducted with the expectation that unitary patterns will characterize all members of families or larger taxonomic groupings seem doomed to failure, even though a gross association between morphology and ecology can be most easily established at this level. A conclusion that emerges from studies of functional morphology (e.g. Leisler and Thaler 1982), size-ratio patterns (e.g. Moulton 1985), or resource relationships (e.g. Schluter and Grant 1984) is that morphological patterns of ecological interest are most likely to be evident within small sets of ecologically and/or taxonomically similar species. Even at this level, however, differences between locations can produce considerable variation in the closeness of the morphology–ecology linkage (Schluter and Grant 1984, Reynolds and Meslow 1984, Tonkyn and Cole 1986).

Despite these difficulties, ecomorphological studies of communities or guilds may have considerable value. They are an important part of any thorough study and may yield insights otherwise unavailable. It is doubtful, for example, that the ecology of *Geospiza* finches could be understood without a consideration of their morphology. A knowledge of morphology cannot substitute for direct ecological information, however. The number of factors influencing the relationship between morphology and ecology is large enough and their effects complex and inconsistent enough that each situation must be evaluated separately. One cannot accept the validity of the basic premise on the basis of faith or a literature citation. Moreover, in the absence of detailed knowledge of other aspects of a system and separate tests, interpretation of ecomorphological patterns must be quite restrained. Such patterns by themselves provide only vague and inferential evidence of ecological processes, such as competition.

8

Distributional patterns of species

The Boreal Chickadee (*Parus hudsonicus*) breeds in coniferous forests from eastern Canada west across the North American continent nearly to the Pacific coast. In the coastal mountains of the west it is geographically replaced by the Chestnut-backed Chickadee (*P. rufescens*), which ranges from southeastern Alaska to mid-coastal California. The two species meet in southern British Columbia and Washington. There, *hudsonicus* occurs inland at higher elevations in boreal and subalpine forests, while *rufescens* occurs at lower altitudes in wet coastal forests. In the interior of northern Alaska and Yukon, the Siberian Tit (*P. cinctus*) occurs. Where it overlaps with *hudsonicus*, it occurs primarily at forest edges and along rivers rather than in the dense, dark coniferous forests favored by *hudsonicus*. The three species thus differ distributionally in range, elevation, and habitat (Lack 1971).

Examples of such distributional patterns among ecologically similar bird species abound. They are usually interpreted as evidence that competition has produced exclusions in either space or habitat. In fact, the influence of competition and niche thinking has been so pervasive in community ecology that the search for patterns has been conducted largely in this context (e.g. Lack 1971, MacArthur 1972, Cody 1974a, Cody and Diamond 1975). Whether or not these or other distributional patterns are consistent with niche theory or provide evidence suggestive of competition or of some other processes, the patterns are of interest in their own right. In this chapter I consider some features of the broad-scale geographical distributions of species and some of the factors that may influence these distributions. In the following chapters I continue this examination of niche patterns, considering broad-scale habitat distributions of species (Chapter 9), ecological relationships among species coexisting in local areas, or 'niche partitioning' (Chapter 10), and changes in these patterns that occur with changes in community composition or other factors, so-called 'niche shifts' (Chapter 11).

Range boundaries and relationships

The ecological tolerances of species dictate the environmental situations they can occupy, and how these conditions are met in space determines where the species may occur (Fig. 8.1). In turn, the distributional patterns of species specify the potential membership of communities and guilds in any locality. As one moves from one location to another across the range boundaries of species, the composition of the species pool and communities changes. Depending on the scale of one's movements, these changes are reflected in the beta or delta diversity of communities (Whittaker 1977; Table 5.1).

Embedded in this spatial turnover may be particular patterns of distributions of ecologically similar or closely related species, such as the *Parus* species noted above. These distributions may show a variety of configurations: disjunct, abutting, partially overlapping, or inclusive (Pielou 1979). Such distributional patterns are determined not only by the geographical ranges of the species but by the evenness or patchiness of occurrence of the species within their ranges. Documenting which pattern exists may not be easy. Distributions are based on presence–absence information, but the absence of a species from an area may reflect a distributional fact or a simple failure to record its actual presence in censuses. The latter is especially likely for low-density and/or cryptic species or if censusing is superficial. Range maps may be used to describe distributional patterns at a biogeographical scale, but accurate maps are not available for many parts of the world, especially the tropics. In addition, the crisp edge of a standard field-guide range map often translates in the real world into a very fuzzy distributional boundary, where neither the presence nor the absence of a species is certain (MacArthur 1972).

Despite these difficulties, the accuracy of the distributional information on which ecological analyses are based is rarely questioned (but see Austin *et al.* 1981). If a distributional pattern has been incorrectly described, any interpretation of that pattern will be erroneous. On the basis of distributional surveys in the Galápagos Islands, for example, Lack (1947) stated that the finches *Geospiza difficilis* and *G. fuliginosa* are altitudinally segregated where they co-occur, as on Isla Pinta. He interpreted this as evidence of competitive exclusion between the morphologically and ecologically similar species, and Bowman (1961) later interpreted the same pattern in terms of species-specific responses to differences in food availability. Detailed surveys indicate that the elevational ranges of the two species on Pinta in fact overlap considerably (Schluter 1982b, Schluter and Grant 1982),

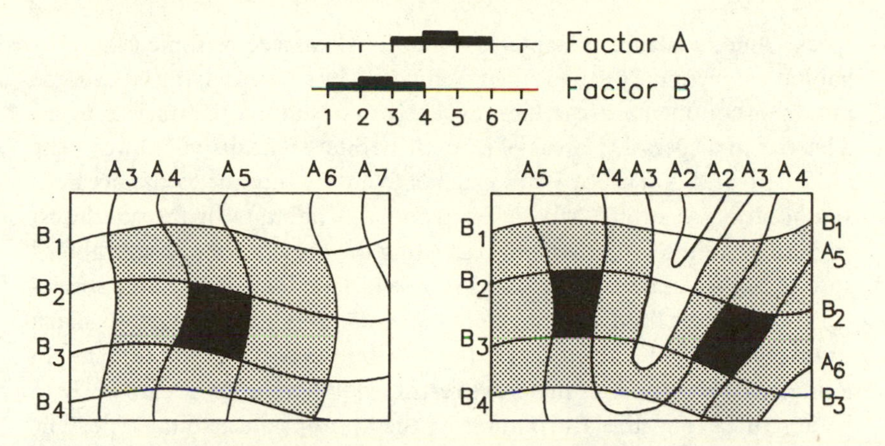

Fig. 8.1. Diagrammatic representation of the relationship between the ecological response of a species to variations in environmental factors (top) and the patterns of distribution and abundance of that species (below). The distribution of the species maps onto environmental gradients of the limiting factors, with abundance being greatest where the combination of factors is most favorable (solid). Because environmental features may not vary symmetrically, multiple peaks of abundance or disjunct distributions may occur (right). After Root (1988c and personal commmunication).

prompting a somewhat different interpretation (see below). Patterns of distribution require the same care and accuracy in their documentation as do other ecological patterns.

Factors influencing range boundaries

What determines the boundary of a species' range? Why is it present in one area and absent in another? Cody (1983a) listed as possible reasons (1) the lack of suitable habitat, (2) barriers that prevent dispersal to the area, (3) competitive interactions with a closely related or ecologically similar species, (4) diffuse competition with several somewhat similar species, (5) the absence of critical resources such as a specific food type, and (6) an area too small to contain an individual territory or support a viable population. To these must be added (7) climatic factors producing physiological stress, (8) historical factors, and (9) chance. Each of these represents a potential process explanation for the observed pattern, and much of the business of the geographical ecology and biogeography should involve testing such hypotheses.

In some situations the factors determining the range boundary may be reasonably clear. Dispersal barriers, for example, restrict the distribution of species among islands (Chapter 5), and the small size of some islands or habitat patches may preclude their occupancy by species having large home ranges. In continental areas, large expanses of unsuitable habitat may act as a barrier to dispersal to areas of suitable habitat some distance away. The extremely arid Nullarbor Plain extends from the interior to the ocean in southcentral Australia, separating woodland habitats in the southwest from those in the southeast and restricting the distribution of several bird species to the southwest (Keast 1961, 1981, Schodde 1982). Some species with specialized food requirements, such as crossbills (*Loxia*) or Black Brant (*Branta bernicla*), are tied to the distribution of their food supplies. In a similar fashion, species that have restricted habitat affinities are likely to be absent beyond the distribution of the appropriate habitat type. The breeding range of Kirtland's Warbler (*Dendroica kirtlandi*) is confined to a small area of northern Michigan containing jackpine (*Pinus banksiana*) of the appropriate age and stature (Mayfield 1960, 1978), and Red-cockaded Woodpeckers (*Dendrocopus borealis*) are confined to open pine woodlands of southeastern USA (Mengel and Jackson 1977). In Australia, the Rock Warbler (*Origma solitaria*) occurs only in a small area of coastal New South Wales, where it is associated with rocky ravines in Hawkesbury Sandstone and adjacent limestone areas (Blakers *et al.* 1984). On a broader scale, many species have affinities for general habitat types such as coniferous forest, grassland, or desert, and the distributions of these species often roughly coincide with the extent of a particular biome type (see Pitelka 1941).

Community ecologists have most often interpreted distributional patterns as consequences of competition. Lack (1971), for example, thought that situations in which ecologically similar species have range boundaries coinciding with the distributions of different habitat types were due to competitive exclusion rather than habitat restriction *per se*. The range limits of many other species do not coincide with the distributions of particular foods or habitats or of other bird species; both Lack (1971) and MacArthur (1972) suggested that these ranges were probably limited by diffuse competition with the complex of other bird species present. MacArthur (1972) admitted that the northern range limits of north-temperate species might be established by climatic rigors but offered as a generalization the view that southern range limits are determined by competition. As an example, MacArthur noted that Belted Kingfishers (*Ceryle alcyon*) breed south on the North American mainland to the Gulf Coast, Texas, and California, although they winter in the tropics as far south as Panama. Why do they stop where they do? After ruling out various other factors, MacArthur observed that two other kingfisher species appear about where the range of *alycon* stops. One of these, the Green Kingfisher (*Chloroceryle americana*), is smaller, the other (the Ringed Kingfisher, *Ceryle torquata*) larger, and they thus 'sandwich the belted kingfisher in the most uncomfortable kind of competitive squeeze' (MacArthur 1972: 133). MacArthur concluded that 'a wintering adult belted may be able to gather enough food for itself in their presence, but to gather food for a brood of hungry young is probably out of the question'. Such conclusions are based not on evidence but on Kipling-like assertions derived from a belief in the competition paradigm. Most of the examples of distributional patterns discussed in this chapter have also been interpreted, often by means of equally *ad hoc* arguments, as consequences of interspecific competition.

The effects of other factors on bird distributions have generally received less attention, and climate has been especially neglected. Lack (1971) argued that, because birds are endothermic, it is doubtful that ranges are ever directly limited by climatic factors, and Cody (1974a) contended that physiological differences between species associated with differences in their geographical ranges are consequences rather than causes of the distributions, which are determined by competitive interactions. Nonetheless, climatic stresses may exert a strong and direct influence on the distributions of some species. This is especially evident in harsh, high-altitude situations. The northern distribution of the Goldcrest in Norway coincides with the limits of the coniferous forests it occupies, but in northern Sweden, Finland, and adjacent areas of the Soviet Union the species is not found in the northernmost spruce or pine forests. On the basis of a detailed investigation of the species' incubation energetics, Haftorn (1978) suggested that its northward distribution in these areas is limited by mean summer air temperatures. Another species, the Siberian Tit, is a well-established resident in northern Finnish Lapland, where it breeds early in the spring. Adults are capable of withstanding severe cold, although chicks may suffer during late cold spells (A. Järvinen 1983). Other species have expanded their ranges into this region during the past few decades. Great Tits are partial residents and breed late but successfully. They are not resistant to winter cold, however, and suffer considerable overwinter mortality. Pied Flycatchers (*Ficedula hypoleuca*) are migrants, suffer severe nesting losses, and undergo substantial annual fluctuations in density and breeding success. The species differ in their tolerance of harsh breeding and winter conditions, and small climatic changes would alter their distributional patterns.

Physiological adaptations may also constrain the distributions of bird species in desert regions (Dawson 1981, 1984). In the western United States,

Sage Sparrows occur throughout the Great Basin and on the western fringes of the Mohave Desert and in chaparral in southern California. Their breeding range overlaps that of Black-throated Sparrows (*A. bilineata*), but this species is absent from the cold deserts of the northern Great Basin and extends much farther into the hot deserts of the southwest and Mexico. Water stress may be severe in such areas, and *bilineata* is physiologically capable of surviving for long periods without free water on a diet of dry seeds, whereas *belli* is not (Smyth and Bartholomew 1966, Moldenhauer and Wiens 1970).

Physiological tolerances may limit the distribution of birds more broadly than only in harsh or extreme environments, especially during winter. From an examination of the wintering distributions of 148 species in the United States and Canada in relation to the geographical distributions of several climatic factors, Root (1988a) found that northern limits were often coincident with isoclines in mean January minimum temperature, duration of the frost-free period, and vegetation composition (habitat) (Fig. 8.2). Both western and eastern range limits were frequently concordant with patterns in annual precipitation and vegetation, but western distributions were also often associated with elevation whereas one-fifth of the eastern species had distributions that closely corresponded with minimum January temperatures. Figure 8.3 provides an examples of these patterns for one species, the Pine Warbler (*Dendroica pinus*).

Such patterns are derived over a broad geographical area and are of necessity correlative only. Consideration of other environmental variables or combinations of variables might produce different patterns (although Root considered the most obvious factors). Nonetheless, it is apparent that the distributional limits of a large number of wintering species in North America may be associated with climatic factors (see also Hennemann 1985). These associations raise the possibility that, like the Goldcrest, some of these species may be distributionally limited by energetic demands. Detailed metabolic studies have been conducted on 14 passerine species whose northward winter distributions were coincident with mean minimum January temperature in Root's analysis. Root (1988b) used this information to calculate the metabolic rate of these species at the northern limit of their wintering ranges, and found a highly significant fit to the value 2.48 BMR (basal metabolic rate). This relationship suggests that the northern winter ranges of several passerine species are restricted to areas where the ambient temperature on average does not dictate that they increase their metabolism by roughly 2.5 × BMR. Why this value should represent a distributional threshold is not apparent, but the consistency of the relation-

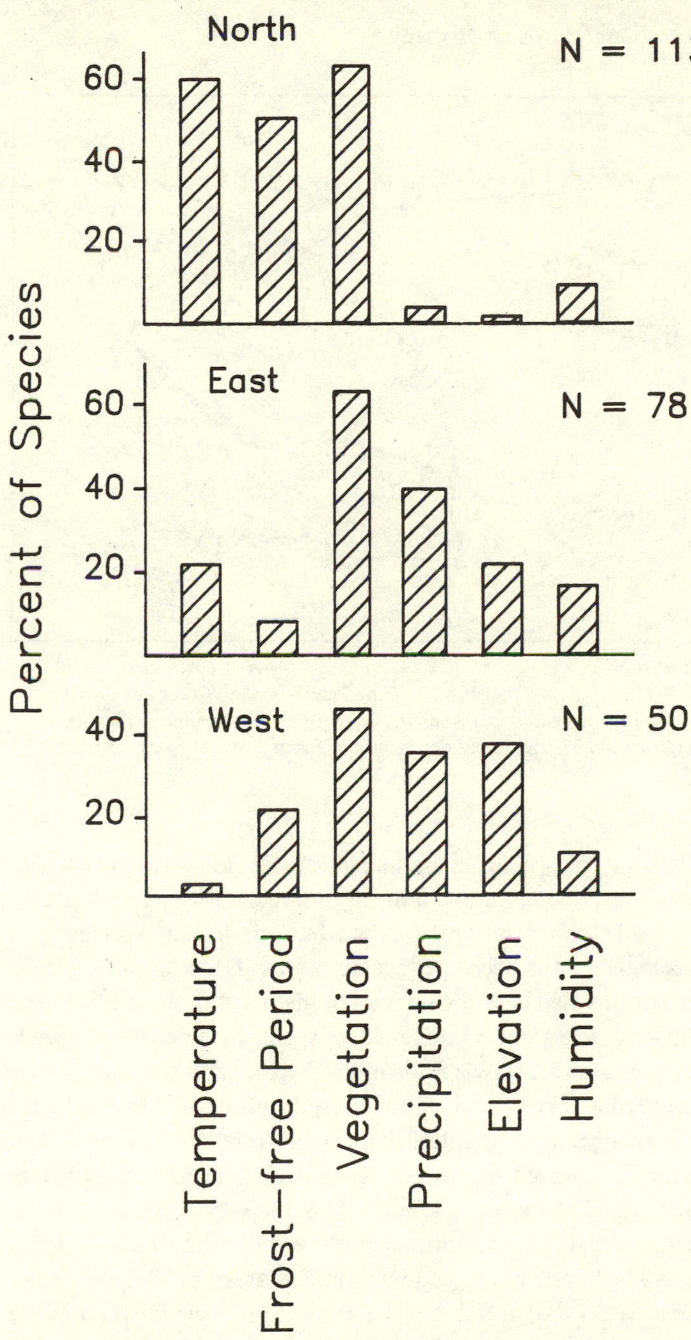

Fig. 8.2. The percentage of species with northern, eastern, and western range boundaries in North America associated with each of several environmental measures. N = number of species considered in each direction. Modified from Root (1988a).

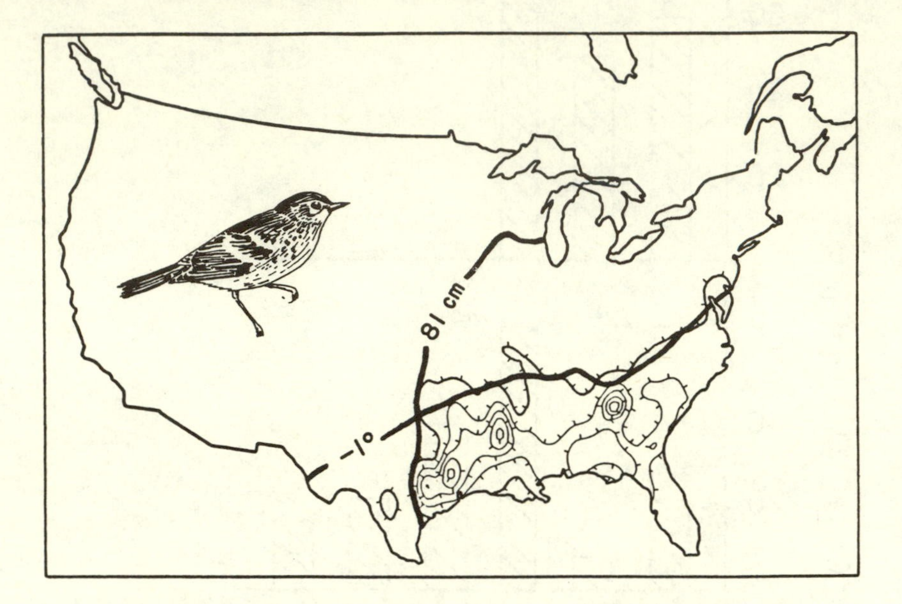

Fig. 8.3. The wintering distribution of the Pine Warbler (*Dendroica pinus*) in North America in relation to isopleths for − 1°C mean minimum January temperature and 81 cm annual precipitation. From T. Root (pers. comm.).

ship across a variety of species suggests something possibly quite basic. Whatever the causes, it is apparent that physiological factors may play a more important role in determining the distribution of many species than community ecologists have been inclined to admit (Bartholomew 1958).

Range distributions also bear the imprint of history, but these effects are especially difficult to determine. Our records of species distributions in well-studied areas cover at most only a few hundred years, and for many areas the historical record is very brief. Following equilibrium thinking, we tend to regard current ranges as fixed unless recent records tell us otherwise. The records of extinctions in island archipelagos during historic times (e.g. Olson and James 1982, Moulton and Pimm 1986, Lack 1976, Steadman and Olson 1985) suggest that interpretations of current distributional patterns of species on islands that do not include such effects are possibly incomplete (Chapter 5). Moreau (1966) noted that the small cormorant *Phalacrocorax africanus* disappeared from the northernmost portion of Africa shortly after its occurrence there had been documented by ornithologists. Interpretations of biogeographic and distributional patterns in this area might have been quite different had those investigations been initiated a few years later.

Range determinants in regional avifaunas

It is relatively easy to produce isolated examples that seem to indicate clearly the role of one or another factor in limiting the range of a species. Determining what factors influence the ranges of a number of species in a taxonomic group or regional avifauna, however, is much more difficult. Consider, for example, the biogeographic patterns of birds of piñon-juniper habitats in the Intermountain Region of the American west. Johnson (1978) compared the breeding assemblages of the cool, moist woodlands of spurs of the Sierra Nevadas with those of the warm, arid woodlands on montane islands some 90 km away. Although nine species have range boundaries that lie between the two areas, there is considerable overlap in the species composition in the avifaunas. The densities of 19 species change along the gradient, however, and several species change habitat affinities between the regions. Species with Boreal affinities predominate in the northern stands, whereas those of the southern stands are primarily of Austral derivation. On this basis, Johnson suggested that climatic factors might play a major role in determining range boundaries on the gradient, southern species adapted to warm and relatively arid conditions being replaced by northern species adapted to cooler, moist environments.

The distributional patterns of several species in this region are of particular interest. Downy and Ladder-backed (*Picoides scalaris*) woodpeckers, for example, are allopatric, their ranges separated by a considerable gap in which apparently suitable habitat for either species remains unoccupied. Johnson suggested that this pattern might reflect differing sensitivities of the species to temperature and moisture regimes, although MacArthur (1972) argued that gaps may occur between the ranges of competing species if resource levels in the intervening area are low. Neither of these suppositions has been tested. Western and Mountain bluebirds (*Sialia mexicana* and *S. currucoides*), on the other hand, occupy parapatric breeding ranges in this region. The species have well-defined differences in habitat preferences; these, together with the approximate coincidence of the zone of parapatry with the northern limits of the Mojave Desert, led Johnson to suggest that bluebird distributions are probably determined by temperature regimes rather than by competition. Finally, Grace's Warbler (*Dendroica graciae*) breeds commonly and exclusively in forests of ponderosa pine (*Pinus ponderosa*) in this region. Its range stops short of the distribution of seemingly suitable habitat in some areas and no other species of warbler or pine-foliage insectivore can be identified as a potential competitor in these areas. The perimeter of this species' distribution is unstable, however, and it has recently been recorded breeding in some well-surveyed areas of southern Nevada from which it was previously absent.

Collectively, the distributions of these species, and of riparian woodland species in the same region, show little adherence to any general pattern and little concordance with one another. In this region, as in most, distributional patterns are complex. Some of the patterns are suggestive of possible underlying factors, but Johnson's explanations are really only inferential suggestions, hypotheses awaiting formal testing.

How important are the various determinants of geographical distributions in birds? Any attempt to answer this question is likely to be influenced by one's preconceptions and, accordingly, one's interpretations of particular patterns. Bearing this in mind, two analyses provide some interesting insights. Terborgh and Winter (1982) examined the range limits of 155 Colombian and Ecuadorian species having small (<50000 km^2) ranges. Because ranges are limited to the east by the Andes and to the west by the Pacific Ocean in this region, Terborgh and Winter emphasized northern and southern distributional limits. Roughly a quarter of the species are locked into closed pockets of topography or habitat, but the remainder are seemingly free to expand their ranges within apparently suitable habitat in at least one direction. A great many of these species have distributions that terminate parapatrically at boundaries with congeneric species; 49 species abut congeners at both ends of their ranges. Some species contact a congener in the north and habitat barriers (dry zones or desert) in the south. Many of the unexplained boundaries are associated with species of the Pacific lowlands, several of which are monotypic or occupy ranges within presumed Pleistocene forest refuges. Overall, Terborgh and Winter attributed 43% of the range boundaries to parapatry with a congener and 36% to ecological and/or topographical gradients; 21% remained unexplained. In a somewhat similar analysis, Hall and Moreau (1962) found fewer instances of range limitations associated with the presence of congeners, perhaps because more of the species were limited to discrete habitats on montane islands. In both studies, many of the parapatric congener distributions involved members of superspecies complexes. If speciation within such complexes is allopatric, semispecies will almost by definition segregate geographically (Terborgh and Winter 1982). As the species evolve further, they may overlap geographically but segregate by elevation, perhaps later achieving full sympatry by virtue of differentiation in habitat use or foraging behavior, as Diamond (1973, 1986b) has suggested occurs among montane birds in New Guinea. Historical factors contribute to the inter-

Table 8.1. *Percentages of contacts between congeneric bird species in several groups that can be attributed to separation by geographical range (including parapatry as well as disjunct distributions), habitat, or feeding behavior.*

In some groups values do not total 100% because assignments to an 'unknown' category have not been included.

Data set	*N*[a]	Geographical range	Habitat	Feeding
European birds				
Passerines	325	25	55	19
Other land birds	88	25	20	43
Water/wading birds	191	28	18	48
European transequatorial migrant passerines				
Breeding season	163	26	62	10
Wintering area (Africa)	161	64	23	2
West Indies passerines[b]	81	59	21	17
Parus species				
Europe	36	25	32	43
Asiatic mountains	91	26	56	18
Subsahara Africa	45	67	30	3
North America	45	84	11	5
Zosterops species				
Mainland	15	80	20	0
Islands[b]	93	46	44	8

Notes:
[a] Possible contacts between congeners.
[b] Including each geographically isolated species only once.
Source: From Lack (1971).

pretation of these distributional patterns in terms of past speciation events rather than distributional changes *per se*.

Unlike Terborgh and Winter, Lack (1971) did not specifically consider the various factors that might determine range boundaries but rather asked how important geographical isolation among congeners is in relation to other forms of ecological segregation. He analyzed the patterns among several large sets of species (Table 8.1). According to Lack's analysis, roughly a quarter of the congeneric species among the breeding birds of Europe are parapatric or have separated geographical ranges, regardless of whether they are passerines, nonpasserine land birds, or aquatic species. Where the distributions of passerine congeners overlap, however, separation by habitat seems quite important; the other groups apparently differentiate more often by feeding location or method. Of the European passerines that are transequatorial migrants, a large proportion of the congeners winter in geographically separate areas, despite their substantial breeding distributional overlap. Geographical separation of congeners is also frequent among the passerines of the West Indies, where it is associated with the distribution of species onto different islands. Again, many of the congeners are members of the same superspecies, and speciation history becomes important.

Lack (1971) considered the distributional patterns of several avian genera in detail. Among tits (*Parus*), species in Europe and the mountains of central Asia generally do not segregate a great deal by range but rather are differentiated ecologically within a common range. On the other hand, in Africa south of the Sahara and in North America many of the *Parus* species occupy parapatric or disjunct ranges, and few species co-occur locally. Lack (1969b, 1971) speculated that this difference might reflect the more recent evolutionary derivation of the African and North American species, as well as the presence of ecologically similar passerine species in other families in these areas. Among the white-eyes of the widespread genus *Zosterops*, species co-occur only infrequently in the mainland areas of Africa, Asia, and Australia, and geographical separation is frequent. On very small and remote islands, only one species is usually present, but fairly small and fairly remote islands may often contain two to seven species, among which habitat (including altitudinal) separation is frequent. Interestingly, unusually large white-eyes (placed in different genera) have evolved independently perhaps 12 times, each where a single *Zosterops* species of normal size is also present, and usually on islands containing few other passerine species.

In these analyses, distributional patterns were ascribed to single factors, although Lack admitted some difficulty in making assignments. This is not surprising, for the controls of species' distributions are often complex (Grinnell 1917). Moreover, given Lack's emphasis on competition, the focus in these surveys was on congeneric species. It would be more instructive to consider the distributional patterns of species in broadly defined guilds.

Range dynamics

Biogeographers and ecologists often treat distributions as if they were fixed in space over time. Distributional patterns are dynamic, however, and change on several scales of time. The temporal dynamics of species and communities will be considered more fully in Volume 2, but a consideration of two examples of range changes is instructive at this point. Using data gathered in Finland by line transect surveys from 1910–29, 1936–49, 1952–63, and 1973–77, Järvinen and Väisänen (1979) examined changes in population densities and in the location of the zone of equal abundances of two pairs of congeneric species. *Parus cinctus* has a northern distribution and breeds chiefly in conifers, whereas *P. cristatus* is a more southern species that occurs in mixed forests; both species are sedentary. The two species are generally similar in their feeding ecology, and Lack (1971) considered the species to be competitors in some areas. During the time period considered by Järvinen and Väisänen, *cinctus* greatly decreased in abundance, first in the southern and then in the middle portion of its range. At the same time, *cristatus* expanded northward and onto the Åland Islands but then decreased, particularly in the north. In the zone of overlap, both species decreased in abundance by 90–95%. The 1:1 zone has shifted northward substantially (Fig. 8.4).

Järvinen and Väisänen also considered the range dynamics of two migrant species, *Fringilla montifringilla*, which breeds mainly in sparse forest, and *F. coelebs*, a habitat generalist. The two species are similar in morphology and diet. Except for a short-term decrease in abundance during the 1940s and a southward expansion of 100–200 km in 1955, *montifringilla* appears to have been stable in both distribution and abundance over the time period considered. On the other hand, *coelebs* increased substantially from the 1920s to the 1940s but has changed little since then. The ranges of the two species now overlap, although the density gradient of each is quite steep (Fig. 8.5). The location of the 1:1 zone advanced northward during the 1940s in association with the decrease in *montifringilla* abundance but since has shifted south, although not to its position early in the century (Fig. 8.5).

The ranges of these pairs of congeners have thus exhibited considerable flux. Other things being equal, the complementarity of the distributional changes of the species is suggestive of competition (e.g. Merikallio 1951, Lack 1971). Other things, however, have not been equal over this period. The climate of Fenno-Scandia has ameliorated over the past century, shifting spring and summer isotherms northward substantially. At the same time, humans, through changes in both grazing and forestry practices, have altered the forest habitats used by these species. Järvinen and Väisänen attempted to relate these changes in climate and habitat to the range dynamics of *Parus* and *Fringilla*. The results (Table 8.2) do not exclude the possibility that competition between the species may also have contributed to the patterns. At the very least, they indicate the complex fashion in which distributional dynamics may be linked to environmental variations.

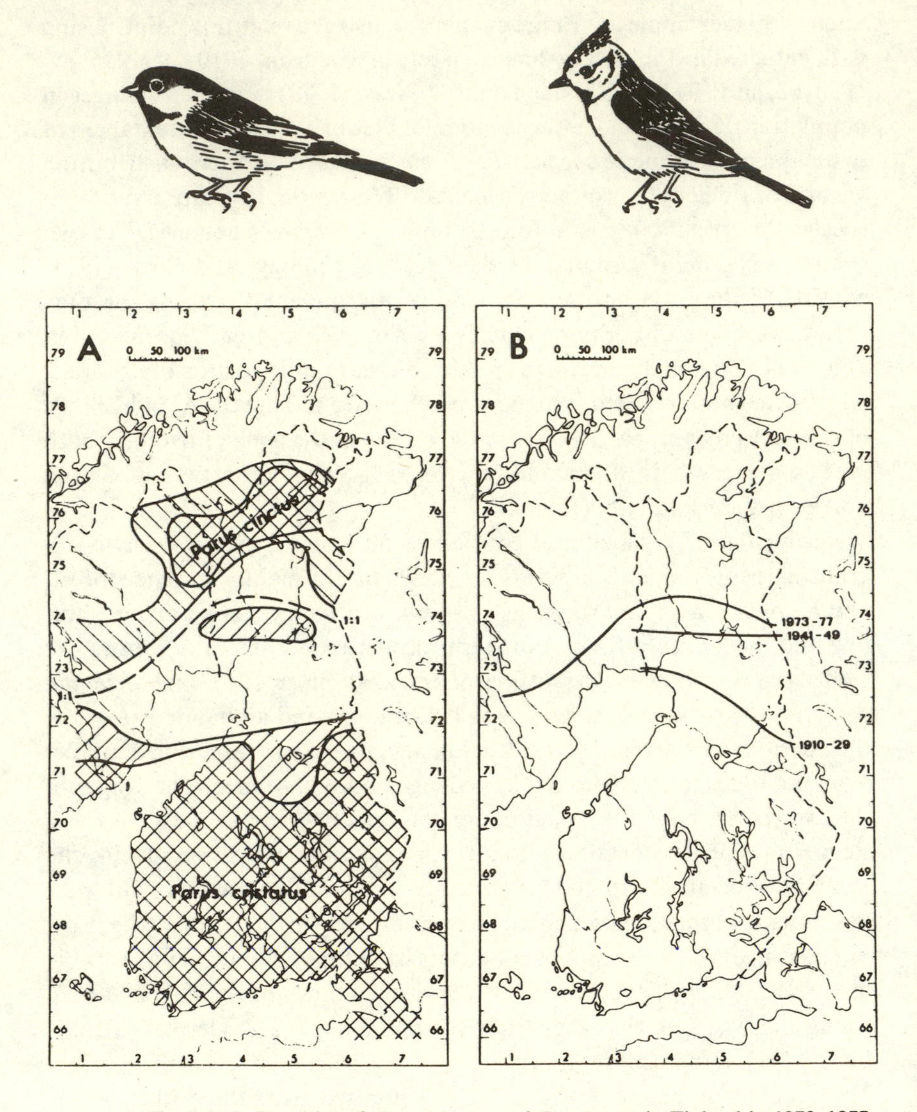

Fig. 8.4. A. Densities of *Parus cinctus* and *P. cristatus* in Finland in 1973–1977. Hatching = 0.25–0.5 pairs km^{-2}, cross-hatching = >0.5 pairs km^{-2}. The zone of equal abundances (1:1 ratio) is also shown, although, as this zone occurs in an area where both species are rare, its position may not be accurate. B. The location of the 1:1 zones in three study periods. From Järvinen and Väisänen (1979).

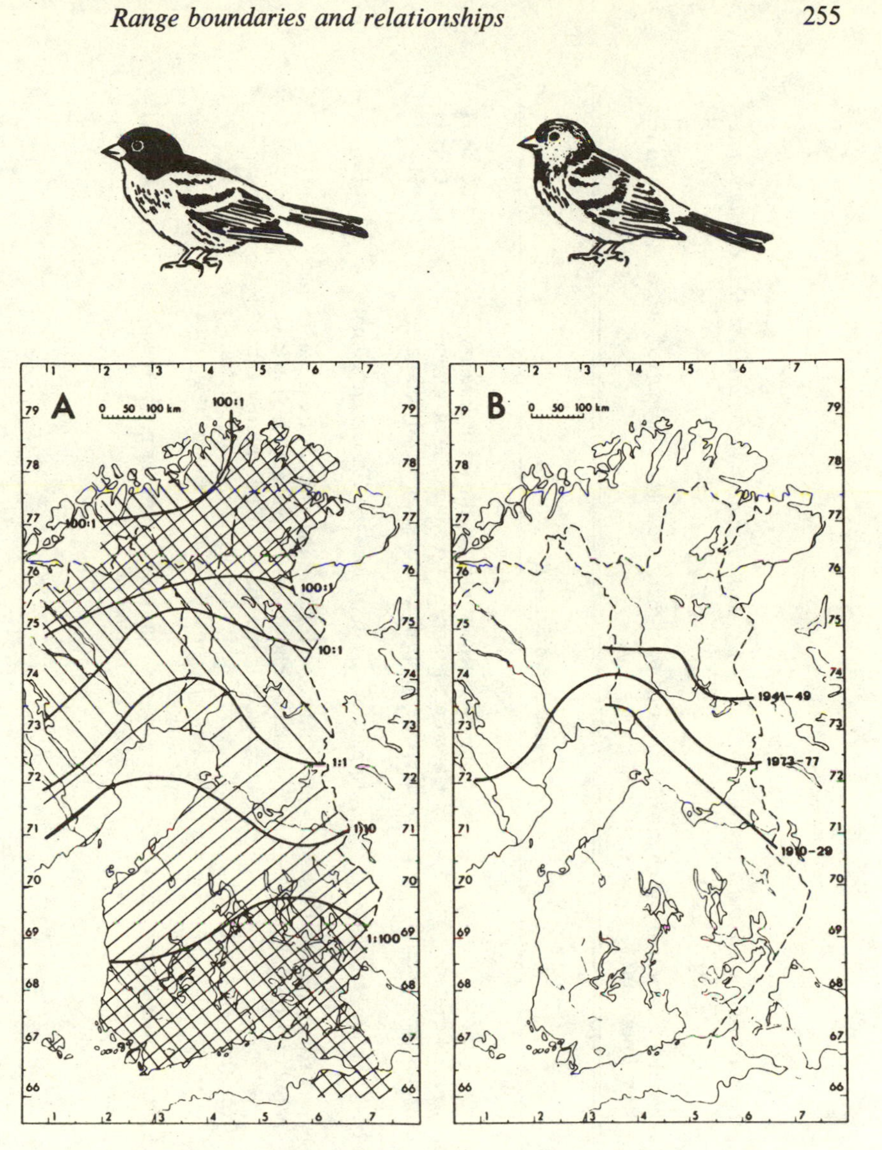

Fig. 8.5. A. The ratio of the densities of *Fringilla montifringilla* and *F. coelebs* in Finland in 1973–1977. B. The locations of the 1:1 zones of the species during three study periods. From Järvinen and Väisänen (1979).

Table 8.2. *The changes in populations of* Parus *and* Fringilla *species observed in Finland since early in this century and possible explanations for each.*

In the third column, C indicates that the theory that the changes are related to climatic changes is supported, while H refers to the theory that habitat changes have been important.

Population changes	Suggested explanation	Theory supported
Parus		
(1)Long-term decrease of *cinctus*	Climatic amelioration, or impact of forestry	C or H
(2)Long-term increase of *cristatus* in N Finland	Climatic amelioration, or increase of spruce	C or H
(3)Colonization of Åland by *cristatus* (and possibly early increase in S Finland	Increase of spruce, or chance; hardly climate	H
(4)Recent drastic decrease of both *cristatus* and *cinctus* in the overlap zone of the species	Impact of forestry (since the 1950s); *not* climate	H
(5)Less drastic recent decrease in northern populations of *cinctus* and in S Finnish populations of *cristatus*	Less intensive forestry than in the areas where the populations have catastrophically declined; increase of spruce in S Finland; hardly climate	H
(6)Drastic decrease of *cinctus* from the 1940s to 1955	The very cold spring of 1955, or population fluctuations, or failure of census in 1955; *not* forestry	C?
(7)Northward change of the 1:1 zone of the species from 1910–29 to 1973–77	Climatic amelioration, or greater negative impact of forestry on *cinctus*	C or H

Fringilla		
(8) Long-term stability of *montifringilla* in Finland	Wide habitat amplitude in N Finland; *not* climate	H
(9) Temporary decrease of *montifringilla* in the 1940s	Cold winters of the early 1940s, or population fluctuations; *not* forestry	C?
(10) Southward shift of *montifringilla* in 1955	The very cold spring of 1955; not forestry	C
(11) Long-term increase of *coelebs* in Finland	Climatic amelioration, or increase of edge effect, spruce and forests of high grade in S Finland, or decreased winter mortality	C or H
(12) Changes of the 1:1 zone of the species	Temporary decrease of *montifringilla* in the 1940s (point 9) and southern breeding in 1955 (point 10)	—

Source: From Järvinen & Väisänen (1979).

Distributional patchiness

Within their geographical ranges, species often have patchy distributions. Although the ecological suitability of a site from which a species is missing is not easy to determine, species often are absent from seemingly suitable sites within their geographical range. To some degree this should be expected, as habitats are usually patchy (Wiens 1976b, 1985a). Given careful censusing, distributional patchiness may be easier to document than range boundaries, especially if the environmental patches are distinct.

The most distinct patches are islands, and the clearest examples of patchy distributions are found when a species is present on some islands in an archipelago but absent from others. In some cases, a closely related or ecologically similar species occurs on some of the islands from which the species is missing. Diamond (1975a) described several such 'checkerborad distributions' among the birds of New Guinea and its associated archipelagos, and Lack (1976) documented similar patterns among West Indies species. Whether or not the frequency of such distributional patterns departs significantly from that generated by some null model (see Chapter 4), the patterns (presuming they are accurately determined) are of interest and invite explanation.

Fourteen species of hummingbirds occur in the West Indies, but few islands contain more than three (in fact, two or more species inhabit 36 islands, there are no islands with only one species, and at least four islands have no hummingbirds, a highly nonrandom distribution of species numbers; Brown and Bowers 1984, 1985). In general, lowlying islands support one small and one large species, which are syntopic. Mountainous islands have a single small species, which occurs in both lowlands and highlands, and two large species, which are elevationally segregated with little overlap (Lack 1976, Kodric-Brown *et al.* 1984, Brown and Bowers 1984, 1985). The composition of the hummingbird assemblage differs among islands, however – the small species may be one of five species, none of which co-occur, while four large lowland species and five large highland species replace one another on different islands. For each species, there is considerable distributional patchiness.

What factors might account for these patterns? In general, the factors are the same as those that influence range boundaries, listed above (see also Table 4.1). For the hummingbirds, Lack attributed the distributional patterns to minor differences between similar-sized species that give one or another species a competitive edge on islands with different ecological conditions. Kodric-Brown *et al.* and Brown and Bowers more specifically ascribed the patterns to close coevolution between hummingbirds and hummingbird-pollinated flowers, coupled with intense competition over floral resources among birds of similar size. Diamond (1975a) interpreted the patchy distributions of birds among islands in the New Guinea region almost exclusively in terms of competition and said little of any other factors possibly influencing distributions.

Some indication of the complex interactions among factors influencing distributional patchiness is afforded by Abbott's (1981) study of the distribution of three passerine species, *Lichenostomus virescens*, *Zosterops lateralis*, and *Sericornis frontalis*, on islands adjacent to the coast of southwestern Australia. The three species differ somewhat in morphology and diet. *Sericornis* (11 g, bill length = 14 mm) feeds primarily on arthropods gathered from leaf litter. Both *Zosterops* (9 g, 12 mm) and *Lichenostomus* (27 g, 21 mm) are versatile feeders, taking fruits, nectar, and arthropods (usually by gleaning foliage). Of the 100 islands Abbott surveyed, *Zosterops* was recorded on 47, *Lichenostomus* on 35, and *Sericornis* on 11; 29 of the islands lacked any of the three species. Their distributions among islands are thus patchy, but in a distinctly nonrandom way: *Zosterops* and *Lichenostomus* occur together less often, and more islands have either species alone, than expected by chance. Island area affects the patterns, however, as *Zosterops* occurs on over five times as many small (< 10 ha) islands as does *Lichenostomus* or *Sericornis*. On islands of 10–100 ha, however, *Zosterops* occurs on less than half as many islands as does *Lichenostomus*. *Sericornis* is infrequent on islands of less than 100 ha.

This pattern is suggestive of an area-dependent competitive relationship between *Zosterops* and *Lichenostomus*, the former dominating on small islands, the latter on larger islands. Information on habitat features gathered in detailed studies on 20 of these islands adds another dimension to this pattern. *Lichenostomus* is generally absent from islands on which the foliage density at a height of 1.5 m (a measure of brushiness) is high and the extent of such habitat is on the order of 1–3 ha; there, *Zosterops* occurs alone. *Zosterops*, on the other hand, is generally absent and *Lichenostomus* is present on islands where brushiness is low and the extent of such habitat is 8–40 ha. On two islands adjacent to a portion of the mainland from which *Lichenostomus* is absent, however, *Zosterops* occupies habitat that would seem to be more suitable for *Lichenostomus*. *Sericornis* occurs only on islands containing more than 60–70 ha of vegetation, regardless of its brushiness. The distributional pattern of *Zosterops* and *Lichenostomus* among the islands thus appears to be influenced by (at least) island size, habitat area, vegetation brushiness, and the presence or absence of the other species, whereas that of *Sericornis* seems to be largely independent of the other species but critically dependent on island and habitat area.

Bird distributions are often patchy in continental areas as well, although because the patches are less clearly bounded the distributional patterns are not so easily recognized. Many tropical species, for example, are patchily distributed on both local and regional scales (e.g. Willson *et al.* 1973, Karr and Freemark 1983, Graves 1985), as are some temperate species (e.g. Fritz 1980). Perhaps the most detailed analysis of distributional patchiness in a continental setting has been undertaken by Vuilleumier and Simberloff (1980). They considered bird distributions in two zones of alpine-like vegetation of the high Andes of South America: páramo, a wet, discontinuously distributed habitat in Colombia, Venezuela, Ecuador, and northern Peru; and puna, the drier, continuously distributed zone of north-central Peru southward to northern Chile and Argentina. Both zones are quite young, dating at the most from the Plio-Pleistocene. The puna supports a large avifauna, which includes many of the páramo species. Each zone, however, contains several species specialized to that region. Vuilleumier and Simberloff based their analysis on surveys of 23 páramo and 17 puna locations. Some of these sites were studied fairly intensively, but many were visited only briefly and the observations were supplemented with information from the literature or museum collections. The data base is thus of uneven quality and composition, a problem that, although explicitly addressed by Vuilleumier and Simberloff, means that inadequate sampling must be seriously considered as a cause of apparent distributional patchiness.

Of the 55 páramo species amenable to analysis, 16 (29%) have patchy distributions, and another 11 may (although this may reflect inadequate knowledge of their distributions) (Table 8.3). Considering just the 25 species ecologically specialized to the páramo, 40% exhibit certain distributional patchiness. Because it is more continuous than the páramo, one might expect patchiness to be less pronounced in the puna zone, but it is significantly greater in the set of all species (although no different for the puna specialists). Both zones contain a substantial number of species with patchy distributions.

Vuilleumier and Simberloff offered several possible explanations for this distributional patchiness. Incomplete sampling has already been mentioned, and it was thought to contribute to the patterns of perhaps one-third of the species in each zone (Table 8.3). The distributional patchiness of over half of the species was probably associated with habitat patchiness; interspecific competition could be implicated as a possible cause of an equal

Table 8.3. *The extent of distributional patchiness and its possible causes among bird species of páramo and puna vegetation zones in the high Andes of South America*

Category	Páramo	Puna
Distributional pattern		
Total species	55	115
Patchy	29%	41%
Not patchy	51%	31%
Uncertain	20%	28%
Restricted species[a]	25	63
Patchy	40%	36.5%
Not patchy	44%	36.5%
Uncertain	16%	27%
Possible causes of patchiness[b]		
Incomplete sampling	3	8
Habitat patchiness	6	15
Competition	6	15
Pleistocene history	6	4
Number of species with:		
Only one explanation	1	8
Two possible explanations	7	11
Three possible explanations	2	4

Notes:
[a] Species ecologically specialized to páramo or puna.
[b] Number of patchy restricted species for which a particular factor possibly influences the patchiness ($N = 10$ for páramo, 23 for puna).
Source: Summarized from Vuilleumier & Simberloff (1980).

number of distributions. For most species, more than one cause seemed likely, and for all of the species chance may also have been involved in producing the observed distributions. On the basis of an analysis of individual cases, Vuilleumier and Simberloff concluded that three páramo species and five in the puna have patchy distributions that cannot be ascribed in whole or in part to environmental patchiness or inadequate sampling. Of these, two of the páramo species and three of the puna species seem most likely to be restricted by competitors; for the remainder, both competition and Pleistocene history may be involved.

Several conclusions emerge from this analysis. First, the effect of sampling errors should not be ignored in distributional analyses, for they may be important. Second, distributional patchiness may be common, even in vegetation types that are more or less continuously distributed. Third, many factors are likely to affect the distributions of most species, suggesting that single-factor categorizations of species (e.g. Table 8.1) may be superficial. Nonetheless, some species do seem to be patchily distributed primarily because of single factors. Although habitat patchiness is the most important of these, competition is clearly of major importance in some situations, but not to the degree that some (e.g. Lack 1971, Terborgh 1971, Terborgh and Weske 1975) have claimed. Fourth, these analyses are based on the presence or absence of species at survey locations. If dispersal rates are high, however, some species may be represented by at least some individuals in virtually all patches. In these situations, an examination of patterns in abundance over a geographical scale rather than presence or absence may be instructive. Finally, because each distributional situation is to a degree idiosyncratic, it is not likely that statistical treatments of the distributional patterns of entire assemblages will demonstrate the existence of situations involving competition, habitat patchiness, history, or some other factor, nor will they reveal which species are involved in them.

Distributions on elevational gradients

In mountainous areas, species that overlap geographically may have differing distributions on elevational gradients. In the páramos of Venezuela, for example, several species occur over broad altitudinal spans, some are found only at low elevations, and others occur only above 3000 m. If one examines the altitudinal distributions of all 25 páramo species, however, they form a more or less continuous series (Fig. 8.6). There are no sudden increases in the rate of species turnover at certain elevations, as might be suggestive of major habitat or community changes (Vuilleumier and Ewert 1978). Two species of wrens (*Cistothorus platensis* and *C. meridae*), however, appear to replace each other altitudinally. In the Mérida Andes, *meridae* is restricted to elevations above 3000 m, whereas *platensis* reaches up only to 1700 m. At another location where *meridae* is absent, a different subspecies of *platensis* ranges from 2200 to 3275 m (Fig. 8.6).

It is tempting to interpret such patterns of altitudinal replacement as evidence of competitive exclusion, as has often been done (e.g. Diamond 1973, 1975a, 1978, 1986b; Cody 1974a; Lack 1971, 1976; MacArthur 1972). Lack explained the distributions of large hummingbird species on West Indies islands in this way. Diamond attributed the abutting elevational ranges of two species of *Crateroscelis* warblers in New Guinea to direct competition and the variations in the altitudinal range of *Turdus poliocephalus* on different islands to diffuse competition with some unknown complex of species. In order to assess both the frequency with which such patterns occur and the plausibility of this interpetation, we can

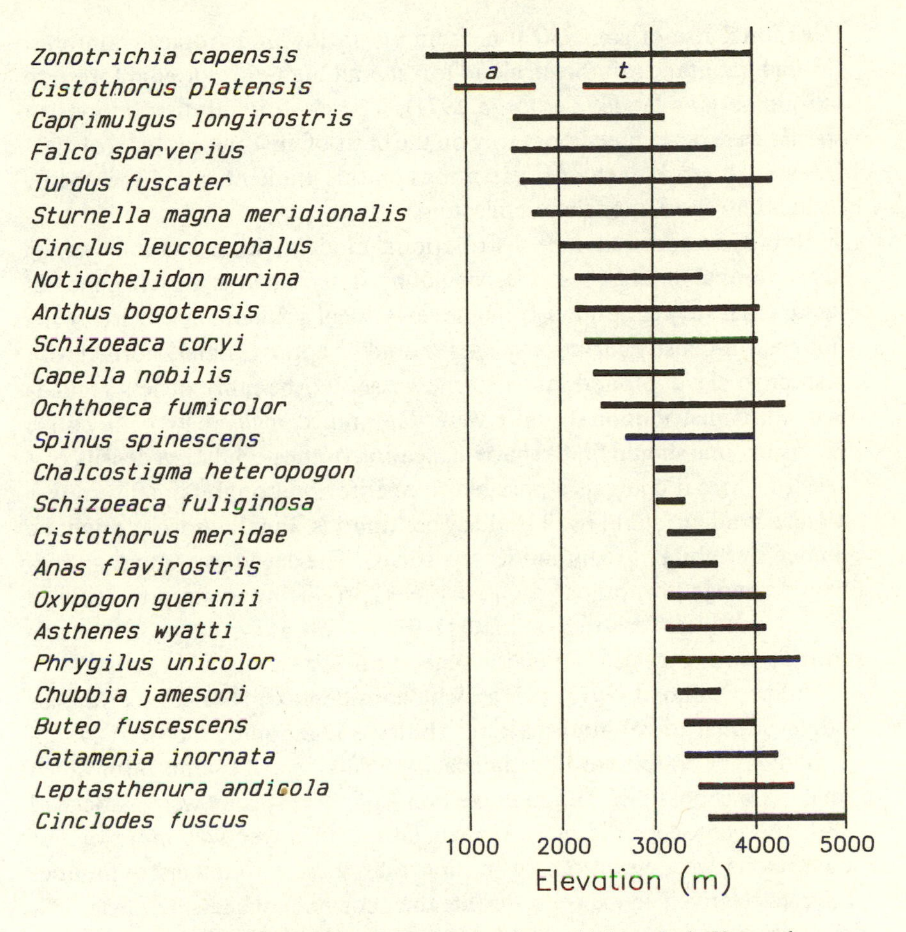

Fig. 8.6. Elevational distributions of 25 páramo bird species in Venezuela. *Cistothorus platensis* has only one population (*tamae*, t) in páramo, but the entire range of the species is included to illustrate altitudinal differences between *C. platensis* and *C. meridae*. After Vuilleumier and Ewert (1978).

consider the findings of several more intensive studies conducted in continental and insular settings.

Elevational distributions on continents

In what must rank as one of the most arduous community investigations ever undertaken, Terborgh and his colleagues (Terborgh 1971, Terborgh and Weske 1975, Terborgh 1985a) documented the distributions of a large number of species over elevational gradients in the Peruvian Andes. The first gradient, on the Cordillera Vilcabamba, included an elevational rise of some 3000 m from virgin lowland tropical rainforest through montane rainforest, cloud forest, and high-altitude elfin forest, to tall alpine grasslands (Terborgh 1971). The elevational distributions of species were determined primarily on the basis of mist-net captures of 200–300 individuals at each of 11 elevations on this gradient, supplemented by incidental observations and collecting.

Terborgh addressed three distributional models with his data. First, the distributional limits of species could be determined by environmental factors that vary continuously on the elevational gradient, in which case the population-density curves of species should be approximately normal with respect to elevation and faunal turnover should be more or less regular. Second, if distributional limits were determined instead by competitive exclusion, one should find a sharp truncation of the population density of a species where it contacts a possible competitor, although faunal turnover along a gradient should still be fairly continuous. Finally, species might be limited by habitat discontinuities (ecotones). The densities of many species would change abruptly at such ecotones, producing a sharp increase in faunal turnover. Terborgh estimated habitat ecotones visually, on the basis of transitions between the major vegetation zones.

Unlike Diamond (1973, 1975a), who considered only selected examples of elevational distribution patterns (but see Diamond 1986b), Terborgh evaluated the patterns of all of the species occurring in his samples for which analysis was possible. Although the limits of some species were associated with the ecotones (Fig. 8.7), overall faunal turnover was more or less continuous with elevation, suggesting that these habitat discontinuities affected relatively few species, despite their conspicuousness. Evidence of a competitive determination of distributions was found in the mutual exclusion of several congeners in linear series along the gradient, a tendency for the elevational range occupied by such species to be reduced in larger suites of congeners, and the truncation of abundance curves at the area of contact between congeners (Fig. 8.7). Terborgh considered the distributional limits of 261 species; of those cases in which the lower or upper distributional limit did not coincide with a terminus of the transect, 20% had limits associated with major habitat ecotones, 33% met the criteria for the competition model, and the remaining cases were assigned (some by default) to limitation by continuously varying environmental factors.

Terborgh considered only situations involving elevational replacements among congeners to be evidence of competition. Terborgh and Weske (1975) suggested that these categorizations underemphasized the role of competition and overestimated the effects of ecotones and environmental

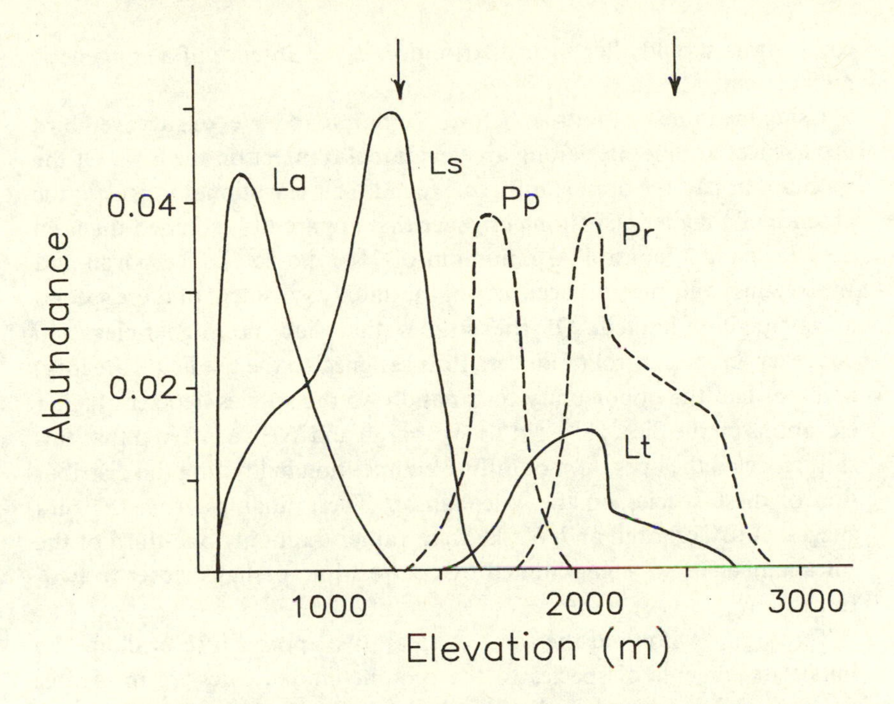

Fig. 8.7. Distribution–abundance curves for several flycatcher species on an elevational gradient in the Peruvian Andes. Species La (*Leptopogon amaurocephalus*) and Ls (*L. superciliaris*) overlap broadly, although *superciliaris* reaches its maximum abundance in the absence of *amaurocephalus*. Both of these species appear abruptly above the lowland-montane rainforest ecotone. Truncation of the curves is apparent in the transitions between *superciliaris* and *L. taczanowskii* (Lt) and *Pseudotriccus pelzelni* (Pp) and *P. ruficeps* (Pr). Arrows indicate the positions of habitat ecotones. After Terborgh (1971).

gradients. To test this, they compared the distributional patterns from the Cordillera Vilcabamba with those on an isolated massif, the Cerros del Sira, some 360 km away. At higher elevations, the Sira is exposed to colonization primarily from below rather than along continuous ridges, as in the Cordillera. As a consequence, roughly 80% of the species that would be expected to occupy the summit zone of the Sira (on the basis of their distribution on the Vilcabamba) are absent. In addition, the elevational positions of several of the ecotones between vegetation zones are shifted in relation to their position on the Vilcabamba. These differences provide an opportunity to determine how various bird species change in elevational distribution on the Sira. Specifically, those whose limits are set by ecotones should shift to match the new ecotone positions, whereas species limited by competitors should alter their distribution in the absence of a congeneric replacement.

Using the same methods as before, Terborgh and Weske surveyed bird occurrences at nine sites along an elevational transect on the Sira. Of the species that had the opportunity to expand their elevational ranges in the absence of a higher-elevation congener that apparently excluded them on the Vilcabamba 'control', a minimum of 71% did so. To Terborgh and Weske, this confirmed the accuracy of the initial assessment of these species as competition-limited. Of the species that had no higher-elevation congener on the control (and were thus assigned to the gradient category) and that had the opportunity to expand into the species-deficient higher elevations on the Sira, 58% did so. Terborgh and Weske inferred that this shift reflected the operation of diffuse competition in limiting the distribution of these species on the Vilcabamba. These findings from the Sira suggested to Terborgh and Weske that, rather than only one-third of the Vilcabamba species being limited by competition, perhaps closer to two-thirds were.

Terborgh (1985) used the same comparative approach to evaluate the initial assignments of species to the ecotone-limited category in greater detail. For 47 species whose limits coincided with the montane rainforest–cloud forest ecotone on the Vilcabamba 'control' gradient, Terborgh determined whether their ranges expanded or contracted on the Sira and two other elevational transects where this ecotone was displaced upward or downward. Because the ability of a species to alter its distribution may be related to how close its center of elevational distribution is to the ecotone, Terborgh considered the patterns relative to the species' center of distribution. When the ecotone was displaced away from the center of distribution (e.g. downward for species occurring above the ecotone), 36 of 41 species (88%) expanded their distributions, confirming the initial categorization of these species as ecotone-limited. When the ecotone was shifted toward the center of distribution, however, 43 of 44 species (98%) failed to contract fully; most exhibited some range contraction, but not enough to maintain the coincidence of their distributional boundary with the ecotone. Terborgh attributed this failure to the tendency of species to occupy a greater variety of habitats close to their center of distribution than far from it. This may be so if the breadth of habitat occupancy of a species is density-dependent, but the evidence of this is generally inconclusive (Chapter 9) and does not exist for any of the species that Terborgh encountered. Terborgh concluded that ecotones might account for only one-sixth of the elevational limits of Andean birds.

How reliable are these patterns and their interpretations? The logistical difficulties of the surveys on the transects cannot be overemphasized (Terborgh nearly lost his life on one expedition), and the data are probably as good as can be obtained under such circumstances. The distributional patterns were established by a combination of mist-netting and unsystematic observations. Information is thus probably adequate for the species occurring within 2 m of the ground that are readily captured in nets but is less reliable for other species. It was also assumed that the different elevational gradients in the Peruvian Andes are ecologically similar except for the shifts in ecotones and the high-elevation species impoverishment of the Sira (a *ceteris paribus* assumption). Although dawn temperatures at a given elevation did not differ significantly between the Vilcabamba and Sira areas, no other measures are available to assess this assumption (Terborgh and Weske note in passing, however, that the vegetation of the cloud forest and elfin forest on the Sira contains many endemic plant species). On the basis of their experience in conducting many elevational transects on Hawaii, Mountainspring and Scott (1985) observed that, unless care is taken to ensure habit uniformity between elevational zones, the observed distributional patterns may represent responses by the species to unrecorded habitat features that differ between sites. The habitat zones that Terborgh and his colleagues used to define the ecotones were also coarse, each encompassing a broad range of habitat variation. Thus, responses of species to habitat changes occurring in locations other than at recognized ecotones could be incorrectly assigned to a gradient explanation. Finally, in order to compare distributional shifts in response to changes in ecotone locations or species diversity on different elevational gradients, one must not only make the *ceteris paribus* assumption noted above but must assume as well that the distributional patterns are stable and have reached an equilibrium determined by whatever factors control the limits. Because of the difficulty of documenting distributional patterns in these areas, Terborgh had no information on their temporal stability and could not assess this assumption.

Thus, although the distributional patterns that Terborgh and his colleagues recorded may be a reasonably accurate representation of the actual elevational ranges of the species that are fairly common and amenable to netting, the conclusions about the general determinants of distributional limits among Andean birds must be regarded with some doubt. These conclusions also stand in marked contrast to those of Vuilleumier and Simberloff (1980), who could find convincing evidence of competitive determination of distributional limits in only 8% of the páramo and puna species they considered. Vuilleumier and Simberloff used more stringent criteria for accepting a competition explanation, however, and they did not consider the possibility of diffuse competitive effects. Graves (1985) also noted that many of the congeneric species that were nonoverlapping on the Vilcabamba gradient and that were cited as evidence of competition by Terborgh and Weske do in fact overlap elevationally at other locations in the Peruvian Andes, so the patterns observed by Terborgh and his colleagues may not be general in this region.

These studies were conducted in species-rich tropical settings. How do the distributions of birds on elevational gradients in temperate locations compare? In the mountains of the northeastern United States, Able and Noon (1976) surveyed bird distributions at 100-m intervals on several elevational transects by means of informal walking censuses covering a specified time period (a procedure of questionable accuracy). They found that the species occurring at higher elevations had the broadest altitudinal amplitudes. This finding agrees with Terborgh's (1971) observations from the Vilcabamba but contrasts with the pattern in páramo species described by Vuilleumier and Ewert (1978) (see Fig. 8.6). Three major ecotones between vegetation zones occurred in the northeastern mountains, and slightly more than half of the species' distributional limits coincided with these ecotones (e.g. Fig. 8.8). Able and Noon found no differences between the elevational amplitudes of species having congeners present and those lacking congeners or between species within the same guild and those in different guilds, nor were there any convincing cases of elevational replacements among potential competitors. Here, then, the primary determinant of distributional patterns appears to be habitat rather than the presence or absence of other species. Sabo (1980) documented similar patterns on another elevational gradient in the same region. Working in the mountains of southeastern Arizona, Miles (personal communication) also found that most of the elevational turnover he observed was associated with habitat change (although his sampling was focused on obvious ecotones and thus may have been predisposed to detect such patterns). There was especially great turnover in the transition from oak to pine woodlands, and this was accompanied by a major shift in community ecomorphological patterns. This ecotone represents a change from broadleaf to needleleaf trees, and the different morphological patterns may be associated with the different foraging tactics that are called for in the different vegetation types. The flora of the pine and fir zones is also of boreal derivation whereas that of the lower oak and desert scrub zones reflects a Madrean biogeographic influence, raising the possibility that biogeographic and historical factors may also be involved in the elevational distributions of the bird species.

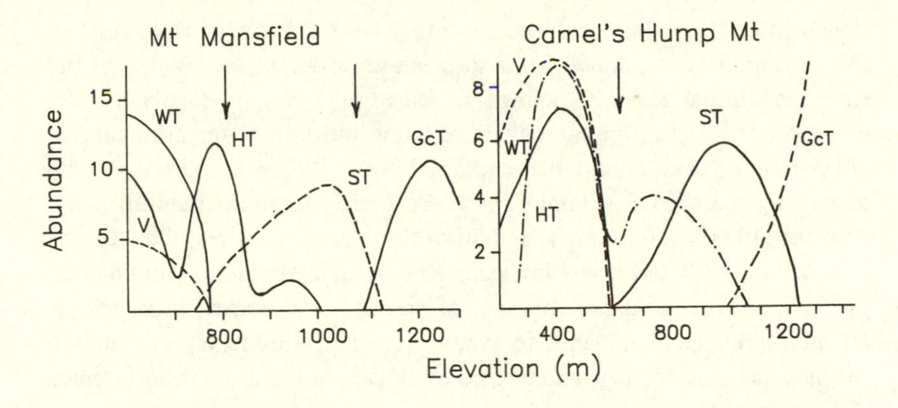

Fig. 8.8. The elevational distributions of thrush species on two northeastern United States mountains. The thrush species are Wood Thrush (WT), Veery (V), Hermit Thrush (HT), Swainson's Thrush (ST), and Gray-cheeked Thrush (*Catharus minimus*, GcT). Arrows indicate the positions of habitat ecotones. After Able and Noon (1976).

Why are the distributional patterns from these temperate elevational transects so different from those in the tropical Andes? Able and Noon suggested that the temperate species, most of which are seasonal migrants, may segregate more easily by habitat than by space (elevation). Spatial separation may require the long-term residency in an area typical of tropical individuals and species. In addition, the greater species diversity of the tropical communities may prompt greater intrahabitat spatial separation (Noon 1981, Diamond 1973) Terborgh (1985) has proposed that the tremendous diversity of plant species in the tropics discourages preferential foraging by bird species. They therefore may not respond to elevational changes in plant species composition in the same manner as birds in temperate vegetation, where ecotones are determined by changes in a few dominant plant species on which the birds may forage selectively (Robinson and Holmes 1984). All of these notions remain untested. In any event, a close comparison of the temperate patterns observed by Able and Noon with those documented in the tropics by Terborgh and Weske is unwarranted, given the vast differences in the methods of the two studies.

Elevational distributions on islands

If species interactions are important in determining elevational distributions, one might expect species on mountainous islands to have broader elevational ranges than their mainland counterparts, as they respond to the reduction in species number and the absence of presumed competitors. Such seems to be the case for Jamaica, where Lack (1976) documented that many of the species range from the lowlands through to the mountains and relatively few species are restricted to any one of the three elevational zones he defined. Species thus overlap broadly in their elevational distributions and there are few altitudinal replacements of congeneric or even confamilial species. In contrast, in Honduras (a presumed source area for Jamaica), where there are more habitat types, Monroe (1968) found more than half of the species to be restricted to one type and only 6% of the species occurred in both lowlands and mountains. This comparison is a bit strained, of course, as the area, topography, climate, biotic environments, food resource distributions, level of human disturbance, past history, and bird communities of Honduras and Jamaica differ vastly.

Although most comparisons of elevational patterns have left other key variables unmeasured, assuming *ceteris paribus*, this is not the case for the Galápagos Islands. There, Schluter (1982b, Schluter and Grant 1982) related the altitudinal distributions of *Geospiza* species during one year on Isla Pinta to quantitative measures of habitat and food supplies. Although relatively small (60 km^2), Pinta rises to an elevation of 650 m and contains a habitat gradient from arid, rocky lowlands through an intermediate zone with scattered trees to a more humid highland zone in which trees form a nearly closed canopy. Five *Geospiza* species occur on the island, but one (*scandens*) is rare and was not considered. During Schluter's study, the altitudinal distributions of the other four species were broad, overlapped considerably, and changed between seasons. During the early wet season in February, for example, *difficilis* was most abundant (as gauged by mist-netting and transect censuses) at higher elevations, *fuliginosa* at intermediate altitudes, and *fortis* at lower regions, although the species overlapped considerably. By the end of the wet season in May, all four species were most abundant at low to intermediate elevations, while toward the end of the dry season in November, the species all reached maximum densities in the higher regions of the islands. As habitat features changed little during this period, it seems unlikely that the elevational distributions of the species were governed to any great degree by the habitat changes with altitude (although *difficilis*, which forages extensively for arthropods and gastropods in litter, tended to be more frequent at the higher elevations where litter is more plentiful; see Table 8.4). The elevational distribution of the availability of the seed and arthropod foods of the finches varied seasonally, however, and the distributions of the birds tended to parallel food abundance, especially during the dry season (Table 8.4). At this time, food may be limiting and diet overlap among the species is reduced as the species

Table 8.4. *Correlations between densities of* Geospiza *finches and measures of food supplies and litter volume (*G. difficilis *only) at several times of year.*

Coefficients in parentheses are based on finch density estimates derived from census-walk data; other values are from mist-netting. Underlined coefficients are significant ($P < 0.05$); $n = 6$ in each case.

Species	Correlation with finch density			
	February	May	August	November
	Litter volume			
G. difficilis	0.93	−0.67	0.72 (0.75)	0.93 (0.72)
	Food availability			
G. difficilis	0.92	−0.75	0.84 (0.93)	0.57 (0.92)
G. fuliginosa	0.08	0.85	0.58 (0.92)	0.88 (0.79)
G. fortis	−0.01	0.94	−0.06 (−0.11)	0.62 (0.36)
G. magnirostris	−0.41	0.46	0.91 (0.92)	0.71 (0.94)

Notes:
From Schluter (1982a).

opportunistically exploit somewhat different food types or switch to different feeding positions. There was no evidence of negative distributional associations among potential competitors. Although this cannot be taken as evidence that interspecific competition is absent in this system, it suggests that other factors, primarily variations in the food suppy, are the major determinants of the altitudinal distributions of the finches.

Two of the *Geospiza* species on Pinta, *difficilis* and *fuliginosa*, have a patchy distribution on lowlying islands in the Galápagos. Of 25 such islands, 20 contain only *fuliginosa*, 3 only *difficilis*, and 2 have neither, a distinctly nonrandom distributional pattern (Grant and Schluter 1984). Schluter and Grant (1982) used the information from the studies on Pinta to explore three hypotheses in an effort to explain the distributional patchiness of these species on lowland islands, testing them with similar observations gathered on Marchena (*fuliginosa* only) and Genovesa (*difficilis* only).

First, the distribution of each species might be determined by fixed food requirements that do not vary between populations. A species would thus be missing from islands not meeting these food requirements. Areas containing *difficilis* alone (Genovesa), then, should be similar in food supply to the higher elevations on Pinta where *difficilis* in most common, and the diet and population densities of *difficilis* should be the same in the two areas. In terms of seed-producing plant genera and litter composition and abundances, Genovesa is actually more similar to lowland than to highland areas on Pinta. Densities of *difficilis* on Genovesa are much greater than would be expected from the density–food supply relationship on Pinta, and birds on Genovesa forage extensively on seeds during the dry season, when Pinta individuals continue to feed largely on invertebrates.

A second hypothesis relates the distribution of the finches to variable food requirements. If the food requirements of a species differ among islands because of local adaptation to food supplies, one might expect the diet and density of the species to differ among islands, although because the suitable ranges of foods of *difficilis* and *fuliginosa* are different, islands containing only one or the other should differ in food supplies. Specifically, Genovesa should be unsuitable for *fuliginosa* and Marchena should be unsuitable for *difficilis*. This prediction is not met. The two islands are quite similar in seed-producing floras, and the food types used by *fuliginosa* on Marchena are at least as plentiful on Genovesa. The reverse also holds: Marchena seems by and large to provide an abundance of suitable foods for *difficilis*.

This leaves the hypothesis that competition plays a part in determining the patchy distributions. If this is true, we should find that the *difficilis* population on Genovesa is similar in diet and density to the population of *fuliginosa* on Marchena. This seems to be the case. Populations of *fuliginosa* occupy similar lowland situations on Marchena and Pinta and not surprisingly have similar diets and densities; this confirms the adequacy of the interisland comparison. The population of *difficilis* occupying Genovesa exhibits changes in diet and density-food supply relationships from the patterns in the population occupying higher elevations on Pinta, and these increase its similarity with *fuliginosa*.

For these *Geospiza* species in the Galápagos, then, altitudinal distributions on islands with highland zones may be variable and associated largely with the distributions of food supplies, especially during the dry season. Patterns of distributional patchiness among lowland islands, on the other hand, may be more closely related to the presence or absence of another species than to food supplies, at least for *difficilis* and *fuliginosa*. This investigation stands in marked contrast to most studies of elevational or patchy distributions by virtue of the careful way in which both the distributional patterns and the associated environmental variables were quantified. Such detail may not be attainable in all situations (e.g. Terborgh's Andean transects), but it permits a level of analysis that fosters increased confidence in the conclusions.

One other island situation indicates the importance of recent habitat changes in determining elevational distributions of species. Native forest habitat has been progressively destroyed on islands in the Hawaiian archipelago, first through clearing and burning by Polynesians (Olson and James 1982) and more recently by grazing, cutting, and clearing for cultivation by European settlers (Moulton and Pimm 1983, 1986; Scott *et al.* 1986). As a consequence, little native vegetation remains below 600–900 m elevation on most islands, and the habitats there are comprised almost entirely of exotic plant species or cultivated crops. As a consequence of these profound habitat changes (as well as heavy predation by rats, dogs, and Polynesians), many of the native bird species have suffered extinction and few of the remaining native species are ever recorded below 900 m. A variety of introduced bird species occupy the disturbed lowlands, but these are largely absent from the remaining native vegetation at higher elevations. Diamond and Veitch (1981) documented a similar pattern on New Zealand islands, where exotic bird species have become established only following the disruption of habitats and the disappearance of native bird species. In Hawaii the situation is complicated by the distribution of avian malaria, to which many native species are susceptible and exotic species largely immune. The mosquito vectors of the parasite are most abundant at lower elevations, and this may also contribute to the elevational limitation of the native species (van Riper *et al.* 1986).

Distribution–abundance patterns

Generally, distributional patterns are determined by a consideration of where species are present and where they are absent. The abundance of species varies within the areas occupied, however, and patterns of geographic variation in abundance may provide additional insights into the distributional relationships among members of communities.

When one examines the geographical distribution of densities of a species, a wide range of patterns is possible. For example, the distribution of Emus (*Dromaius novaehollandiae*) in Australia is discontinuous, with concentrations in semiarid regions of the eastcentral and western portions of the continent (Grice *et al.* 1985) (Fig. 8.9A). In the east, densities are greatest in the center of the range and decrease more or less steadily toward the periphery; this pattern is less clearly expressed in the west. Abundance generally decreases from areas used for sheep grazing to grain-growing regions to cattle-grazing areas and is least in areas receiving no commercial use (deserts). Grice and his colleagues attributed this pattern to climatic factors that determine the availability of food during breeding, the avail-

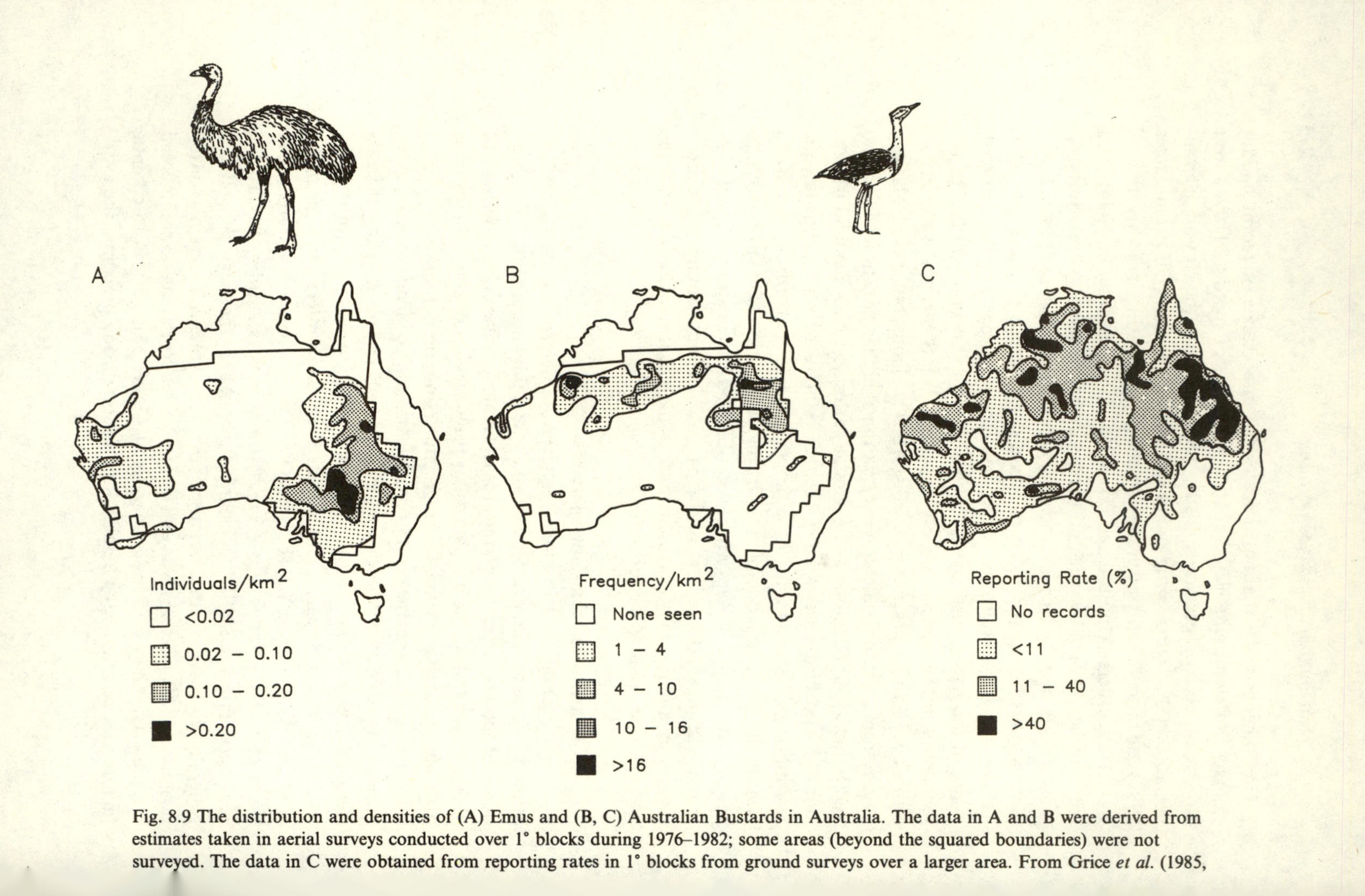

Fig. 8.9 The distribution and densities of (A) Emus and (B, C) Australian Bustards in Australia. The data in A and B were derived from estimates taken in aerial surveys conducted over 1° blocks during 1976–1982; some areas (beyond the squared boundaries) were not surveyed. The data in C were obtained from reporting rates in 1° blocks from ground surveys over a larger area. From Grice *et al.* (1985,

ability of surface water, and the prevalence of dingos. Australian Bustards (*Ardeotis australis*), on the other hand, have a more-or-less continuous distribution in arid regions of northern Australia but show multiple peaks of abundance (Fig. 8.9B); Grice *et al.* (1986) suggested that this pattern is related to the distribution of low-stature vegetation, especially mitchell grass (*Astrebela* spp.). The distributional patterns of both Emus and bustards were derived from aerial surveys of 1-degree blocks. That such patterns may be sensitive to the area surveyed and the methods used is evident from Fig. 8.9C, which shows the distribution of bustards determined by ground surveys in both wooded and open regions of the continent (Blakers *et al.* 1984). According to these surveys, bustards reach greatest densities in the eastern tablelands, where recent clearing of woodlands and planting of pastures has created apparently favorable habitat. Whether the accompanying increases in bustard abundance are temporary or permanent is uncertain (Grice *et al.* 1986). Additional examples of distributional patterns may be found in Blakers *et al.* (1984), Scott *et al.* (1986) and Root (1988d).

The proposition that densities of a species should be high in the central portion of its range and decrease toward the periphery was central to Grinnell's thinking (1922), was termed a 'general biogeographical rule' by Hengeveld and Haeck (1982), and is a basic prediction of Brown's (1984) 'general theory' of distributions. Brown based his argument on two assumptions: (1) the abundance and distribution of species are limited by a combination of physical and biotic variables that collectively define their multidimensional niches, and (2) spatial variation in these variables is autocorrelated – that is, nearby sites tend to have more similar environments than do distant sites. Together, these assumptions led Brown to conclude that ecological conditions should be most favorable for a species (and density correspondingly greatest) in one area and that abundance should decrease with increasing distance in any direction from that location (see Fig. 8.1). The decline is gradual because densities are determined by the additive effects of many factors that are distributed independently of one another in space. Under these conditions, the Central Limit Theorem predicts normal abundance-distribution curves. Brown's argument also rests on a third, unstated assumption, that there is an equilibrium between niche conditions at the locations and population densities – in other words, that habitats are fully saturated.

Exceptions to the predictions of the theory occur when these assumptions are not met. Thus, environmental patchiness may lead to multimodal abundance patterns (e.g. Figs. 8.1, 8.9) or abrupt environmental changes may produce truncated distributions in which abundance is high close to the edge of the range. In both instances, the assumption of spatial autocorrelation has been violated. Brown suggests that such anomalous patterns may often disappear when the data from many sites are averaged or when the distribution is viewed over a sufficiently large area. This, of course, is an inevitable consequence of any averaging procedure; it does not imply that centered distributions of abundances are necessarily more general, biologically, than are patchy or truncated distributions.

Some avian distributions do appear to agree with the predictions of Brown's model. In a survey of the density distributions of 41 landbird species along a 1200-km transect through bottomland deciduous forest in the central United States, for example, Emlen *et al.* (1986) found that the densities of most species tended to be greatest in the range center and to decline toward the boundaries. In some cases the distributional patterns of one species were truncated in association with sudden increases in the densities of presumed competitors (Fig. 8.10), although other, noncompetitive explanations of these patterns are also possible (see Volume 2). In other situations, the distribution of abundances seems to vary considerably among species. Scott *et al.* (1986), for example, produced detailed distributional maps for all of the species occurring in forest and woodland habitats on the Hawaiian Islands; some species, such as the Elepaio (*Chasiempis sandwichensis*) or the Maui Creeper (*Paroreomyza montana*), have maximum abundances centered in their ranges, whereas others, such as the Common Amakihi (*Hemignathus virens*) and the Apapane (*Himatione sanguinea*) show multimodal or sharply truncated abundance distributions. In another study, Root (1988c) plotted the distributional abundances of 48 species wintering in North America. If areas in which density reaches at least 60% of the maximum abundance recorded are considered as 'peaks', then 85% of the species had multiple peaks of abundance within their distributional range. According to this criterion, only 4% of the species exhibited the sort of unimodal, Gaussian distribution of abundances emphasized by Brown.

Brown's theory also offers another prediction, based on another assumption. If closely related or ecologically similar species differ in no more than a few niche dimensions, the more widely distributed species should also be expected to have the greatest average abundance where they occur. There is an intuitive appeal to this idea: one might expect rarer species to be less frequent in their occurrence in a series of localities sampled, whereas very common species are also likely to be widespread and occur in most locations where the habitat is suitable. Hanski (1982) suggested that this pattern is a 'general rule' of nature.

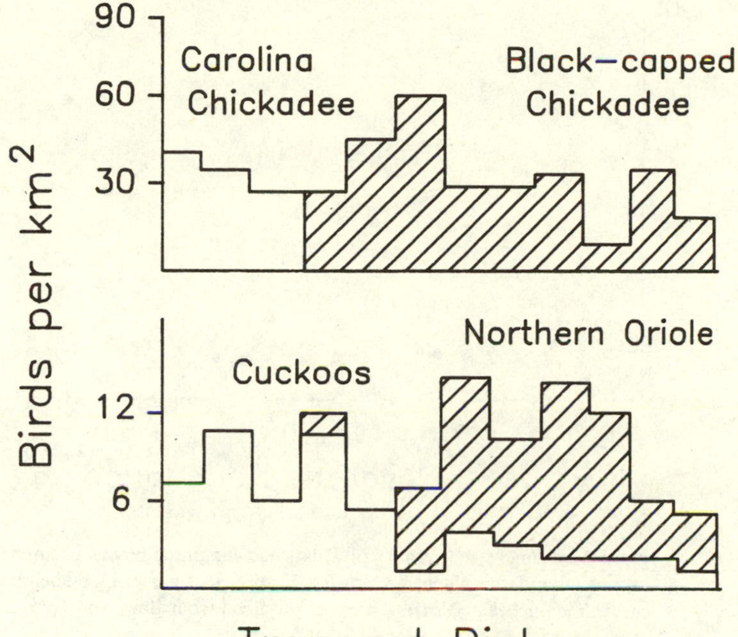

Fig. 8.10. Density profiles of species along a 1200-km transect in bottomland deciduous forest in North America. Above: patterns for Carolina (*Parus carolinensis*) and Black-capped chickadees; below: patterns for Northern Orioles (*Icterus galbula*) and cuckoo (*Coccyzus spp.*) species that are ecologically similar to the orioles. The abscissa is oriented from south (left) to north (right) on the transect. After Emlen *et al.* (1986).

In many instances it does seem that the more abundant bird species in a region also occur most frequently at various locations. Using information from Christmas Bird Censuses, Bock and Ricklefs (1983) determined that emberizid and cardueline finch species with large wintering ranges in North America had significantly greater mean local abundances than more restricted species. If species are most abundant at the centers of their ranges and one then samples an arbitrarily defined geographical area, however, species that by chance have their centers of distributions in the study area will be recorded as being widespread and abundant but equally widespread and abundant species whose centers of distribution lie outside the study area will be tallied as rare and infrequent because only the periphery of their distributions will be sampled. The correlation may thus be spurious. Bock and Ricklefs controlled for this source of error by analyzing only those

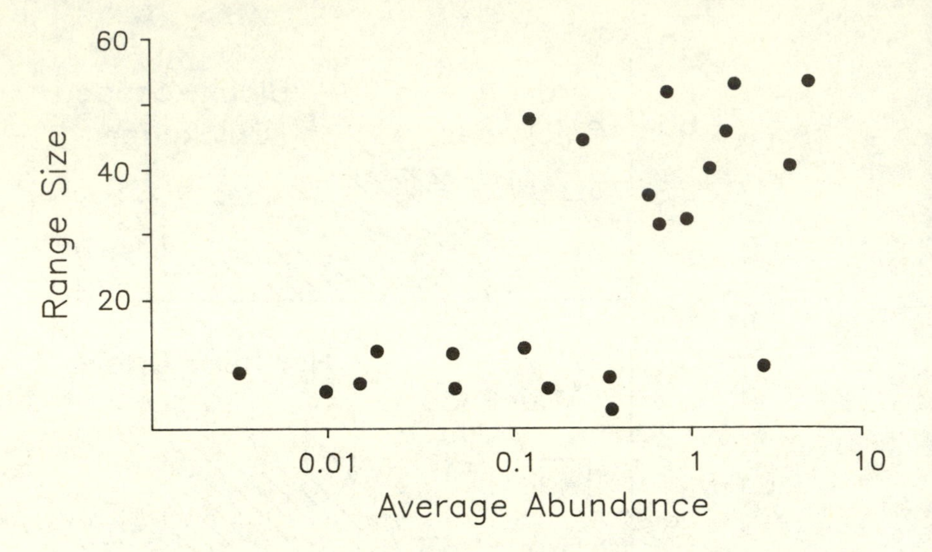

Fig. 8.11. Range size (number of 5° latitude–longitude blocks occupied) versus average abundance where present for 22 species of wintering emberizid and cardueline finches in North America. Modified from Bock and Ricklefs (1983).

species with at least 90% of their wintering range in the study area; again, there was a significant positive correlation between range size and mean local abundance (Fig. 8.11), although the species separated rather sharply into two distributional groups (widespread and restricted). All of the widespread species are abundant. Most of the restricted species are much less common where they occur, but some reach densities equivalent to those of the widespread species.

Other studies of birds (e.g. Bock 1984, 1987, Lacy and Bock 1986) have also found a general relationship between range size or frequency of occurrence and average local abundance. There are exceptions, however. In British woodlands, for example, some low-density species are also infrequent regionally but other equally low-density species occur at most sites (Fuller 1982) (Fig. 8.12). Haila *et al.* (1980) showed that, although the abundance of landbird species on Åland is positively related to the variety of habitats occupied (and thus frequency of occurrence), several species are restricted in habitat occupancy but abundant and other species that occur at low densities have very broad habitat amplitudes. In North America, abundances are not correlated with distributional breadth when ecologically and/or taxonomically dissimilar species are combined in the analysis (Root 1988c) or when one compares habitat-distribution breadth and

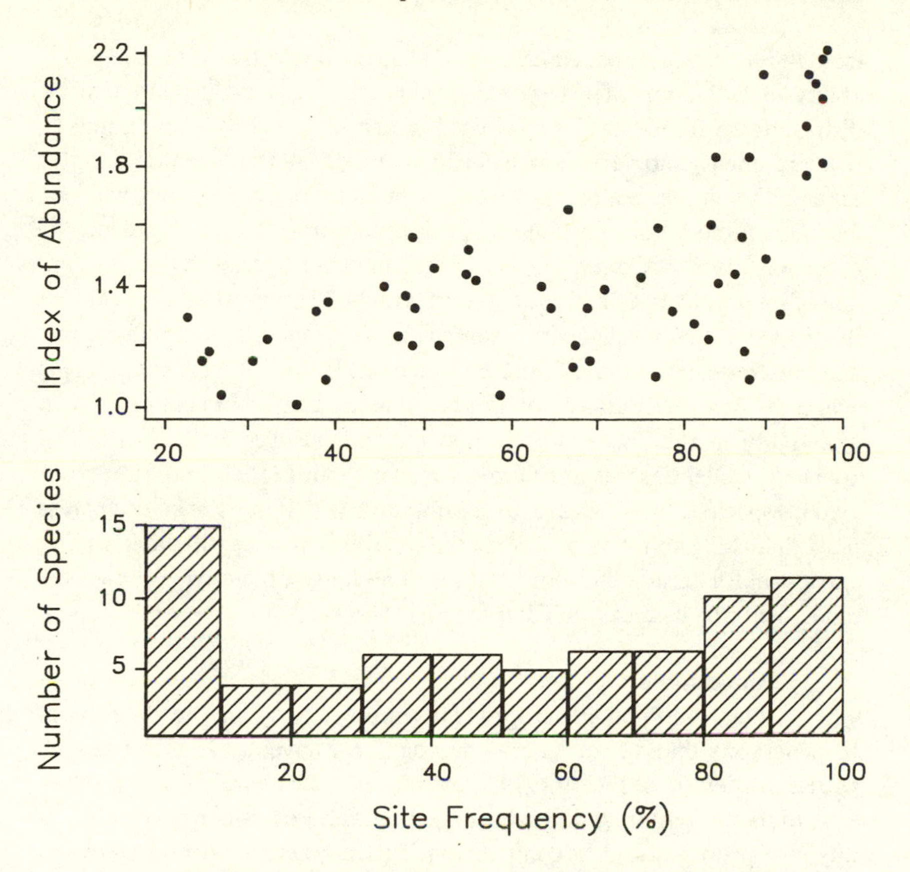

Fig. 8.12. Above: The relationship between the abundance of breeding woodland birds and the frequency with which they occur in British woods. Site frequency is the percentage of sites at which each species was recorded (frequencies < 20 are not shown). Below: Frequency-histogram showing the number of species occurring at sites at a given frequency level. From Fuller (1982) and data therein.

abundance at a local scale (Seagle and McCracken 1986). In Australia, Ford (MS) found no relationship between measures of range size or habitat breadth and mean local density for eucalypt woodland birds. The absence of a clear relationship in this case may be due to the distribution of habitats in Australia. As the environment becomes drier toward the interior of the continent, habitat types occupy progressively larger areas. Species that are rainforest specialists will therefore have relatively small ranges, regardless of their abundances, whereas savannah or desert specialists will have much larger ranges, again regardless of their abundances.

One additional prediction about distributional patterns has been derived from these relationships. Hanski (1982) has proposed that if (1) the probability of extinction of a regional population is inversely related to its distributional frequency (as visualized in den Boer's (1981) 'spreading of risk' argument) and varies stochastically, and (2) all the sites one samples are equally suitable to all of the species in an assemblage, then there will be a distinctly bimodal distribution of species according to their frequency of occurrence at sites. Communities will contain 'core' species that are locally abundant and regionally frequent, and 'satellite' species that occur at low densities and low frequencies among sites. The separation of widespread and low-frequency species found by Bock and Ricklefs (1983; Fig. 8.11) is suggestive of such a pattern, but there is only the faintest indication of such bimodality in the data for British woodlands (Fuller 1982; Fig. 8.12). Hanski's model has been criticized (Brown 1984, Gotelli and Simberloff 1987), especially with regard to assumption (2). If bird species express individualistic habitat associations, the suitability of sites will differ among species and a basic condition for the bimodality predicted by Hanski's model will not be met (see Chapter 9).

Conclusions

Distributional geography falls more clearly in the domain of biogeography than of community geology, so I have not considered such patterns in detail (see Pielou 1979, Udvardy 1969, Brown and Gibson 1983). Distributional patterns, however, have been used as evidence of community-level processes such as competition. This may account for the emphasis given to the distributions of congeners in many studies. Our preconceptions about what factors may be important has led us to examine distributional patterns in certain kinds of ways, almost to the exclusion of others.

Competition does seem to contribute to determining some distributional patterns. The patterns, however, merit our attention quite beyond any role that competition may play. A great many other factors may act to determine distributional limits and abundance patterns, and knowledge of the influence of these factors may produce some valuable insights into the process of community assembly. In particular, the effects of history, as expressed in speciation events, recent extinctions, Pleistocene changes in climate and habitat distributions, and the like, is almost certain to be important. Perhaps because such effects are difficult to document, they are often ignored. Climatic or physiological influences on distributional patterns have also been neglected, although their effects may be easier to establish.

Usually the determinants of distributional patterns are complex. In the absence of detailed, quantitative information on several factors, any attempt to attribute a particular distributional pattern to any single cause must be regarded with skepticism, as a provisional hypothesis to be tested rather than a confirmed explanation. Any such attempts, of course, rest on the presumption that the patterns have been documented accurately. Conclusions based on patterns derived from informal surveys involving a few individual sightings at one place at one time are fragile. The absence of a species from an area often has an uncertain meaning, and unless censusing has been extremely careful and thorough, 'inadequate sampling' must be considered as a possible explanation of distributional patterns.

9

Habitat distributions of species

Differences in the habitat affinities of species figure prominently in interpretations of the broad distributional patterns discussed in the previous chapter. Studies of the habitat relationships of species were popular before the ascendancy of niche theory (e.g. Grinnell 1917, Lack 1937, Pitelka 1941, Kendeigh 1945, Svärdson 1949), but the ideas of MacArthur and his colleagues provided a fresh impetus and fostered a profusion of publications. Evaluation of habitat relationships has become an important part of wildlife and resource management as well (e.g. Capen 1981, Van Horne 1983, Verner *et al.* 1986).

All bird species are restricted, to varying degrees, in the range of habitats they occupy. By definition, the members of a local community share at least portions of their habitat ranges, but as one moves between habitats in a local or regional landscape mosaic, some species disappear and other species appear; this turnover is expressed in beta diversity (Table 5.1). Such patterns are seen most clearly by following the changes in bird populations that accompany habitat succession through time. As abandoned cultivated fields or pastures change from grassland through open shrubland to young forest, for example, bird species enter and leave the community in sequence, in accordance with species-specific habitat requirements related to nest sites, cover, food, foraging sites, and so on (Lanyon 1981, Johnston and Odum 1956, Diehl 1986, Törmälä 1980, Mehlhop and Lynch 1986; Fig. 9.1). Similar changes occur during the succession that follows clear-cutting in forests (e.g. Moss *et al.* 1979, Titterington *et al.* 1979).

In general, analyses of bird habitats have been conducted to address three interrelated questions: (1) Do species sort into well-defined assemblages on habitat gradients, quite apart from the geographical distribution patterns? (2) Do ecologically or taxonomically similar species segregate by habitat? (3) Are variations in the distribution or abundance of particular species closely associated with habitat characteristics? In this chapter I examine a sampling

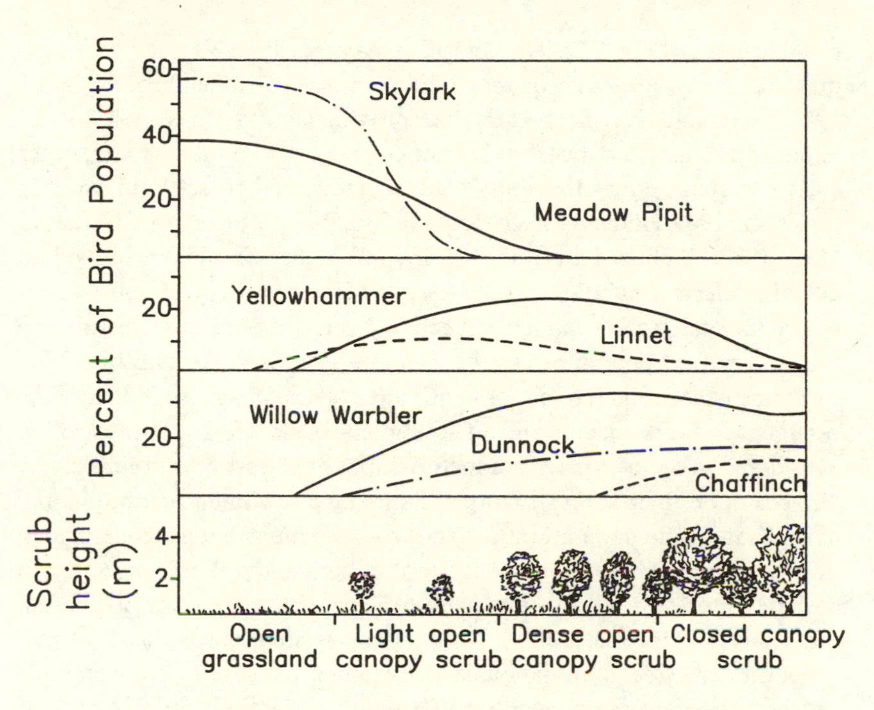

Fig. 9.1. Changes in the relative dominance of several bird species in British habitats arrayed on a successional sequence from open grassland to closed-canopy scrub woodlands. After Fuller (1982).

of bird-habitat patterns related to these questions, emphasizing several analytical approaches that have been especially popular in habitat studies.

The basis of habitat distributions: habitat selection

The patterns of habitat distribution that we observe are a consequence of decisions made by individual birds in selecting a place in which to establish a breeding territory or to overwinter. A great many factors influence these decisions, and differences in the effects of these factors produce variation in habitat occupancy patterns both within and between species. These differences among species are the focus of most studies of bird–habitat relationships, but the variations frequently complicate the correlative comparisons on which they are based. This is not the place to consider these factors and effects in detail (see Partridge 1978, Cody 1985), but some of the basic determinants of individual habitat selection should be mentioned briefly.

We may view individuals of a species as possessing an internal image or template (genetically determined and/or learned) of what constitutes a suitable habitat (Fig. 9.2). This template may specify habitat requirements quite narrowly, although in most species it seems rather broad (Wiens 1985b). Habitats that fit the template provide various 'cues' – aspects of habitat structure, floristics, food resources, edge area, microclimate, other species present, etc. – that may lead an individual to settle in an area (Svärdson 1949, Hildén 1965, Slagsvold 1980, Wiens 1969, 1985b, Helle and Helle 1982, Ambuel and Temple 1983). Whether or not an individual actually selects a habitat on the basis of these cues depends on various environmental factors, which collectively may constrain or distort the 'optimal' pattern determined by the template. Population density may have profound effects, and interactions with other species may also influence the occupancy of a particular area. These interspecific effects are most often considered solely in terms of competitors, but predators or mutualists may also be important in some situations. Time lags also complicate the picture, as individuals and populations may not respond immediately to changes in habitat features, population density, or the abundances of other species. All of these features are not constant in time or space but vary, often in different ways on different scales. The consequence is that the realized habitat selection expressed by individuals is dynamic, the result of an ongoing process. The patterns that result from summing over the individuals of one or several species in one or several areas at one or several times are thus likely to be quite variable, scale-dependent, and often fuzzy. Some of this inconsistency will be apparent in the examples discussed in this chapter.

Multivariate analysis of habitat distributions

At a qualitative level, the general habitat associations of many species are known to any good birdwatcher, and the work of ecologists might be regarded as expressing (or sometimes obscuring) that knowledge in detailed quantitative analyses. These analyses of bird–habitat relationships have been conducted in several ways, but nearly all are based on correlating the presence or absence of species or their densities with measures of habitat features (Wiens and Rotenberry 1981a, Verner *et al.* 1986). In the simplest analyses, variations in the distributions or densities of birds have been related to variations in single habitat features, such as tree density, grass coverage, or foliage height distribution (e.g. Wiens 1969, Flack 1976, Emlen 1972, 1977b). This approach has the advantage of being straightforward and revealing clear, readily interpretable patterns (if such exist). If a large number of bird species is being related to a large number of habitat variables, however, this approach can lead to 'fishing', exploring many possible relationships in the hope of finding some that seem to make

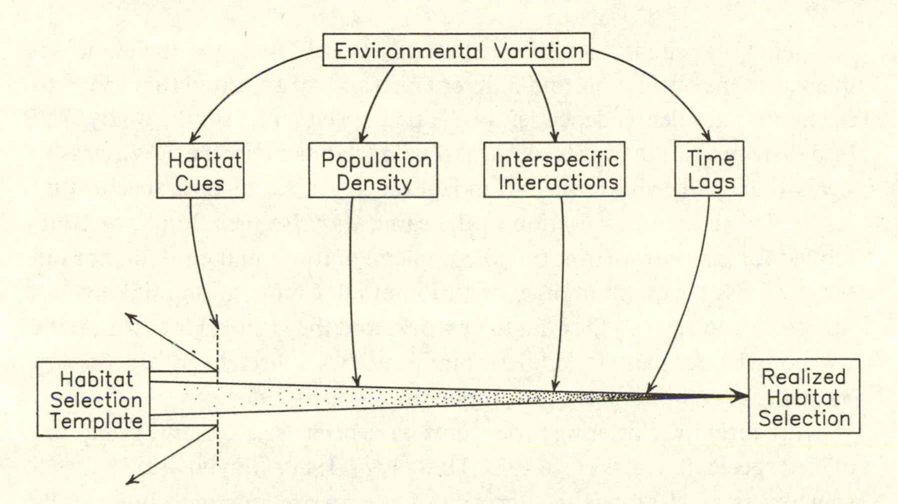

Fig. 9.2. A conceptual model of the influences of various factors on the expression of a basic habitat-selection template of a bird as its realized habitat selection. From Wiens (1985b).

sense (Chapter 3). Of a large number of bivariate correlation tests, some will be significant by chance alone, and distinguishing the spurious correlations from those that have some ecological foundation may be difficult. In single-factor approaches, the effects of intercorrelations among habitat features are also ignored. Significant correlations may then lead to the conclusion that the species responds to a habitat variable in isolation rather than to an interacting complex of habitat features, a 'niche-gestalt' (James 1971). For these and other reasons, multivariate statistical analyses of bird–habitat relationships have become especially fashionable.

Species similarities and groupings

Similarities and differences in the habitat responses of species are best seen when their distributions or abundances are displayed with respect to a continuous environmental gradient, ordered so that the response curves of the species are unimodal (ideally, Gaussian; Gauch *et al.* 1974, Austin *et al.* 1984). Such gradients may actually exist in space (as habitats change with elevation or along an alluvial fan), or they may be synthetic, derived from habitat measures taken at spatially separate locations. Bond (1957), for example, arrayed 64 forest stands from scattered areas in Wisconsin on a continuous ordination according to the importance values of tree species in the stands and then asked whether or not discrete communities of birds could be recognized in relation to this ordination. They could not; avifaunal similarities between stands closely matched the stand sequence on the ordination, and different bird species appeared to respond to the habitat gradient independently of other species. In a similar analysis of bird distributions in forests on the Apostle Islands of northern Wisconsin, Beals (1960) likewise found no indication of suites of bird species that responded to habitat variation in the same way. Neither Bond nor Beals looked for pairwise distributional complementarities in their data, nor did they address the question of competition at all. Competition thinking was not yet in vogue when they did their work, and they instead focused on the issue of whether plant (and bird) communities are discrete entities or vary more or less continuously in composition (Curtis 1959).

More recently, clustering procedures have been used to define groupings of bird species (e.g. Kikkawa 1982, Haila *et al.* 1980, Grzybowski 1982). In some cases no clearly defined groups of species are apparent, but usually several clusters can be detected. After all, clustering is a procedure designed to produce clusters and, given any degree of unevenness in the habitat-similarity matrix of the birds, it will. Rotenberry and I (1978), for example, used cluster analysis to examine the patterns of similarities in the bird-species composition of communities surveyed at a number of locations in the northwestern United States. At one level, three distinct groupings of sites emerged, based on species distributions: those of deciduous and coniferous forests of the mountains and coastal areas, those of the more arid interior ponderosa pine parkland and high desert plateau, and another group of shrubsteppe sites. At another level, several distinct avifaunal regions could be recognized within the shrubsteppe. The agreement of the groupings of sites based on avifaunal similarity with the distribution of general habitat types indicates that bird species within groups are restricted in their habitat distributions in similar ways.

Our analysis included sites that were scattered over a large geographical area containing quite different vegetation formations, so it is not surprising that the clustering procedure defined clear groupings. Such analyses may be more informative when they are applied at a more local scale and based on similarities in the habitat occupied by the species rather than general similarities in bird-community composition. Thus, Collins *et al.* (1982) based their analysis of the habitat distributions of 16 warbler species in northcentral Minnesota on similarities and differences in the species' habitats, as described by 13 features of habitat structure. The analysis defined three groups of species on the basis of their association with open shrubby fields, shrub-forest edge habitats, or mature forest.

In cluster analysis, the distributions of species are not explicitly related to either natural or derived environmental gradients. Instead, species are simple aggregated on the basis of their similarity with respect to some array of habitat features. Of several multivariate procedures that may be used to derive gradients from habitat measures, principal components analysis (PCA) has been most widely used. Measurements of various habitat features are often correlated with one another, and PCA is used to construct synthetic, orthogonal axes that account for the variation present in a matrix of habitat measurements taken at a number of locations. Bird species may then be related to the multivariate space defined by the PCA dimensions according to their correlations with the factor scores of sites on each dimension.

In some instances, the bird species are widely and rather uniformly scattered in the PCA space (e.g. Whitmore 1977, Smith 1977). This pattern indicates both the continuity of habitat variation portrayed by the PCA and the absence of discrete suites of species associated with specific portions of the habitat gradients. In other situations, however, groupings of species in the PCA space are evident (e.g. Sabo 1980). This suggests that real-world habitats are distributed discontinuously in the synthetic space and/or that several bird species share responses to similar habitat conditions. Cody (1983c), for example, used PCA to define gradients of variation in habitat structure in South African veld. The first two components of the PCA accounted for some 80% of the variation present in the original data matrix (which is not surprising, as Cody based his analysis on only six habitat variables, all of which were derived from the vegetation-height profile). Cody determined the positions of his survey plots in the PCA space according to their habitat characteristics and then calculated the similarity in bird species composition between plots. If many bird species do not respond to the gradient represented by the PCA components, the distribution of similarity values in the PCA space should be haphazard. If the species respond to the gradients but do so independently of one another, however, similarity values should vary more or less evenly. In fact, there were distinct breaks in the contours of avifaunal similarity (Fig. 9.3), suggesting that habitat changes at some points on the gradient are accompanied by major shifts in the composition of the bird community. Distinct suites of species appear to be associated with certain habitat types, even though the PCA gradients scale the habitat variation continuously.

Rotenberry and Wiens (1980) also used PCA to assay variation in 22 habitat variables measured at 26 grassland and shrubsteppe locations in North America. The three significant PCA dimensions corresponded with variation in the horizontal structure and patchiness of vegetation (compon-

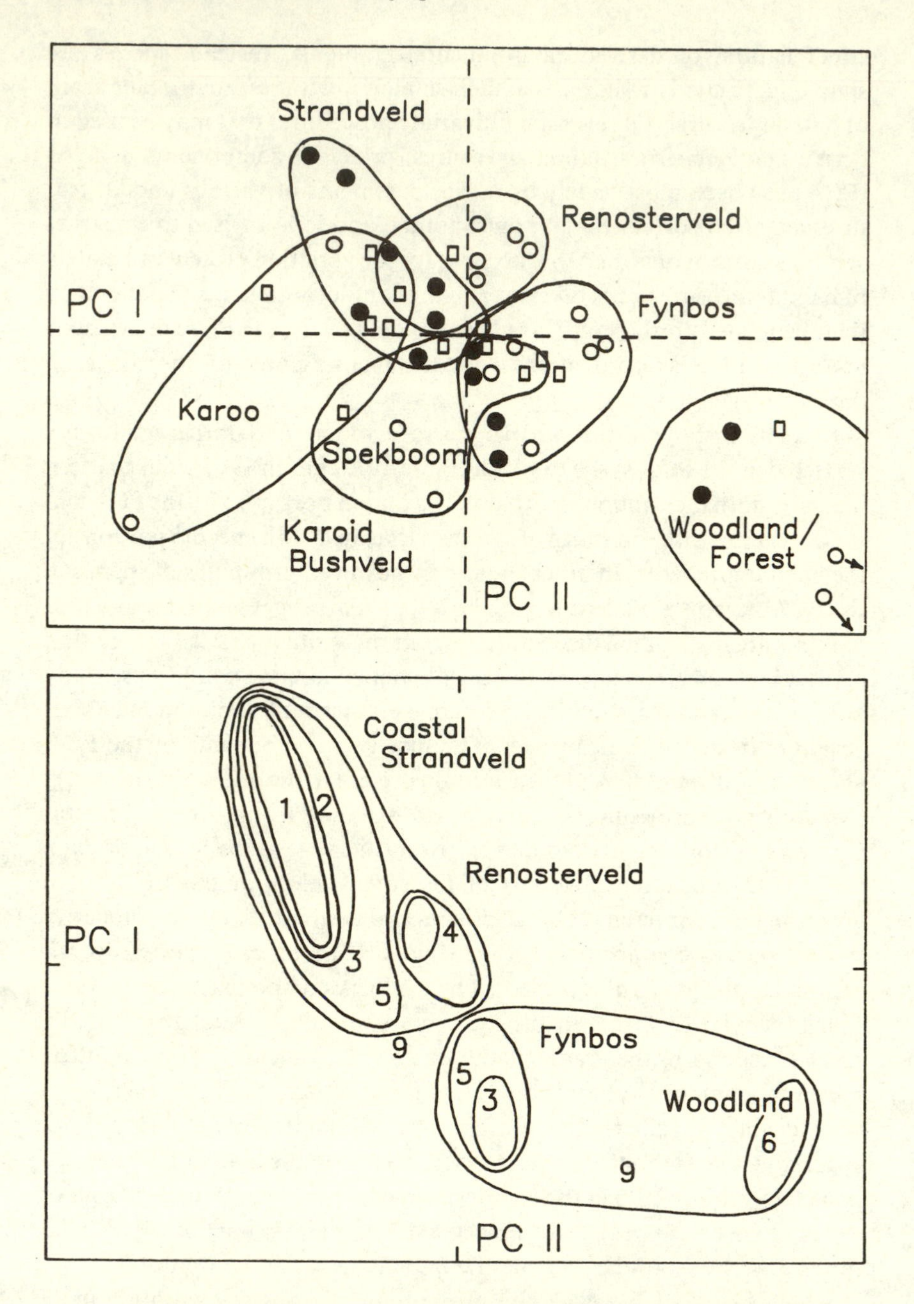

Fig. 9.3. Top: Distribution of 44 South African veld sites in a habitat space defined by the first two components of a PCA based on vegetation structure. Vegetation becomes more closed and continuous from the left on PC I to the top on PC II and taller from the top of PC II to the right of PC I. The sites

ent I), differences in vertical structure (II), and variation in the coverage and density of low forbs (III). Of the 26 bird species that bred on one of more of the sample plots, 15 varied in abundance in ways that were significantly correlated with the habitat gradients portrayed by the PCA components (Fig. 9.4). These species were not evenly scattered through the PCA space, but formed groups that were recognizable by their association with habitat conditions characteristic of tallgrass prairies (low horizontal heterogeneity, high vertical structuring), shortgrass prairies (low vertical stature and intermediate horizontal patchiness) and shrubsteppe (high vertical and horizonal heterogeneity). Three species were correlated with variation in component III, reflecting their occurrence on forb-rich montane meadows. Of the 11 species that were not correlated with any of these components, only one (the Western Meadowlark) was present on more than one plot.

One of the species groups that was defined in this analysis is associated with shrubsteppe habitats. When the same approach was used to examine how this group and other bird species respond to habitat variation at a within-shrubsteppe scale, the patterns that were clear in the broader analysis disappeared and were replaced by entirely different relationships (Wiens and Rotenberry 1981a). Three of the 'shrubsteppe' species (the two sparrows and the thrasher), together with the Horned Lark (a member of the 'shortgrass' group; Fig. 9.4), comprised a set of 'core' species that were present in $>80\%$ of the plots surveyed in the shrubsteppe. These species evidenced relatively few significant correlations with the shrubsteppe PCA habitat gradients, in contrast to their strong correlations with the gradients defined in the broader analysis. On the other hand, several 'peripheral'species, which occurred sporadically or had patchy distributions in the shrubsteppe, showed stronger associations with the shrubsteppe PCA components. The patterns of associations of species with multivariate habitat gradients are thus dependent on the scale at which they are viewed.

Procedures such as PCA, factor analysis, or canonical correlation analysis can be quite effective in depicting the relationships of variations in bird densities to habitat structure, but they may also produce synthetic gradients that are ecologically uninterpretable. Moreover, if the species exhibit

Fig. 9.3. (*cont.*)
were located on three geographically separate transects (open circles, solid circles, and open squares). Bottom: Avifaunal similarity contours for sites on one of the transects (solid circles). The contour intervals indicate groupings of sites among which beta diversity is at a specified level; censuses are similar within contours but are increasingly different across contours. Modified from Cody (1983c).

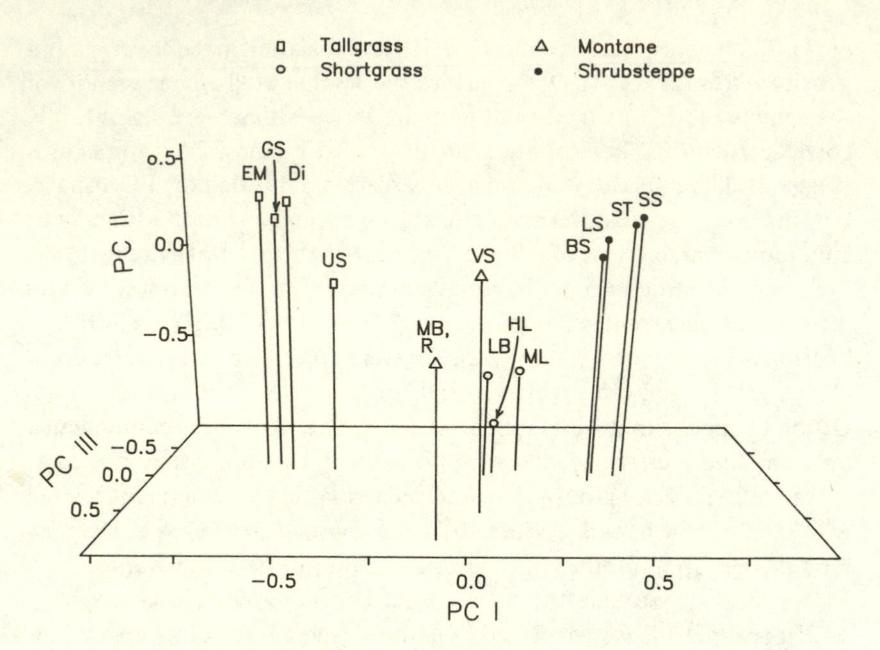

Fig. 9.4. Mean positions of several bird species in a PCA space defined by vegetation structure for North American grassland and shrubsteppe locations. Only species that are significantly correlated with at least one factor are shown. Species codes: EM = Eastern Meadowlark, GS = Grasshopper Sparrow, Di = Dickcissel, US = Upland Sandpiper, MB = Mountain Bluebird (*Sialia currucoides*), R = American Robin (*Turdus migratorius*), VS = Vesper Sparrow (*Pooecetes gramineus*), LB = Lark Bunting (*Calamospiza melanocorys*), HL = Horned Lark, ML = McCown's Longspur (*Calcarius mccownii*), BS = Brewer's Sparrow, LS = Loggerhead Shrike (*Lanius ludovicianus*), ST = Sage Thrasher, SS = Sage Sparrow. After Rotenberry and Wiens (1980a).

nonmonotonic distributions on the gradients, the relationships of the species to one another and to the multivariate space may be distorted. In such situations, indirect ordination procedures such as reciprocal averaging or correspondence analysis may minimize these problems and provide a better portrayal of the distributions of birds on environmental gradients (Sabo and Whittaker 1979, Prodon and Lebreton 1981). Rather than deriving habitat gradients *a priori* and then matching bird distributions to them, the habitat gradients are defined *a posteriori*, by determining how the birds scale on environmental factors. Prodon and Lebreton's (1981) application of reciprocal averaging (RA) to the analysis of a grassland-forest habitat gradient in the Pyrenees showed that the habitat varied continuously and that bird species were scattered more or less evenly on the gradient, although groups of species characteristic of the major habitat types could be discerned. In a New Hampshire subalpine forest area, RA revealed that a major guild of several foliage-gleaning insectivore species occupied one portion of the ordination, but members of several other guilds were more widely scattered over the habitat gradient (Sabo and Whittaker 1979). Other techniques, such as canonical correspondence analysis (Ter Braak 1986), nonmetric multidimensional scaling (Orloci and Kenkel 1985), and multidimensional plexus diagrams (Moskát, personal communication) may also be useful in such analyses.

Species differences

The existence of clusters of species defined by habitat similarities does not necessarily mean that species within a cluster do not differ in habitat occupancy, nor do differences in the mean positions of species in a multivariate gradient space necessarily imply that they cannot be found together in the same habitats in nature. Among the warblers studied by Collins *et al.* (1982), for example, the species groupings revealed by cluster analysis were also evident in RA and PCA ordinations. Each species actually occupied a broad range of habitat conditions, however, and there was substantial overlap among species belonging to different as well as the same clusters. These differences between the multivariate patterns and reality stem in part from an emphasis on the mean position or centroid of a species' distribution on habitat gradients in procedures such as PCA or RA (e.g. Fig. 9.4). Although such statistical procedures are useful in searching for patterns in complex matrices of bird species and habitat variables, they may not provide a reliable basis for detailed between-species comparisons unless the variation in habitat occupancy about the species' centroids is also considered. Folse (1979) suggested using multivariate contrast analysis to evaluate the effects of variation within species and covariation among species on such comparisons.

Analyses aimed at determining how species differ in habitat distributions more often have employed stepwise discriminant function analysis (DFA) (e.g. Cody 1968, James 1971, Whitmore 1975, Noon 1981, Ambuel and Temple 1983). In this procedure, the habitat variable that maximally distinguishes the distributions of bird species is first determined, then another habitat variable that has a low partial correlation with the first and combines linearly with it to create the best two-variable discriminant function is incorporated, and so on. The analysis thus produces a series of multivariate discriminant functions. These are then used to generate canonical variates, uncorrelated linear combinations of the original variables that in sequence account for as much of the overall interspecific separation as possible. Habitat separation among the species is generally interpreted with regard to these canonical variates. Discriminant function analysis has also been used to analyze the habitat–occupancy patterns of single species, by contrasting features of occupied and unoccupied study sites (e.g. Rice *et al.* 1983a, 1984). In either case, the discriminant functions may be used in a classification role to predict the probability that a bird occupying a particular habitat will belong to a given species or that sites with certain habitat characteristics will contain particular species. Although DFA may sometimes produce groupings of bird species in the canonical variate space (e.g. Morrison and Meslow 1983), the use of DFA is inappropriate if one's primary goal is to determine species groupings based on habitat (e.g. Crawford *et al.* 1981).

This methodology, like most multivariate approaches, rests on the assumptions that the species distributions on the canonical variates are normal multivariate and that the conditional covariance matrices are equal. It also requires that the sample size of observations be substantially greater than the number of variables considered. Violation of these requirements distorts the relationships of species in the multivariate space, rendering their interpretation problematic (Green 1979, B.K. Williams 1981, 1983). Thus, when Morrison *et al.* (1986) used DFA to differentiate patterns of habitat use by species in Sierra Nevada coniferous forests, they found that the classification performance of the functions was poor. They noted, however, that their data rarely met the assumptions of the analysis, which may account for the failure of the model to classify species correctly. Capen *et al.* (1986) also reported a poor classification performance by DFA, which they attributed to problems in their study design and violation of the statistical assumptions. It seems that little is to be gained by applying a statistical model to a system that violates the assumptions of the model, especially if the procedure is as sensitive as DFA is.

One example illustrates the power and the pitfalls of using DFA in habitat analysis particularly well. In his studies of sylviid warblers in England and Sweden, Cody (1978) first determined the approximate center of an individual's territory by observing the bird's activities. He then measured the foliage profile (divided into 10 height intervals) in each of four 'haphazardly selected' directions from this point. The area under the profile curve in each of the height intervals was used as a variable in DFA. The habitat measures that Cody used were thus recorded within the territories of individuals (using a rather loose sampling design) rather than for entire study plots and pertained only to aspects of the vertical structuring of the vegetation.

Three species of *Phylloscopus* warblers bred in Cody's study site in England. These species occur in somewhat different positions in the space defined by the first two canonical variates of the DFA (Fig. 9.5, top), *trochilus* occupying a wide habitat range from dense low shrubs through tall open forest, *collybita* generally avoiding the low habitat, and *sibilatrix* occurring chiefly in woodlands with open understory. There is clearly considerable overlap in their habitats, however, and this impression is confirmed by the relatively poor classification performance of the discriminant functions for each species. Of the 55 habitat samples in which *trochilus* was actually found, only 37 could be classified as *trochilus* habitat on the basis of the DFA; according to their habitat features, the remaining 18 territories should instead have contained either *collybita* (11) or *sibilatrix* (7).

It is also possible to determine confidence limits and equal-frequency ellipses for each species in the canonical space (Fig. 9.5B). These contours define the likelihood that a point in the habitat space will be occupied by a given species (it should be noted that the form of these ellipses is especially sensitive to both small sample size and inequality of the covariance matrix; B.K. Williams 1981). Where the equal-frequency ellipses for species meet, Cody defined a 'separation function', which divides the habitat space into nonoverlapping regions most likely to be occupied by one or another species.

The regions and probability ellipses defined in this way can be used to make predictions about which species should be present in additional habitat samples that were not used to generate these patterns. Cody used the functions of Fig. 9.5B to test whether or not warblers from southern Sweden segregate with respect to the same habitat criteria contained in the English canonical variates and to assess whether or not the species occurred in the same habitats (positions in the canonical space) in Sweden. *Phylloscopus collybita* was absent from the Swedish area, but another sylviid warbler, *Hippolais icterina*, was present. For the two *Phylloscopus* species, the separation pattern of the Swedish birds is similar to that of the English birds. This led Cody to conclude that habitat selection is based on the same variables in the two areas, although it really means only that the species separate in a similar manner in the two regions with respect to the habitat features measured. The positions of species in the habitat space changed, however (Fig. 9.5C). Because the directions of these shifts are associated with the absence of *collybita* and the addition of *icterina*, Cody interpreted them as responses to a changed complex of competitors. This is a possibility, of course, but in the absence of information on the densities of the species in each area, the availability of various habitat types, the long-term stability of these habitat patterns, or measures of habitat features other than

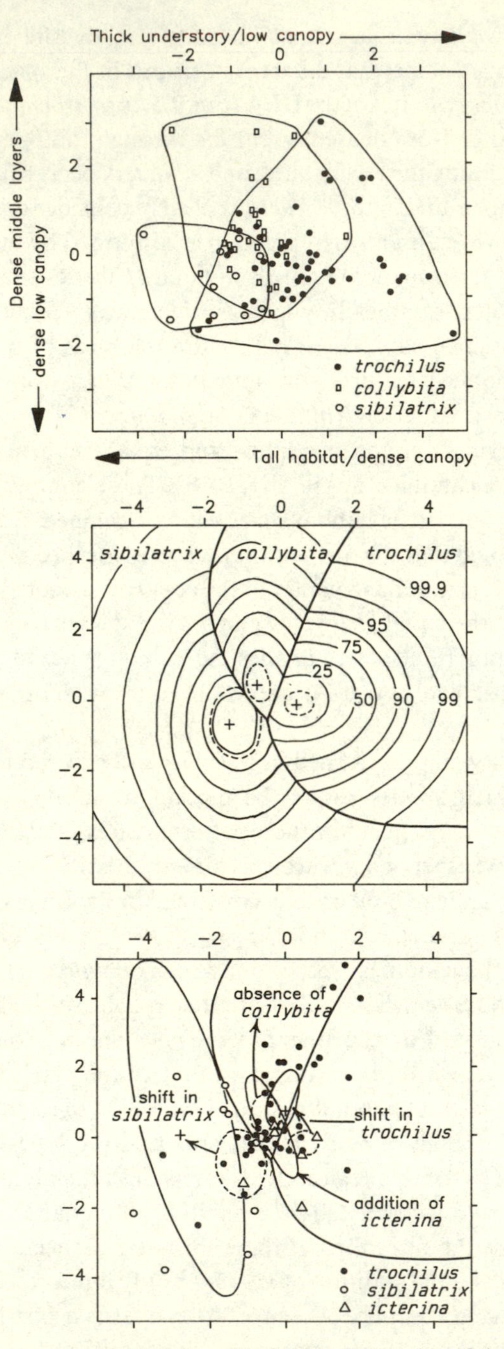

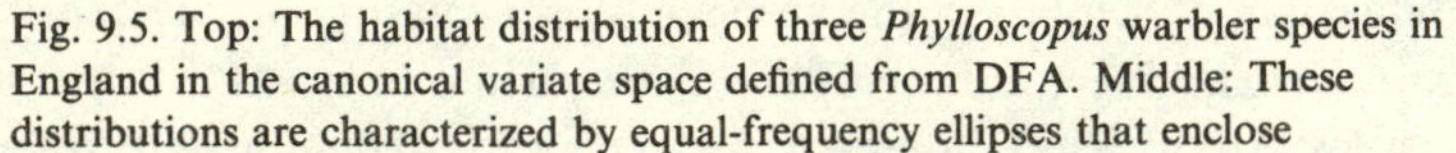
Fig. 9.5. Top: The habitat distribution of three *Phylloscopus* warbler species in England in the canonical variate space defined from DFA. Middle: These distributions are characterized by equal-frequency ellipses that enclose

foliage-height profile, one can really only conclude that the habitat patterns of the species vary geographically.

Cody also conducted a reverse analysis, deriving canonical variates for both *Phylloscopus* and *Sylvia* warblers on the basis of Swedish data and testing how well the English birds fit them. Occupancy of breeding habitat differed among coexisting congeners in both areas, although there was considerable overlap. In *Sylvia*, for example, only 55% of the predictions of territory occupancy based on the DFA classifications were correct, whereas 65–70% of the *Phylloscopus-Hippolais* cases were correctly predicted. When species from one area were related to the predictions generated from the other area, less than 25% of the *Phylloscopus–Hippolais* and 45% of the *Sylvia* territories were correctly classified. The species thus appear to have broad habitat tolerances, and segregation by habitat is generally poor.

This example indicates some of the difficulties in determining the extent of habitat separation between species and assessing whether the differences reflect independent, species-specific responses to environmental variation (e.g. Bond 1957, Folse 1982, Sabo and Holmes 1983, Grzybowski 1982, Rotenberry and Wiens 1980, Wiens and Rotenberry 1981a) or consequences of species interactions (e.g. Cody 1968, 1974a, 1978, Noon 1981). When the habitat distributions of individuals are considered, overlap may be extensive (Fig. 9.5, top), even though at a population level such detail can disappear (Fig. 9.4, 9.5, middle). Cody accounted for individual variation in habitat occupancy by establishing confidence limits of equal-frequency ellipses in DFA space, and other workers (e.g. Rotenberry and Wiens 1981, Sabo 1980) have plotted density contours of species' distributions in multivariate PCA habitat space. Sherry and Holmes (1985), for example, derived PCA axes separately from data on the vegetational composition of the tree and herb-shrub strata of a hardwoods forest in New Hampshire and then calculated 95% confidence ellipses for all of the vegetation samples and for the samples falling within territories of each of seven bird species.

Fig. 9.5. (*cont.*)
concentrically from 25% to 99.9% of the points in each species' distribution. The + indicates the mean and the dashed line the 95% confidence limit. The equal-frequency ellipses are used to generate 'Separation Functions' (solid lines) that indicate the regions in which habitat will be occupied with maximum likelihood by one of the three species. Bottom: The English *Phylloscopus* separation functions are used to predict the habitat occupancy of two *Phylloscopus* species and *Hippolais icterina* in Swedish woodlands. The data points are for habitats measured and censused in Sweden, the dashed lines are the confidence limits about the English means, and the solid ellipses are the 50% equal-frequency ellipses for the new Swedish data. Shifts in the habitat occupancy of *trochilus* and *sibilatrix* are indicated. Modified from Cody (1978).

The confidence ellipses for five of the species did not contain the origin of the PCA space defined by components I and II (Fig. 9.6), and the ellipses of several species did not overlap one another. Sherry and Holmes interpreted these patterns as evidence of habitat selection by the five species and of selection of different habitats by the species with nonoverlapping ellipses (e.g. Least Flycatchers vs. Blackburnian Warblers, Fig. 9.6). Similar patterns emerged in the analysis of the herb-shrub data set.

Sherry and Holmes were cautious in their interpretations of the 95% confidence-ellipse patterns they described, but others (e.g. Noon 1981) have interpreted such contours or ellipses as symbolizing habitat niche widths, and the overlaps of ellipses for different species as measuring niche overlap. Such ellipses are a handy way of expressing quantitatively the variability in habitat responses of species with respect to the derived gradients, but they may or may not translate into similar patterns of habitat breadth or overlap in the real world. More importantly, there are serious methodological problems associated with the forms of such ellipses (Smith 1977, B.K. Williams 1981, Carnes and Slade 1982, Van Horne and Ford 1982). They must therefore be interpreted with discretion.

The sort of pattern that is obtained also depends on the form of the analysis. If one uses a procedure such as DFA that is explicitly designed to maximize differentiation between species, the results will be markedly different from those obtained for the same data using an approach that is neutral to species relationships (such as PCA or RA). A procedure that emphasizes similarities (e.g. cluster analysis) will produce still different patterns. The methodology should be tailored to the questions being asked, of course. One should not be too surprised, however, if DFA reveals differences between species, and ecological interpretations of these differences should be made cautiously (Wiens and Rotenberry 1981a, B.K. Williams 1981). Moreover, the patterns revealed by any of these procedures will change with changes in the data set used. When Titterington *et al.* (1979) applied DFA to a set of bird–habitat data from areas of clearcut and uncut 'control' spruce-fir forest, the major pattern was a separation of those species using clearcuts from those using the controls. When the uncut controls were excluded from the analysis, however, a new DFA not surprisingly provided better discrimination among the species using the clearcuts.

Temporal and spatial variations in habitat occupancy

Because resource levels, shelter, and many other features of habitats undergo large seasonal changes in most environments, seasonal variation in habitat occupancy is not unexpected, and it has been documented in

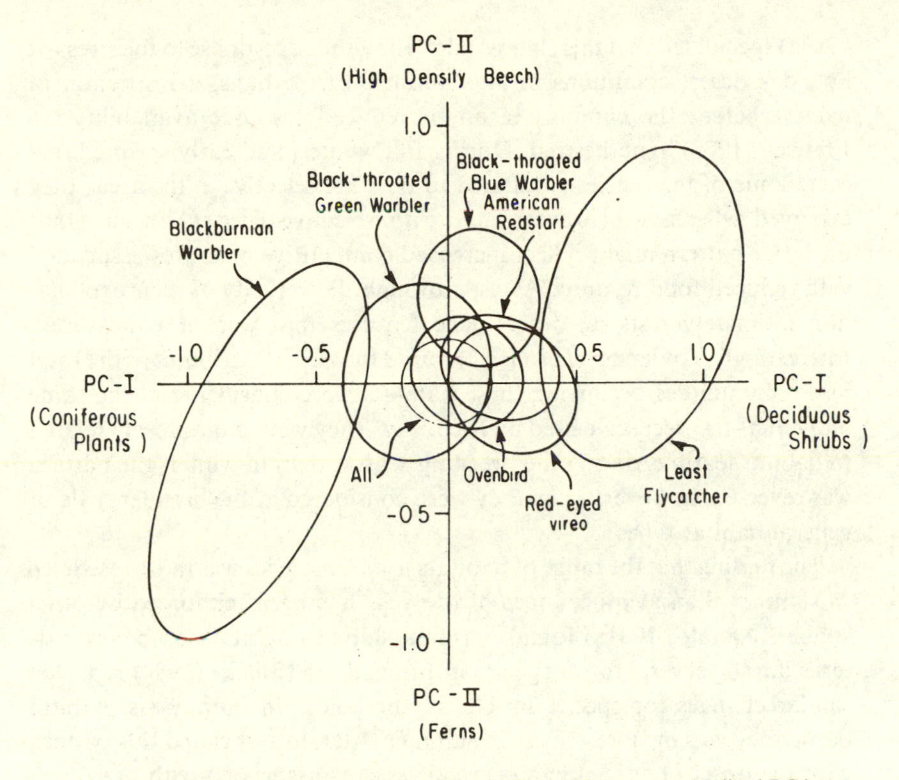

Fig. 9.6. Ninety-five percent confidence ellipses of mean standardized PCA scores (based on tree-strata variables) for all sample locations ("All") and for subsets of plots occupied by each of seven passerine species in a northeastern hardwoods forest in the USA. From Sherry and Holmes (1985).

several studies. In one particularly intensive investigation, Rice *et al.* (1983a, 1984) used DFA to distinguish the features of habitats in areas occupied by the bird species occurring in the riparian zone along the Lower Colorado River from those of areas in which the species were absent. In this zone, both the degree of habitat selectivity of most species (i.e. the degree to which the presence or absence of the species is correctly predicted by the habitat-based discriminant functions) and the habitat measures contributing to the DFA varied between seasons. More variables were required to discriminate used from unused areas in spring and early summer than in late summer, perhaps indicating greater habitat selectivity in late summer. Interspecific competition does not seem a likely contributor to this increased habitat specialization, as the habitat-occupancy patterns of the species at that time were largely independent of those of other species in the community and resources were seemingly superabundant. Rice *et al.* (1980,

1983a) speculated that this change might be either a response to the stress of hot, dry desert conditions in late summer or, perhaps, a restriction in habitat before the coming season of reduced resource availability, as Fretwell (1972) hypothesized. During fall, winter, and early spring, however, some of the species continued to be quite selective in the areas they occupied, whereas others became more nonselective. Rice *et al.* argued that the latter pattern might reflect increased competitive pressures associated with reduced food resource levels, although the veracity of their explanation ultimately rests on other niche features that were not measured. Interestingly, Anderson *et al.* (1983) found that when attributes of the total bird communities occupying these habitats were considered at the same individual-transect scale used by Rice *et al.*, they were more closely related to habitat features during the breeding season than in winter; the pattern was reversed, however, when they were considered at the broader scale of general habitat types.

The finding that the range of habitats used by species was more restricted in summer than at other times of the year has been reinforced by other studies. Alatalo (1981a) found seasonal shifts in habitat-occupancy patterns among several forest species in Finland, and Bilcke (1984) reported similar changes for species in The Netherlands. In both areas, habitat occupancy was most restricted in summer. Alatalo associated this pattern with the onset of special habitat requirements associated with breeding.

Characteristics of the habitat occupied by a species may vary geographically, as Lack (1971: 255) noted anecdotally for several species. Collins (1983b) explored such variation quantitatively by comparing the habitat features within individual territories in populations of several warbler species in Maine with those within territories of the same species in northern Minnesota. All of the species differed significantly between areas in their patterns of correlation with individual habitat measures, and clear geographic shifts in a multivariate habitat space defined by RA were also evident (Fig. 9.7). A more detailed analysis of geographical variation in the habitat occupancy of one of these species in five areas of the northern United States (Collins 1983a) reinforced these findings (see also Shy 1984, James *et al.* 1984). On the other hand, Noon *et al.* (1980) found that the habitat features that entered into stepwise discriminant functions for species in several regions analyzed separately were similar, leading them to conclude that the habitat requirements of most forest bird species 'apply generally throughout their breeding ranges'. The data base used by Noon *et al.* contained information gathered by different observers using different census procedures and different methods of habitat sampling; given this

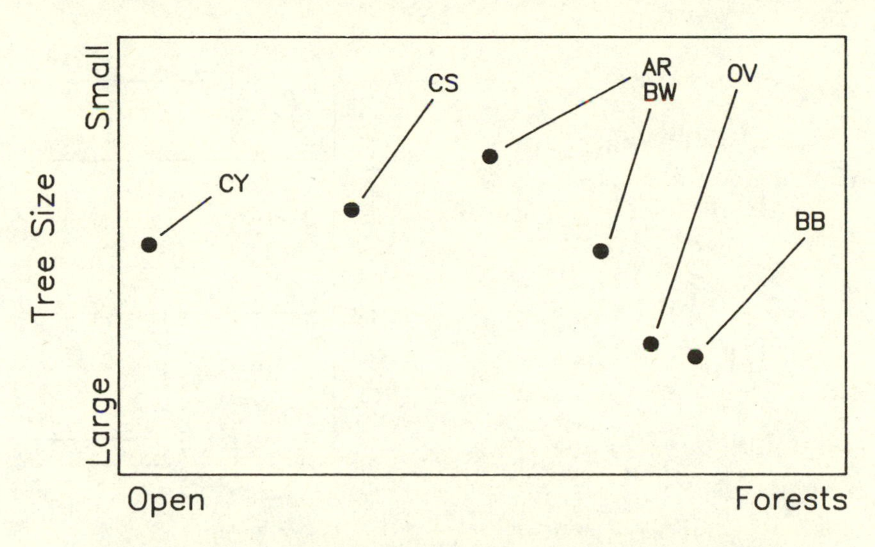

Fig. 9.7. Shifts in the habitat occupancy of woodland warblers in a habitat space defined by reciprocal averaging ordination. Letters indicate habitat occupancy in Maine, connected for each species to its habitat occupancy in Minnesota (solid circle). Species codes: CY = Common Yellowthroat (*Geothlypis trichas*), CS = Chestnut-sided Warbler (*Dendroica pensylvanica*), AR = American Redstart, BW = Black-and-white Warbler (*Mniotilta varia*), OV = Ovenbird, BB = Blackburnian Warbler. Modified from Collins (1983b).

heterogeneity and the rather general habitat categorizations they employed, such a sweeping conclusion is not justifiable. The observation of James *et al.* (1984: 35), that 'regional variation in habitat is common in birds . . . and it deserves more attention', seems closer to the mark.

The effects of density

Variation in habitat occupancy by a species can arise from many factors, but the effects of the population density of that species may be especially important. Svärdson (1949) suggested that increasing density should be associated with an increase in the range of habitats occupied by a species because the intensified intraspecific competition forces some individuals to occupy marginal habitats (Fig. 6.1). This thinking was formalized in the models of Brown (1969) and Fretwell and Lucas (1969). Brown proposed that habitats differ in their suitability to a species and that individuals will preferentially select the most suitable habitat. As density increases, the available area of this habitat type becomes saturated with individuals, at which point there are no further increases in density in that habitat but additional birds 'spill over' into a second, less suitable habitat type (Fig.

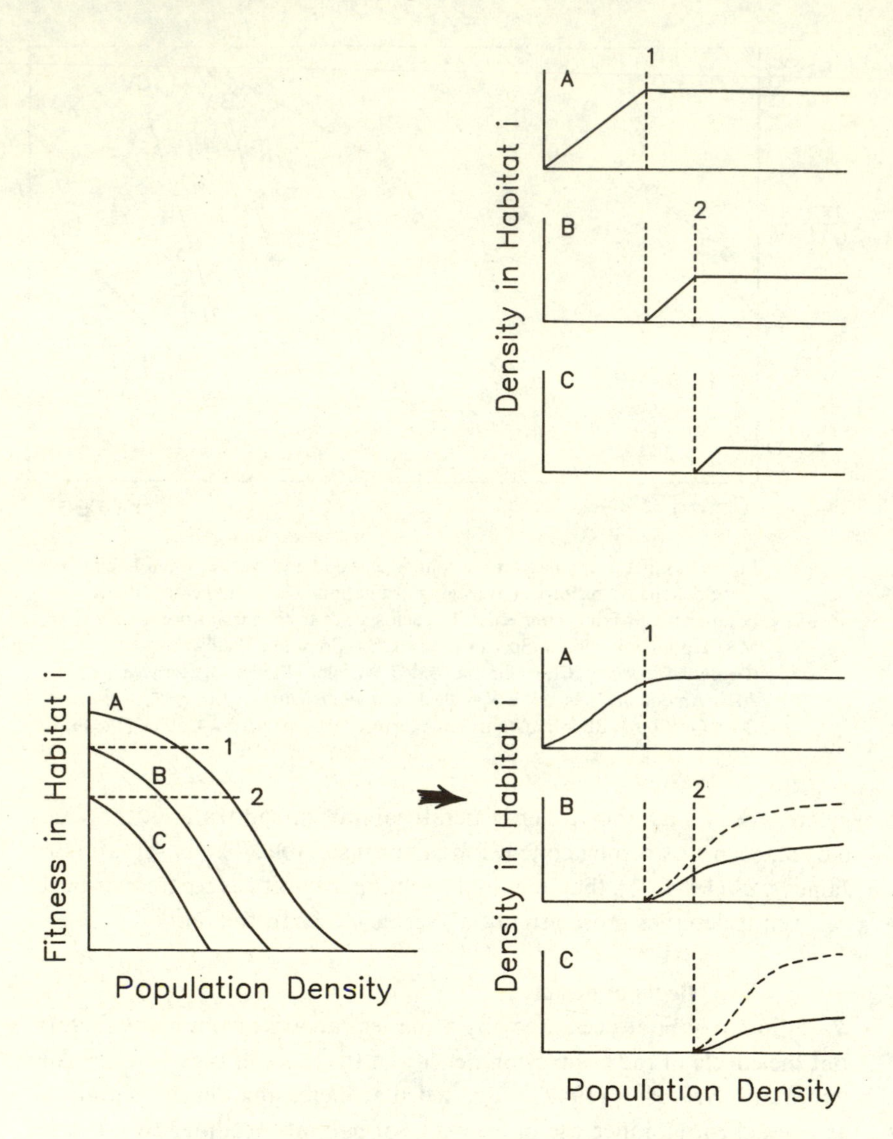

Fig. 9.8. Models of density-dependent habitat occupancy. Top right: Brown's model predicts that the preferred habitat (A) will be occupied first, densities building to threshold 1; when that habitat is saturated, individuals begin to occupy a second, less suitable habitat (B). At threshold 2, habitat B is saturated and individuals will spill over into habitat C. Below: The Fretwell–Lucas model proposes that individual fitness potential in a habitat decreases with increasing density in that habitat but that different habitats differ in fitness potential (left). At thresholds where fitness prospects in two habitats are equal, individuals will occupy both habitats. Density in the 'preferred' habitat will continue to increase after threshold 1 is passed (right). The solid lines in the

9.8). The scenario may be repeated when the available area of the second habitat is saturated. Fretwell and Lucas modeled habitat occupancy as a function of the fitness potential ('quality') of habitat types. Some habitats are of better inherent quality than others, and these will be occupied first. As density within that habitat increases, however, the fitness potential of the habitat declines (Fig. 9.8). Eventually, a point is reached at which an individual may realize equivalent reproductive success by occupying instead another habitat of slightly poorer quality. As densities in both habitats continue to increase, a threshold will be passed at which the fitness prospects in each are now equal to those in a third 'marginal' habitat, and birds will begin to settle there. The factors producing these habitat expansions may involve assessment of habitat quality by individuals according to various cues (Fig. 9.2) or the despotic eviction of subordinate individuals to the lower-quality habitats by dominant individuals (as is suggested by morphological differences between birds in different habitats; Fig. 7.9; see also Lundberg *et al.* 1981). According to either model, the range of habitats occupied by a species increases with increasing density, although in a nonlinear manner. The Fretwell–Lucas model also predicts that reproductive success within a given habitat will decrease with increasing density.

There is some support for these models, at least in terms of the patterns they predict. To occupy new habitats, individuals must disperse, and there is evidence for some resident passerines that dispersal into such areas is density-dependent (e.g. Greenwood *et al.* 1979, Berndt and Henss 1967, Hildén 1982, Ekman 1979, O'Connor 1980). Dense populations of a species frequently (but not invariably) do occur over a greater variety of habitats than do sparse populations (e.g. Emlen 1977b, Tiainen *et al.* 1983). The best way to determine these relationships is to chart the dynamics of density and habitat occupancy over time. This has been done for several species by the use of long-term data gathered in the Common Birds Census of the British Trust for Ornithology (Williamson 1969, O'Connor 1980, 1981, 1985, 1986). For example, Yellowhammer (*Emberiza citrinella*) populations were severely reduced in Britain by the extraordinarily severe winter of 1962–63. In subsequent years, densities first increased in farmland habitat, but levelled off after a period of recovery. Densities in woodland, however,

Fig. 9.8 (*cont.*)
right panels indicate the population-density patterns expected from the 'ideal-free' version of the Fretwell–Lucas model; the dashed lines indicate that populations in lower-quality habitat may actually increase to levels greater than those in the 'preferred' habitat if there is excess dispersal from and/or aggressive exclusion of individuals from habitat A.

continued to increase, suggesting a spillover from farmland. O'Connor (1980) found that clutch sizes decreased with increasing densities in both habitat types, although more steeply in farmland. Thus, although farmland is seemingly the preferred habitat, the density-dependent clutch-size depression erodes the advantage of breeding there and makes it reproductively advantageous for individuals to move into the 'inferior' but less crowded woodland habitat (O'Connor 1981). Other species showed similar density-dependent habitat-occupancy patterns, but the form of the relationship differed among the species (O'Connor 1985, 1986). Chaffinches, for example, used much the same range of habitats at all recorded densities, but the degree of habitat dominance (the proportion of nests in the most frequently used habitat) decreased sharply with increased density (Fig. 9.9; O'Connor and Fuller 1985). As the population size increased, abundance in some of the less suitable habitats increased disproportionately in relation to those in 'preferred' habitats, reducing habitat dominance.

If individuals select and use habitats that best meet their requirements (i.e. are 'optimal'), differences in the magnitude of use of habitat types or in densities in those habitats should be directly related to the quality of those habitats (Schamberger and O'Neil 1986). This linkage between local densities and habitat suitability is a fundamental premise underlying most interpretations of bird-habitat correlations: the habitat variables positively correlated with density variations are presumed to signify high-quality, 'preferred' habitat. In fact, this density–quality relationship may frequently not hold. It may be distorted, for example, if patterns of individual habitat occupancy are influenced by site tenacity (Wiens 1985a, b; O'Connor 1985, 1986). Thus, the habitat (territory) occupancy of Painted Buntings (*Passerina ciris*) is determined both by territory quality (as gauged by the distribution of resources and the frequency of polygynous matings) and by site fidelity: birds generally settle first in the highest-quality territories but individuals that have bred previously in less suitable areas may return there even when there are vacancies in better areas (Lanyon and Thompson 1986).

The assumption that high densities are associated with high-quality habitat may also be violated if a large component of the population (e.g. subordinate individuals) is forced to emigrate into low-quality 'sink' habitats, where densities may build up to high levels (Wiens and Rotenberry 1981c, Lidicker 1975). Van Horne (1983) considered these problems in detail and proposed that density is most likely to be decoupled from a strong association with habitat quality in environments that are strongly seasonal, temporally unpredictable, or spatially patchy and in species that are eco-

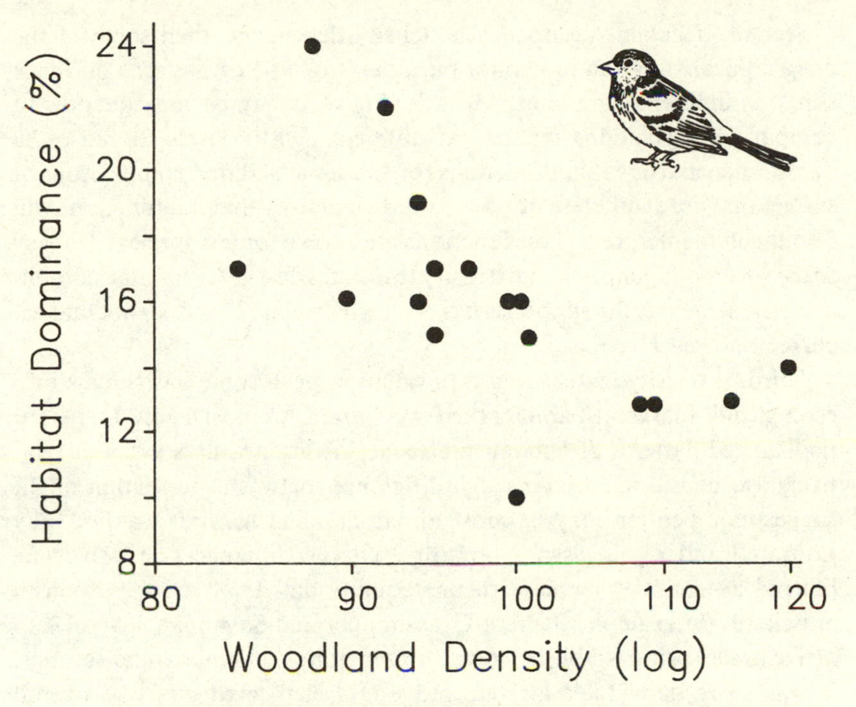

Fig. 9.9. The degree of habitat dominance (proportion of nesting records in the most frequently used habitat) in relation to densities of Chaffinches in British woodlands, as measured by Common Bird Census surveys conducted during 1964–1980. After O'Connor and Fuller (1985).

logical generalists and have a social dominance structure and a high reproductive capacity. Except for the latter, these criteria apply to many birds and most environments.

All of this has several important implications. First, it is improper to refer to habitat as 'selected', 'preferred', 'suitable', or 'unsuitable' on the basis of correlational patterns alone, especially if the density–habitat occupancy dynamics in a local setting are not known. Areas considered 'unsuitable' because they are not occupied at a certain density may become 'suitable' with a small increase in density, and this may have little to do with their capacity to support reproductive activities or permit long-term survival of individuals. Habitats that have recently become unsuitable may continue to be occupied by site-tenacious individuals (see e.g. Wiens and Rotenberry 1985). Our conceptions of habitat selection are generally couched in optimization thinking (e.g. Cody 1974a,b, Rosenzweig 1979, 1981), but variation in the factors shown in Fig. 9.2 can easily deflect nature from the theoretical optima.

Second, if habitat occupancy is density dependent, then some of the geographical variation in habitat patterns (Fig. 9.7) or patterns of 'niche shifts' in different community settings (Fig. 9.5) may be consequences of comparing populations that are at different density levels. Because the density-dependent habitat functions for species may differ, comparisons of species that are at different positions on their density–habitat functions will be difficult to interpret. These functions are known for few species. It is easy to see why the assumption that density translates directly to habitat suitability in the same way for all species has been so popular. That does not make it correct, however.

Third, if density increases in a population push some individuals into ecologically marginal habitat, then we should be most likely to record undistorted patterns of habitat 'preference' from measures taken at relatively low densities. Paradoxically, differences between species that might suggest competition may be most apparent when habitats are not fully saturated and resources not limiting, but the differences may become blurred as densities increase to an 'equilibrium' level. In a Wisconsin grassland, for example, the first Grasshopper and Savannah sparrows to arrive in the spring established territories in quite different habitat settings, but as more individuals arrived and established territories, the overall differences between the species began to diminish. At peak densities, both species blanketed the entire area with territories and their habitat-occupancy patterns were indistinguishable (Wiens 1973a; see Fig. 10.14).

Fourth, in correlating density with habitat features, one assumes that a given density means the same thing in different habitats. A density that in one habitat represents saturation and forces spillover into other habitats may in another habitat be well below a saturation level (Fig. 9.8). To the extent that this occurs, statistical relationships between densities and habitat variables that are actually important may be destroyed.

Fifth, analytical procedures that use regression or correlation tests of density variation (e.g. PCA, multiple regression) not only yield different patterns from methods (such as DFA) that use presence–absence information but also differ in the scale on which sampling is usually conducted and the sensitivity of the patterns to temporal variation. The use of regression-based techniques is usually directed toward documenting relationships of habitat features to variations in abundance where the species is present, and clearly unsuitable habitat is rarely sampled. If samples in which the species is absent (abundance = 0) are included in the analysis, the patterns will be different from those obtained in analyses of density variations in occupied areas only (Rice *et al.* 1986; Fig. 9.10). The discriminant approach, on the

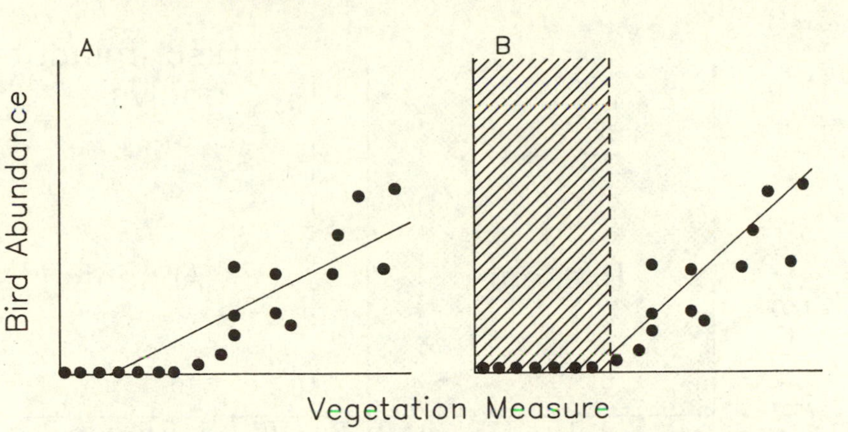

Fig. 9.10. A. Linear regression of bird abundance on a vegetation feature in which samples lacking the species are included in the analysis. At intermediate values on the vegetation axis the regression overestimates the relationship, whereas it is underestimated at high values. B. The regression is limited to the portion of the vegetation gradient actually occupied by the species; DFA is used to separate occupied (clear) from unoccupied (hatched) areas. After Rice *et al.* (1986).

other hand, explicitly requires information from unoccupied portions of a habitat gradient and thus may lead one to sample over a broader environmental scale (Fig. 9.11). Moreover, substantial changes in the patterns of habitat associations revealed by regression/correlation procedures may be produced if density at some locations varies independently of habitat conditions, but these variations will have little influence on the power of a discriminant function to predict presence or absence (Rotenberry 1986; Fig. 9.11).

Finally, the scale at which bird/habitat relationships are evaluated affects the inferences one may draw about habitat quality to the birds. 'Quality' is probably best measured in terms of factors influencing individual reproductive success (e.g. hatching success, fledging weights, nestling growth curves; Van Horne 1983, Maurer 1986), and analyses in which habitat conditions are averaged over areas much larger than individual territories may not reveal such factors.

Do these considerations mean that habitat analyses based on density measures are useless? No, but they do indicate that such analyses should be conducted and interpreted with care and restraint. A statistically verified relationship between the density of a species and actual or derived habitat variables may be used, for example, to predict densities or habitat occupancy in other circumstances, to suggest new hypotheses for testing, or to

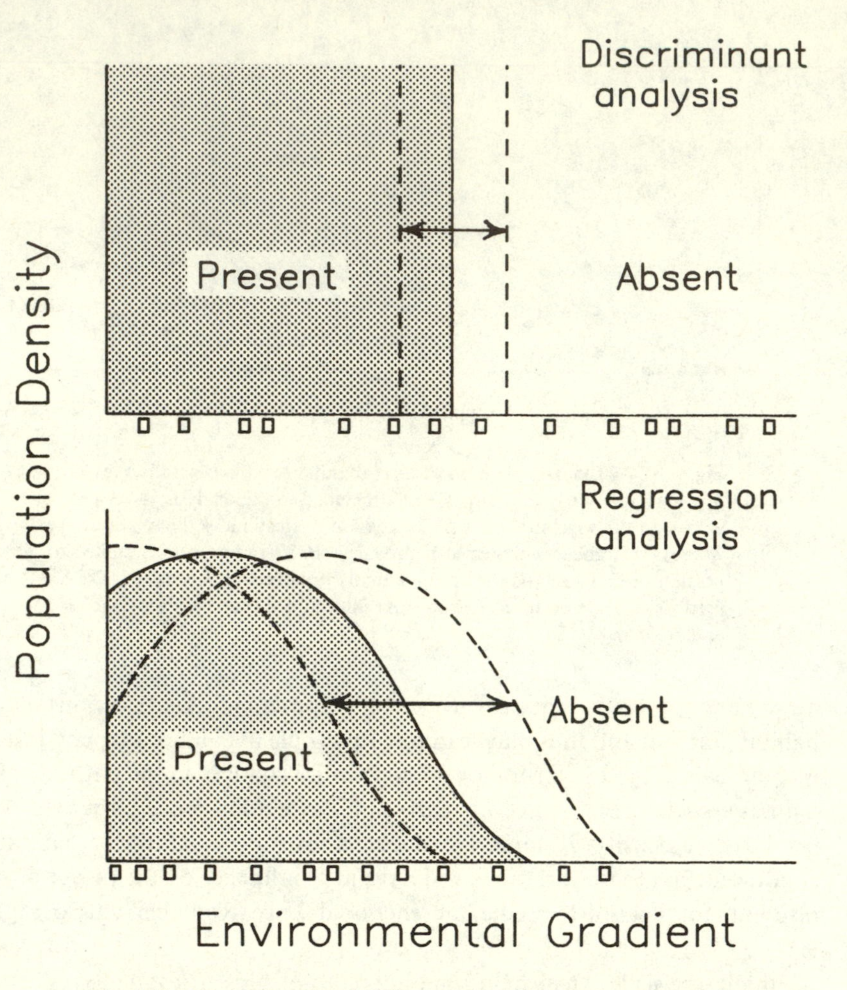

Fig. 9.11. The relationship of different analytical procedures to sampling scale and sensitivity to habitat-independent variation in population density. Dashed lines and arrows represent annual variation in density independent of habitat features; squares below the abscissa indicate sampling sites. After Rotenberry (1986).

point to profitable directions for experimental manipulations. It permits the conclusion that density variations in the species are associated with variations in certain habitat variables. This finding, however, does not warrant the conclusion that these variables are directly important to the individuals, that they are the cues used in habitat selection, or that they measure habitat quality.

Some methodological comments

Investigations of the relationships between birds and their habitats are subject to all of the methodological concerns discussed in Chapter 3, but several specific aspects of these studies merit additional consideration. Correlative analyses require careful censusing and accurate density estimation, of course, but habitat measurement requires similar attention. One must consider where, how, and when samples should be taken, which and how many variables should be measured, and whether or not nonlinear relationships may be important.

Where and how to sample?

In our studies of grassland and shrubsteppe birds, we measured habitat features at random points within study plots (Wiens 1969, Rotenberry and Wiens 1980, Wiens and Rotenberry 1981a; see also James and Shugart 1970, Willson 1974). Our analyses are thus based on comparisons of values of habitat measures and bird densities among plots. We have also used the random sample points located within mapped territories in the plots to characterize habitats in individual territories, analyzing differences between occupied and unoccupied areas within plots (Wiens 1969, 1973b, Wiens, Rotenberry and Van Horne 1987). Other workers have investigated habitat relationships at this scale by first mapping territories and then taking habitat measures systematically within these boundaries (e.g. Ambuel and Temple 1983, Cody 1978, Tiainen *et al.* 1983, Shy 1984, Crawford *et al.* 1981). In still other studies, habitat features have been measured at points or in circular quadrats centered on specific activity areas of individuals, such as singing perches, foraging areas, or nest sites (e.g. Noon 1981, Smith 1977, Morrison and Meslow 1983, Collins 1983a, b, Morrison *et al.* 1986, Larson and Bock 1986).

Each of these procedures measures habitat on a different scale and with varying biases. Sampling in the plot-based approach is unbiased with respect to the distribution or habitat use of individuals within a plot, and the patterns that emerge relate populations (densities) to habitat conditions averaged over the plot area. At the territory level, on the other hand, the patterns relate to areas occupied by individuals, which are generally contrasted with unoccupied areas. Sampling centered on individual activity sites indicates only the characteristics of habitats used for specific activities, and these may differ from the habitat used for other activities or averaged over entire territories (Mehlhop and Lynch 1986). Sampling is also more likely to be biased and inadequate at the finer scale of activity sites (Holmes 1981). If one measures the habitat features of locations in which individuals are first seen (e.g. Larson and Bock 1986), for example, the observations may be sensitive to an initial visibility bias (see Chapter 3). Each method indicates something different about the habitat relationships of a species. The choice of method should therefore be based on a careful consideration of one's objectives. Samples taken using different procedures cannot be combined in analysis or compared directly.

When to sample?

Both habitats and habitat selection by birds vary in time, and sampling that is insensitive to these variations may produce misleading patterns. Whitmore (1979) documented how the habitat features in territories of several grassland sparrows change through the breeding season. As a consequence, the habitat at the time of territory establishment differs markedly from that at the time when young are fledged 2 months later. In Colorado montane forests, the dominant breeding species exhibit 'preferences' for particular vegetation types (as gauged by their occurrence at greater densities in a type than would be expected from the areal coverage of that type; Winternitz 1976). The extent and direction of these 'preferences' vary over the breeding season, however, and do so in different ways for different species in the community. These observations mean that an investigator measuring bird habitats of these species at one time might obtain strikingly different results if he or she took the measurements a few weeks earlier or later. If one includes in correlative analyses habitat measurements taken at different times in relation to the phenology of the birds in different locations, the comparisons might be distorted. Seasonal or yearly variations (e.g. Rice *et al.* 1980, 1983b, Karr and Freemark 1983, Wiens 1981b) can produce similar complications. Once again, when one measures habitat depends on one's objectives, but in any study the 'when' must be standardized.

Which variables to measure?

Assuming that a satisfactory sampling program can be established, one must then decide which habitat variables to measure (Wiens and Rotenberry 1981a, Karr 1981, Whitmore 1981, Noon 1981, Anderson 1981). Most often, the variables are selected arbitrarily, on the basis of what *we* perceive to be interesting or find easy to measure rather than their potential importance to the birds (Wiens 1969, Karr 1981, Vuilleumier and Simberloff 1980, Holmes 1981). Traditionally, the emphasis in studies of bird habitats has been on features relating to the structural configuration or physiognomy of habitats. Indeed, consideration has often been restricted to such structural variables on the premise that structure alone is important (e.g. MacArthur and MacArthur 1961, Wiens 1969, Karr and Roth 1971, James 1971, Willson 1974, Anderson and Shugart 1974, Cody 1968, 1978). Several rather detailed systems for describing habitat structure have been developed (Emlen 1956, Dansereau *et al.* 1966, Wiens 1969, James and Shugart 1970).

Certainly aspects of habitat structure *are* important to birds, but there is also mounting evidence that the presence or abundance of particular plant species in an area may be important features of the habitat as well (e.g. Holmes *et al.* 1979b, James and Wamer 1982, Rice *et al.* 1983a, Beedy 1981, Holmes and Robinson 1981, Morse 1978, Morrison and Meslow 1983, Wiens and Rotenberry 1981a, Anderson *et al.* 1983, Abbott and Van Heurck 1985). For example, Airola and Barrett (1985) found that breeding species in mixed coniferous forests in the Sierra Nevadas were strongly associated with certain tree species in their foraging. In particular, incense cedar (*Calocedrus decurrens*) was consistently avoided by all species. Interestingly, Morrison *et al.* (1985) reported *increased* use of incense clear by wintering birds in the same study area, which they attributed to the high availability of an insect overwintering under the loose bark of small trees. In shrubsteppe areas in North America, models based on the floristic composition of the habitat generally predict bird abundances better than do models based on vegetation structure (Rotenberry and Wiens MS; see Table 9.1). In North American grasslands, over half of the variation in bird community composition was associated with floristic variables, whereas only a third of the variation was accounted for by structural measures (Rotenberry 1985). When floristic variation was controlled by partial correlation analysis, the contributions of structural variables were halved (Fig. 9.12). Rotenberry speculated that the greater degree of association of bird species with the floristic composition of habitats might be related to the distribution of specific food resources that different plant taxa provide. Rice *et al.* (1984) also reported a stronger association with tree species than with vegetation structure for riparian woodland birds on the Lower Colorado River, and they suggested that 'many of the commonly found correlations of bird community relationships to vegetation profiles may be the result of combining analyses of many different bird species with many different tree or plant species associations' (1984: 895).

Several of these studies attributed the importance of floristic variables to their associations with food resources. In some cases, food availability has

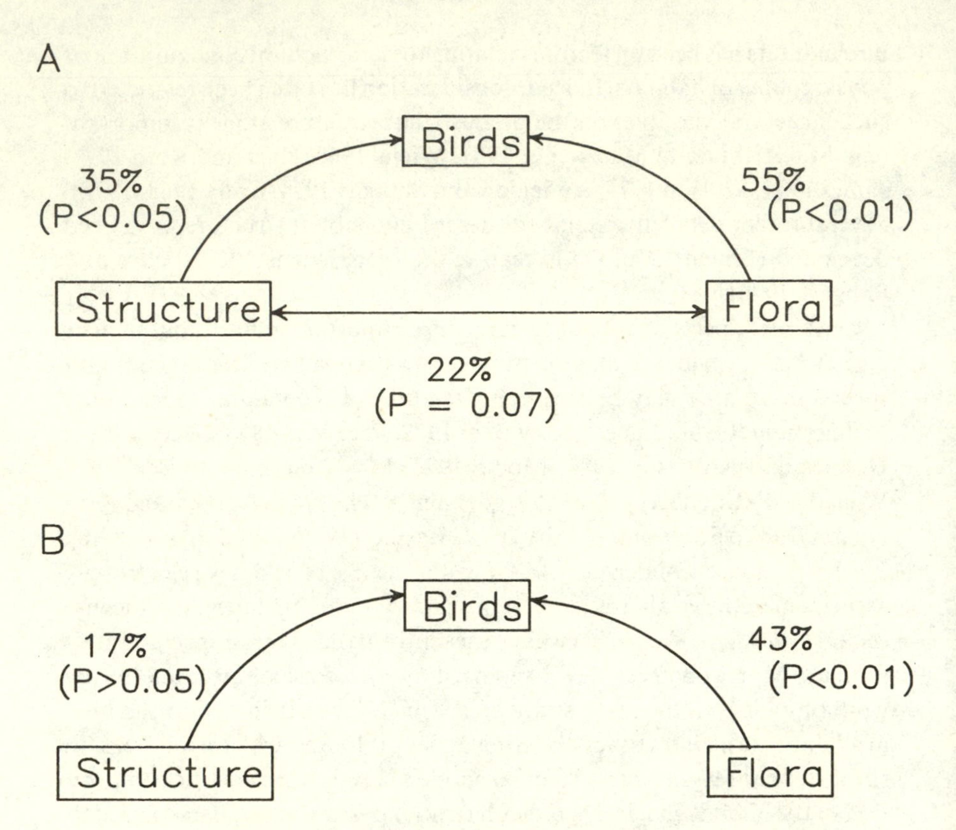

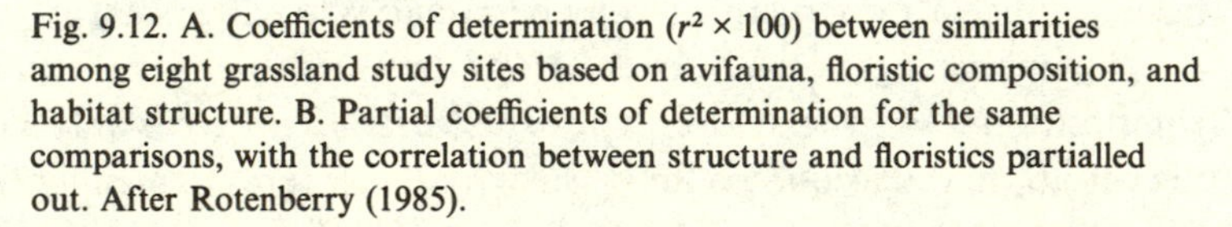
Fig. 9.12. A. Coefficients of determination ($r^2 \times 100$) between similarities among eight grassland study sites based on avifauna, floristic composition, and habitat structure. B. Partial coefficients of determination for the same comparisons, with the correlation between structure and floristics partialled out. After Rotenberry (1985).

been considered directly (albeit not in quantitative detail) as a determinant of habitat–occupancy patterns. The distributions of several native and introduced bird species among habitats in New Zealand appears to be associated largely with the differing availability of food such as fruit or honeydew (Clout and Gaze 1984), and the amount of arthropod biomass present is a better predictor of avian habitat use than are measures of foliage diversity or volume in the New Jersey Pine Barrens (Brush and Stiles 1986). In some situations, microclimate or water availability may also be critical features of habitats (Karr and Freemark 1983, Weathers 1983). The array of environmental factors that may possibly contribute to habitat selection and occupancy is quite large (Fig. 9.2), and to consider them all would pose formidable sampling and analytical difficulties. To reduce this problem, Gray and King (1986) suggested the use of multidimensional scaling, a technique that allows one to determine which of a series of resource dimensions actually influences animals in selecting from an array of choices (see also Kenkel and Orlóci 1986). Unfortunately, this procedure requires information on the degree to which organisms perceive choices as similar or different, which is difficult to obtain even in carefully controlled laboratory experiments.

Related to the problem of determining which variables to measure is the question of how many variables to consider. Cody (1974a) argued that one need measure only a few variables, those associated with the key niche dimensions on which competing species are segregated, and this 'few variables' emphasis is characteristic of the approach to habitats used by MacArthur and his associates. The fact that multivariate analyses often include a relatively small number of variables in the explanatory functions seems to reinforce this view (e.g. Rice *et al.* 1983a). This finding is an almost inevitable consequence of the application of multivariate procedures to any data set containing very many intercorrelated variables. Other workers (e.g. Bond 1957, Wiens 1969, James 1971, Emlen 1977b) have measured a great many habitat variables, perhaps in the hope of including among them those that are of critical importance to the species.

The 'few variables' approach has the advantages of ease and simplicity. It is not likely to produce deep insights into the habitat relations of species, however, especially if different features influence individuals to varying degrees at different places or different times and these effects vary among species (all of which seem likely). The 'many variables' approach is more difficult, but it offers the potential of documenting habitat patterns in quantitative detail. Multivariate procedures may be used to condense the mass of variation among these variables into some comprehensible pattern, but this defines only what is 'important' statistically, not necessarily biologically. One cannot rely on multivariate magic to tell one what the measurements mean. Moreover, the patterns obtained by either of these approaches relate only to the domain of the original measurements. Analyses based on consideration of several features of vegetation height profiles (Cody 1978; Fig. 9.5) reveal patterns related to the vertical structuring of habitats, nothing more. The ways in which data are restructured to produce derived habitat gradients by a multivariate analysis depends entirely on the variables considered. If one includes a great many irrelevant variables, these will fade into the background as 'noise' in the analysis only if they vary randomly with respect to one another; to the extent that they are intercorrelated, they will enter into the analysis, contributing to patterns that are nonsensical. Karr and Martin (1981) report an interesting example of this effect (see also Stauffer *et al.* 1985).

Nonlinearity

Viewing the literature, one might conclude that we (and birds) live in a world totally dominated by linear effects and relationships, as virtually all ecological analyses rest on linear procedures. In fact, the patterns of variation in habitat features are often nonlinear on continuous environmental gradients (e.g. Austin *et al.* 1984). Because the responses of birds to many of the factors shown in Fig. 9.2 frequently involve thresholds, the distributions of birds with respect to habitat features may be nonlinear as well. If several habitat variables are interrelated in nonlinear ways, habitat gradients constructed as the best linear combination of such variables (as in PCA or DFA) may be quite misleading (Johnson 1981). Indirect approaches such as RA (Sabo and Whittaker 1979) may reduce such problems, but the effects of nonlinearities are not eliminated, as these procedures also assume a Gaussian model (Ter Braak 1985). Canonical correspondence analysis procedures (Ter Braak 1986) may provide a way to deal with these problems.

The problems of dealing with nonlinearities are exacerbated if one has not considered the entire range of habitat occupancy by a species is an area but has arbitrarily truncated the data set in space or time (Johnson 1981, Wiens and Rotenberry 1981a; see Fig. 3.5). Analyzing portions of a nonlinear response curve with linear procedures may reveal almost any sort of pattern, depending on which portion of the curve is considered.

Meents *et al.* (1983) approached the problem of nonlinearity by using PCA to condense the variation in a large number of habitat measurements into four derived variables. The component scores for each of these four factors were then transformed by squaring and cubing. This produced a new set of 12 variables that were then used in stepwise multiple regression analyses of the abundance distributions of the bird species. Although the linear (untransformed) PCA components were clearly dominant in these regressions, many of the polynomial factors also entered into the expressions, suggesting the importance of nonlinearity in the birds' responses to habitat variation. Rotenberry and I (Rotenberry 1986, Wiens 1985b, Rotenberry and Wiens MS) used response-surface methods (Myers 1971) to evaluate the habitat relationships of shrubsteppe birds. In this analysis, the habitat measures are used as terms in polynomial regression equations rather than transformed to polynomials that are then subjected to linear regression. We found that the nonlinear models generally fit the data much better than did the linear equations, indicating the importance of nonlinear relationships between the birds and habitat features.

This analysis also provides an instructive example of the difficulty of using equations relating bird densities to habitats as predictive models. In our response-surface analysis, we used information from 14 shrubsteppe sites sampled over 3 yr (1977–79) to derive multi-term equations relating densities of several breeding species to PCA components based on measures of either habitat structure or floristic composition. Most of these models produced a significant fit to the data (Table 9.1), especially when floristic measures were used. For Sage Sparrows over 70% of the variations in abundance were described by the floristic response surface, although the birds appeared to vary more or less randomly with respect to habitat structure (confirming our earlier analyses; Wiens and Rotenberry 1981a). Horned Larks and Sage Thrashers (*Oreoscoptes montanus*) exhibited similar patterns. Brewer's Sparrow (*Spizella breweri*) densities, on the other hand, were significantly associated with both the structural and floristic response surfaces.

Models that fit the observations this well should be useful in predicting the abundances of these species in additional samples not included in the original analysis. We tested the predictive capacity of the models in three ways. First, a subset of 5 of the original 14 sites was surveyed during 1980–83; these provide a way of validating the model performance in time. We used the model derived from the 1977–79 data to predict the bird densities expected on these five sites, given the 1980–83 habitat measures, and then related the expected densities to those observed by correlation analysis. In general, the models predicted abundances of these species about as well for the 1980–83 validation sample as they had in the original analysis (Table 9.1), implying a fair degree of temporal consistency in bird–habitat relationships. The observed abundances of Sage Sparrows and Sage Thrashers, however, differed significantly from those predicted by the floristic model. What were splendidly tight relationships during 1977–79 eroded during the following 4-year period, despite the fact that coverages of the shrub species changed little during this period.

Second, we applied the models to data generated from several 'secondary' sites not included in the original model derivations to test the geographical consistency of the models. Two of these sites (Owyhee and Jack

Table 9.1. *The fit of nonlinear response–surface regression models based on PCA components of habitat structure and floristic composition to the abundances of breeding bird species.*

The models were derived from data gathered at 14 original sites during 1979, and then applied to independent data from 5 validation sites (temporal comparison) and 5 secondary samples (geographical comparison). Because different satistical procedures were required for these comparisons, values of correlation coefficients are shown for the original and validation sites, whereas probabilities based on *t*-test of the hypothesis of no difference between observed and expected values are shown for the secondary sites.

			Secondary sites				
			Owyhee		ALE		
Species and model	Original Sites, 1977–79	Validation Sites, 1980–83	1978	1979	Plot 1	Plot 2	Jack Creek
Sage Sparrow							
Structure	0.60	0.16	0.59	0.49	0.28	0.03*	0.09
Floristics	0.84***	−0.07	0.03*	<0.01**	0.09	0.09	0.02*
Brewers's Sparrow							
Structure	0.68**	0.52*	0.70	0.04*	0.24	0.19	0.07
Floristics	0.82***	0.53*	0.32	<0.001***	<0.001***	0.06	0.89
Sage Thrasher							
Structure	0.47	0.03	0.46	0.80	0.54	0.46	0.81
Floristics	0.76**	0.40	0.92	0.42	0.02*	0.35	0.37
Western Meadowlark							
Structure	0.67**	0.81**	0.96	0.65	0.28	0.02*	0.49
Floristics	0.74*	0.59**	0.18	0.79	<0.001***	0.09	0.81
Horned Lark							
Structure	0.40	0.31	0.02*	0.05*	0.32	0.74	0.26
Floristics	0.87***	0.66**	<0.001***	<0.001***	0.04*	0.46	0.21

Notes:
$* = P < 0.05$, $** = P < 0.01$, $*** = P < 0001$.
Source: From Rotenberry and Wiens (MS).

Creek) are located in southeastern Oregon, within a few kilometers of sites used in the original analysis; the other (ALE) is in shrubsteppe habitat in southeastern Washington, some 450 km north. The observed densities of most species at most of these sites differed significantly from those predicted by the nonlinear structural or floristic models (Table 9.1). For example, the floristic model described over 75% of the variation in Horned Lark numbers for the original sites, but it failed dismally in predicting lark abundance at the Owyhee site (both years) and at one of the ALE plots. It produced adequate predictions, on the other hand, for larks at the other ALE plot and at the Jack Creek site. Overall, only two of the seven habitat-association models that were statistically significant over the original sites (thrashers/floristics and meadowlarks/structure) performed well when applied to the secondary sites. Indeed, the three best-fitting original models (all floristic) performed worst. Clearly, there is a substantial geographical component to the patterns of habitat association of these bird species. Because the secondary sites did not differ markedly in habitat structure or floristic composition from the original sites, this variation appears to reflect geographical differences in the responses of bird populations to habitats.

The third test involved using the models to predict densities of the species at a site (Guano Valley) in which the shrubsteppe habitat was dramatically altered in both structure and floristic composition as part of a massive 'range improvement' program of state and federal agencies in this region (Wiens and Rotenberry 1985; see Volume 2). Here also the densities of the birds predicted by the response-surface models on the basis of the changed habitat configuration failed to approximate the observed densities at all closely (Rotenberry 1986).

Several reasons can be suggested for the generally disappointing performance of the response-surface models when they are applied to other data sets. For the ALE site this might be a consequence of geographical variation in habitat selection by the species, but this seems unlikely to have contributed to the poor fit at the Owyhee or Jack Creek sites. At Guano Valley, time lags produced by site tenacity may have destroyed the anticipated relationship, and at all of the locations variations in density that are independent of habitat features may have important effects (Wiens and Rotenberry 1981a, Wiens 1985b, Wiens and Rotenberry 1985, Rotenberry 1986). We are left, however, with the disquieting conclusion that even 'good' models of bird–habitat relationships that incorporate nonlinear effects may be quite limited in their capacity to be extrapolated to predict densities in specific situations. Maurer (1986) has noted similar difficulties in extrapolating bird/habitat predictions from linear correlations.

Conclusions

Some of the most obvious patterns in bird communities are those relating species to habitats. Patterns of differences between ecologically similar species have often been emphasized because they can be interpreted as evidence of adjustments to competitive pressures. Although these patterns are of considerable interest, a documentation that two or more species differ in their distributions in relation to real or derived habitat gradients does not justify any conclusions regarding the role of competition (or any other factors) in producing this pattern, any more than a recognition that several species group together by shared habitat affinities permits the conclusion that the species represent a tightly-knit, integrated assemblage. In addition, habitat features represent only one aspect of species' niches, and to regard a space or volume defined by multivariate analysis as some sort of total Hutchinsonian 'niche space' simply because it is multidimensional is short-sighted.

Like any ecological patterns, habitat relationships are sensitive to the methods used to derive them. This is particularly troublesome in habitat analysis, as several of the more popular methods have distinct (and intended) biases toward discerning either similarities or differences among species. Given the same set of habitat information, these methods will produce quite different results, prompting different interpretations and conclusions. A determination of which method may be appropriate and which inferences are warranted depends upon a clear conception of one's objectives (which is best accomplished if the hypotheses being tested are clearly stated) and an intimate understanding of the biology of the species. Multivariate analyses cannot substitute for knowledge, nor will they disguise ignorance.

We usually determine patterns of habitat occupancy at the broad scale considered in this chapter by using some measure of density as the response variable to habitat variation on gradients. The premise that habitats supporting high densities of a species are of better quality than those in low-density areas cannot be used as a basis for reaching conclusions or framing management policies without independent verification of the relationship. In the same way, it is incorrect to conclude that the habitat variables most closely associated with high-density populations are the most important variables to the species or those that individuals use in making habitat-selection decisions. The density stability of populations in different habitats might be a more appropriate measure of habitat suitability than density alone, as fluctuations would be expected to be greater in marginal habitats subject to density-dependent spillover than in 'optimal' habitats. One test of this notion, however, yielded ambiguous results (O'Connor 1986).

If everything in nature were strictly optimal, one might have reason to expect variations in population densities or habitat use to be closely related to habitat suitability, and patterns of similarities or differences among species in communities might then be easy to interpret. This ideal state of affairs is complicated by a large array of factors, and these in turn make it difficult to construct models of bird–habitat relationships that have much predictive power or lead to an understanding of habitat selection (Best and Stauffer 1986, O'Neil and Carey 1986, Van Horne 1986). Both our understanding and our models suffer when we fail to include the effects of:

1. responses by the species to predators, prey, parasites, or competitors that may modify habitat responses;
2. events elsewhere in the range of individuals or populations on habitat use or density (e.g. migrants);
3. harsh or sporadic environmental factors, such as weather, that act in a stochastic manner in relation to any bird–habitat linkages;
4. temporal or geographical variations in the habitat responses of species;
5. events occurring at times when the system has not been studied (e.g. winter, for breeding studies, or previous years, for short-term studies);
6. time lags in response to habitat changes (e.g. site tenacity);
7. nonlinearities or thresholds in the variation of habitats or in the responses of individuals or populations to this variation;
8. variations among habitats in the degree to which local areas are fully saturated with individuals;
9. incomplete sampling of environmental gradients (e.g. Fig. 3.5);
10. using habitat measures that are inappropriate, incorrectly scaled, or too few, or using so many variables that actual relationships are submerged in a mass of irrelevant detail; or
11. measuring habitats at an incorrect scale with reference to the question being asked.

Some of these factors are attributes of one's methodology, and their effects may be minimized by proper attention to the design of a study. Others are attributes of the species being investigated and are less easily remedied. Greater insight into the effects of all of these factors is likely to emerge, however, if studies of avian habitat associations are conducted at several locations over several years, and if they are founded on careful consideration of features of the natural history of the birds and their environments.

10

Resources and their use

Species that differ from one another in distribution or broad-scale habitat occupancy are, in effect, removed from common membership in local communities. A large number of species *do* co-occur to form local communities, however, and by virtue of their shared occupancy of the same general habitat type they are at least generally similar in ecology. How do such species relate ecologically to one another? Classical competition theory dictates that they will differ in some essential ways, as ecologically identical species cannot coexist.

Consider a well-worn example. In broadleaved woods in England, one may find four or five species of tits together: Blue (*Parus caeruleus*), Great, and Marsh (*P. palustris*) tits occur regularly, Coal Tits somewhat less so, and Willow Tits are sparse and irregular. Synthesizing information from several sources, Lack (1971) presented the following synopsis of their ecology (Fig. 10.1). The smallest species, the Coal Tit, feeds chiefly on the branches of oak and ash trees. Most of the insects it captures are tiny. The slightly larger Willow Tit feeds primarily on birch, somewhat less on elder, and avoids oak trees. During winter, it uses branches even more extensively than the Coal Tit, but, like all of the species, shifts its attention to leaves in summer. By contrast, the Blue Tit feeds primarily on oak throughout the year, foraging toward the tops of trees on twigs and buds or, in summer, almost entirely on leaves. It is an agile species, often hanging from twigs or foliage to forage. Most of the insects it captures are small. The Marsh Tit is the same size but has a longer bill. If feeds extensively in the shrub layer or on twigs and branches in the lower canopy of oaks or in the understory herb layer. It captures somewhat larger insects than the other species and also eats various seeds and fruits. Finally, the larger Great Tit feeds mainly on the ground in winter, although it shifts to the leaf canopy in summer when gathering caterpillars for its young. Most of the insects it eats are relatively large, and it eats more seeds (especially hard seeds) than the other species.

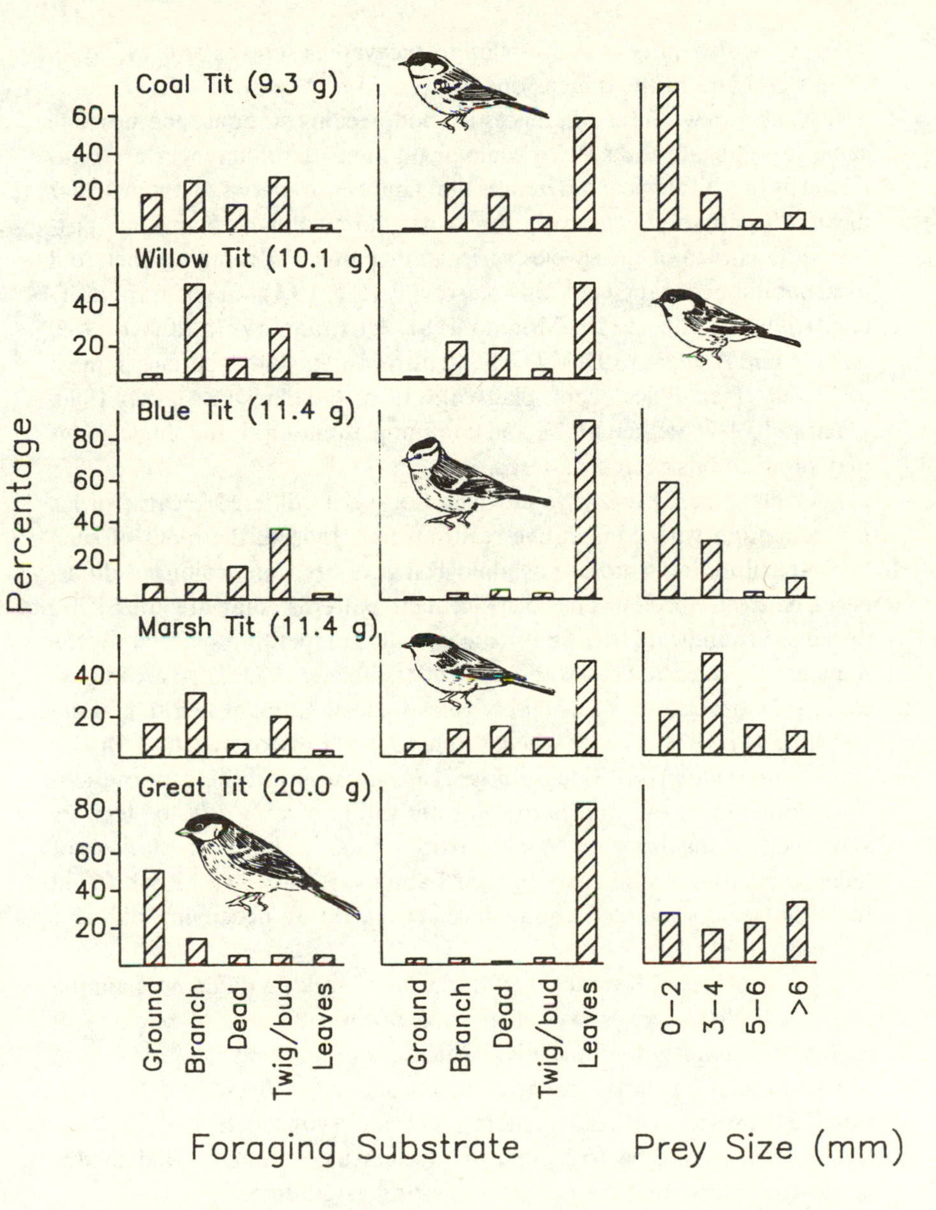

Fig. 10.1. Use of foraging substrates during winter (left) and summer (center) by five tit species in England. Observations of the use of other substrate types are omitted. At the right are shown the frequency distributions of sizes of insect prey in the diets of the birds, obtained at another location where the Willow Tit was absent. Modified from Gibb (1954) and Betts (1955), after Lack (1971).

The species also differ in their ability to excavate nest holes or use holes in the ground and in the dimensions of favored nest cavities.

In Lack's view, these differences in food, feeding stations, and nest site serve to segregate the species ecologically, and the differences are reinforced by morphological differences that adapt each species to its particular niche. The situation is actually somewhat more complex than Lack portrayed it, as each of the species varies in its pattern of foraging, diet, and local habitat occupancy not only seasonally (Fig. 10.1) but geographically (e.g. Alatalo 1980, 1982b, Moreno 1981, Ulfstrand 1977, Sæther 1982, Nilsson and Alerstam 1976). The guild also contains other species in most areas (e.g. Pied Flycatchers, Slagsvold 1975; Goldcrests, Alatalo 1980, Ulfstrand 1977, Sæther 1982), and confining attention to the single genus may produce misleading patterns.

Lack emphasized the ways in which the species differed because of his preoccupation with competitive relationships. Indeed, the focus of such 'niche-partitioning' studies has almost always been on ecological differences between species. The 'displacement patterns' that are found are presumed to indicate 'the limits interspecific competition place(s) on the number of species that can stably coexist' (Schoener 1974a) and are important in the generation of assembly rules for communities (e.g. Diamond 1975a, Cody 1978). Thus, when Fischer (1981) discovered that Brown (*Toxostoma rufum*) and Long-billed (*T. longirostre*) thrashers in southern Texas differed in bill and body sizes by ratios of only 1.06 or less, he surmised that the thrashers should partition food resources on the basis of habitat. He found differences in their habitats, from which he concluded that 'the species avoided competition primarily by occupying different habitats' (1981: 340).

Because of the well-established tradition of seeking differences among species, the literature is skewed toward documentations of patterns of ecological separation. Similarities, which are often more impressive (e.g. Fig. 10.1), are frequently ignored or are considered bothersome details that obscure the features of critical interest. Our main concern in studying local assemblages *should* be with how the species use resources, and in this context similarity may be just as interesting as difference.

Resources

In order to consider patterns of similarities or differences in resource use or niche relationships among species, it is necessary to determine exactly what 'resources' really are. Hutchinson's (1957) concept of the multidimensional niche has become translated over the years into theoretical models in which the niche dimensions are formally equivalent to resource dimensions (or, more precisely, utilization functions of species on resource dimensions) (MacArthur 1972, Diamond 1975a, Cody 1978). One might expect, then, that careful attention would be given to the identification and measurement of these resource dimensions. Unfortunately, this is rarely done. At one time or another, almost any environmental factor that correlates with variations in the distribution and abundance of species has been termed a resource, even including derived multivariate axes that have no apparent reality in nature. In 'resource partitioning' studies, whatever variables separate species ecologically are often termed resources, and the coexistence of the species is, in turn, explained by the partitioning of those resources. There is a certain circularity to such arguments.

The definition of resources

What is or isn't a resource? The critical features of a definition are that (a) it must be used (but not necessarily consumed) by the organism of interest, and (b) it must be at least *potentially* limiting (but not necessarily always limiting) to individual fitness and/or population dynamics (Tilman 1982, Wiens 1984b, Andrewartha and Birch 1984; see also Abrams 1988). Food (and the essential nutrients, energy, and materials that it provides), therefore, is clearly a resource. Space is a resource to sessile organisms, and adequate living space may be a resource to territorial or social animals. More often, however, space may affect the distribution of factors that are resources than be a resource itself. Some components of habitats, such as nest sites, roost locations, singing perches, or cover from predators, qualify as resources, as they may be limiting and are clearly used by the organisms. Other aspects of habitats, such as foliage profile or 'patches' in a heterogeneous habitat, are not resources *per se* but rather reflect the spatial distribution or accessibility of things that are resources. 'Time' is generally not a resource either, but rather is something that constrains resource availability (e.g. seasonal phenology) or use (e.g. prey handling time) (see Tracy and Christian 1986). In the northern waterfowl assemblages that Toft *et al.* (1982) considered, for example, 'time available for breeding' is not a resource, as they claimed; rather, there are temporal restrictions on food abundance or availability that influence breeding success. Aspects of foraging behavior are not themselves resources but have to do with how resources are obtained or used.

It is not always easy to determine whether or not something is a resource. Is the individual plant or its flowers or the nectar they produce the resource to a hummingbird? The nectar is used, but the entire plant or patches of

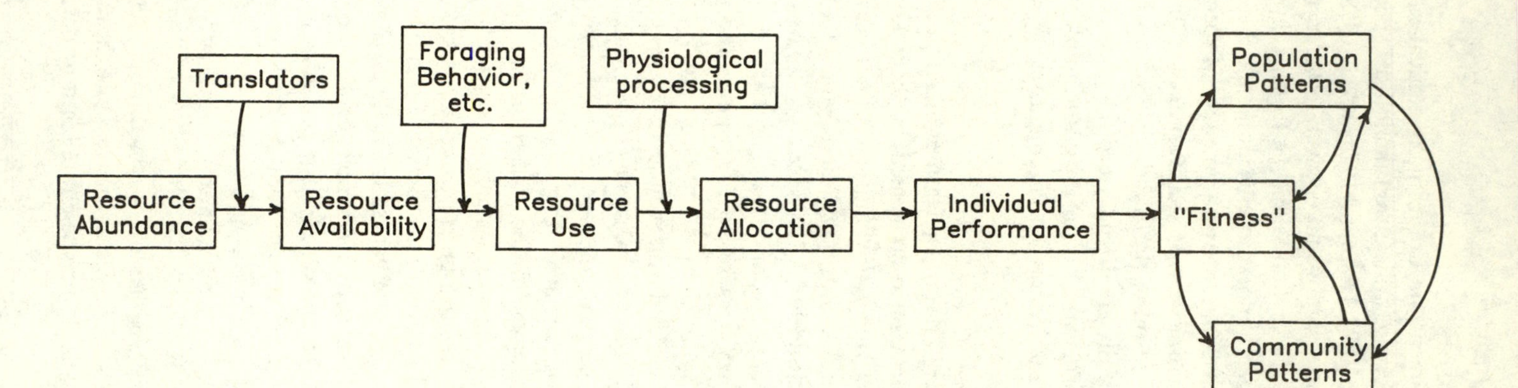

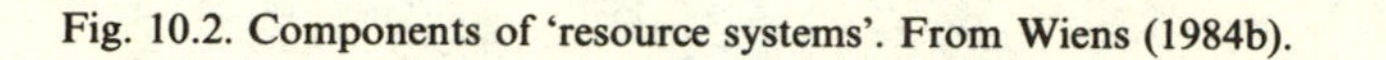

Fig. 10.2. Components of 'resource systems'. From Wiens (1984b).

plants may be defended territorially. If birds consistently prefer trees of a particular species for foraging or nesting, is the tree species the resource, or is it simply a correlate of particular distributions of food types or nest sites? The observation that individuals use some features of the environment nonrandomly in relation to measures of their abundance or availability (e.g. Szaro and Balda 1979, Alldredge and Ratti 1986) does not necessarily mean that one has correctly identified the resource (although it may be suggestive).

Thinking clearly about resources is critically important to achieving any understanding of niche relationships in communities. Some of the ambiguity of niche patterns in communities and the argument over their meaning probably stems from a failure to define resource dimensions correctly. Different definitions or measures of resources may produce quite different patterns of niche breadths or niche overlaps among species (Case 1981, 1984, Greene and Jaksić 1983), and, if one combines factors that are resources with those that are not in multidimensional niche analyses, real patterns may be obscured while the observed patterns are interpreted incorrectly. Without a precise definition of the resources in a system, it is not possible to derive accurate patterns of resource or niche relationships among organisms in the community, much less to measure the resources so that hypotheses resting on assumptions of resource limitation may be tested fairly.

Resource systems

Defining what factors are or are not resources, however, is only part of the problem. Some of the fuzziness of resource patterns in communities derives from a confusion of aspects of resource use (e.g. foraging position) or its consequences (e.g. resource partitioning) with the actual resources themselves and a failure to recognize these as different elements of 'resource systems' (Fig. 10.2; Wiens 1984b).

Resources exist in an environment, independently of whatever uses them, in a certain *abundance* that varies in both space and time (P. Price 1984). The abundance of a resource in an environment, however, is of less importance to an organism than its *availability*. If prey are cryptic or occupy microhabitat shelters, only a portion of those actually present in an environment will be available to a consumer at any given time. Parasitized prey individuals, on the other hand, may be more available to predators than unparasitized individuals (Moore 1983, 1984). Widely scattered seeds or nectar sources may be less available to a foraging bird than the same quantity of resource clustered in patches. Flying insects may not be readily available to birds that glean inactive or resting insects from foliage, leading individuals to forage at those times of day when insect activity is reduced (Hutto 1981). Intra- or interspecific competitors that exploit (consume) resources used by other individuals alter resource abundance, but if the competitors instead aggressively preclude access to resources, resource availability is affected. A variety of biotic and abiotic factors may therefore act as 'translators' in resource systems, changing a particular pattern of resource abundance into a different pattern of resource availability.

The set of available resources forms the foundation for resource *use* by individuals. Resource use, however, is influenced by constraints or predispositions associated with the morphology, physiology, and behavior of the species and by external factors such as the availability of alternative food types, the spatial relationship among resources, attributes of the resources themselves (e.g. seed hardness, prey palatability), or competition with other species. Once used or consumed, resources are allocated to various functions by individuals (Fig. 10.2 emphasizes physiological allocation of food resources, but a tree hollow may be allocated in a parallel way to incubation or to roosting). How resources are allocated influences the performance (e.g. reproduction, survival) of individuals and thus their fitness.

An example may clarify the distinction between resource abundance, availability, and use. Clams and mussels are important resources to shorebirds feeding in intertidal mudflats in many parts of the world. In The Netherlands, the clam *Mya arenaria* is an important prey of both Oystercatchers (*Haematopus ostralegus*) and Curlews (*Numenius arquata*) (Zwarts and Wanink 1984). The clams are distributed through the mud over a range of depths, larger clams occurring deeper (Fig. 10.3). This size–depth distribution specifies the abundance of the resource. The birds capture prey by probing in the mud with their bills, so only those clams occurring within the depth zone determined by the birds's bill lengths (7 cm for Oystercatchers, 12–14 cm for Curlews) are actually available to the birds (Fig. 10.3). Small clams occurring at shallow depths are available to both species but are generally not taken as prey because they fall below a size threshold at which prey can be handled efficiently. This threshold is higher for the Curlew than for the Oystercatcher (Fig. 10.3). The birds also select the larger clams occurring within the substrate-depth zone they probe; Oystercatchers, for example, take many clams of 3–4 cm length, even though most clams of that size class occur at depths beyond its bill reach. Thus, aspects of resource use shift the diet from that expected on the basis of availability alone.

Zwarts and Wanink experimentally manipulated the clam distribution in

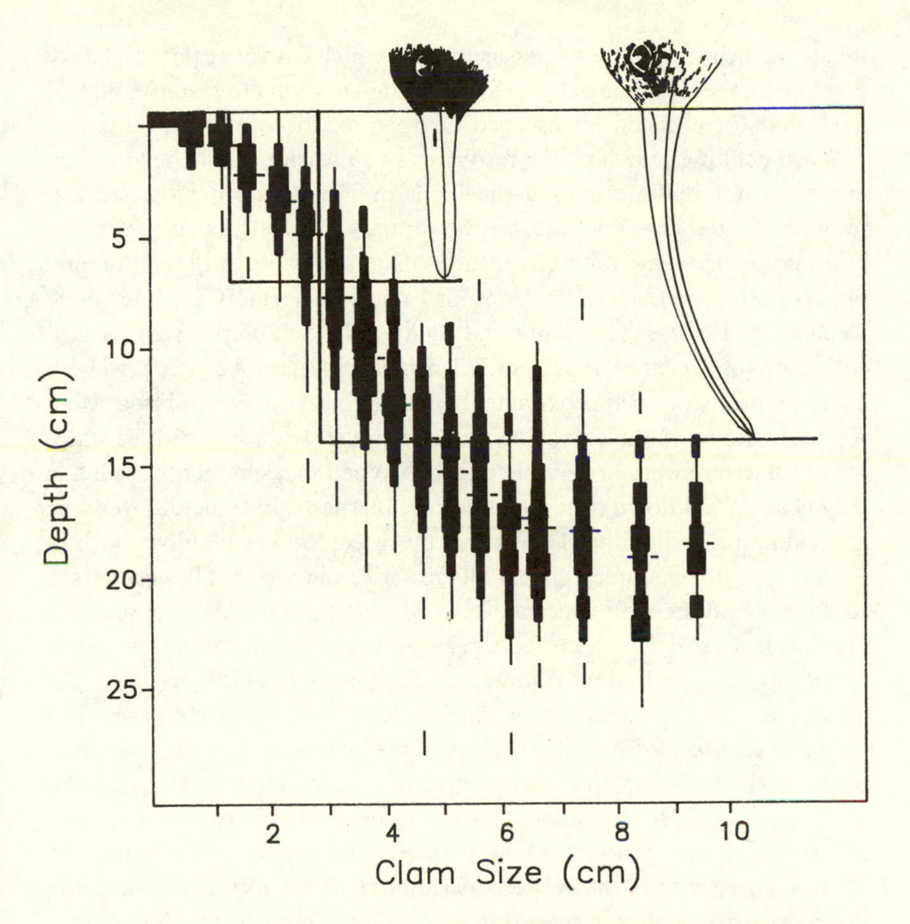

Fig. 10.3. The frequency distribution of depths at which clams of different size classes are found in intertidal mudflats in The Netherlands. The lines show the portion of the clam distribution available to Oystercatchers and to female Curlews, as determined by their bill lengths (probing depths) and the minimal clam sizes they can handle efficiently. Modified from Zwarts and Wanink (1984).

one area by burying hundreds of large clams just below the surface. An Oystercatcher occupied the site, preyed upon the large clams, and defended the sites against congeners as well as Curlews. Normally, however, differences in resource use by the species are nearly complete and interactions are rare. Because the size–depth distribution of clams also reflects an age-class distribution, the two species affect the age structure of the clams differently. Zwarts and Wanink's calculations indicated that predation by Oystercatchers might deplete the clam populations substantially in the winter following their second growing season, after which Curlews exerted a heavy pressure on the remaining clams. Those clams surviving to reach lengths of more than 7 cm effectively escaped further predation by these birds.

When defining or measuring resources as dimensions in niche analyses, one must differentiate among the levels in Fig. 10.2. In most studies, resource availability and use are of primary interest, as they provide information necessary to test hypotheses of competition, optimal foraging, opportunism, and the like. Some resources, such as nectar (Carpenter 1978, Heinemann 1984), seeds (Smith and Balda 1979, Schluter 1982a), or nest cavities (Saunders *et al.* 1982, van Balen *et al.* 1982), can be measured with sufficient precision to approximate their availability to certain birds. Most resources are more difficult to measure, however, so often resource states are not measured but are simply inferred. When measurements are made they bear an unknown (but also inferred) relationship to actual resource abundance or availability. To estimate the insect food available to birds in alder forests, for example, Stiles (1980) recorded the extent of insect damage to leaves of different tree species, from which he projected insect biomass per unit leaf area. He then used a relationship of total leaf area for each species to the distribution of foraging activities by the birds to justify the assumption that insect abundance is directly proportional to leaf area. The estimates that result from all of this are of dubious value. Cody (1978, 1981) used 'stickyboards' (small squares coated with insect tanglefoot) to measure insect numbers in several habitats. He considered these to provide reasonable estimates of the food available to insectivorous birds, since he found a relatively high correlation between variation in bird densities and variation in stickyboard catches in several habitat types (Cody 1983c). When standardized by exposure time, the plate captures may provide a useful index of the relative numbers of certain types of insects active in different areas or microhabitats at different times, but, as Cody himself (1980: 1072) observed, it would be foolhardy to claim that such stickyboard captures are a direct measure of the actual availability of insects eaten by insectivorous birds.

The difficulty of documenting resource states has not stifled investigations into resource use or partitioning. Schoener (1974a) calculated that the number of such studies grew exponentially at a rate four times that typical of scientific work in general during the period 1959–72. The rate of growth has probably decreased some since then, but the topic is still quite popular (Schoener 1986c). These studies have been focused on how birds use food resources and microhabitats (especially in foraging), how coexisting species use space, and how species differ in nesting locations. An immense literature has accumulated, which I will dip into only at scattered points to indicate something of the diversity of resource-use patterns that has been found.

Diet-niche relationships

Some patterns of food resource use

'Food' is seemingly a clearly identifiable resource that is often thought to be limiting (see Volume 2). Similarities or differences among species in diet composition are thus especially relevant to tests of niche or guild concepts. Generalizations, however, are hard to come by. The *Parus* species that coexist in British woodlands, for example, differ in the sizes and types of prey consumed (Fig. 10.1), although the differences are sometimes slight. The species of foliage insectivores that Root (1967) studied in California oak woodlands overlapped considerably in prey sizes and types, and Wooller and Calver (1981) found similarly high overlap among some (but not all) of the small birds feeding in the understory of kari (*Eucalyptus diversicolor*) forests in Western Australia. Birds in both the California and Australian communities differ in morphology and foraging behavior (e.g. Fig. 6.3), so the diet similarity is not directly related to morphological or behavioral similarity or difference.

In other groups, diet is more closely related to morphology and/or behavior. Many seabirds, for example, feed only at the surface, whereas others dive for prey, sometimes to considerable depths (see Fig. 10.8). Among the large assemblage of seabirds breeding on South Georgia, the surface-feeding species (e.g. albatrosses, petrels) feed primarily on squid, copepods, and krill, whereas the pursuit-divers (penguins, shags) feed on fish and exploit krill to an even greater extent than do the surface feeders (Croxall and Prince 1980). Within each group, species differ in diets according to foraging location (inshore vs. offshore waters) and bill morphology. In the northwestern Hawaiian islands, most seabird species feed opportunistically, taking nearly any organism of appropriate size (determined by the bird's morphology) that occurs in the feeding zone (Harrison *et al.* 1983). Nonetheless, dietary differences among species do exist, and these can be related to feeding technique or location, time of day, morphology, or the seasonality of breeding. Similar patterns of dietary differences that are related to differences in morphology, behavior, or foraging zones have been reported for other seabirds as well (Furness and Barrett 1985, A.W. Diamond 1984, Trivelpiece *et al.* 1987, Croxall 1987). Interpreting these patterns in the absence of information on prey resource abundance or availability, however, is difficult. High dietary overlap has been reported for seabirds of quite different morphology and behavior where prey such as

Table 10.1. *Designations of the food resources of 16 species of insectivorous, neotropical flycatchers inhabiting lowland habitats in Costa Rica.*

The food resource defines the lowest common denominator of prey characteristics as reflected by stomach contents.

Species	Food resource characteristics
Contopus cinereus, C. virens	A variety of aggregated, probably ephemeral flying prey (inc. many social Hymenoptera)
Contopus borealis	Mostly aggregated, probably ephemeral social Hymenoptera (ants and bees)
Colonia colonus	Predictable patches of flying stingless bees, *Trigona* species, perhaps near bee nests
Terenotriccus erythrurus	Mostly fulgoroid, and some cercopoid, Homoptera that jump to evade predators
Myiobius sulphureipygius	A wide variety of active insects, many of which were fleeing from members of mixed-species bird flocks
Myiornis atricapillus	Cryptic, relatively immobile and homogeneously distributed arthropods and some Homoptera (primarily Membracidae)
Todirostrum sylvia	Cryptic, relatively immobile and homogeneously distributed arthropods
Todirostrum cinereum	Many insects including relatively evasive Diptera and parasitoid wasps, perhaps at extrafloral nectaries
Todirostrum nigriceps	Many insects including evasive Diptera and some parasitoid Hymenoptera, perhaps around flowering trees
Oncostoma cinereigulare	Cryptic, relatively immobile arthropods
Rhynchocyclus brevirostris	Almost any arthropods within vegetation, except relatively small and/or more evasive Diptera and parasitoid Hymenoptera
Platyrinchus coronatus	Cryptic, relatively immobile and homogeneously distributed arthropods, and some jumping Homoptera
Tolmomyias assimilis, T. sulfurescens	Relatively active, conspicuous, or

Table 10.1. (*cont.*)

Species	Food resource characteristics
	easily catchable prey, perhaps in productive vegetation zones
Empidonax virescens	A vareity of active, probably ephemeral prey

Source: After Sherry (1984).

anchovies, squid, or sandeel are seemingly superabundant (Pearson 1968, Crawford and Shelton 1978, Baltz and Morejohn 1977, Furness and Barrett 1985). Croxall *et al.* (1984) found that several South Georgian seabird species consumed krill that were considerably larger than those captured in net hauls, suggesting prey selection, but the birds may have been feeding in areas that differed in krill size distribution from that in the limited net-sampling areas.

In other cases, species that use a generally similar and seemingly constrained foraging mode, such as the neotropical flycatchers studied by Sherry (1984), may differ markedly in diets (Table 10.1). In this situation, the diets of the species reflect differences in their responses to the detectability, location, and ease of capture of prey, which in turn are associated with the morphology, foraging behavior, and microhabitat position of the species. Sherry suggested that these species-specific patterns may reflect phylogenetic constraints imposed during the adaptive radiation of neotropical flycatchers in the Tertiary.

Variation in Diets

Any consideration of diet relationships among species is complicated by the fact that diets of species may change as prey abundance and availability vary in both space and time. Most owl species in middle and northern Europe feed upon voles, which are abundant prey, and dietary overlap may be high (Herrera and Hiraldo 1976, but see Korpimäki 1986). In the Mediterranean region where microtines are less plentiful, owls diversify and differentiate in diets. In the Australian kari forest mentioned before, many of the species have much the same diet in early winter and in spring but most species broaden their diets during the breeding season. Some species also shift their diets, changing the form of the diet–overlap matrix (Calver and Wooller 1981).

The nature of these variations in diet composition and interspecific

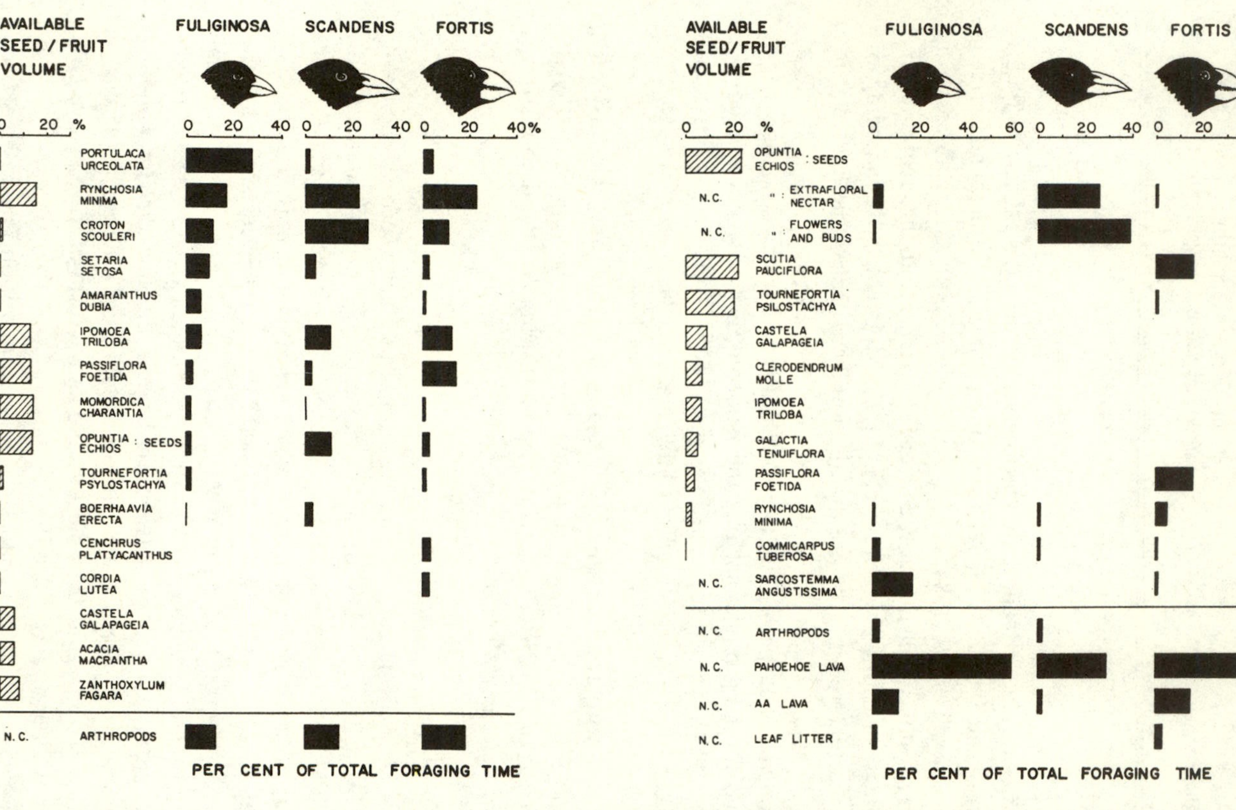

Fig. 10.4. Diets of three common finch species at B. Academia in the Galápagos Islands during the wet (left) and dry (right) seasons. The hatched histograms to the left of the plant species names indicate the proportion of that plant species in the total available food volume. The horizontal line divides specific foods taken (above) from foraging categories classed by general food type (below). N.C. = food availability not censused. From Smith *et al.* (1978).

overlap is especially clear among the finch species of the Galápagos Islands. Most of the *Geospiza* populations have generalized diets, and many sympatric pairs overlap in diet by > 50% (Abbott *et al.* 1977). Diet composition varies between areas and between seasons on a given island as the overall spectrum of available foods changes. Smith *et al.* (1978) documented such changes between the end of the wet season and the dry season 6–8 months later. At Bahía Academia on Santa Cruz Island, for example, three sympatric finch species fed upon generally similar foods during the wet season, taking some of the abundant food types, ignoring others, and feeding on several rare species as well. In the dry season, the species fed extensively on the ground on unidentified prey, but for the prey types that could be identified by observation the species specialized differently (Fig. 10.4). *Geospiza fuliginosa*, for example, fed extensively on nectar of the vine *Sarcostemma* and on the fruits of *Commicarpus*, whereas *fortis* fed on the medium-sized seeds and fruits of *Scutia*, *Rynchosia*, and *Passiflora*. In contrast to the wet season, birds only rarely fed on arthropods during the dry season, when insects were scarce. The same sorts of seasonal shifts were recorded at other sites: all species fed chiefly on the same soft, easily handled seeds and fruits during the wet season but shifted to different diets during the dry season, in accordance with the morphological specializations of each species. Dietary overlap thus decreased during the season of relative resource scarcity at some locations, but at other sites the shift occurred even when resource levels remained relatively high. Schluter (1982b) also documented patterns of seasonal shifts in diets and reduction of overlap during the dry season among the finches in Isla Pinta, although the changes were more closely associated with changes in patch-use by foraging birds than with food selection *per se* (Schluter 1982a).

In the studies at Bahía Academía, *Geospiza scandens* fed extensively on *Opuntia*, especially during the dry season (Fig. 10.4). On Daphne Major, *scandens* fed extensively on flowers and buds of *Opuntia* in both a dry year and a wet year (Grant and Grant 1980; see also Grant and Grant 1981, Price 1987), but the diet of *fortis* changed. During the dry year, it foraged more on small seeds and on *Tribulus* fruits; it avoided the seeds but fed more on the floral and extra-floral parts of *Opuntia*. The result was an increase in diet overlap with *scandens* during the dry year. Grant and Grant attributed this diet convergence to the appearance of an abundant resource (*Opuntia*) against the background of general resource scarcity; if observations had been taken a few weeks earlier, before *Opuntia* began flowering, the relationships might well have been different.

These studies illustrate the dynamic nature of feeding patterns of species (and the patterns of diet overlap that result) in both space and time and the importance of associated measures of resource levels. In both the finches and the neotropical flycatchers, there are species-specific aspects to food use that are determined in part by the birds' morphology and behavior. Within these constraints, however, the diet may still be diverse and variable, producing patterns of interspecific relationships that reflect the effects of a great many factors (see Ellis *et al.* 1976, Krebs *et al.* 1983, Kamil and Sargent 1981). Thus, although documentation of the dietary patterns among members of a guild may be reasonably straightforward, interpreting the results of such studies is difficult, especially if the contributions of patchiness in resource distributions or of temporal changes in resource levels are not considered.

Methodological comments

Actually, *accurate* determination of diet patterns may not be so easy. In the Galápagos studies, for example, the feeding habits of the birds were determined by direct observation of food consumption by foraging individuals. Usually this is not possible, and various other methods are used to document or infer diets. Sherry's (1984) work was based on the examination of stomach contents of collected specimens, whereas the data of Wooller and Calver (1981) were derived from fecal samples from captured individuals that were then released. Other investigators have used emetics to induce the vomiting of stomach contents, fitted pipe-cleaner collars to nestlings to prevent swallowing of prey delivered by adults (e.g. Orians and Horn 1969, Török 1986), estimated prey consumption from feather pluckings or regurgitated pellets at feeding sites (e.g. Opdam 1975, Reynolds and Meslow 1984), or simply inferred diet from morphology or feeding behavior (Cody 1974a).

Each of these procedures has drawbacks. Prey taxa differ in the rates at which they are digested and disappear from stomach samples. Stomachs therefore contain a mixture of prey types ingested over different lengths of time before collection (Custer and Pitelka 1975). Pellets contain hard parts, such as bones, hair, or exoskeleton, but fail to indicate any soft-bodied prey that may have been consumed. Collars are suitable only for immobile young in accessible nests. Direct observation is possible only in open situations with relatively fearless birds; even then, a great many prey captures may remain unidentified. Neither morphology nor behavior provides a detailed picture of food consumption, and in some situations these features are not closely related to diet at all (Wooller and Calver 1981, Hespenheide 1975). As a consequence of these biases, diets documented by different methods bear different relationships to what the birds actually do consume, and they cannot be combined in analyses or even compared directly. In any single study the patterns of resource use indicated by a given procedure must be judged with regard to the limitations of that procedure before interpretations can be offered.

Foraging behavior and microhabitat use

Some patterns

Because it is far easier to watch what a bird is doing than to sample its diet, many more studies have documented how and where birds forage or use portions of the habitat. The operational assumption of these studies is that differences in foraging behavior among species reflect differences in the ways they encounter and use food resources (MacArthur 1958, 1972). Although this assumption is rarely tested, it is probably valid in a general sense – surface-feeding seabirds encounter different prey than do divers, and canopy-feeding insectivores are exposed to resources that differ from those of ground-foragers. Unless one defines resources carefully, however, it is easy to conclude (incorrectly) that features of the habitat such as height above ground, position in a tree canopy, or substrate type are the resources partitioned by differences in behavior among species.

Members of a guild usually do differ in foraging tactics or microhabitat use, although the differences tend to be quantitative rather than qualitative. Among the five *Dendroica* warbler species that MacArthur (1958) studied in Maine coniferous forests, for example, Cape May Warblers (*D. tigrina*) fed in the peripheral portions of the tops of trees, moving vertically through the canopy and often hawking prey. Blackburnian Warblers also fed in the upper canopy but used the interior as well as the periphery of the trees and moved radially about the tree, usually peering for prey from a perched position. Black-throated Green and Bay-breasted (*D. castanea*) warblers both concentrated their foraging activities in the middle canopy, especially in the darker interior of the tree, but the former moved tangentially to the canopy in rapid short flights and the latter moved radially in a much more deliberate fashion. Black-throated Greens often hovered for prey, whereas Bay-breasteds usually searched from a stationary position, at times hanging to peer beneath the foliage. Yellow-rumped Warblers exhibited the most varied feeding habits, moving radially, tangentially, and vertically in the canopy in nearly equal proportions, searching for prey by both hawking and peering, and using most tree locations (although concentrating activities in the lower canopy). Morse's (1968, 1980) studies in a nearby spruce forest confirmed these basic patterns but also showed that males of each species tended to forage higher in a particular zone than did females (Fig. 10.5). This is because males normally forage in the vicinity of singing perches whereas females forage close to the nest sites, which are lower in the trees (see also Franzreb 1983, Holmes 1986).

MacArthur combined a small number of observations of varying durations from two summers and several study locations to derive the patterns he reported. Their accuracy is thus uncertain, and neither temporal nor spatial variations in foraging behavior can be determined. In MacArthur's study, some of the species (Blackburnian and Black-throated Green warblers) appeared to be rather stereotyped in their foraging behavior. Similar patterns have been reported for these species in other studies in the northeast (Morse 1968, 1971a; Holmes *et al.* 1979a; Sabo 1980), although the behavior of Black-throated Greens is considerably more diverse in the southern Appalachians, where Rabenold (1978) studied them. Black-throated Green Warblers are socially dominant to other warblers (Morse 1971a), and in the subalpine forest studied by Sabo (1980) they aggressively displaced both Yellow-rumped and Magnolia (*D. magnolia*) warblers, both generalized foragers with rather plastic behavior (Morse 1968, 1971a). Cape May and Blackburnian warblers were absent from this forest, and Sabo attributed this to the low stature of the trees and the absence of the favored foraging zone of these species.

Many other examples of niche differences among coexisting species could be described, but one more may be sufficient to illustrate some representative patterns. Three species of treecreeper (*Climacteris*) are locally sympatric in portions of eastern Australia. They all feed primarily on ants, and at times one may see individuals of all three species foraging in the same trees at the same time (Noske 1979). The Brown Treecreeper (*C. picumnus*) feeds extensively on the ground or fallen logs and avoids the outer portions of trees away from the trunk, whereas Red-browed (*C. erythrops*) and White-throated (*C. leucophaea*) treecreepers both rarely forage on the ground or on logs but use boughs, limbs, and branches more extensively (Fig. 10.6). The distributions of the species still overlap considerably, and much the same pattern is seen in the height distributions of the species. All of the species use tree species nonrandomly: *leucophaea* and *picumnus* concentrate their activities on rough-barked stringybarks (*Eucalyptus caliginosa* and *E. laevopinea*), and *erythrops* feeds more extensively on smooth-barked tree species (*E. blakelyi*, *E. amplifolia*, *E. viminalis*, *E. melliodora*, and *E. cypellocarpa*) (Fig. 10.6). The species also differ somewhat in foraging tactics, *leucophaea* and *picumnus* pecking and excavating pieces of bark from rough-barked trees, *erythrops* peering and probing

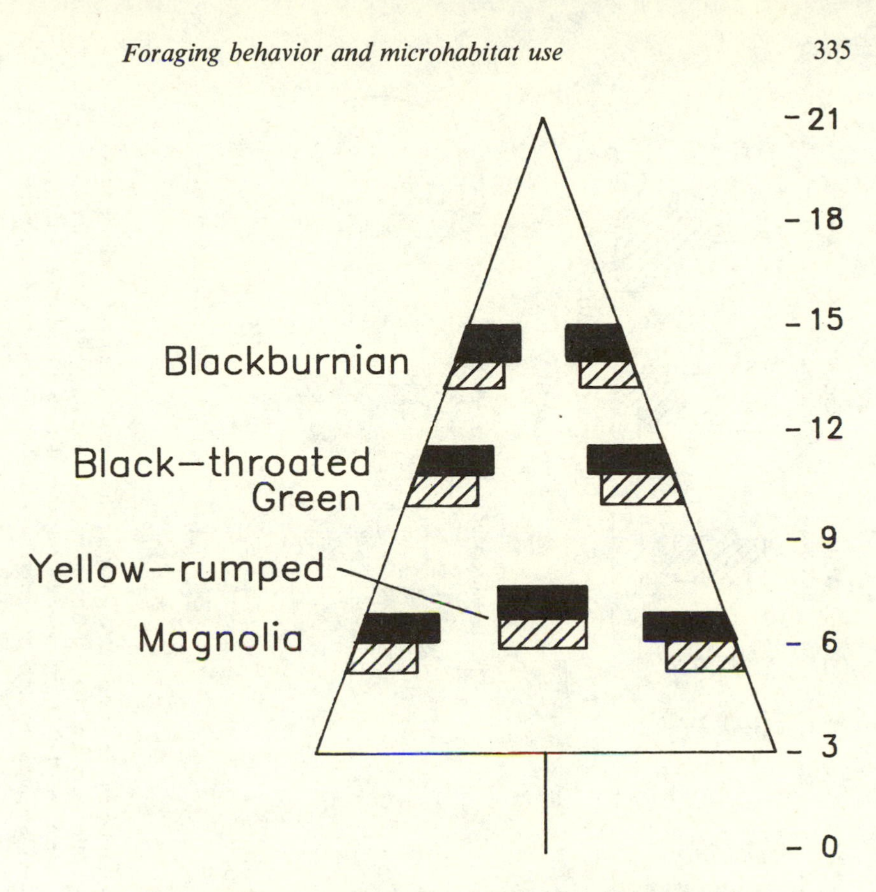

Fig. 10.5. Centers of foraging areas of warblers inhabiting spruce forests in Maine. The solid bar represents males, the hatched bar females. Numbers to the right indicate height (m). Data from Morse (1967, 1968), after Morse (1980).

under strips of bark hanging from smooth-barked trees. Individuals of one species sometimes aggressively supplant individuals of other species from trees, and in these interactions the heavier species (*picumnus*, 37 g) displaces either of the other two species (both 23 g), while *erythrops* consistently displaces *leucophaea*. In southeastern Australia, a fourth bark-foraging species, the Varied Sittella (*Daphoenositta chrysoptera*) co-occurs with *leucophaea* and *erythrops*. It also forages primarily on stringybarks, but focuses its attention on dead branches, excavating flakes of bark with its wedge-shaped bill (Noske 1985). Unlike the treecreepers, it rarely eats ants. Both the treecreepers and the sittella exhibit intersexual differences in foraging behavior (Noske 1986).

In both the warblers and the treecreepers, differences between the species emerge when several features of foraging behavior and microhabitat use are

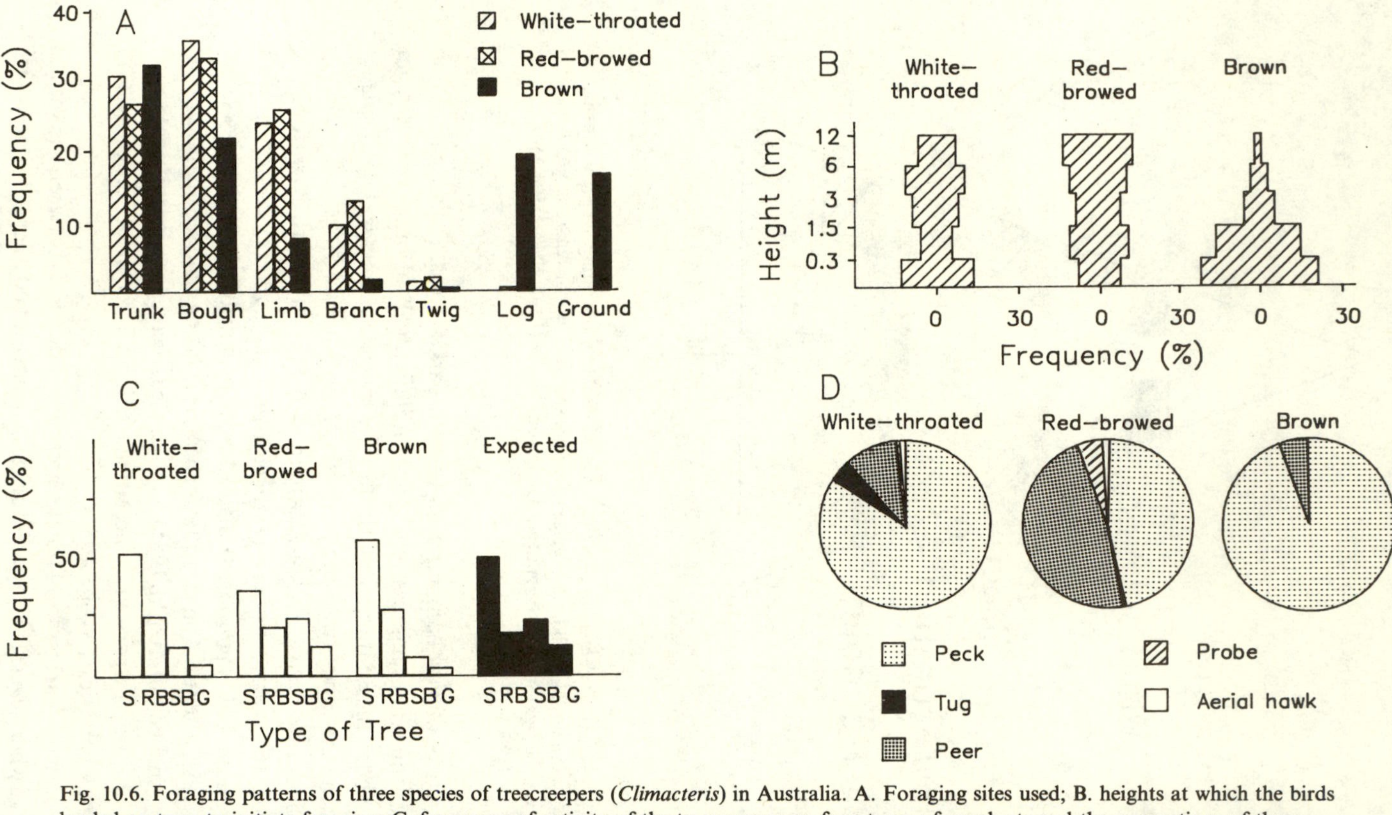

Fig. 10.6. Foraging patterns of three species of treecreepers (*Climacteris*) in Australia. A. Foraging sites used; B. heights at which the birds landed on trees to initiate foraging; C. frequency of activity of the treecreepers on four types of eucalypts and the proportions of these types on the study area ('expected'), which gives the distribution expected if the birds selected trees at random (S = stringybark, RB = rough box, SB = smooth box, G = gum); D. proportion of foraging activities of the treecreepers devoted to each of several movement types. Modified from Noske (1979).

examined. Because niche separation may involve a number of aspects of behavior and habitat use, multivariate analyses may be useful in synthesizing general patterns from the array of separate features. Sabo and Holmes (1983) used detrended correspondence analysis (a variant of PCA) to ordinate the species of a subalpine forest in New Hampshire (in which the species were predominantly of boreal affinities) and a nearby hardwoods forest (populated by species of more southern, deciduous forest affinities). Overall, the subalpine community was less diverse, occupied a smaller total foraging niche space, and included species with greater overlap in foraging behaviors. Sabo and Holmes attributed this, at least in part, to the greater climatic severity of the subalpine zone. The ordination separated the species on the basis of foraging position (ground vs. arboreal) and the coarseness of foraging substrate (bark vs. fine foliage) (Fig. 10.7). In the latter dimension, three woodpeckers and a nuthatch of the hardwoods forests that forage primarily on tree trunks were separated from the other species, which then followed a gradient from the Red-breasted Nuthatch and the Solitary Vireo (which used both branch and foliage substrates), through species using progressively finer resources, to the Blackburnian Warbler and Ruby-crowned Kinglet, which hover at leaves in the upper crown. The distribution of subalpine species was skewed toward the fine-substrate end of the gradient. In general, however, the species in both communities were relatively evenly spread with respect to the two 'niche axes'.

Several habitat and behavioral features have emerged as important dimensions of species differentiation in various studies (see Table 10.2). Features of the tree species (particularly bark texture) were especially important to the treecreepers, and plant taxa have also contributed to species separation in several other studies (e.g. Szaro and Balda 1979, Franzreb 1983, Balda 1969, Ulfstrand 1977, Nilsson and Alerstam 1976, Holmes and Robinson 1981). In northern Finland, eight species of foliage gleaners were differentiated primarily by their use of tree species and the portions of trees in which they foraged (Alatalo 1982c). Balda (1969) and Franzreb (1983) found that species foraging in montane conifers differed in the heights at which they foraged as well as their position in the tree. In a neotropical location, nine morphologically similar species occurring in antwren feeding flocks differed with respect to foraging substrate, foraging method, vertical foraging height, foraging position, or the foliage density of foraging sites (Jones 1977). Sæther (1982) found that species in central Norwegian forests generally differed in foraging height, position on branches, or use of tree species. Most investigators have measured aspects of foraging height, position, substrate, and movements, and most patterns

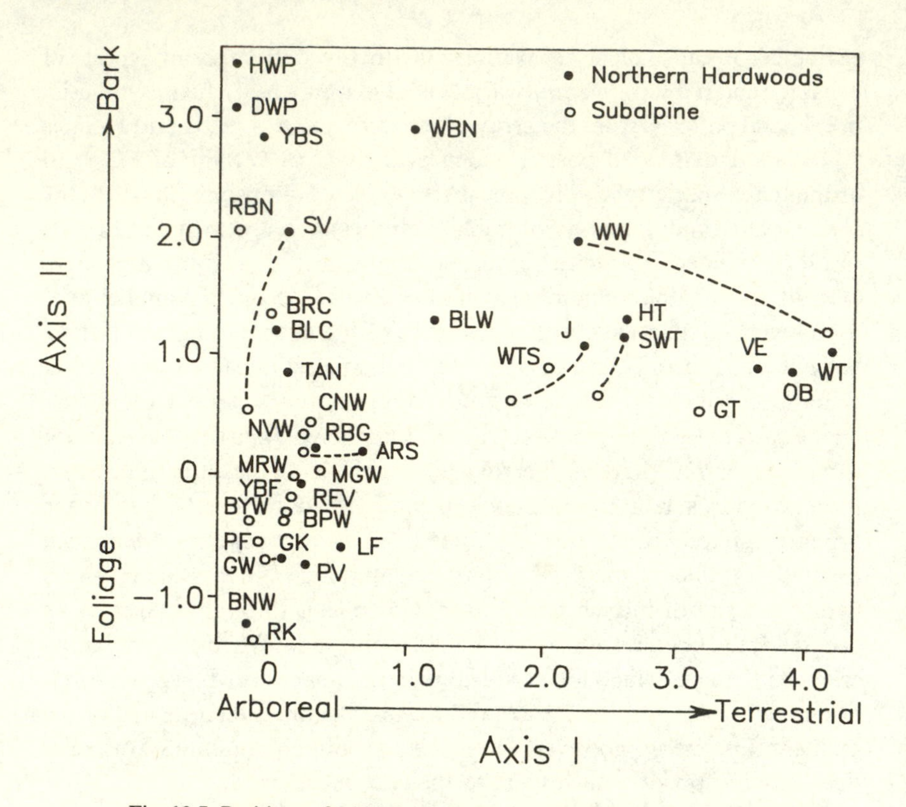

Fig. 10.7. Positions of 36 bird species breeding in subalpine and hardwoods forests in New Hampshire along the first two DCA axes of a joint ordination. Axis I represents an arboreal to terrestrial gradient, Axis II a gradient from foliage to bark substrate use. Dotted lines connect the positions for those species occurring in both forest types. Species codes: HWP = Hairy Woodpecker, DWP = Downy Woodpecker, YBS = Yellow-bellied Sapsucker, WBN = White-breasted Nuthatch (*Sitta carolinensis*), RBN = Red-breasted Nuthatch (*Sitta canadensis*), SV = Solitary Vireo (*Vireo solitarius*), BRC = Boreal Chickadee (*Parus hudsonicus*), BLC = Black-capped Chickadee, BLW = Black-throated Blue Warbler, WW = Winter Wren, HT = Hermit Thrush, SWT = Swainson's Thrush, VE = Veery, WT = Wood Thrush, OB = Ovenbird, GT = Grey-cheeked Thrush, J = Dark-eyed Junco, WTS = White-throated Sparrow (*Zonotrichia albicollis*), TAN = Scarlet Tanager, CNW = Canada Warbler (*Wilsonia canadensis*), NVW = Nashville Warbler (*Vermivora ruficapilla*), RBG = Rose-breasted Grosbeak, MRW = Yellow-rumped Warbler, ARS = American Redstart, MGW = Magnolia Warbler, YBF = Yellow-bellied Flycatcher (*Empidonax flaviventris*), REV = Red-eyed Vireo (*Vireo olivaceus*), BYW = Bay-breasted Warbler (*Dendroica castanea*), BPW = Blackpoll Warbler, PF = Purple Finch (*Carpodacus purpureus*), GK = Golden-crowned Kinglet, LF = Least Flycatcher, GW = Black-throated Green Warbler, PV = Philadelphia Vireo, BNW = Blackburnian Warbler (*Dendroica fusca*), RK = Ruby-crowned Kinglet. From Sabo and Holmes (1983).

therefore relate to these variables. Whether these actually are the critical dimensions of niche separation among terrestrial birds or are simply the most fashionable variables to measure is uncertain.

The use of habitat or microhabitat is not restricted to feeding (even though one might sometimes get that impression from the literature). Birds use these features for nesting, courtship and mating, singing, resting, and so on. In some cases, how microhabitat is used may be influenced by potential predation risks as well. In grasslands and other open or scrubby habitats, predation risk may increase with greater distance from cover and/or lower vegetation height, and species differ in the distances from tree or shrub cover at which they normally forage, especially during winter. In Arizona grasslands, for example, Pulliam and Mills (1977) found that White-crowned Sparrows rarely occurred more than a few meters from cover, whereas Grasshopper and Savannah Sparrows occupied more open areas farther from shrubs and trees and Chestnut-collared Longspurs were not encountered close to trees at all. Working with different finch species in British Columbia, Lima *et al.* (1987) observed that most individuals foraged at intermediate distances from cover, which they interpreted as a trade-off between the risks of feeding too close to or too far from cover. In these situations, 'cover' is presumably the resource, and 'distance from cover' is the means of resource partitioning (Pulliam and Mills 1977). In another study, Nilsson (1979b) calculated seed density and cover in grid blocks in Swedish beech forests during two winters. During one winter, the species were concentrated in patches where seed density was high, as might be expected. Seed distributions were similar during the second winter, but densities of four of the six species were positively associated with patches of low-stature cover and were inversely related to high seed density patches. Nilsson attributed this to a greater abundance of predators in the woods during the second winter.

Aquatic birds may be differentiated ecologically on somewhat different dimensions. In both freshwater and marine habitats the primary separation is between species that feed on or near the surface (e.g. dabbling ducks, storm-petrels) and those that dive and pursue prey (e.g. grebes, murres). The differences have been particularly well documented among several seabird assemblages (Fig. 10.8) (Ashmole 1971, Harrison *et al.* 1983, Croxall and Prince 1980). These differences in feeding behavior define the water zones in which prey are available to foraging individuals. Surface-plunging species can reach prey only close to the surface, but by increasing their search height they may enlarge the area that is searched. A diver searching for prey while swimming on the surface with its eyes underwater,

Table 10.2. *Summarization of the extent of differences between congeneric species of birds occupying eucalypt forests and woodlands in Australia*

	Morphology	Food	Foraging method	Foraging site	Habitat	Plant species	Height	References
Rosellas *Platycercus*	--	+	--	+ + +	+	--	+ +	Wyndham & Cannon 1985
Lorikeets *Trichoglossus*	--	+	--	--	+	+	--	Wyndham & Cannon 1985
Thornbills *Acanthiza*	+	+	+	+ + +	--	+ +	+ +	Bell 1985
Pardalotes *Pardalotus*	+	--	+	--	+	+	+	Woinarski 1985
Fantails *Rhipidura*	+	+	+ +	+ +	+ + +	+	+	Cameron 1985
Treecreepers *Climacteris*	+	--	+ +	+ +	--	+ +	+	Noske 1985
Honeyeaters *Lichenostomus*	+	+ +	+	+ +	+ +	+	+	Wykes 1985
Honeyeaters *Lichenostomus*	+	+	+	+ +	+ + +	+	+	Ford and Paton 1977
Honeyeaters *Melithreptus*	+	+	+	+ +	+ +	+	--	Keast 1968
Honeyeaters *Phylidonyris*	--	--	+	+	+ +	+	--	Recher 1977
Robins *Petroica*	+	?	+ +	+ +	+ +	+	+ +	Fleming 1980
Whistlers *Pachycephala*	+	--	+ +	+	+ +	+	--	Bridges 1980

Notes:
-- negligible; + slight; + + moderate; + + + major.
Source: From Ford (1985).

on the other hand, has a quite different prey detection/availability volume (Fig. 10.9; Eriksson 1985). Differences in feeding behavior among species may also lead them to adopt different roles when they participate in mixed-species foraging flocks, in which the effects of their interactions may be mutualistic rather than competitive (Hoffman *et al.* 1981).

In addition to these differences in feeding tactics, however, seabirds may also differ in the distances from breeding colonies at which they feed (Fig. 10.10; Croxall and Prince 1980, Harrison *et al.* 1983, Ford *et al.* 1982) or the physical characteristics of water masses selected for foraging (Abrams and Griffiths 1981, Haney 1986). Cody (1973b), for example, argued that the six alcid species breeding on the northern Washington coast differed only slightly in breeding seasonality, diving depth, and diet but coexisted because of differences in foraging distances from the colonies. The foraging-distance zonation among most species, however, is not nearly so absolute as Cody portrayed it (or as Fig. 10.10 suggests). Bédard (1976) noted that the species Cody considered differ considerably in morphology and probably in diet as well (Cody's diet conclusions were based on 54 samples of fish loads carried to nestlings). Bédard also found that Cody's distributional data actually indicated substantial overlap in foraging distances among the species. Further, the relatively few observations were gathered from 'an unspecified number of transects over an unspecified number of days in unspecified sea conditions and in an unspecified number of localities' (Bédard 1976: 178). The data may have been confounded by the inclusion of individuals loafing at sea as well as foraging birds, and no consideration was given to the patchy distribution of food resources at sea or the varying distances from colonies to upwelling zones or deep-water areas. Our own observations (Wiens and Scott, unpublished) in the area where Cody conducted his studies indicated considerable daily variations in the flight patterns and distances of species. These variations were related to such factors as weather conditions and the location of the regional fishing fleet. It is clear that a large variety of factors may influence the distributional patterns of seabirds at sea (Brown 1980, Bourne 1981).

It is apparent from these examples that locally coexisting species may differ from one another in many ways (Table 10.2). The number of potential 'resource dimensions' that one may consider is large, and it is therefore inevitable that differences will be found among species in some of them. Those that emerge as important in differentiating species may often be those that the investigator subdivides (often arbitrarily) into a large number of categories, permitting finer resolution of apparent species differences.

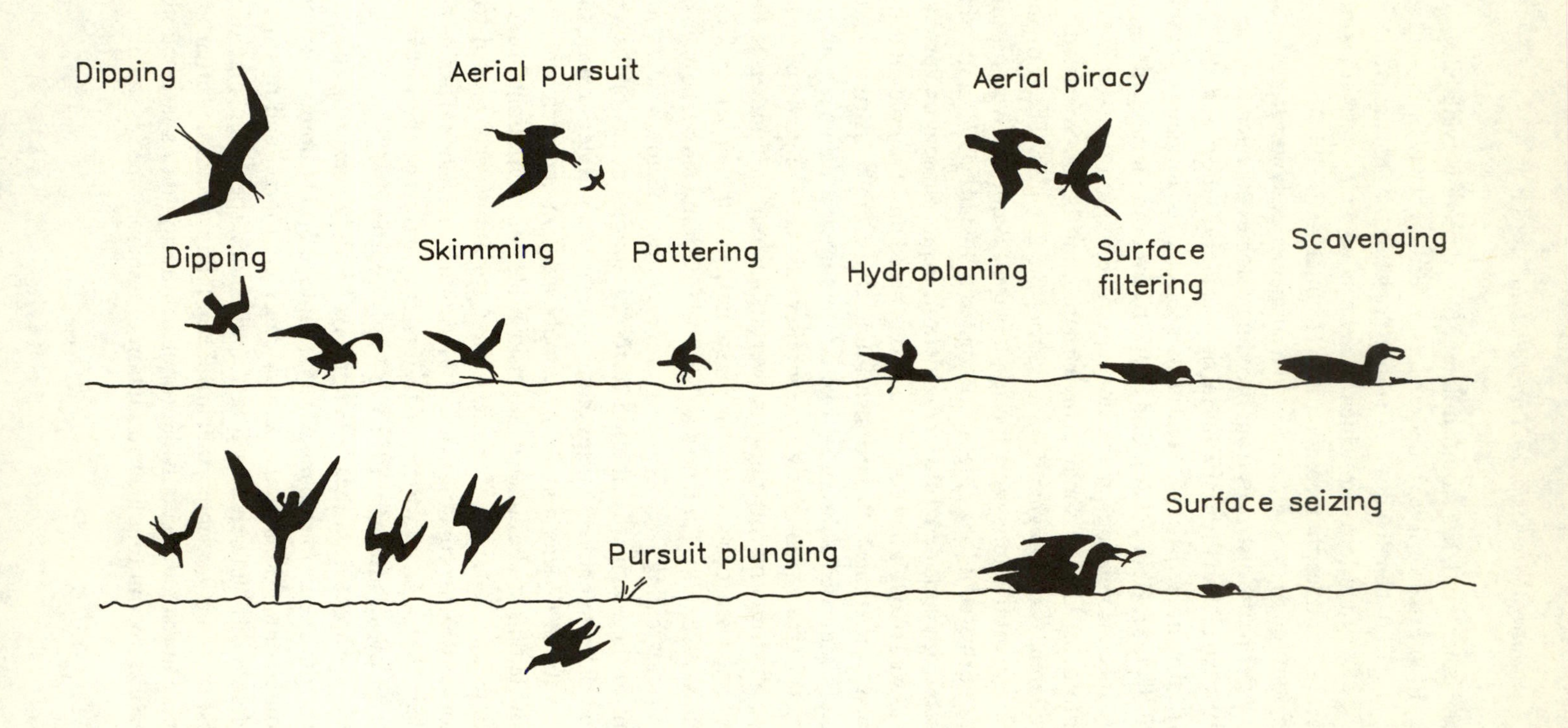
Dipping
Aerial pursuit
Aerial piracy
Dipping
Skimming
Pattering
Hydroplaning
Surface filtering
Scavenging
Surface seizing
Pursuit plunging

Fig. 10.8. Feeding methods of seabirds. The genera shown, from left to right, are (top): *Fregata*, *Stercorarius* pursuing *Phalaropus*, *Catharacta* harrying *Larus*; (second row), *Anous*, *Larus*, *Rynchops*, *Oceanites*, *Pachyptila*, *Daption*, *Macronectes*; (third row), *Sterna*, *Pelecanus*, *Phaethon*, *Morus*, *Diomedea*, *Phalaropus*; (under water), *Puffinus*, *Uria*, *Pelecanoides*, *Pygoscelis*, *Phalacrocorax*, *Melanitta*. Modified from Ahlquist in Ashmole (1971).

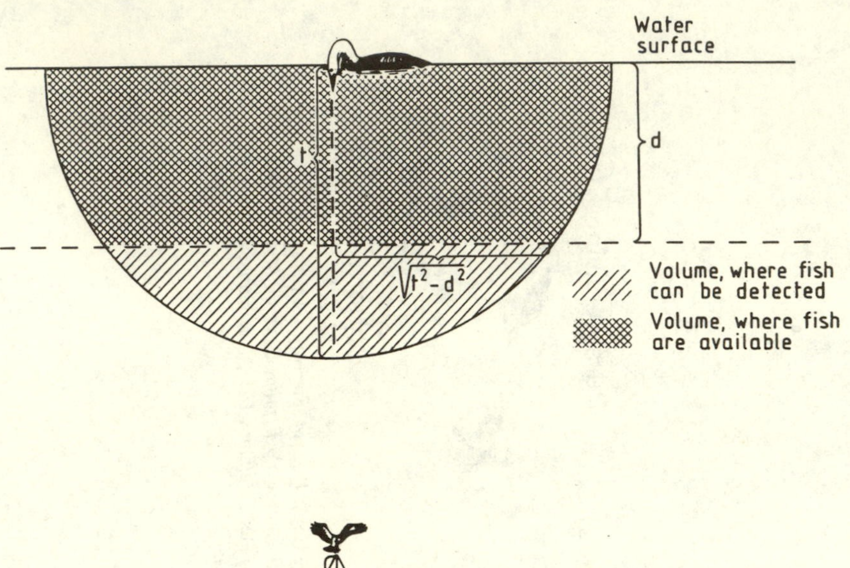

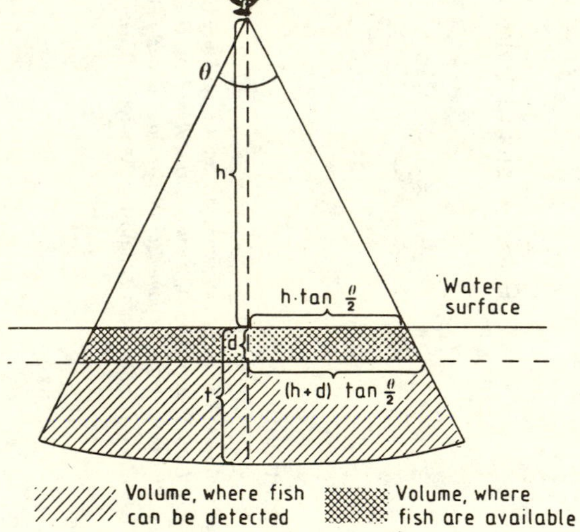

Fig. 10.9. Top: A 'pursuit diver', which searches for prey with its eyes below the water surface, can survey a volume of water equivalent to that of a half sphere with a radius given by the water transparency (t). If the maximal diving depth (d) is less than t, fish are available in a volume given by a spherical segment of depth d, with a top surface with the radius t and a bottom with the radius $\sqrt{t^2 - d^2}$. Bottom: A 'surface plunger', which searches for prey while flying or hovering over the water, can survey a volume with a top surface given by the search height (h) and the visual angle (θ) and a depth given by the water transparency (t). When d is smaller than t, fish are available only in a portion of this volume. From Eriksson (1985).

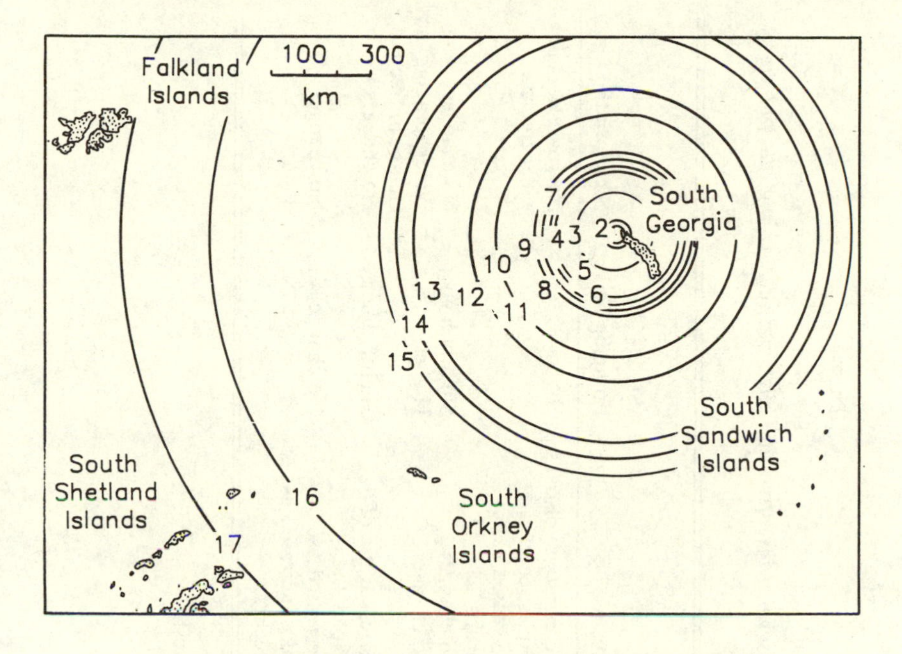

Fig. 10.10. The mean maximum foraging ranges of seabirds breeding at South Georgia during the chick-rearing period. Species codes: 1 (not numbered) = Blue-eyed Shag (*Phalacrocorax atriceps*), 2 = Gentoo Penguin (*Pygoscelis papua*), 3 = Macaroni Penguin (*Eudyptes chrysolophus*), 4 = Northern Giant Petrel (*Macronectes halli*), 5 = Southern Giant Petrel (*Macronectes giganteus*), 6 = Wilson's Storm-Petrel (*Oceanites oceanicus*), 7 = South Georgia Diving-Petrel (*Pelecanoides georgicus*), 8 = Common Diving-Petrel (*Pelecanoides urinatrix*), 9 = Antarctic Prion (*Pachyptila desolata*), 10 = Cape Pigeon (*Daption capense*), 11 = Snow Petrel (*Pagodrama nivea*), 12 = Black-browed Albatross (*Diomedea melanophris*), 13 = Grey-headed Albatross (*Diomedea chrysostoma*), 14 = Blue Petrel (*Halobuena caerulea*), 15 = Light-mantled Sooty Albatross (*Phoebetria palpebrata*), 16 = White-chinned Petrel (*Procellaria aequinoctialis*), 17 = Wandering Albatross (*Diomedea exulans*). Modified from Croxall *et al.* (1984).

Variations in foraging and microhabitat use

Foraging behavior and microhabitat use vary over time and between locations, and this also complicates the search for general patterns. Shifts may be most apparent when resource levels change dramatically. During an outbreak of western spruce budworm in Montana conifer stands, for example, budworms were most plentiful in the upper canopy, and most of the bird species (including insectivores and granivores that normally feed on the ground as well as foliage-feeding insectivores) concentrated their foraging there (Manuwal 1983). Maurer and Whitmore (1981) documented changes in aspects of the foraging behavior of several species

Table 10.3. *Changes in the foraging behavior and substrate use of several insectivorous birds between young and mature deciduous forests in West Virginia.*

Species	Foraging behavior				
	Manoeuvres	Locations and substrates	Tree use	Foraging heights	Niche width
Empidonax virescens	No changes	Decreased use of woody substrates, increase in use of peripheral foliage	Increased use of oaks, maples, beeches. Decreased use of shrubs	Higher	Narrower foraging, wider tree use
Vireo olivaceus	No changes	No changes	No changes	Slightly higher	No changes
Dendroica virens	No changes	Decreased use of peripheral foliage	No changes	Higher	Wider foraging, no change in tree use
Setophaga ruticilla	More flycatching	Less use of foliage	Increased use of maples	Slightly higher	No changes
Piranga olivacea	No changes	No changes	Decreased use of oaks, increased use of beeches	No change	No changes

Notes:
*Changes are those that occur in mature forest relative to the young forest.
Source: From Maurer and Whitmore (1981).

when they compared young and mature deciduous forest stands in West Virginia, although the nature of the changes differed between species (Table 10.3). In Sabo and Holmes' (1983) study, six species were present in both the subalpine and hardwoods communities. One (the Black-throated Green Warbler) foraged in virtually the same way in both communities, but others (Winter Wren, Solitary Vireo) shifted foraging patterns considerably (Fig. 10.7). The stereotypy of the warbler is consistent with the observations of Morse (1968, 1971a) and MacArthur (1958) from this area. The wrens, on the other hand, apparently modified their foraging in accordance with changes in the structure of the forest-floor litter and fallen logs on which they forage. The shift of the vireo may have been in response to habitat differences and/or the presence in the hardwoods of two other vireo species (PV and REV in Fig. 10.7).

Similar variations occur on a regional scale as well. Birds in coniferous forests in northern Finland used ground and low shrub layers more and peripheral parts of the trees less than did individuals of the same speces in similar habitats in southern Sweden, and some of the species also shifted in their use of tree species (Nilsson and Alerstam 1976). Similar sorts of changes in foraging behavior have been reported in comparisons of species occupying oak woodlands in California and Sonora (Landres and MacMahon 1983) or forests in the southern Appalachians and in Maine (Rabenold 1978). In the latter comparison, the six species present in both areas were more generalized in their foraging tactics and locations in the southern area. Foraging overlap was nonetheless greater among the 11 species occurring in Maine, where the guild as a whole occupied a smaller total foraging niche volume than did the six species in the south. At this scale of comparison, then, increased foraging niche overlap was associated with increased species diversity; Sabo and Holmes (1983), however, recorded the opposite relationship in their more local hardwoods–subalpine forest comparison.

Food and microhabitat conditions also change seasonally and between years, and these changes may produce variations in foraging behavior and patterns of species similarities or differences. Where Abraham and Ankney (1984) studied Arctic Terns (*Sterna paradisaea*) and Sabine's Gulls (*Xema sabini*) in the North American arctic, the species foraged opportunistically and occupied similar habitats during the prelaying period when snow was melting and melt pools formed. As the season progressed, the availability of foraging habitat and prey increased and the species diverged in their use of habitats, foraging behavior, and diets. In another study, Alatalo (1982c) found that most of the foliage-gleaning species in northern Finland changed their frequency of use of tree species and foraging positions in trees and altered their foraging tactics between summer and winter. As a consequence, the pattern and amount of overlap in foraging behavior and microhabitat use among these species varied seasonally. The species were rather evenly spread in the multidimensional space defined by the six foraging variables in both seasons, but this may simply reflect the averaging effects of combining six partially interdependent variables in the analysis. In fact, overlaps between species on different axes were positively correlated in most instances, providing little evidence of niche complementarity. Sæther (1982) reported similar seasonal and annual variations in foraging behavior and use of microhabitat and tree species among the passerines occupying an alder forest in Norway; he likewise found no evidence of complementarity between species on different resource-use dimensions.

A methodological comment

Patterns of resource use are sensitive to how measurements are made. In some studies, the methods used to obtain behavioral data are not even mentioned (Edington and Edington 1983, Alerstam and Ulfstrand 1977, Maurer and Whitmore 1981, Craig 1984). Most often, however, the form and location of behavior are recorded by opportunistic spot observations (e.g. Wooller and Calver 1981, Baker and Baker 1973, Franzreb 1983, Moreno 1981, Ulfstrand 1977). In some situations, such as the dense tropical rainforests in which Crome (1978) conducted his studies, this may be the only feasible procedure, but it is still subject to the biases discussed in Chapter 3. Sometimes the investigators have recognized the possible biases and errors of this procedure and have attempted to minimize them (Root 1967, Sæther 1982, Wagner 1981). Others (e.g. Abraham and Ankney 1984, Noske 1979, Rabenold 1978, Landres and MacMahon 1980, 1983, Hertz *et al.* 1976) have employed focal-animal sampling (Altmann 1974), continuous observations, or interval sampling to record behavior. Each procedure is likely to yield a somewhat different image of the patterns of foraging or microhabitat use, and any given pattern (such as those mentioned above) must thus be interpreted in the context of the methodology employed.

Individual spatial relationships and interspecific territoriality

Individuals of different species may also differ in their occupancy of a local area, especially if the area contains a mosaic of microhabitat patches. In the context of competition theory, these differences may be viewed as enabling potential competitors to coexist by virtue of spatial segregation, even if other features of their use of resources are identical (Cody 1974a). Interspecific aggression, for example, may produce a short-term spatial separation of individuals. Edington and Edington (1983) observed that several Nigerian species with generally similar foraging behavior were often separated into different habitat zones by aggressive interactions. Each of several sunbird species was most likely to prevail in the interactions occurring in the feeding zones that it used most heavily, and the outcome of aggressive encounters was therefore habitat-dependent. In other interactions, the relative densities of the species were important. On several occasions, either White-throated Bee-eaters (*Merops albicollis*) or Ethiopian Swallows (*Hirundo aethiopica*) altered their foraging patterns following the arrival in an area of individuals of the other species; which of the species changed, however, was related to the number of individuals of the other species present (Fig. 10.11).

Complete microspatial segregation among species is often associated with interspecific territorial defense and exclusion, but in patchy environments it may also be produced by differences in habitat selection. Distinguishing between the contributions of habitat differences and species interactions to patterns of spatial occupancy, however, is often difficult. Cody and Walter (1976) mapped the locations of individual territories of several *Sylvia* species on small plots in Sardinia during a breeding season and then related the spatial overlap between species to that expected if each species were to occupy the same proportion of the plot at random. In their most intensively studied location, *S. cantillans* and *S. melanocephala* were distributed independently of each other, but overlap among the other species was less than expected by chance. Cody and Walter separated the effects of behavioral exclusion from those of differences in habitat preferences by using the overlap in vegetation height profiles of areas occupied by different species to predict the probability of spatial overlap on the basis of habitat alone. The difference between the predicted habitat overlap and the observed spatial overlap was attributed to behavioral interactions, after adjustments to reflect the amount of habitat that each species potentially could occupy in the plot. Cody and Walter concluded that some of the species were separated largely by habitat differences, but for others (e.g. *undata-melanocephala*), behavioral exclusion seemed especially likely. This approach is interesting and merits broader use, although it is hampered by several unrealistic assumptions (e.g. that patches with differing height profiles are uniformly or randomly distributed in space) and some circularity (e.g. determination of potentially occupiable habitat from measures of

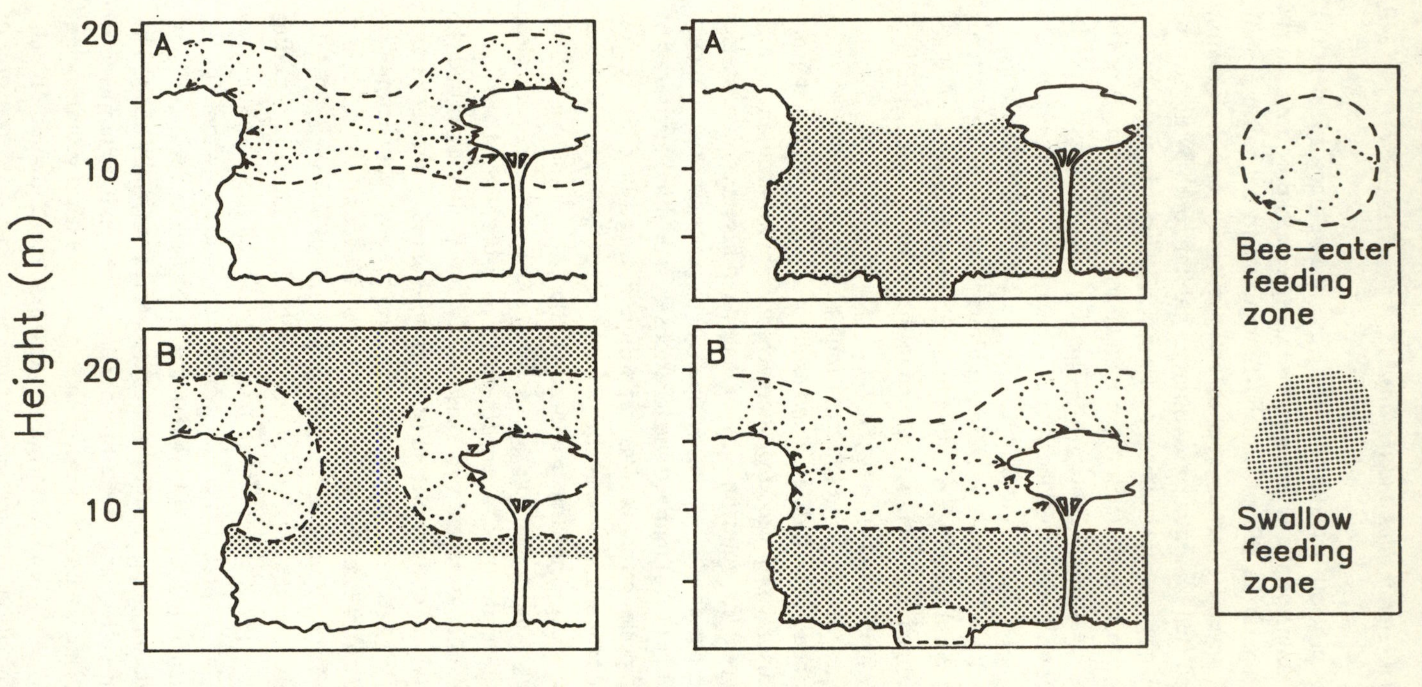

Fig. 10.11. Left: The effect of the arrival of a flock of Ethiopian Swallows on the feeding pattern of a smaller group of White-throated Bee-eaters in Nigeria. (A) Feeding patterns observed when the bee-eaters were present alone; (B), change in pattern after the arrival of the swallows. Right: The effect of the arrival of a flock of bee-eaters on the feeding pattern of a smaller group of swallows. (A) Swallows present alone; (B), after the arrival of the bee-eaters. Modified from Edington and Edington (1983).

habitat occupied at that time, with no consideration of the effects of population density, site fidelity, or chance on the observed habitat-occupation patterns).

Long-term investigations may provide additional insights into the complexities of microspatial relationships among species. After nearly a decade of work in Norwegian forests, Hogstad (1975) concluded that breeding densities of Willow Warblers and Bramblings (*Fringilla montifringilla*) fluctuated inversely with respect to each other from year to year, that the species exhibited partial interspecific territorial exclusion, and that the degree of avoidance between the species was proportional to their combined densities. He found no evidence of differences in microhabitat occupancy, and a territory occupied by a Willow Warbler in one year might well be defended by a Brambling the next. Fonstad (1984), however, noted that the inverse density relationship reported by Hogstad for 1966–74 did not continue in the same area during 1975–82, nor was it found in a 19-yr study in Swedish Lapland. Fonstad found that the two species were quite similar in foraging height and diet, but when he removed males of either species from territories the sites were reoccupied by males of the same species. A reanalysis of Hogstad's data revealed that an area occupied by one species in one year was equally likely to be occupied by an individual of either species in the following year. Fonstad concluded that, although the territories of the species do overlap less than expected, this is probably a consequence of different microhabitat preferences rather than interspecific territorial aggression. In alder woodlands in central Norway, however, Willow Warblers and Chiffchaffs, which are quite similar to each other in foraging behavior and microhabitat occupancy, occupy almost mutually exclusive territories (Sæther 1983). Edington and Edington (1972) reported interspecific territoriality between Willow Warblers and Wood Warblers (*Phylloscopus sibilatrix*) in England, but Tiainen *et al.* (1983) found no evidence of spatial interactions between these species in Finland.

Part of the difficulty in documenting interspecific spatial exclusion is that the period of actual interaction between the species may be quite limited. Noon (1981), for example, found no evidence of interspecific aggression or response to song playbacks between Wood Thrushes and Veeries or Hermit Thrushes in northeastern USA. Because he did not map their territories, he could not assess their spatial relationships accurately. Where the three species co-occur in Maine forests, however, Wood Thrushes aggressively dominate the other species and establish interspecifically exclusive territories (Morse 1971b). Overt aggression between the species becomes quite infrequent within a week of the Wood Thrush's arrival, however, and studies conducted later in the breeding seasons (such as Noon's) would therefore fail to find it.

Interspecific territoriality

Under some conditions, interspecific aggression may lead to reciprocal territorial exclusion, in which individuals of each species respond to members of the other species as they would to conspecifics when defending territories. Although situations of completely balanced interspecific aggression are rare (Orians and Willson 1964, Cody 1974a), examples of one species behaviorally excluding an ecologically similar and/or closely related species from portions of a habitat are well-documented. Red-winged Blackbirds establish breeding territories in marshes in many areas of western North America in early spring, occupying a broad range of marsh habitats. Yellow-headed Blackbirds (*Xanthocephalus xanthocephalus*) are ecologically more restricted, favoring relatively large, open marshes with deeper water. When yellow-heads arrive at breeding marshes later in the spring, they aggressively displace red-wings from such habitats within marshes. The territories of the two species abut but do not overlap (Miller 1968). In central California marshes, red-wings establish territories in winter, but they may abandon some of these territories after large numbers of Tricolored Blackbirds (*A. tricolor*) arrive in spring (Fig. 10.12). In this case, individual red-wings aggressively dominate individual tricolors, but the tricolors simply overwhelm the red-wings by force of numbers (Orians and Collier 1963).

Another example illustrates the complexities of spatial relationships between species and how these relate to ecological similarity in other features. Sedge Warblers and Reed Warblers are locally sympatric during the breeding season in many areas of Britain and Europe. The species usually nest in somewhat different habitats, the Reed in marshes (*Phragmites*) over water, the Sedge in drier areas such as *Carex* close to water (Lack 1971, Catchpole 1973, Henry 1979). Reed Warblers also place nests higher in the vegetation than do Sedge Warblers, and Sedge Warblers may breed somewhat earlier (Catchpole 1973). Where Catchpole studied the species in southern England, however, their nesting distributions overlapped in ungrazed fallow fields. There the species interacted strongly and defended mutually exclusive territories. These territories were primarily nesting areas, as both species foraged exclusively in other areas, the Reed mainly in willow woods and the Sedge in *Glyceria* marsh. Despite these differences, Catchpole found substantial overlap in the feeding habitats and types of prey taken by the two species. In France, the species also forage

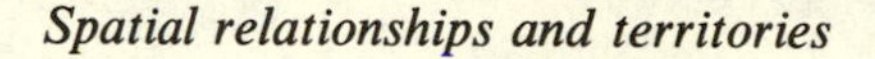

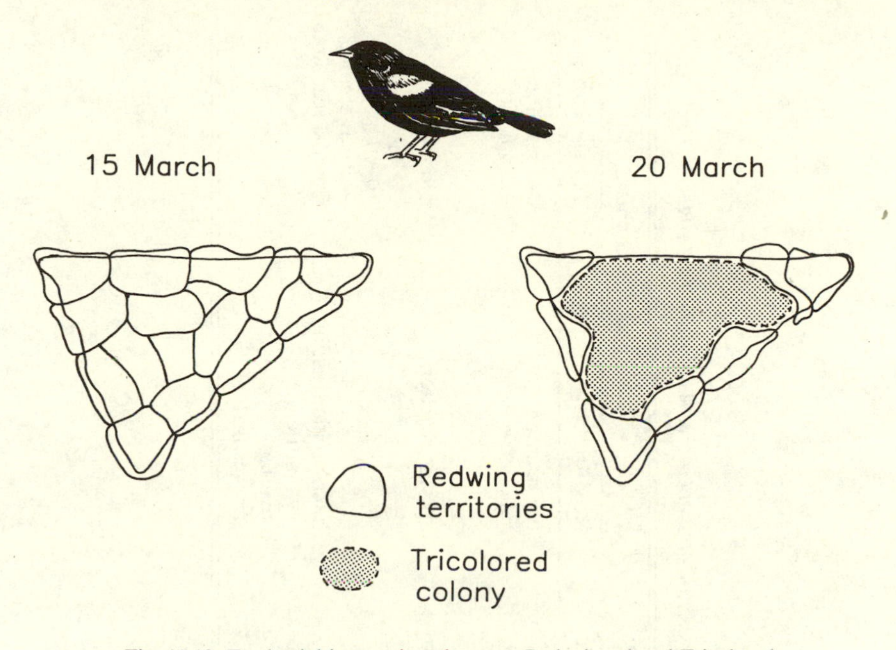

Fig. 10.12. Territorial interactions between Red-winged and Tricolored blackbirds at a California marsh. From Orians and Collier (1963).

extensively outside their breeding territories, in willow groves and agricultural fields (Henry 1979). Henry found that, although both species captured some prey types not consumed by the other, their diets were extremely similar in both taxonomic composition and prey–size distributions.

Catchpole (1978) examined the behavioral correlates of the territorial relationships of the two species in detail. He presented playbacks of conspecific and allospecific songs to individuals of each species in England, where they are sympatric, and in Germany, where the Sedge Warbler is absent but another species, the Marsh Warbler (*Acrocephalus palustris*), is sympatric with the Reed Warbler. The results (Table 10.4) were rather complex. Each species responded strongly to playbacks of its own song, as expected. In England, Sedge Warblers responded strongly to playbacks of Reed songs, but the reverse experiment produced no responses. In Germany, on the other hand, Reed Warblers were responsive to playbacks of Marsh Warbler songs, but the reverse did not occur. The common feature of these results is that the first species to arrive and establish territories in the spring is behaviorally dominant over the species that arrives later and defends territories interspecifically in shared habitats, forcing the spatial

Table 10.4. *Responses of Sedge, Reed, and Marsh warblers to playbacks of conspecific and allospecific songs.*

	Sympatric in England		Sympatric in Germany	
Playback songs:	Sedge warbler	Reed warbler	Reed warbler	Marsh warbler
Sedge warbler	Specific reaction	No reaction –sympatric –overlap –arrives second	No reaction –allopatric	No reaction –allopatric
Reed warbler	Interspecific reaction –sympatric –overlap –arrives first	Specific reaction	Specific reaction	No reaction –sympatric –overlap –arrives second
Marsh warbler	No reaction –allopatric	No reaction –allopatric	Interspecific reaction –sympatric –overlap –arrives first	Specific reaction

Source: From Catchpole (1978).

separation. It is the first-arriving species that is responsive to allospecific songs. (This contrasts with the sequence exhibited by blackbirds and by some of the *Sylvia* warblers, in which the later-arriving species displaces the established territory holders.)

Interspecific territoriality is likely to occur only in certain ecological situations. Using optimization arguments based on food density, foraging travel time, and the costs of territorial defense, Cody (1974a) proposed that species should be interspecifically territorial when diet overlap is high (regardless of their territory sizes). As diets diverge, interspecific territoriality may be advantageous only if territories are very large. The empirical support that Cody offered for this model was less than convincing (diet overlap was gauged by bill-length ratios, and the data exhibited considerable scatter). After reviewing the occurrence of interspecific territoriality among birds, Orians and Willson (1964) concluded that it seems most likely to occur between ecologically similar species in structurally simple habitats with low vertical relief, especially if the habitat is suboptimal for one of the species.

Both of these sets of predictions are based on the premise that interspecific territoriality is adaptive because it lessens competition between species that overlap in resource use. Murray (1971, 1976), however, argued that interspecific territorial defense is largely a consequence of misdirected aggression toward individuals of other species that are similar in behavior, appearance, and ecology and that it is therefore nonadaptive. Rice (1978a, b, c) examined Murray's ideas with reference to the vireos he studied and concluded that his results were more consistent with the arguments of Cody and of Orians and Willson. Gochfeld (1979), on the other hand, concluded that the interspecifically territorial meadowlarks (*Sturnella*) that he studied in the Argentinian pampas conformed more closely with Murray's model. The vireos and meadowlarks, like the blackbirds, warblers, and most of the other species for which interspecific territoriality has been reported, are phylogenetically closely related. This makes it difficult to disentangle the effects of similarities in behavioral systems due to relationship from those associated with ecological similarities. Tests of Cody's hypothesis using such groups are therefore not definitive.

Convergence in territorial signals

Cody (1973a, 1974a) also suggested that convergence in the social signals used in territorial defense may lend stability to the spatial separation of ecologically similar species. The songs of the *Acrocephalus* warblers give no evidence of convergence in sympatry, but Rice's (1978a, b, c) studies of interspecifically territorial Red-eyed Vireos and Philadelphia Vireos in Ontario are consistent with Cody's views. Although on a regional scale *olivaceus* occupies a broader range of habitats than does *philadelphicus*, Rice (1978a) recorded no differences in habitats occupied at a local scale. The species were also quite similar in the frequently used foraging behaviors, although they differed in the use of some uncommon foraging methods. Individuals responded to playbacks of both allospecific and conspecific songs (Rice 1978b) and, on the basis of differences in song discrimination between the species, Rice (1978c) concluded that *philadelphicus* has converged toward mimicry of *olivaceus* songs.

The notion of signal convergence is also supported by studies of Thrush Nightingales (*Luscinia luscinia*) and Nightingales (*L. megarhynchos*) in Europe (Sorjonen 1986; see also Murray 1987, Sorjonen 1987). The habitat occupancy of the species is more similar where they are sympatric than in allopatry, and where they co-occur they occupy mutually exclusive territories. Both species responded to playbacks of the other species' song, although *luscinia* did so significantly more frequently than did *megarhynchos*. Comparisons of song patterns of *luscinia* in allopatric and sympatric populations showed convergence toward *megarhynchos* songs by sympatric birds. This was due to the inclusion of both conspecific and heterospecific elements in the songs. In both sympatry and allopatry, individual Thrush Nightingales shared more song features with their nearest neighbors than with distant birds. Where the individuals are interspecifically territorial, in sympatry, some of these neighbors are Nightingales, and learning of elements from the songs of neighbors (conspecifics or heterospecifics) may produce the pattern of vocal convergence.

Cody (1978, 1979, Cody and Walter 1976) supported his ideas with observations of *Sylvia* warblers, which differ between regions in both ecological (habitat) overlap and the degree of interspecific territorial exclusion. Cody reported high ecological overlap between species in Sweden. There, direct behavioral interactions between several pairs of species resulted in partial or complete territorial exclusion. These interactions were asymmetrical, one species dominating the other in a habitat-dependent fashion similar to the sunbirds observed by Edington and Edington (1983). In England there was greater habitat separation among the species, behavioral interactions were more restricted and symmetrical, and the songs of the interactive species were apparently more closely convergent than in Sweden. Habitat separation was more precise in Sardinia, where only certain species pairs interacted behaviorally and occupied nonoverlapping territories. The interactions were primarily vocal rather than aggressive and involved song convergence among some species. Finally, Cody's observations in Morocco indicated large-scale habitat separation among the warbler species, such that only a few were syntopic. Territories of these species frequently overlapped, no interspecific aggression was observed, and the songs of the *Sylvia* species were all distinctive. Cody hypothesized that the regional differences may reflect the relative ages of the communities, species in the younger, northern assemblages partitioning resources primarily by direct aggressive encounters, those in somewhat older systems developing differences in habitat preferences and adopting more formalized conventions of behavioral interactions, and those in the oldest communities in the south having achieved a degree of habitat and microhabitat partitioning that obviates the need for any behavioral interactions. Although Cody's documentation of song convergences is based on anecdotal observations rather than quantitative analysis of vocalization patterns, the scenario he outlines is intriguing.

Use of nest sites

Nesting locations have received much less attention than food or habitat parameters in studies of niche partitioning, perhaps because for many cup-nesting species suitable nest sites are presumed to be readily available and therefore not a resource over which competition is likely to occur. In groups such as hole nesters or seabirds, however, nesting requirements are specialized and suitable locations may be difficult to obtain. Species with such specialized requirements may comprise a substantial portion of local and regional avifaunas. In Australia, for example, 18% of the terrestrial species are known to use tree hollows for nesting, and 21% of the nonpasserine species are obligate hole nesters (none of which are capable of excavating cavities by themselves) (Saunders *et al.* 1982, 1985). Cavity-nesting is less important in the avifaunas of southern Africa or North America (10% of the species, 7% and 4% of the nonpasserines obligate, respectively).

The abundance and availability of nest sites for hole-nesting species may be quantified with reasonable accuracy. Studies in several areas have involved detailed comparisons between the attributes of nest cavities occupied by various species and those of all cavites present (Saunders *et al.* 1982, van Balen *et al.* 1982). In a salmon gum (*Eucalyptus salmonophloia*) woodland in Western Australia, 47% of the natural tree hollows were occupied by nesting individuals of eight species (Saunders *et al.* 1982). Hollows occupied by the four most common breeding species differed in nest-entrance size, internal diameter, depth, height above the ground, and severa ıttributes (Fig. 10.13). In mixed woodlands in The Netherlands, van Balen *et al.* (1982) found occupancy of natural nest cavities ranging from 54% to 93% in different areas and years. There, the most common hole-nesting species were passerines (Starlings and several *Parus* species). There was considerable overlap among the species in the attributes of nest holes used, and some cavities were occupied by different species in different years. Nonetheless, there were significant differences among some of the species, and most species used the spectrum of potentially available nest cavities nonrandomly.

The finding in both of these studies that a considerable portion of the nest cavities in an area may not be occupied during a given breeding season might be taken as evidence that nest cavities are not a limited resource. In some situations this may be true, but other factors may affect the actual availability of nest sities that are present in an area. Galahs (*Cacatua roseicapilla*) in Australia occupy and defend a nest site throughout the year, and the requirements of individual spacing may preclude the use of some cavities that would otherwise be suitable. Similar factors may affect the availability of nest holes to some other cockatoos, although for other species (e.g. Red-tailed Black Cockatoos, *Calyptorhynchus magnificus*) such social factors seem unimportant (Saunders *et al.* 1982).

Seabirds usually nest on isolated islands where native predators are absent or rare. Often, they occupy narrow ledges on cliff faces or nest in sheltered burrows. On large islands where predators are absent, such as South Georgia or islands in the Benguella upwelling region off southern Africa, suitable nesting sites may not be limited (Croxall and Prince 1980, Duffy and La Cock 1985), although the species still exhibit distinct nest-site preferences. In other areas where nesting habitat and space are restricted and breeding populations are quite large, aggression related to nest sites may be frequent (Duffy 1983, Belopol'skii 1957). Seabirds that share similar nest-site requirements tend to be positively associated in their distributions among colonies (Whittam and Siegel-Causey 1981b), enhancing the likelihood of interactions. Within a group of species with generally similar nest-site requirements, however, there may be more subtle, quantitative interspecific differences. Thus, the six species of cliff-nesting seabirds on St. George Island, Alaska, for example, differ in the size, shape, slope, and degree of overhang of nesting ledges used, although there is considerable overlap among species and exchanges of particular nest sites between species in successive breeding seasons are not uncommon (Squibb and Hunt 1983). Duffy (1983) also recorded clear differences in mean nest-site occupancy among three species nesting on Peruvian guano islands, although

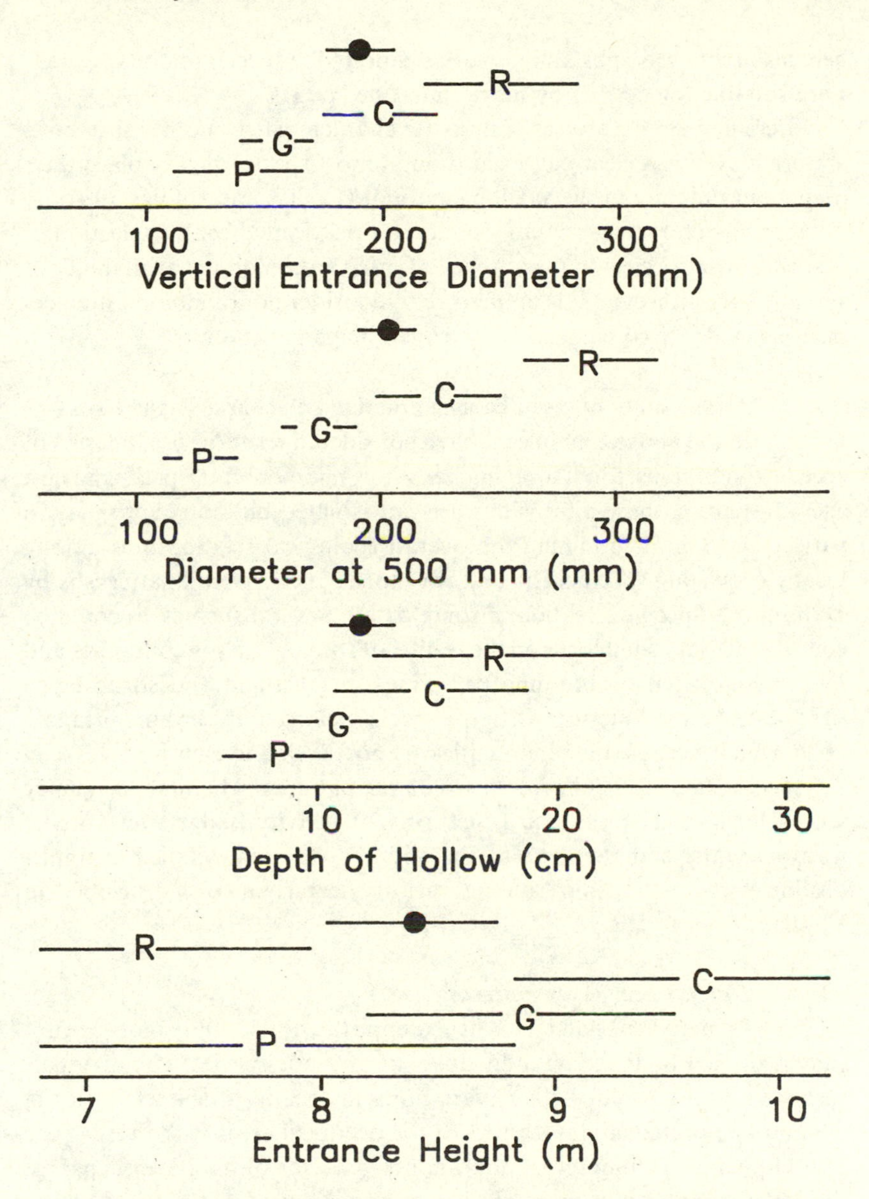

Fig. 10.13. Dimensions and height of nest holes available (solid circle) and occupied by Red-tailed Black Cockatoos (R, 664 g), Little Corellas (*Cacatua sanguinea*) (C, 550 g), Galahs (G, 305 g), and Port Lincoln Parrots (*Barnardius zonarius*) (P, 127 g) in a salmon gum woodland in Western Australia. Symbols denote mean values, lines +/− 2 standard errors of the mean. Calculated from data in Saunders *et al.* (1982).

here also there was substantial overlap among the species and many areas were suitable for nesting by more than one species.

Differences among species in nest-site characteristics in other situations are often even less clear-cut, even though most species clearly place their nests nonrandomly in the available habitat (Collias and Collias 1984). A number of factors (most notably predation risk) can affect the selection of nest sites by a species. In the absence of information on the availability of suitable sites, however, it is premature to consider nonrandom nest placement as evidence of competitively driven niche partitioning.

A case study of resource use: grassland and shrubsteppe birds

In the above examples, I have considered resource use in terms of specific dimensions: food, foraging behavior, microhabitat, space, and nest sites. In nature, these dimensions are not isolated but come into play in various ways in determining the overall ecological relationships among locally coexisting species. The best way to see these overall patterns is by examining a specific situation in some detail. Several such cases could be considered, but I shall focus on the results of studies that my colleagues and I have conducted on breeding passerines in grassland and shrubsteppe habitats in North America. All of the species belong to the ground foraging guild, which, because the communities are not diverse, usually involves only 2–7 species. Our studies have focused on the patterns of habitat occupancy within local study plots, the structuring of activity budgets and use of microhabitats, and the diets of the species. Another detailed example, dealing with relationships among various nectarivores, is developed in Volume 2.

Habitat occupancy patterns

In many grassland and shrubsteppe locations, study plots are not fully occupied by breeding individuals of any one species. If individuals respond to within-plot habitat variations in locating their territories, a comparison of the habitat features in the occupied areas with those in the unoccupied areas may reveal nonrandom patterns of habitat occupancy. On the other hand, if portions of a plot are unoccupied by a species because of interspecific territorial exclusion, low population density, or chance effects in the placement of territories, consistent differences in habitats between occupied and unoccupied areas should not be evident. In a grassland area in Wisconsin (Wiens 1969), the habitat characteristics of occupied areas often differed from those of unoccupied areas. Most species, for example, occupied areas in which the vegetation density close to the ground

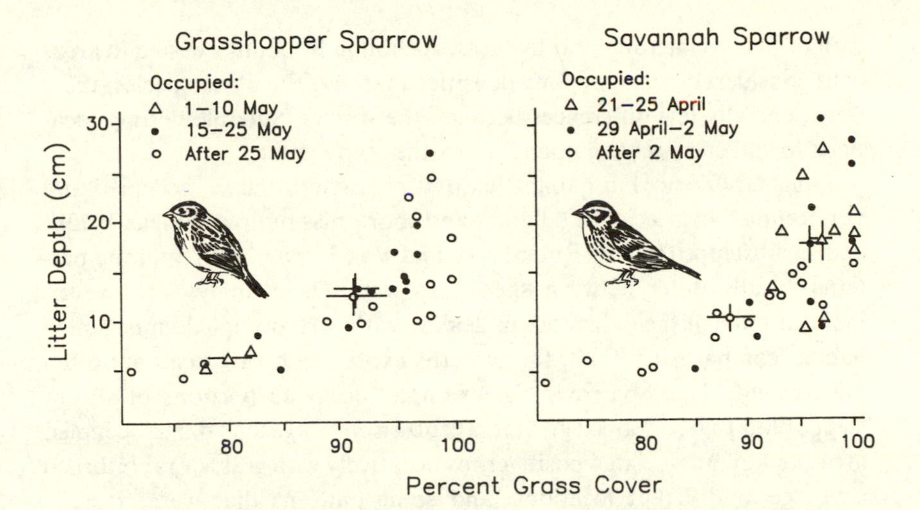

Fig. 10.14. Litter depth and grass coverage characteristics of Grasshopper and Savannah sparrow territories in a Wisconsin field, as measured in mid-June. Symbols indicate mean values of individual territories; symbols with lines indicate mean for a set of territories at a particular time; lines extend one standard error on each side of the overall means. From Wiens (1973a).

was greater than in unoccupied areas, although the reverse characterized the Henslow's Sparrow (*Passerherbulus henslowi*). Emergent forbs were taller in areas occupied by Henslow's Sparrows, Bobolinks (*Doliochonyx oryzivorus*), and Eastern Meadowlarks but did not differ between occupied and unoccupied areas of the other sparrows. There were thus differences between the species in their habitat-occupancy patterns, although for most species these differences were slight. In particular, the habitat measures in areas occupied by Grasshopper and Savannah sparrows were nearly identical. These occupied–unoccupied patterns were measured at the peak of the breeding season, however, after all individuals had established breeding territories. If one considers the temporal progression in habitat occupation by these two sparrow species, differences are more apparent. The first individuals to arrive established territories in quite different portions of the plot, and only as densities built up to peak levels did the average habitat measures for the two species converge (Fig. 10.14).

In this Wisconsin grassland, Bobolinks, Eastern Meadowlarks, and Savannah Sparrows all built nests in areas with greater vertical vegetation density and height and denser, deeper litter than characterized their entire territories, and the same general patterns were evident to a lesser degree in Grasshopper Sparrows. The species thus located their nests nonrandomly with respect to microhabitat features. Although Bobolinks nested in areas with considerably greater forb densities than did the other species, there were generally few differences between the species. Suitable nesting locations for all species were abundant in the study area.

Similar differences in habitat features characterize areas occupied and not occupied by species in tallgrass and shortgrass prairies (Wiens 1973b) and shrubsteppe (Wiens, Rotenberry and Van Horne 1987), and the patterns usually differ between species as well. These analyses, however, indicate some of the difficulties in dealing with such occupied–unoccupied habitat comparisons. First, the patterns expressed by a species are often inconsistent. Sage Sparrows, for example, occupied portions of shrubsteppe plots that contained greater sagebrush coverage than did unoccupied areas, but they associated positively or negatively with grass or rabbitbrush coverage at different locations, and some patterns that were strongly expressed at a given location in one year were absent or reversed in other years. Although a variety of factors might contribute to these variations, they were not related to changes in the densities of Sage Sparrows in the plots. We thus find substantial variability in habitat-occupancy patterns at this local scale, and between-species similarities and differences are correspondingly variable.

The second difficulty has to do with gauging what really constitutes a habitat difference between occupied and unoccupied areas. Like most investigators, we selected our study plots to be relatively homogeneous (i.e. all grassland, rather than a mosaic of grassland and woods). Differences between portions of a plot may therefore be rather subtle. Because many of the habitat measures are coverage values derived from the total sample of points in occupied or unoccupied areas, they generally have no associated variance measures. This may lead one to interpret any habitat difference that seems reasonably large as a real difference. This judgment is subjective, however. If one's goal is to document differences between occupied and unoccupied areas or between species, rather small quantitative differences may be considered important (e.g. Wiens 1969, Cody 1968), and the use of procedures such as DFA (e.g. Rice 1978a) only serves to highlight such differences. The habitat differences should be tested against some null expectation to determine whether or not they depart significantly from random habitat occupancy. We used simulation procedures to generate distributions of differences between occupied and unoccupied areas of a given plot that would be expected from random occupancy (Wiens, Rotenberry and Van Horne 1987), and the results indicate that many of the differences that intuitively seem large and important might reasonably be

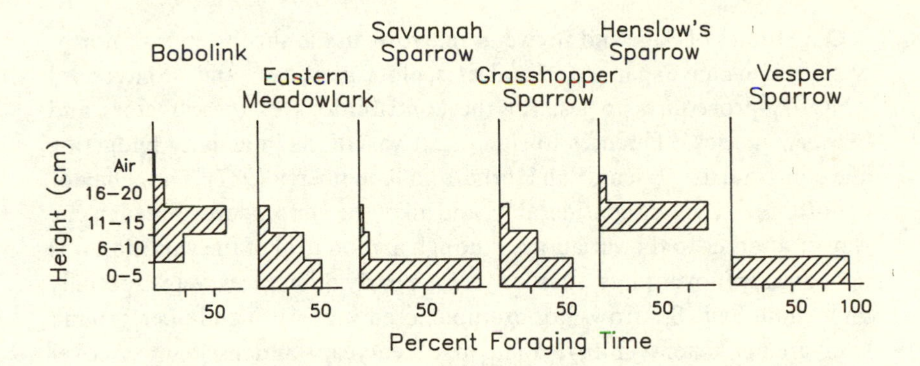

Fig. 10.15. Distributions of foraging activity among height strata of vegetation by birds breeding in a Wisconsin grassland. Modified from Wiens (1969).

expected to occur by chance. If this is so, it is difficult, in the absence of additional information, to attribute the patterns to any active habitat response by the birds or to interactions between species.

Behavioral patterns

Perhaps because of the limited structural complexity of the habitat, birds in grassland and shrubsteppe exhibit a limited range of patterns of habitat use and behaviors. In the Wisconsin grassland (Wiens 1969), continuous recordings of the behavior of individuals of six species indicated that the relative frequencies of activity types were generally similar among all of the sparrows, although Vesper Sparrows spent somewhat more time singing. Bobolinks spent proportionately more time perching, and Eastern Meadowlarks sang relatively little and spent more time foraging. The species differed more in their use of microhabitats for various activities. Most of the species foraged in grass and overlapped considerably in their use of grass of different heights, although the overall patterns of grass use differed somewhat among species (Fig. 10.15). The species also differed in their use of perches for singing.

These differences, like those Cody (1968) reported in his studies of grassland birds, are not dramatic. Although one may interpret them as divergent patterns of resource use (as both Cody and I did), the overall similarities in behavior are perhaps equally impressive. In neither study were the differences in behavior between species tested statistically (although Cody did apply DFA to differentiate among the species he considered), and the patterns were derived largely from data gathered in a single year at a location.

Our studies of Sage and Brewer's sparrows in the shrubsteppe of northwestern USA have spanned several sites, plots, and years, and we have used ANOVA procedures to test for the contributions of these factors and between-species differences to significant variations in activity budgeting and substrate use (Wiens, Van Horne, and Rotenberry 1987). The behavior of both species varied considerably, and all of the factors contributed in one way or another to the variation, although a good deal of the variation was not consistent over plots, sites, or years. Yearly differences were especially important. Sage Sparrows, for example, used sagebrush and open ground to a greater extent in 1976 than in other years and used grass cover significantly less in 1978. Brewer's Sparrows used sagebrush more frequently in 1977 than in 1976 and were not recorded using grass during the 1978 observations at all. One might expect these variations in substrate use to be related to annual variations in sagebrush or grass coverage, but such relationships were neither direct nor simple (Wiens 1985b, Wiens, Van Horne, and Rotenberry 1987). Use of grass and sagebrush by both species increased with increasing grass cover. Both substrates were also used more frequently by Sage Sparrows when their population densities were greater. Rather than varying in a reciprocal fashion, however, Brewer's Sparrows also used sagebrush to a greater extent when Sage Sparrow densities were high.

Both species thus structured their activity budgets and use of microhabitats in a nonrandom fashion, and there were significant differences between the species in the details of their behavior. Brewer's Sparrows spent more time singing then did Sage Sparrows and foraged more in 1976 and 1977 but less in 1978 and 1979. Brewer's Sparrows consistently used sagebrush substantially more than Sage Sparrows, but this difference was associated entirely with their singing behavior; the species did not differ in the use of sagebrush for foraging. Other differences between the species in activity or substrate use were significant in some plots or years but not others, and our search for environmental features that might covary with these differences was unproductive.

In these studies we attempted to relate behavior to possible resource characteristics by means of correlational tests. These relationships may be examined more directly through experimental manipulations. On one of our shrubsteppe study sites we experimentally altered habitat configuration and substrate availability by systematically removing individual shrubs from 25 × 25-m blocks to create a mosaic of blocks in which 0%, 25%, 50% and 75% of the shrubs were removed (Wiens, Rotenberry, and Van Horne 1986). We then recorded the behavior and substrate-use patterns of Sage and Brewer's sparrows in the manipulation area in relation to those in an adjacent unaltered control area over the following 4 yr. The effects of the manipulation on habitat structure were dramatic, but the behavioral responses of the birds were fairly subtle. Sage Sparrows in the manipulated area spent significantly less time singing and were generally less active than in the control area, and they also tended to forage somewhat more in the altered habitat. Brewer's Sparrows foraged less in the manipulation area, but sang more. Despite these differences in activity budgeting, neither species used the substrate types differently in the manipulation versus the control area, although both species used the treatment block from which no shrubs had been removed significantly more often than the altered blocks. The birds thus altered their use of habitat patches and some aspects of their behavior in response to the manipulation, but they did not change the way in which specific substrate types were used for foraging or other behaviors.

In another aspect of our studies we assessed microhabitat use in foraging in greater detail. In one area (containing two study plots) we measured 11 variables for each of 150 randomly selected vegetation patches in 1981. A PCA condensed the variation in these variables into four major components, which accounted for 73% of the variation in the original data set. We also quantified the patch-use by Sage and Brewer's sparrows by following foraging individuals, marking each patch visited, and later measuring the same 11 variables for these patches. By comparing these values to the PCA generated from the randomly selected patches, we determined that both species used the patches nonrandomly. In particular, both species in both study plots showed significant negative factor scores on component II, which contrasted patches dominated by sagebrush with those dominated by green rabbitbrush (*Chrysothamnus viscidiflorus*) (Fig. 10.16). The species also departed from random patch-use on component I, but only on the West plot. This component scaled variation in several attributes related to shrub size. Random patches were somewhat larger, on average, on the East plot, where both species used them in a nondiscriminating manner with respect to size. On the West plot, where shrubs were on average smaller, use was biased toward patches containing larger shrubs. Both species thus used patches in a nonrandom fashion, but there were no significant differences between the species (Wiens 1985b, Rotenberry and Wiens unpublished). When we repeated these studies in 1982, there were changes in both the attributes of random patches and in the use of patches by the birds, but the species were still nearly identical in their (nonrandom) use of patches.

In general, these detailed and lengthy studies of behavior have indicated that the patterns of foraging and microhabitat use by species in these

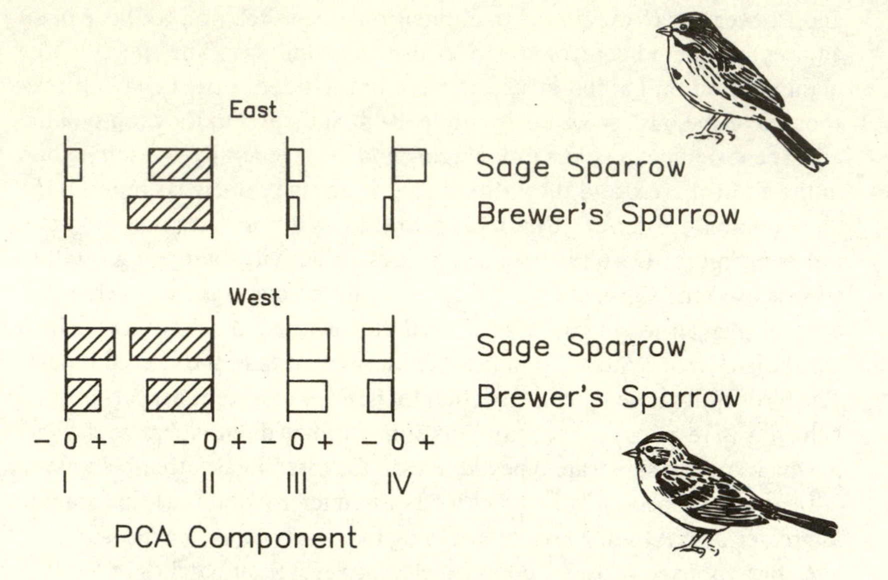

Fig. 10.16. Histograms reflecting the degree of positive (+) or negative (–) departure of features of habitat patches used by Sage and Brewer's sparrows in their foraging on two plots (East and West) of an Oregon shrubsteppe in relation to four dimensions of a PCA of attributes of randomly sampled habitat patches in each plot. Significant departures are hatched. Component I represents variation in size of patches (– = small); component II, variation in contributions of sagebrush and of green rabbitbrush (– = sagebrush); component III, variation from compact, isolated patches (–) to more diffuse patches; and component IV, sagebrush coverage (–) vs gray rabbitbrush. From Wiens (1985b).

environments are quite variable in both space and time and that much of this variation is not readily associated with concurrent variations in resources or environmental conditions. Coexisting species do differ from one another, but the differences are not consistent, and similarities among the species are often more striking than the differences. Some of the differences that might have been considered noteworthy on the basis of the values alone, however, are not statistically significant. This and the importance of yearly variation in behavior caution against attaching great importance to apparent (but untested) differences between species revealed in short-term studies.

Diet relationships

We examined the dietary patterns of coexisting species at three grassland and one shrubsteppe site during our studies. In general, the diet of

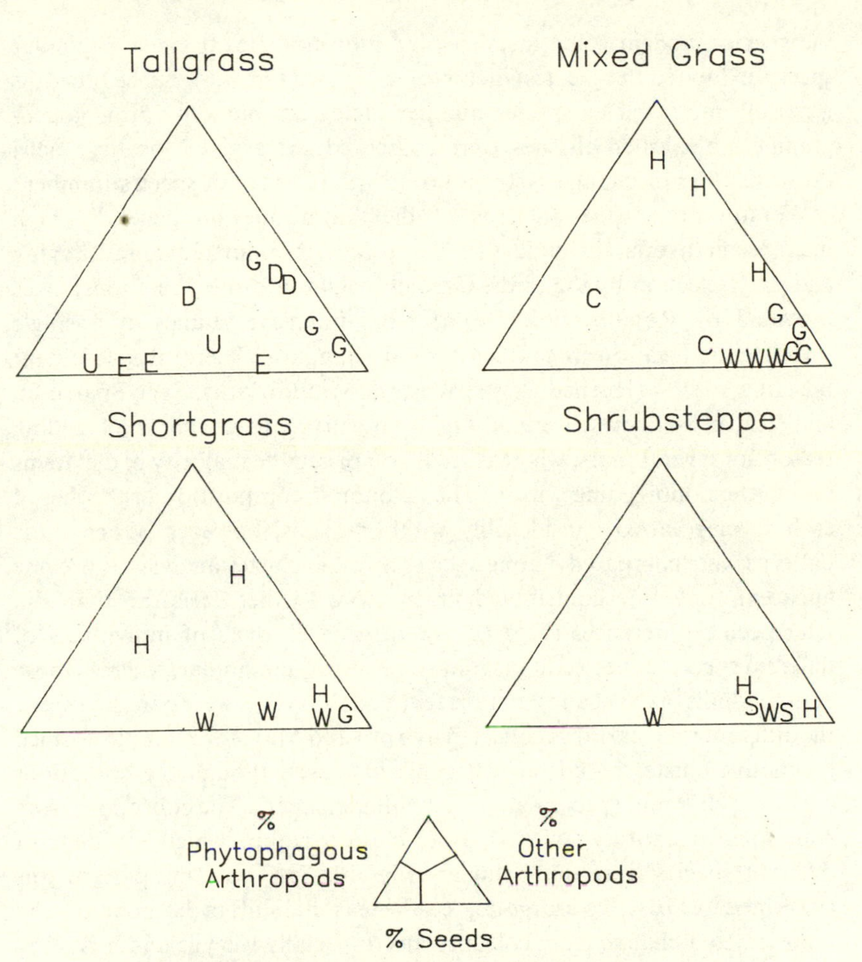

Fig. 10.17. Generalized diet composition of breeding birds in three grassland and one shrubsteppe site in North America. G = Grasshopper Sparrow, D = Dickcissel, E = Eastern Meadowlark, U = Upland Sandpiper, H = Horned Lark, C = Chestnut-collared Longspur, W = Western Meadowlark, S = Sage Sparrow. From data of Wiens and Rotenberry (1979).

each species varied between sites and years, as might be expected in these climatically variable environments (Wiens 1973b, Wiens and Rotenberry 1979). Each species followed a rather different pattern of variation, however (Fig. 10.17), and, as a result, the patterns of diet overlap among species at a site were inconsistent between years. Overlap in prey type and size was generally high, especially among the small finches and longspurs. Overlap and diet niche breadth were unrelated to morphological differences among the species, and in some instances the morphologically most dissimilar species exhibited the greatest dietary overlap (see Fig. 7.6). Niche breadths generally increased as species number increased, but only at a biogeographic scale, when all sites were compared. At a given locality, niche breadths did not change systematically with changes in species number.

We interpreted the patterns of diet variation and generally high interspecific overlap as indicative of substantial opportunism in foraging and prey selection by the birds. Greater insight into these relationships is provided by Rotenberry's (1980a) more intensive studies at a single shrubsteppe location in southeastern Washington. There, the dominant breeding species (Horned Larks, Western Meadowlarks, Sage Sparrows, and Vesper Sparrows) were all highly insectivorous during the breeding seasons of several years, whereas seeds comprised the majority of diet items during the nonbreeding season. The taxonomic composition of the diet of each species varied considerably within seasons, however. When these patterns were compared among species through cluster analysis, it became apparent that time contributed much more to diet variation than did interspecific differences (Fig. 10.18). In general, diets of individuals of different species collected at the same time were quite similar, whereas those of individuals taken at another time revealed shifts that were parallel among the different species. Birds collected in April and May, for example, formed a cohesive cluster based on diet similarity, even though the collections contained different species sampled in different years. The collections from June formed another quite separate cluster, within which the diets of different species were quite similar (Fig. 10.18). The same pattern was confirmed by DFA. To a large degree, the seasonal shift in diet composition represented a change from coleopterans (especially weevils and larvae) in early spring to grasshoppers in summer. This paralleled a seasonal change in the abundance of these prey types, to which all of the bird species apparently responded in a similar, opportunistic fashion.

Conclusions: food limitation, resource use, and the generality of the shrubsteppe system

On the basis of these patterns, bioenergetic calculations (Wiens 1977c, Rotenberry 1980b), observations of growth rates of chicks in enlarged broods (Wiens and Rotenberry, unpublished), and our demonstration that the arthropod faunas of shrubs protected from bird predation do not differ from those of plants exposed to predation (Wiens, Cates, Van Horne, Rotenberry, and Cobb, unpublished), we have concluded that food resources are generally not limiting to breeding birds in these grassland and

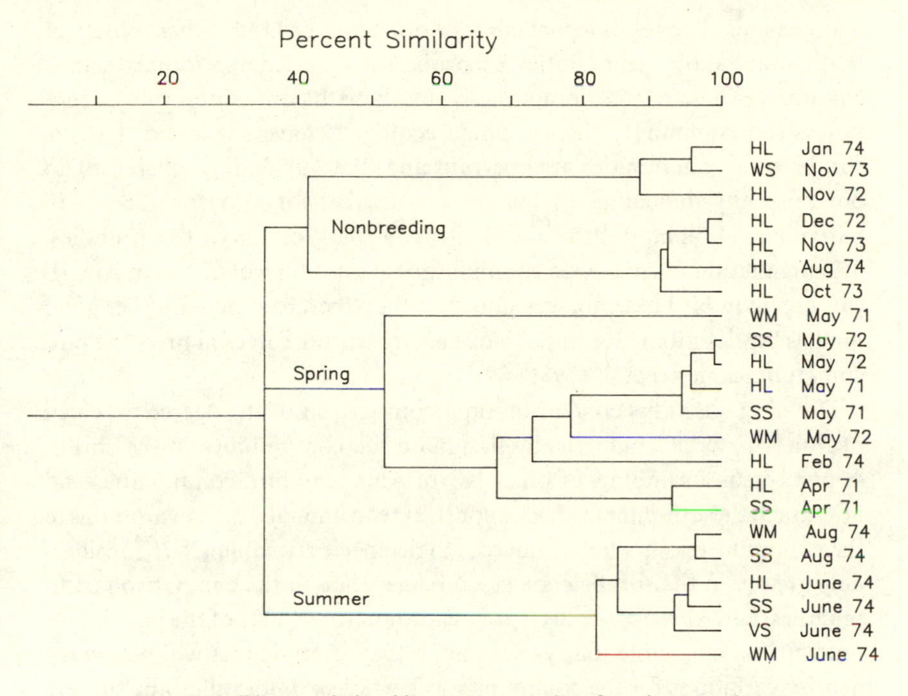

Fig. 10.18. Dietary relationships among samples of species occurring at a southeastern Washington shrubsteppe location. The cluster analysis was based on similarities in the relative biomass of all prey taxa in the diets. HL = Horned Lark, SS = Sage Sparrow, WM = Western Meadowlark, VS = Vesper Sparrow. After Rotenberry (1980a).

shrubsteppe systems. This resource abundance fosters the behavioral and dietary variability and opportunism that we have observed. We lack direct measures of resource abundance or availability, however, and our conclusion thus relies on inferences more than one might wish. We have attempted to relate patterns of dietary differences between sites to general information on vegetation standing crop and primary production at these sites (Wiens and Rotenberry 1979). Although the diets of birds occupying sites with greater vegetation standing crop were more variable between years than in more sparsely vegetated locations, niche breadths were unrelated to any of the general site measurements. Niche overlap among species, however, was greater at sites with low net primary production rates.

On the basis of these results and other findings from our studies in grassland and shrubsteppe systems, we have concluded that there is little evidence that interspecific competition is an ongoing, proximate process in these breeding bird assemblages, that the species respond to environmental features and changes independently of one another (and rather loosely at that), and that the communities are probably not often in resource-defined equilibria. These results do not mesh closely with contemporary competition-based community theory. Some ecologists have suggested that the shrubsteppe communities are aberrant and that our findings therefore do not pose any difficulties to the conventional theory (Wiens 1984a). In particular, Dunning (1986) has challenged the generality of our studies, arguing that the shrubsteppe communities are not typical of North American breeding bird assemblages and that they therefore may not be 'good models for describing the importance of structuring forces in breeding bird communities in general' (1986: 82).

Dunning based his conclusion on a comparison of the degree to which one or two species numerically dominate the communities in the shrubsteppe versus a sampling of other North American bird communities and how much these dominant species contribute to annual density variations in the community as a whole. Indeed, single-species dominance *is* considerably greater in the shrubsteppe assemblages than in the comparison communities, and when one or two species account for > 50% of the individuals present it is inevitable that variations in their abundances will determine density variations for the community as a whole. Almost all of the surveys that Dunning used were from more mesic habitats than shrubsteppe, and they therefore contained more breeding species (exactly how many more is uncertain; the surveys Dunning used were made in areas of different sizes and Dunning did not adjust for species–area effects by rarefaction or other procedures). The shrubsteppe differs from more mesic habitats by containing fewer breeding species per unit area; the greater degree of dominance is an inevitable consequence of this difference in species richness. Dunning therefore showed only that the attributes of species-poor communities are not typical of species-rich communities (or vice versa) (Wiens and Rotenberry 1987). Because there is a wide range of species richness in bird communities in different habitats or locations, it seems doubtful that one can label any one of them as 'typical' of bird communities in general.

Conclusions

To those who seek clear and consistent generalizations about patterns of resource use among coexisting species, the variation in niche relationships described in this chapter must be discouraging. Patterns of dietary overlap, habitat use, and foraging behavior change in both time and space; some of these changes are clearly associated with changes in the environment, but others are not. The predictions of niche theory receive only ambiguous support. The notion of niche complementarity, for example, is supported by the findings of some studies (e.g. Cody 1966, Ashmole 1968, Szaro and Balda 1979, Wagner 1981) but not of others (e.g. Sæther 1982, Landres and MacMahon 1983, Rotenberry and Wiens 1980), and in some cases conclusions of 'striking complementarity' are not supported by the data presented (Hertz *et al.* 1976). A proper test of the niche complementarity prediction requires resource dimensions that are truly independent of one another, and high overlap on several dimensions may reflect interdependence of the dimensions rather than a falsification of the prediction (Schoener 1974b). Even if one documents high overlap on several independent dimensions (as may occur among species opportunistically responding to resource abundances), however, the testability of the notion is further compromised by the possibility that the species segregate on other, yet unmeasured resource dimensions (Wiens 1977b).

The variability of resource-use patterns also means that short-term 'snapshot' studies (e.g. Ulfstrand 1975, Nilsson and Alerstam 1976, Moreno 1981, Edington and Edington 1983) may produce incomplete or misleading results. The overlap in use of *Opuntia* by Galápagos finches recorded by Grant and Grant (1980) would have been missed had they collected their observations a few weeks earlier, and the patterns of diet composition and overlap among shrubsteppe birds recorded by Rotenberry (1980a) differed markedly at different times of year (see also Burger *et al.* 1977, Baker and Baker, 1973, Alatalo 1982c). The documentation of resource-use patterns is thus sensitive to when and where the observations are made. Short-term investigations cannot record the dynamics of resource use, and their results therefore cannot be generalized beyond the time and space boundaries of the study unless one assumes that resources and the ways they are used are stable. This is usually unrealistic. Because of the variability in niche relationships, combining observations from several locations or years (e.g. MacArthur 1958) or comparing results from studies conducted over different scales of space or time (e.g. Schoener 1974a) are not ideal procedures. Unfortunately, decisions about when and where resource-use patterns should be studied are usually made on the basis of logistical factors rather than the biology of the system or a consideration of resource dynamics.

Most investigations of resource use among species in communities have been guided by niche theory, and they therefore contain preconceptions about the importance of ecological differences among species. If an examination of one or two presumed resource dimensions reveals the expected differences, the study may go no further; the required 'niche partitioning' has been demonstrated. Because the differences are rarely tested statistically, almost any degree of interspecific separation may be deemed sufficient to permit coexistence (e.g. Cody 1968, Wiens 1969, Tye 1981, Szaro and Balda 1979). Species are likely to differ from one another in a great many ways for a great many reasons, however, so the discovery that they do differ does not represent a conclusive test of competition-based hyptheses. Studies of the resource-use patterns of species in local guilds or communities should not be conducted solely within the framework of a body of theory that compels us to search for differences. Instead, such studies should focus on how the species functionally relate to resources (e.g. Sherry and McDade 1982) and on the patterns of differences *and* similarities among the species. These patterns may provide the foundation for hypotheses related not only to competition but to history, phylogeny, fluctuating resource levels, predation, physiological capabilities, or other features of the species or their environment (see Volume 2).

In most studies of niche partitioning and resource utilization, little attention has been given to either defining or measuring the resources on which the patterns are presumably based. To avoid documenting patterns that have little to do with actual resources, we should give careful consideration to what the resources for the species in a community really are and distinguish between resource abundance, availability, and use. If we measure foraging behavior, we should be aware that we have recorded an aspect of resource use, not of the resources. If we combine measures of foraging tactics with measures of foliage-height profile in multidimensional niche analyses, we should do so with the knowledge that we are combining quite different things. If we attempt to measure the resources directly, we should be certain that we have considered the definition of resources carefully, and we should ask how closely the measure approximates resource abundance or availability. Moreover, unless it can be determined whether or not resources are in limited supply, developing realistic explanatory hypotheses for the observed patterns will still be difficult. Thus, the determination of accurate resource-use patterns depends on proper measurement, the description of realistic patterns depends on a proper definition of resources, and the development of potential explanations for the patterns requires information on the likelihood that resources are limiting.

11

Density compensation and niche shifts

Many of the examples discussed in the previous four chapters testify to the importance of geographical variation in the morphology, distribution, habitat occupancy, and niche patterns of species. When these variations can be related to the presence or absence of other species, they take on special significance. Competition theory predicts that populations of a species that occur in areas or habitats lacking one or more competing species should exhibit ecological release relative to populations occupying similar situations where the competing species are present (MacArthur 1972, Pianka 1981). The populations change to use some portion of the resources or niche space made available by the absence of the other species. In this way, 'a new state of integration and balance is achieved within the avifauna' (Keast 1970: 61). An 'optimum' community structure is maintained despite the loss of species from the community, at least in theory.

These changes can be expressed in several ways. Population densities may be higher in the absence of putative competitors, leading to *density compensation* in populations or in entire guilds or communities (MacArthur *et al.* 1972). The niche breadth of the population may increase with respect to habitat, elevation, foraging, or diet, either through the expansion of the niche space of individuals or through an increased diversity of individuals within the population (Roughgarden 1979). This *niche expansion* need not involve a change in the mean positions of the population on niche dimensions, although in the absence of competitors populations often undergo *niche shifts* as well, changing niche positions toward those occupied elsewhere by the absent species (Diamond 1978). The populations may also undergo an evolutionary adjustment to the absence of the competitor(s), their morphology shifting in the direction of the missing species (Grant 1972). These patterns have been viewed by many as providing the most compelling and persuasive evidence of the importance of competition in structuring communities and determining species' niches (e.g. Diamond

1978, Diamond and Jones 1980, Noon 1981, Faeth 1984). To cast the patterns so firmly in the context of competition requires acceptance of the assumptions that the species are resource-limited, that the populations and community are saturated and in equilibrium, and that the presence or absence of the presumed competitors is the only ecologically meaningful difference between the areas being compared.

In practice, the different forms of population change in response to the absence of presumed competitors are often not clearly distinguished. 'Niche shifts' may refer to a wide variety of ecological phenomena occurring on different scales of space or time (Herrera 1978b). The notion of 'character release' has been associated with changes in mean body size (Lack 1947), niche expansion (Lack 1969b), increased morphological variation (Van Valen 1965, Arthur 1982), or density increases (Yeaton and Cody 1974). These changes differ in their characteristics, in what is required to demonstrate them, and in the interpretations or hypotheses that can be offered to explain them. It is therefore important to distinguish among them.

Most studies of ecological release have involved comparisons between islands and generally similar mainland environments. Here, it is argued, the reduction in species numbers on islands facilitates tests of the predicted patterns, and several islands may be viewed as 'replicates' in a natural experiment in which the mainland is the 'control' (Diamond 1970, 1978). Because islands may provide sufficient isolation for populations to undergo evolutionary change, it may also be possible to examine ecological changes on several scales, from proximate behavioral shifts through genetically founded changes to the sorts of general evolutionary trends envisioned in the taxon cycle (Diamond and Marshall 1977). The concepts and predictions are equally applicable, however, to comparisons among islands within an archipelago or among different sites on a mainland, although such comparisons have been less popular. In either case, the emphasis is not on *why* species are present or absent from various areas (see Chapter 4), but on the consequences of their distributions.

Community-wide patterns

At the most general level, patterns of ecological release are expressed over entire communities. Most attention has been focused on community-wide density compensation, in which 'the summed population density of individuals of all species on islands [is] equal to the summed mainland density as a result of niche expansions and higher abundances of island species compensating for the absence of many mainland species' (MacArthur *et al.* 1972: 330). This definition is not entirely satisfactory, as it combines elements of pattern with an accompanying process explanation. As a *pattern*, density compensation is characterized by the first portion of this statement, whatever process(es) might have caused it. Actually, unadulterated niche theory predicts that summed densities on faunally depauperate islands should be somewhat lower than summed mainland densities in similar habitats. This is because the niche expansion by the island species leads individuals to occupy ecologically marginal situations, which reduces their efficiency of resource use and results in higher average densities per species but lower overall community density (Wright 1980, Case *et al.* 1979, MacArthur *et al.* 1972, Faeth 1984). Competitive pressures among mainland species, on the other hand, prompt the efficient use of niche space and optimal species packing in the community, at least in theory. Island densities may also be greater if the island species are generally smaller than their mainland counterparts, so that a larger number of individuals will result in the same community biomass (MacArthur *et al.* 1972, Emlen 1979, 1986, Blondel 1985a).

Observations of so-called 'density overcompensation', in which summed island densities exceed mainland densities, are reconciled with the theory by either or both of two postulates (Case *et al.* 1979). First, some of the mainland species may be much more efficient at resource exploitation than others, leading to resource overexploitation and a depression of resource production on the mainland (see Diamond 1975a). If the islands are colonized by the less efficient species, resource levels will be greater than on the mainland, supporting higher population and community densities. On the other hand, the mainland community may be composed of species that have similar harvesting rates, but one or more species may gain competitive dominance by direct interference. If these dominant species are absent from islands, density overcompensation will occur as the other species are freed of the interference, but resource levels will be lower on the islands. The reverse predictions about resource levels should enable one to separate these hypotheses in instances of density overcompensation (Case *et al.* 1979), but this requires an accurate definition of the critical resources as well as information on resource production rates.

Whatever the causes or correlates, it is clear that in some situations summed island densities are as high as or higher than densities in similar mainland habitats. Corsica, for example, supports 109 breeding bird species, whereas areas of equivalent habitat types and diversity on the French mainland contain 170–173 species. Densities across the gradient of habitat types on Corsica averaged 40.4 pairs/10 ha in comparison to 28.9 pairs/10 ha over an equivalent gradient on the French mainland (Blondel 1985a).

Total community biomass, on the other hand, was roughly similar in the different areas. Corsica thus contains smaller species, on average, and the species that are missing from Corsica are those that are generally the rarer (and larger) of the mainland species. Densities of the species occurring in both island and mainland locations averaged 4.2 pairs/10 ha, whereas the species present only on the mainland averaged 2.7 pairs/10 ha.

In censuses confined to spruce forest on very small islands and on the adjacent coast of Maine, Morse (1977) also found overall densities to be greater on the islands. His measures of foliage arthropod abundances indicated that levels were quite similar in island and mainland locations at the time of nesting, although later in the summer arthropod abundance was greater on the islands; the differences in densities could therefore not be attributed to differences in food resource levels (as gauged by Morse's measure), at least at the time of peak breeding. On the Bahamas, pineland habitats support considerably greater densities than are found in similar habitats on the Florida peninsula, and biomass is also greater, at least among foliage gleaners (Emlen 1978). Emlen's admittedly coarse measures of arthropod abundance indicated no systematic differences between the areas, and he hypothesized that the low densities on the peninsula reflected the corrosive effects of gene flow in reducing levels of local adaptation to environmental conditions, thereby reducing efficiencies of niche exploitation and producing lower mainland densities.

Other documentations of community-wide density compensation are more equivocal. MacArthur *et al.* (1972) surveyed the bird fauna of Puercos Island in the Pearl Archipelago off Panama during January of one year by mist netting and conducting song censuses over two routes on two mornings. On the basis of these rather cursory (and possibly biased; see Abbott 1980, Williamson 1981) samples, they concluded that summed density was slightly greater on the island, even though it contained less than one-third as many species as the comparison sites on the Panamanian mainland. They supplied no information on the island habitats, however, so the comparability of the sites cannot be ascertained (Abbott 1980). In another comparison, of 39 small islands in a Swedish lake to two adjacent mainland areas, Nilsson (1977) found that combined densities on the larger islands were similar to the mainland values but densities on smaller islands were greater. Nilsson's conclusions were based on careful censusing (in a single year), but the islands had fewer trees and more saplings than the mainland areas, so the habitats were not strictly similar (Abbott 1980). Schluter (1984) reanalyzed Nilsson's data and concluded that they do not show density compensation, but his analysis was directed at patterns of reciprocal density associations of species rather than overall community densities.

The extent of density compensation may vary among habitats. On Corsica, for example, density compensation is largely restricted to the species occurring in forests, where the reduction in species number in comparison to mainland areas is greatest (Martin 1984, Blondel 1985a). In the Åland Archipelago of Finland, summed densities were similar between the island of Ulversö and the Main Åland 'mainland' in three habitat types but were greater on the island in three other habitats (Järvinen and Haila 1984). On both Corsica and Ulversö, however, the island habitats appear to differ from their mainland counterparts in subtle ways, and this may contribute to the density patterns (Martin 1984, Järvinen and Haila 1984).

Density compensation may also occur in some guilds within a community but not in others. In South Africa, afromontane forests are distributed in a series of increasingly smaller and more isolated habitat 'islands' from east to west, and this is accompanied by a decrease in species number. In the five areas Cody (1983c) sampled along this gradient, species number decreased from 43 to 15, but summed community densities remained amazingly constant (Table 11.1). Cody's measures of insect abundances and vegetation structure varied considerably across the gradient. Using the values at the most depauperate site as a basis for comparison, he calculated the density patterns that would be expected on the basis of species numbers, arthropod levels, and vegetation (Table 11.1). The uniformity of densities over the forest sites could not be accounted for by the patterns of change in food or habitat resource levels, and Cody suggested that the reduction of competition in the more depauperate sites allowed birds access to resources used by other species at the more diverse sites. Cody calculated bird densities by mapping territories on plots of 5–10 ha (without rarefaction) during a $2\frac{1}{2}$-month period over the five sites, and he derived the food abundance measures from a variable number of sticky boards arrayed at each site for an unspecified period of time. The projection of expected bird densities on the basis of insect levels was made using conversions derived from studies in Arizona pine–oak woodlands (Cody 1981). All of these parameters thus have an unknown but probably substantial measurement error, so the degree to which the observed densities approximate those predicted on the basis of arthropod levels is uncertain. A null-model examination of these patterns might prove instructive.

In spite of these methodological problems, it is obvious that the pattern of community-wide density compensation in the South African forests does not occur uniformly across guilds. In the foliage insectivore guild, a change in guild size from 5 to 1 species is accompanied by little overall guild density change, and partial density compensation is also evident in the slow-searching omnivore guild. Among frugivores, granivores, nectarivores, and

Table 11.1. *Avifaunal characteristics of five afromontane forest sites in South Africa, listed from west to east.*

The percentage values (standardized to the situation in the most species-poor site) indicate the density levels observed and predicted from environmental measurements.

	Site				
	Kirstenbosch	Riviersonderend	Grootvadersbos	Kynsna	Alexandria
Number of species	15	22	28	31	43
Density (pairs/ha)	11.7	10.7	11.0	11.2	11.8
Observed density	100%	92%	94%	96%	101%
Predicted from species number	100%	147%	187%	207%	287%
Predicted from arthropods	100%	113%	91%	98%	107%
Predicted from vegetation	100%	71%	87%	67%	49%

Source: Modified from Cody (1983c).

sallying flycatchers, on the other hand, no patterns of density compensation are evident. Of course, these patterns are sensitive to how the guilds are defined and which species they include. Cody used different taxonomic and/or functional levels to define the guilds, and their density responses along the gradient are therefore not directly comparable.

In the foregoing example, a pattern of density compensation is apparent in at least in some habitats or segments of the community. It would be a mistake to conclude, however, that community-wide density compensation has been found by most investigators who have looked for it. Martin (1983a), for example, found variable and inconsistent patterns of density compensation among islands in the Åland archipelago and recorded no evidence of community-wide density compensation in a comparison of 18 small islands near Helsinki with the adjacent mainland (Martin 1983b). In his surveys of several islands in the Sea of Cortez, Cody (1983a) found no relationship between total community density and variations in species numbers on the islands, although Emlen (1979) found island densities to be more than twice the mainland densities in the dominant desert-scrub habitat on four southern islands in the same region. Emlen (1986) later extended this comparative approach to an analysis of summed community densities among several matched habitats on different Hawaiian islands. Contrary to his expectations, he found that summed guild and community densities differed greatly in similar habitat types, often by several orders of magnitude. He suspected that these density differences might be related to inter-island differences in the abundance of non-avian competitors for shared resources, in the patterns of social interference among species, or in equilibrium density levels, but he also concluded that his methods were too coarse and his sample sizes too small to rule out the possibility that the density differences were associated with differences in resource levels.

Community-wide density compensation is usually thought to be associated with niche expansions by populations occupying the more species-impoverished locations. Data to assess this assumption on the scale of entire communities are scarce, although information about the niche patterns of some of the dominant species in the community is often gathered (see below). In their studies of ecological release in Caribbean avifaunas, Cox and Ricklefs (1977, Ricklefs and Cox 1978) used the occupancy patterns of species over nine habitat types (determined by rather coarse census procedures) to determine mean niche breadths on islands of differing species richness and on the Panamanian mainland. The decrease in the mean number of species per habitat type (30 on Panama to 12 on St Kitts) was accompanied by an increase in the mean number of habitat types occupied per species (2.0 on Panama, 5.3 on St Kitts) and in an index of density (the mean number of occurences per habitat; 2.9 on Panama, 5.9 on St Kitts). The larger, more speciose island (Jamaica), however, supported more species that were in late stages of the taxon cycle than did the smaller islands; by definition, these species occupy restricted habitat niches in comparison with the broad-niched colonists of initial stages of the cycle. The colonists, on the other hand, often include a disproportionately large number of the more abundant, widespread, and relatively broad-niched species from the mainland source pool (Ricklefs and Cox 1972, Martin 1983b, Morse 1977, Bengtson and Bloch 1983, Blondel 1985a). This inevitably leads to an expansion of average niche size on islands containing such species in relation to mainland areas or large islands on which evolutionary changes (such as those hypothesized in the taxon cycle) have occurred.

Community-wide density compensation and niche expansion are thus evident in some situations and not in others. Can we detect any general patterns in the likelihood of their occurrence? The above arguments suggest that niche expansion and density increases may be most prevalent on islands containing recent colonists; these may often be relatively small islands. Wright (1980), on the other hand, concluded that summed island densities are generally high relative to the mainland when the island faunas are large and the species occupy familiar habitats but are low when the islands contain few species occupying less appropriate habitats. Nilsson's (1977) study, which indicated that densities were greater only on the smaller, species-poor islands, was not included in Wright's review. There is no obvious consistency to the occurrence of these community-wide patterns, perhaps because many of the studies conducted so far have used rather coarse methods and have therefore produced equivocal results. In order to document density compensation satisfactorily, one must demonstrate that densities have been determined in an accurate or unbiased manner, that the normal variation in densities in the different locations has been taken into account, that the survey areas are of equal size (or the effect of area is removed through rarefaction or similar procedures), and that the areas being compared are indeed matched in climate, habitat, and key environmental characteristics other than species number (Faeth 1984). Otherwise, one must assume that these conditions hold, but this is a fragile assumption.

Density compensation in species populations

The expectation of density compensation in entire communities or guilds is based on the premise that all or most of the species respond to differences among the compared areas in similar ways. Given the individuality of species, this may be expecting too much, and it seems more likely that only a portion of the species in a community or guild should exhibit the predicted patterns. Three of the eight species occurring on small islands off Maine, for example, showed density compensation (Morse 1977). Four of eight species breeding in chaparral habitat on Santa Cruz Island and the adjacent California mainland occurred at higher densities on the island, but densities of the other four species were equivalent (Yeaton 1974). Of 25 species that Emlen (1979) recorded in his comparisons of islands in the southern Sea of Cortez with the adjacent Baja California mainland, 8 were most abundant on the islands, 10 on the mainland (5 of which did not occur on the islands), and 7 were roughly equally common in both areas. Haila (1983) made a similar comparison of densities of species on 41 islands in a large lake in northern Finland, finding that 14 of 24 abundant species were present on the islands in the same densities as on the mainland and 10 occurred at greater densities on the islands. Haila used prevalence functions (Chapter 4) to examine these 10 cases of apparent density compensation more closely and found that all but one could be attributed to habitat differences between island and mainland.

Populations of some species are undeniably denser on islands than in generally similar mainland habitat. Song Sparrows, for example, occupy considerably smaller territories on small islands than on larger islands or the mainland. Yeaton and Cody (1974) attributed this pattern to the presence of fewer potential competitors on the small islands; whether or not one agrees with their interpretation (Wiens 1983), the pattern itself seems secure. On the Faroe Islands, the densities of Rock Pipits (*Anthus spinoletta*), Common Snipe, and Wrens (*Troglodytes troglodytes*) are inversely related to island size (and thus species number) (Bengtson and Bloch 1983). Bengtson and Bloch attributed the density patterns of these sedentary species not to a reduction in pressure from other species, however, but to winter conditions. The birds are territorial then, but they are confined to the shores; this means that smaller islands, with more shore relative to area, support greater densities of birds. Bengtson and Bloch also measured habitat 'quality' with a system of values that the human inhabitants of the islands place on the land according to the grazing or grain-production potential of particular habitats. This evaluation is thus independent of any direct measures of performance or preference by the birds. Densities of both wrens and snipe (but not pipits) were positively correlated with this measure of habitat quality. Densities of the migratory species did not show any relation to island size, although several were positively related to habitat quality. Here, then, the pattern of density compensation is restricted to the

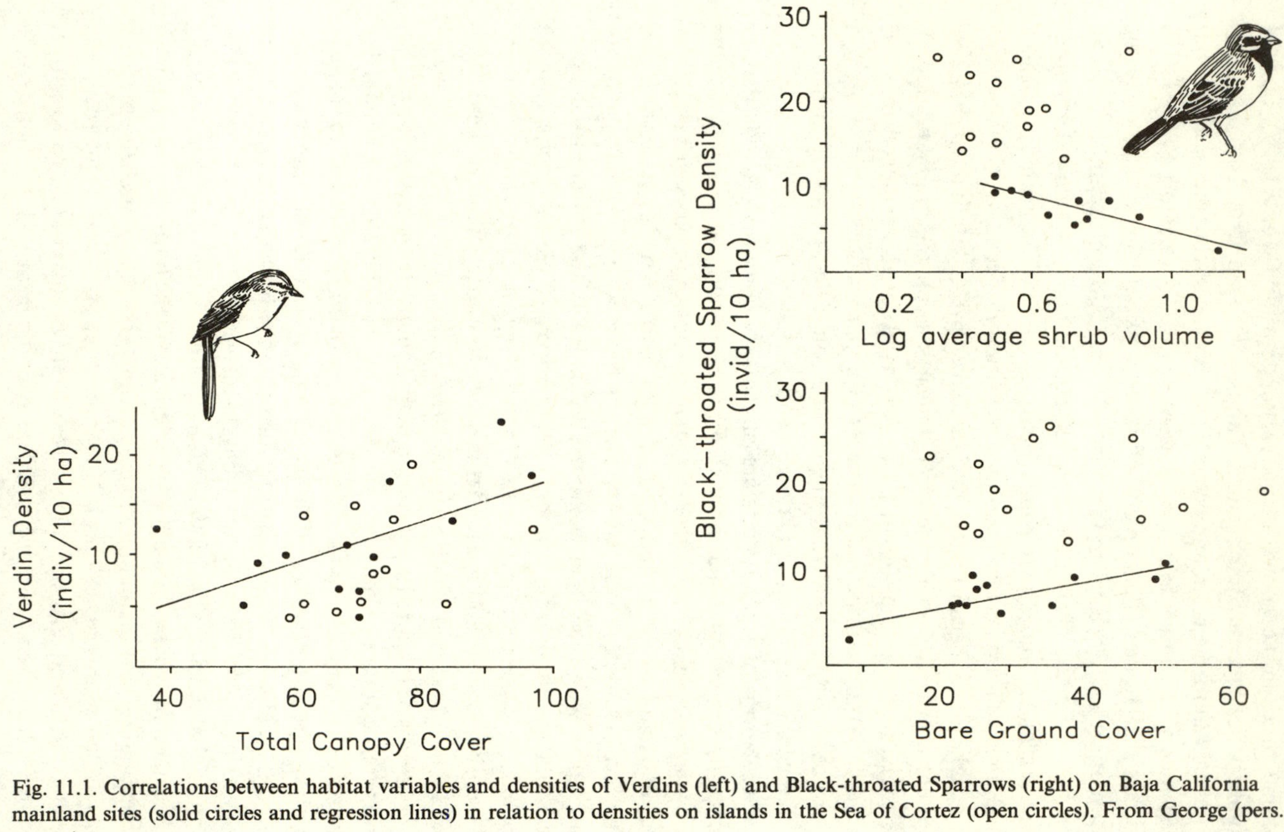

Fig. 11.1. Correlations between habitat variables and densities of Verdins (left) and Black-throated Sparrows (right) on Baja California mainland sites (solid circles and regression lines) in relation to densities on islands in the Sea of Cortez (open circles). From George (pers. comm.).

sedentary species, but the relations of these density patterns to environmental variables differs among the species. In contrast, all of the species that Morse (1977) considered on Maine islands were seasonal migrants, some of which exhibited density compensation and some of which did not.

Bird faunas on islands in the southern Sea of Cortez have been studied independently by several people. There, densities of Black-throated Sparrows by all accounts are substantially greater on the islands than on the adjacent mainland (Emlen 1979, Cody 1983a, George 1987a,b). Emlen and Cody both failed to find evidence of density compensation in Verdins (*Auriparus flaviceps*) and gnatcatchers (*Polioptila* spp.); George confirmed this result for the Verdins, but, when he distinguished between the two gnatcatcher species present in the area, he found that densities of Blue-grays (*P. caerulea*) were substantially greater on the islands whereas Black-tails (*P. nigriceps*) were present only on the mainland. Their combined densities were thus equivalent in island and mainland areas, although the species had strikingly different distributional patterns. George also carefully measured habitat structure and composition in his study plots. Although the habitat types in both areas were generally similar, there were significant quantitative differences between the island and mainland habitats. If these differences contribute to the density patterns, the abundance of a species on the islands should be predictable from the regressions of densities versus habitat features on the mainland. This was indeed the case for the Verdin (which exhibited no density compensation), but Black-throated Sparrow densities were consistently higher than predicted from the mainland regressions (Fig. 11.1). Thus, although some of the increased island densities of this species could be related to habitat differences, other factors are also involved.

Most studies of density compensation have been conducted over relatively short periods, and one must therefore assume that the density patterns observed are relatively stable. This assumption may not always be valid. Nilsson (1977) found that densities of Chaffinches and Willow Warblers were greater on islands in a Swedish lake than on the associated mainland, and Martin (1983b) reported the same pattern for Chaffinches on islands near Helsinki. In neither study was a given island censused for more than 1 yr. Some indication of the effects of temporal variations on such patterns is provided by the results of continuing surveys of the islands in the Swedish lake (Nilsson and Ebenmann 1981). During the period 1976–1978, densities of Chaffinches in the area remained relatively unchanged, whereas Willow Warbler densities decreased by 51% on the islands but only by 10% on the mainland. The density compensation pattern of the warbler that was apparent in 1976 did not persist, and the reduction in warbler densities on the islands was not accompanied by the sort of compensatory shifts by other species that one would expect in a balanced, saturated community.

Attempts to derive patterns of density compensation in individual species must meet the same requirements noted above for community-wide investigations if their findings are to be accepted with confidence. Few studies meet all these criteria. It seems apparent that populations of some species do reach unusually high densities on islands in which total species number is reduced. An accompanying relaxation of competitive pressures is one hypothesis that may account for such patterns, but it is certainly not the only one; several alternative explanations are considered in Volume 2.

Niche shifts and expansion

Changes in niche characteristics or distributional patterns of species between islands and mainland or between different mainland areas are often regarded as clear evidence of the effects of competition. This interpretation is especially favored when the changes are associated with the presence or absence of another potential competitor instead of a more general measure of the species richness of the community. When a single putative competitor species cannot be determined, however, variations in the numbers of a larger set of diffuse competitors may be related to the patterns (e.g. Diamond 1975a, 1978). The idea of niche shifts is so inexorably linked with competition theory that the pattern has come to include both the niche changes *and* the presence or absence of the reputed competitor(s), even though the latter are sometimes only vaguely defined and their actual position as competitors is established intuitively rather than empirically. In this case, the process has become part of the pattern itself.

Some basic patterns

Niche shifts may occur on any or all of the dimensions considered in Chapters 8–10, although most investigations focus on only one particular aspect. Diamond and Marshall (1977), however, attempted to document the frequency and form of niche shifts among all 56 species in the avifauna of the New Hebrides. They compared conspecific populations on different New Hebridean islands and populations in the New Hebrides with conspecific or congeneric populations on another archipelago. Of the 56 species, 35 exhibited niche shifts of some kind: 28 changed in aspects of habitat occupancy, altitudinal distribution, or vertical foraging range, 13 showed shifts in incidence functions (e.g. Fig. 4.3), 5 changed in abundance within the same habitat, and 2 shifted in diet. Of the 21 species that failed to evidence niche shifts anywhere, 2 were quite restricted in range and habitat and 9 were uncommon or of local occurrence. Diamond and Marshall thus found niche shifts to be widespread among New Hebridean birds, and they concluded that 'competition is the proximate cause of distributional limits for most New Hebridean species' (1977: 61). There are of course many sets of conspecific or congeneric populations distributed over the islands of the New Hebrides and the other archipelagos that Diamond and Marshall used in their comparisons, and an advocate of null-hypothesis testing might well object that the frequency of apparent niche shifts has not been demonstrated to be any greater than that expected by chance. This point aside, the comparisons are only qualitative rather than quantitative, and they rely heavily on a *ceteris paribus* assumption that is not likely to hold over such a broad array of islands.

Consider an example of the sort of pattern that Diamond has used to support his arguments regarding niche shifts and competition in the New Guinea region. The Island Thrush (*Turdus poliocephalus*) is widespread over this region but, by Diamond's accounts, it occupies quite different altitudinal ranges in different areas (Fig. 11.2). On New Guinea, it is found only above 2750 m, but on islands in the New Hebrides, Fiji, and Rennell (Solomons) it occurs from high elevations down to sea level. It occupies different habitat types in some of the areas, but Diamond (1975a, 1978, Diamond and Marshall 1977) has attributed the varying distributional pattern to the existence of a tolerable level of diffuse competition. The species is absent from areas containing more than 36 species in the thrush's habitat; on species-rich islands this threshold occurs at higher elevations than on more depauperate islands. The pattern is an intriguing one (assuming that the distributional limits have been accurately determined), but a large number of environmental features are likely to vary with elevation in different ways on the different islands, and Diamond has measured none of these. The conclusion that 'this is a clear example of niche shift related to shifts in diffuse competition from many species rather than to effects of a single dominant competitor' (Diamond and Marshall 1977: 64) is not yet justifiable. In their comparison of the Cordillera Vilcambamba with the Cerros del Sira in the Peruvian Andes (Chapter 8), Terborgh and Weske (1975) also interpreted many of the shifts in altitudinal distributions of species they recorded as evidence of competitive effects, likewise in the absence of detailed measurements of associated environmental features.

Niche shifts are most obvious when a species occupies different elevational ranges or different habitats in different areas. On the island of

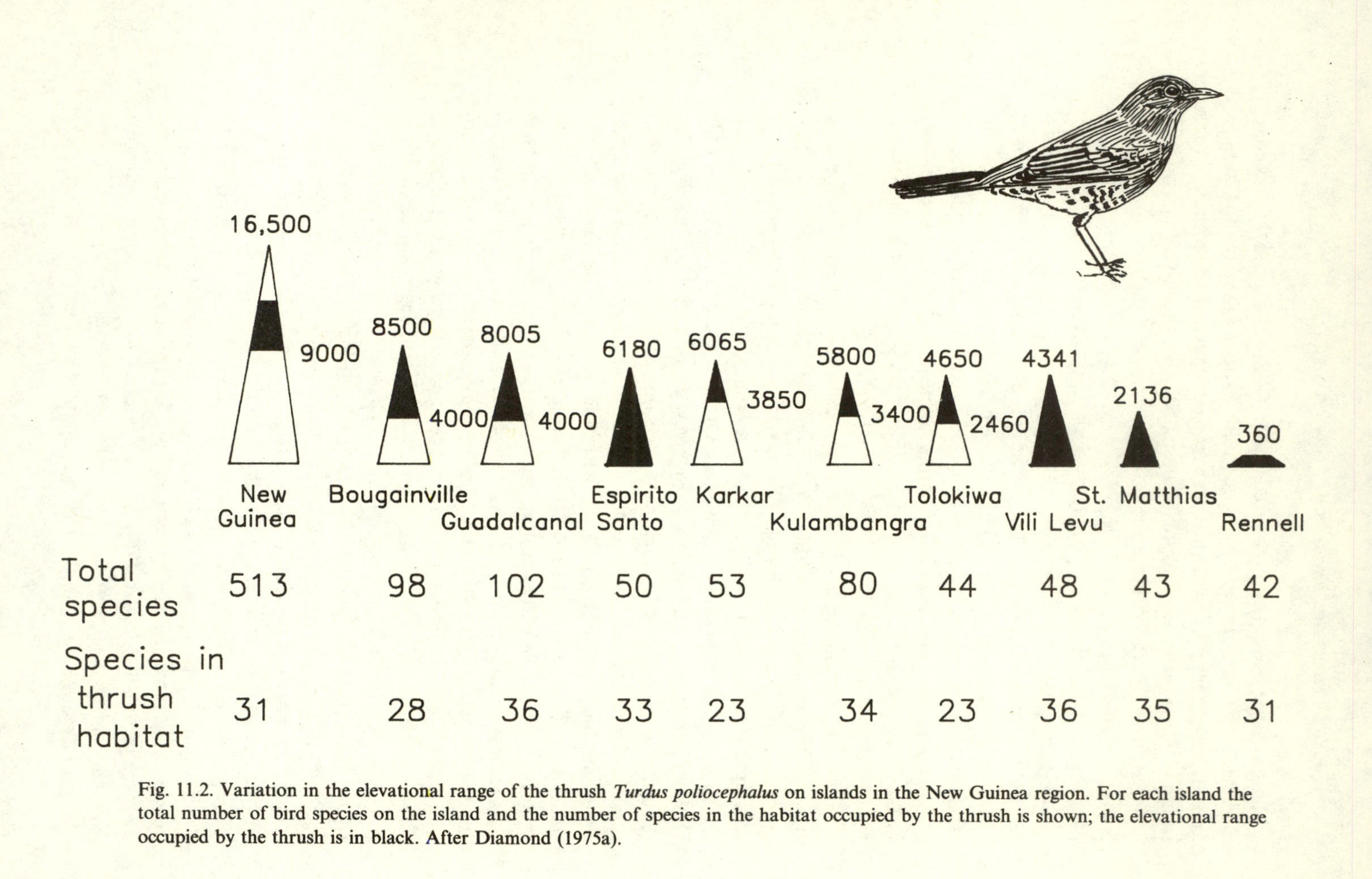

Fig. 11.2. Variation in the elevational range of the thrush *Turdus poliocephalus* on islands in the New Guinea region. For each island the total number of bird species on the island and the number of species in the habitat occupied by the thrush is shown; the elevational range occupied by the thrush is in black. After Diamond (1975a).

Tasmania, for example, the only species of *Lichenostomus* honeyeater that is present, the Yellow-throated Honeyeater (*L. flavicollis*), occurs in a wide range of habitats. This stands in sharp contrast to its restricted habitat distribution on the adjacent Australian mainland, where several congeneric species are found in woodlands of various types (Keast 1968, 1970). Kangaroo Island off South Australia is composed mainly of sclerophyll forest and mallee/heath, and there is only a limited amount of woodland. Judging from their mainland habitat affinities, one would expect to find Yellow-faced Honeyeaters (*L. chrysops*) in the forest, Purple-gaped Honeyeaters (*L. cratitius*) in the mallee/heath, possibly White-plumed Honeyeaters (*L. penicillatus*) in the woodland, and Singing (*L. virescens*) and White-eared (*L. leucotis*) honeyeaters in a broader range of habitats. In fact, only Purple-gaped and White-eared honeyeaters are common on the island, where the former species has expanded to occupy woodland and dry schlerophyll forest in addition to the 'normal' mallee/heath. The White-eared Honeyeater, in contrast, is more or less restricted to tall trees in woodland and is absent from the mallee/heath that it frequents on the mainland (Ford and Paton 1976).

Not all island species exhibit habitat expansion in relation to their mainland counterparts, of course. There are five species of *Sylvia* warblers present on Corsica; six occur on the adjacent mainland. There is thus no within-habitat insular impoverishment for this group (Cody and Walter 1976, Blondel 1985a). The two species that are present only on the mainland occur there at low densities, whereas *S. sarda* occurs at moderate densities on Corsica but is absent from the mainland. Over a similar sequence of habitat types, the three species that occur in both areas have very similar habitat distributions (Fig. 11.3), and their use of the vegetation profile in foraging is similar as well (Blondel 1985a). Among the *Parus* species of Corsica, on the other hand, there are dramatic expansions of habitat occupancy in relation to that found in populations of the same species on the mainland (Fig. 11.3); the island populations also occur at greater densities and exhibit reduced reproductive output in comparison with the mainland populations (Blondel 1985b, Gaubert 1985). Blondel attempted to match habitat types on the sequence as closely as possible, but there are still some differences in the characteristics of equivalent successional stages, and these may have contributed to the differences between island and mainland patterns. For whatever reason, it is clear that Corsican *Parus* species exhibit density compensation and habitat expansion, whereas the *Sylvia* species do not.

Because Blondel measured several habitat features carefully, he was able

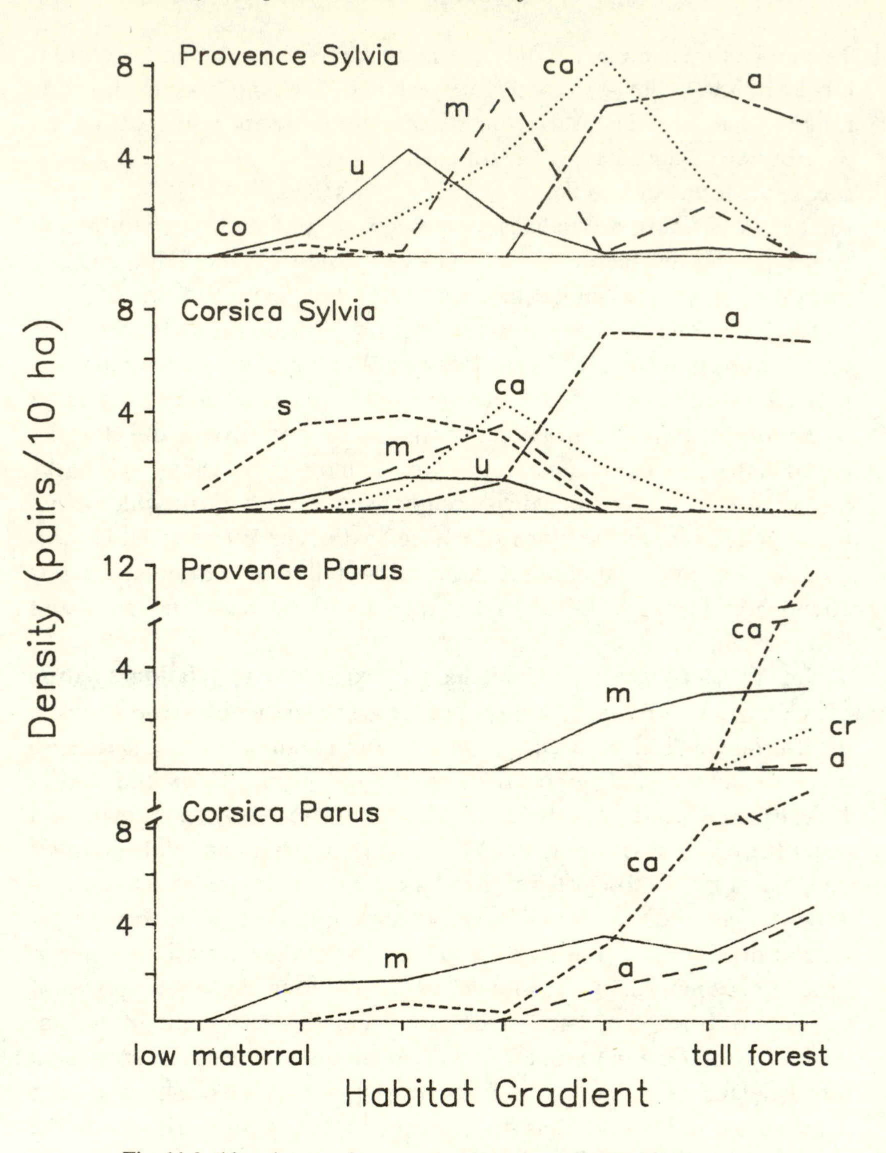

Fig. 11.3. Abundances of several *Sylvia* species (above) and *Parus* species (below) in matched stages of habitat succession on the French mainland (Provence) and the island of Corsica. For *Sylvia*, co = *conspicillata*, u = *undata*, m = *melanocephala*, ca = *cantillans*, a = *atricapilla*, and s = *sarda*. For *Parus*, m = *major*, ca = *caeruleus*, a = *ater*, and cr = *cristatus*. Derived from data in Blondel (1985a).

Table 11.2. *Classification matrix showing the percentage of individual territories of several thrush species that were correctly placed in the proper group (species) as defined by a DFA of habitat characteristics.*

Part A shows the classification performance for the five species occurring in northeastern forests in relation to the DFA for those species in that region. Part B shows the classification performance for individuals of two species occurring in southern Appalachian forests when displayed against the northeastern DFA.

	Predicted group membership				
Actual group	H.m.	C.f.	C.g.	C.u.	C.m.
A. Northeast					
Hylocichla mustelina	78	14	8	0	0
Catharus fuscescens	22	66	8	4	0
Catharus guttatus	22	2	62	14	0
Catharus ustulatus	4	10	0	84	2
Catharus minimus	0	0	0	2	98
B. South					
Hylocichla mustelina	45	49	4	2	0
Catharus fuscescens	9	51	9	32	0

Source: Modifed from Noon (1981).

to quantify his comparison of the island and mainland situations, even though he did not test to determine the extent of similarity or difference of the habitat samples. In one of the more detailed quantitative analyses of habitat niche shifts, Noon (1981) compared the habitat-occupancy patterns of thrushes in forests in the northeastern United States (5 species) with those of thrushes in similar forests in the southern Appalachians (2 species). The two species in the south occupied a considerably broader elevational range than they did in the northeast. Noon tested for an associated habitat expansion by classifying the southern species in relation to a DFA derived from the habitat associations of the five species in the northeast. Even in the northeast, the overlap in habitat occupancy by the species was sufficient to produce misclassifications. *Hylocichla mustelina*, for example, was correctly classified in 78% of the instances but was misclassified (according to habitat) as either *Catharus fuscescens* or *C. guttatus* in the remaining samples (Table 11.2). Overall, 77.6% of the individuals were correctly classified by the DFA. When the two southern species were analyzed using the northeastern DFA, the model failed to classify individuals correctly most of the time, and the overall classification accuracy was only 46.9% (Table 11.2). Clearly, these two species occurred over a considerably broader range of habitat conditions in the south than in the northeast, occupying situations that were associated with other thrush species in the northeast. Noon interpreted this habitat expansion as ecological release, from which he reasoned that interspecific competition played a strong role in determining habitat patterns among thrushes in the northeast. The logic that prompted this conclusion is flawed (see Chapter 2; Wiens 1983, James *et al.* 1984), but the documentation of the pattern of habitat expansion is innovative.

Niche shifts may also be more subtle, involving changes in behavior or microhabitat use where species occupy quite similar elevations or habitat types in the areas being compared. The small forested islands that Morse (1971a, 1980) studied on the Maine coast supported Parula Warblers (*Parula americana*) alone, Parula + Yellow-rumped warblers, or Parula + Yellow-rumped + Black-throated Green warblers. Where the latter species was absent, the other species both shifted foraging activity into zones normally used by Black-throated Greens (Fig. 11.4; see also Fig. 10.5). Where the Yellow-rumped was absent, Parulas expanded to forage over all zones (Fig. 11.4). In this situation, arthropod levels were roughly the same on the mainland and the islands but there were clear differences in the vegetation. It is thus difficult to separate the effects of habitat differences from those of differences in species composition (Abbott 1980).

The importance of species interactions is more clearly indicated in the studies of Grant and Grant (1982) in the Galápagos. There, *Geospiza conirostris* occurs on Espanola in the absence of *Geospiza magnirostris*, *G. fortis*, and *G. scandens*. Its foraging niche is unusually broad there and is more similar to the combined *magnirostris–fortis–scandens* foraging niche than to that of any of these species alone. Its bill features are intermediate between those of the missing three species, and it feeds on the same sorts of seeds used by *magnirostris*, *fortis*, and *scandens* elsewhere. On the island of Genovesa, where *fortis* and *scandens* are absent, the foraging niche and morphology of *conirostris* are similar to those of *scandens* rather than being intermediate between *scandens* and *fortis* combined. On Genovesa, *magnirostris* and *conirostris* occupy dissimilar foraging niches, *magnirostris* exploiting foods that are taken by *conirostris* on Espanola but not on Genovesa. Measurements of seed abundances suggest that food supplies on both islands are sufficient to support populations of the missing species, leading one to surmise that *conirostris* has not only altered its foraging behavior in the absence of the other species but may have actively excluded the other species as well. Grant and Grant conclude that their evidence indeed supports the hypothesis that *conirostris* has excluded *fortis* and

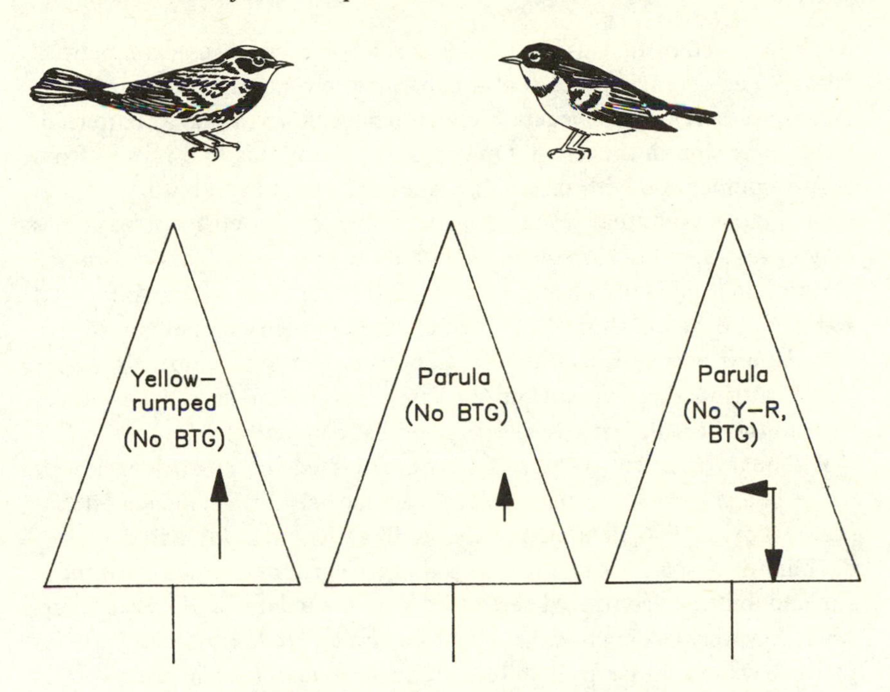

Fig. 11.4. Expansion of foraging ranges of Northern Parula and Yellow-rumped warblers on small islands of Maine when other warblers were absent. The left diagram illustrates the shifts of Yellow-rumped Warblers in the absence of Black-throated Green Warblers (Parula Warblers present); the center diagram, the shifts in Parula foraging when Black-throated Greens were absent (Yellow-rumps present); at the right, shifts in Parula foraging in the absence of both Yellow-rumped and Black-throated Green warblers are shown. The direction and length of the arrows indicate the direction and magnitude of the foraging shifts. Data from Morse (1971a), after Morse (1980).

scandens from Espanola and *scandens* from Genovesa, but other hypotheses are required to explain the absence of *magnirostris* from Espanola or *fortis* from Genovesa.

Other studies of behavioral niche changes have not produced such clear-cut patterns. Of the 14 species present in chaparral or pine-forest habitats on both Santa Cruz Island and the California mainland, for example, 6 expanded, 3 decreased, and 5 showed no change in the range of foraging heights used on the island (Yeaton 1974). Of 6 species that Rabenold (1978) recorded in spruce-fir forests in both the southern and northern Appalachians, 3 were the sole representative of their family in both areas and 3 occurred together with a congener in the north but not in the south. The foraging patterns of those lacking congeners in the south expanded there into zones used by the missing species in the north, suggesting competitive release. Their foraging niches also expanded in other directions, and the three species lacking congeners in either area exhibited a similar pattern of niche expansion in the south. One might attribute this to a release from diffuse competition with unspecified species in the north, but it seems at least equally likely that resource distributions or foraging opportunities may have differed in the south, leading all species to shift their foraging patterns in similar manners. Several of the apparent niche shifts that Landres and MacMahon (1983) documented in their comparison of oak woodland insectivores in southern California and Sonora are also more readily attributable to opportunistic responses to differences in resources than to the presence or absence of potential competitors.

Although divergent niche shifts have attracted the most attention, it would be a mistake to conclude that they are the only form of niche shift that occurs. Pöysä (1986) documented the feeding sites, methods, and depths of dabbling duck species foraging in mixed-species groups in a lake in southern Finland and then compared these measures for members of species pairs when they were feeding together versus separately. In four of the 12 species pairs, there was a divergent shift in these niche features when the species occurred together. These shifts were both symmetrical (both species changing) and asymmetrical; in the latter case, the more abundant and/or larger species did not change. Pöysä noted *convergent* niche shifts in seven of the 12 species pairs, however. In four of these situations, both species shifted toward the other; in three the shift was asymmetrical. Pöysä related the convergent shifts to facilitation in finding food by copying the foraging patterns of another species in the group. The important point, however, is that convergent shifts were more frequent than divergent shifts in these multispecies assemblages, at least over the short time scale of Pöysä's observations.

Temporal variations

The way in which temporal variations may alter patterns of apparent density compensation was noted above, and the same comments apply to niche shifts. Most studies of niche shifts have been conducted over very short periods of time, sometimes only a week or two. If niche characteristics vary, or if the magnitudes or patterns of variation are different in the areas being compared, the niche-shift patterns recorded at one moment in time may be more apparent than real. They may be a consequence of sampling the temporal dynamics of distribution, behavior, or resources at different stages in the different areas. By the same token, actual niche shifts may be missed if sampling happens to coincide with a time at which niche patterns are temporarily similar.

Perhaps the clearest demonstration of temporal effects on niche-shift patterns comes from a 13-month study of hummingbirds on Trinidad and Tobago (Feinsinger and Swarm 1982). Sixteen hummingbird species occur on Trinidad; Tobago supports only five. Feinsinger and Swarm compared the foraging patterns of *Amazilla tobaci* in sites with similar climates and sets of hummingbird flowers. At both sites, nectar was superabundant from December through August but became scarce during the latter part of the wet season, from September through November. At the time of scarcity, *Amazilla* individuals in Trinidad shared the floral resources with six other hummingbird species, whereas the Tobago birds faced only two other nectarivores, both of which fed in very different ways. The Trinidad individuals fed upon a restricted range of flowers whose attributes matched their bill morphology, but individuals in Tobago fed upon a broad range of flowers. The apparent 'ecological release' of the Tobago birds was apparent only during the 3-month period of nectar scarcity, however; over the year as a whole there were no consistent patterns of niche shifts between the areas. Had their study been restricted to the early wet season in June through August or the dry season in December–January (when ecologists from temperate-zone institutions usually visit the tropics), Feinsinger and Swarm would have missed the critical period of diminished resource levels and might have reached misleading and incomplete conclusions. Similar temporal variations in patterns of niche shifts have been observed among waterfowl in Spain (Amat 1984) and between migrant Willow Warblers and resident members of the same foraging guild in Kenya (Rabøl 1987), and they would seem to be likely in almost any system.

The flexibility of niche patterns

Competition theory leads us to expect niche shifts or expansion to occur on an island because traits can be expressed there that are present but suppressed by competitors in mainland populations. We thus assume that both mainland and island populations have equivalent fundamental niches and that the behavioral flexibility, physiological plasticity, and genetic variation of populations will facilitate changes in niche patterns in different competitive settings. If this is so, niche shifts should be commonplace (Williamson 1981).

Some indication that the niche patterns of species may not be so flexible as we expected is provided by Emlen's studies in the Bahamas (Emlen 1981, Emlen and DeJong 1981). On Andros, the only members of the arboreal insectivore guild are the Pine Warbler and Blue-gray Gnatcatcher, but on nearby Grand Bahama five other members of the guild are present, two of which are bark-feeding specialists. The Pine Warblers and gnatcatchers forage almost entirely in foliage on Grand Bahama, but on Andros they forage extensively on bark as well (Table 11.3). Using an index of arthropod abundance, Emlen (1981) was unable to detect any major difference in potential food levels of a given substrate type between the two islands. This suggests that the warbler and gnatcatcher may have been prevented from using the bark substrate on Grand Bahama by the bark-specialist species and that they expanded their foraging niches on Andros in the absence of these competitors. This hypothesis is compromised, however, by the observation that Pine Warblers and gnatcatchers make little use of bark in locations on Grand Bahama in which the bark-specialist species are locally rare or absent. Emlen suggested that the species might possess innate preferences for foraging substrates – foliage on Grand Bahama, foliage and bark on Andros – and that the Grand Bahama birds were therefore incapable of responding to localized reductions in pressures from bark specialists.

If this hypothesis is correct, Andros birds should forage on both foliage and bark in any suitable location, while Grand Bahamas birds should restrict their activity to foliage. Emlen and DeJong (1981) tested these predictions by transporting Pine Warblers between the islands and monitoring their foraging behavior in large enclosures in natural habitat containing both substrate types. True to expectations, birds from Grand Bahama spent little time foraging on bark, whether the substrates were unaltered or had been sprayed to remove arthropods, while Andros individuals spent significantly more time foraging on bark (Table 11.3). There are intrinsic differences (either genetically determined or learned traditions) between the populations in their foraging behavior, and these limit the birds' potential to respond to variations in competitive pressures (or any other factors) in the different areas.

The assumption that flexibility in distributional or niche patterns is sufficient to produce niche shifts may therefore not always be justified. For this and other reasons, niche-shift patterns are far from ubiquitous. In a comparison of two lowland areas in Amazonian Peru, for example, Terborgh (1985b) found that habitat-occupancy patterns of the 190 species common to both areas were both similar in most cases. Terborgh recorded only 12 clear cases of habitat shifts, but 10 (or 11, depending on taxonomy) of these involved congeners. Shifts were thus uncommon, but those that did occur might indeed reflect competitive processes. In Diamond's (1970) survey of niche shifts among species on islands in the southwest Pacific,

Table 11.3. A. *Distributions of foraging activity (% observations) of arboreal insectivores in pine substrates on Grand Bahama and Andros islands.* B. *Distribution of foraging activity of Pine Warblers (% seconds of observation) in experimental enclosures of natural pine substrate and substrate sprayed to remove arthropods. In the experiments, Grand Bahama birds were transported to Andros island for testing, and vice versa.*

	Grand Bahama		Andros	
	Foliage	Bark	Foliage	Bark
A. Observed behavior				
Brown-headed Nuthatch	16	84	—	—
Yellow-throated Warbler	40	60	—	—
Olive-capped Warbler	79	21	—	—
Pine Warbler	93	7	57	43
Blue-gray Gnatcatcher	75	25	30	70
B. Experimental (Pine Warbler)				
Untreated Substrate	97	3	77	23
Treated Substrate	99	1	81	19

Source: After Emlen (1981) and Emlen and DeJong (1981).

roughly half of the species failed to evidence any niche change at all. Habitat and niche shifts by island birds may more often be the exception rather than the rule (Abbott 1980).

Morphological character release

Emlen's observations in the Bahamas suggest that populations on different islands may have diverged evolutionarily, and the occurrence of island subspecies of many widely distributed species (e.g. Keast 1968, Lack 1976) reinforces this conclusion. The changes in niche patterns in depauperate faunas may thus be accompanied by morphological change, as noted above for *Geospiza conirostris* on Genovesa in the Galápagos. Grant (1972) formally defined such situations, in which a population derived from another population that is sympatric with some congeneric or other potential competitor undergoes morphological change in the absence of the other species, as *character release*. In fact, island populations may undergo morphological divergence for a variety of reasons (isolation, founder effects, environmental differences), so a morphological change provides no more direct or convincing evidence of the influences of competition than do more proximate changes in niche features or habitat distributions. If anything, evolutionary divergence complicates island–mainland comparisons (Morse 1980).

Several instances of morphological changes of island populations that coincide with ecological shifts have been reported, but the 'classic case' of character release has to do with *Geospiza fortis* on the island of Daphne in the Galápagos. Lack (1945a, 1947) pointed out that on many islands in the Galápagos, *fortis* occurs together with *G. fuliginosa* and *G. magnirostris* and is morphologically intermediate between these species. On Daphne, where *fuliginosa* and *magnirostris* are absent (or so Lack thought), *fortis* has shifted in bill size toward the smaller *fuliginosa*. Lack (1945a) initially interpreted the pattern as nonadaptive but by 1947 he had changed his view to embrace a competitive explanation; later (1969a) he admitted the possibility that differences in food supplies between the islands might also be involved.

The situation was subjected to renewed scrutiny by Boag and Grant (1984a), who demonstrated that the Daphne *fortis* population is indeed morphologically intermediate between *fortis* and *fuliginosa* found on the nearby island of Santa Cruz (Fig. 11.5). Moreover, known immigrants of *fortis* and *fuliginosa* to Daphne (squares in Fig. 11.5) were generally atypical of Daphne birds, and they differed behaviorally and in plumage characteristics as well. The situation is somewhat more complicated than Lack depicted, however. First, both *fuliginosa* and *magnirostris* were consistently present on Daphne during Boag and Grant's study, although they contributed only 0.5–5% of the total finch density. *Geospiza fortis* apparently does not differentiate behaviorally between conspecifics and *fuliginosa* on Daphne, although it does so on Santa Cruz (Ratcliffe and Grant 1983a,b). As a consequence, *fortis* may hybridize with *fuliginosa* at low frequencies on Daphne, producing offspring of intermediate phenotype (Fig. 11.5). There are insufficient data to determine whether or not such matings are selected against (Boag and Grant 1984a).

Another complication is created by temporal variation. Between May 1976 and March 1978, the Daphne *fortis* population decreased by 85% in association with the scarcity of small, soft seeds caused by a prolonged drought. This produced strong directional selection on *fortis* morphology, favoring larger individuals and shifting the population distribution (Fig. 11.5). The morphological intermediacy of Daphne birds may thus change through time.

Geospiza fortis is not the only abundant finch on Daphne, however; *G. scandens* also occurs there. During the drought cycle, *fortis* specialized on small seeds when they were abundant but increased its diet niche breadth

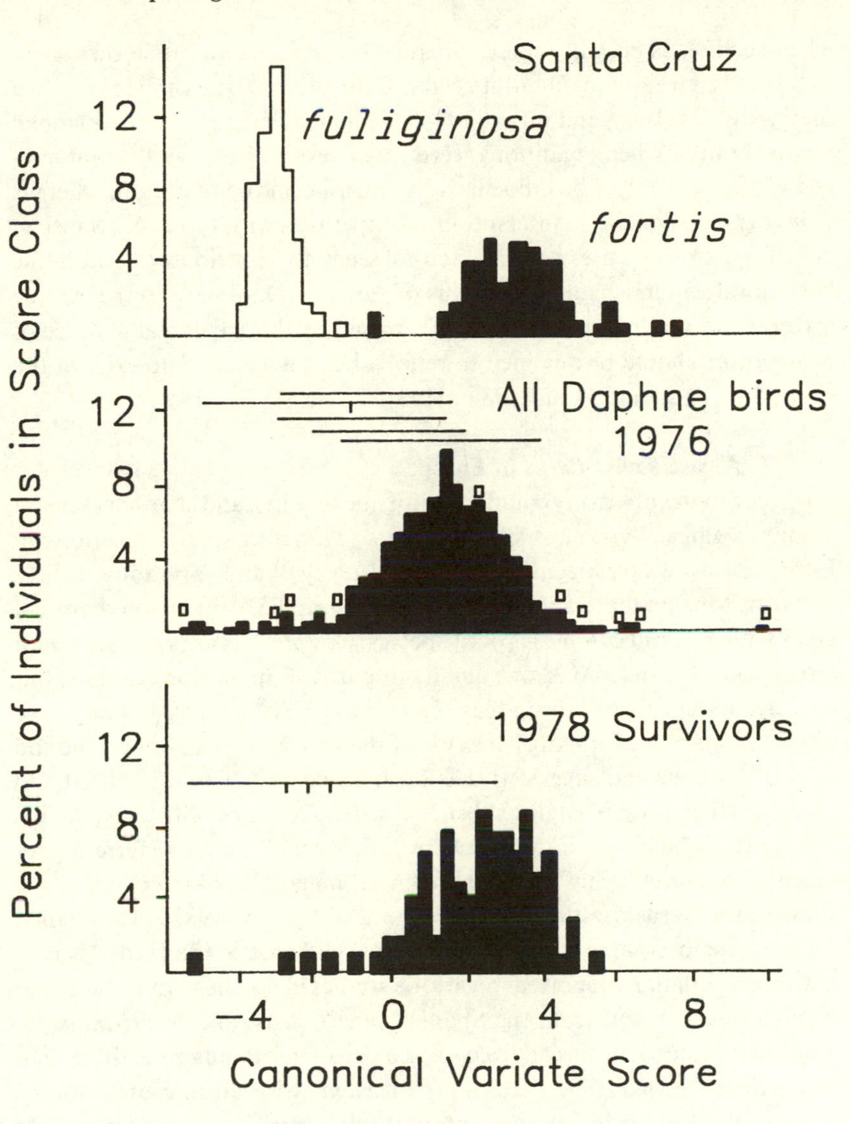

Fig. 11.5. Frequency distributions of canonical variate scores of finch morphology on Santa Cruz and Daphne Major islands in the Galápagos Islands in 1976 and 1978. The horizontal lines represent matings between *fortis* and *fuliginosa*; the vertical lines below these lines indicate the morphological scores of hybrid offspring produced by these matings. The open squares indicate the morphology of birds known to immigrate to Daphne from Santa Cruz in 1977. After Boag and Grant (1984a).

when such seeds became scarce, whereas *scandens* had a broad diet when small seeds were abundant but specialized on *Opuntia* fruits and seeds when they were not (Boag and Grant 1984b). The two species thus overlapped relatively little when conditions were most severe. Boag and Grant suggested that *fortis* may be influenced by intraspecific competition, whereas *scandens* is responsive to interspecific competition with *fortis* (c.f. Smith *et al.* 1978). In any case, several factors seem to contribute to both the behavioral and morphological status of *fortis* on Daphne, and the 'simple textbook account of a character shift caused by the accidental absence of competitors should be qualified to reflect the ecological complexity of the situation' (Boag and Grant 1984a: 243).

A case study: *Parus* in Europe

Studies of density compensation, niche shifts, and character release among ecologically similar species of tits (*Parus*) in several locations in Europe provide a perspective on the consistency of and variations in these patterns. Most of the studies have drawn geographic comparisons between areas with different complexes of tit species. Crested Tits are sympatric with Great and Blue tits in oak woodlands in parts of Spain, for example, but they are absent from other areas. Herrera (1978b) gathered continuous observations of the foraging behavior of the species over a 3-month period at two locations and suggested that both Great and Blue tits shifted and narrowed their use of foraging substrates when sympatric with Crested Tits. The shifts in behavior, however, were rather small, and, as Herrera provided no quantitative information on environmental characteristics of the sites, the patterns are difficult to evaluate. Moreno (1981) also studied *Parus* in Spain, comparing his observations (gathered over a 15-day period in winter) with those reported in various studies in northern Europe. *Parus montanus* was absent from the Spanish pine forests, and *P. cristatus* occurred at greater densities and occupied a broader foraging zone there than in northern Europe. The pattern is consistent with an interpretation of ecological release from competition with *montanus*, but the locations are several thousand kilometres apart and contain very different kinds of environments. Moreno was careful to point out that the niche changes of the Crested Tit were not obviously related to the absence of the Willow Tit.

The most detailed studies have been conducted in England and northern Europe during winter, when food resources are frequently likely to be limiting (Gibb 1954, Betts 1955, Askenmo *et al.* 1977, Jansson *et al.* 1981). In northern Finland, *P. ater* is absent and Goldcrests are uncommon, but both *cristatus* and *montanus* are present. There, the birds forage extensively on outer twigs and needles, shifting their pattern from that found in southern Sweden, where Coal Tits and Goldcrests (both of which forage primarily on needles) are abundant (Fig. 11.6; Alatalo 1981a, 1982b). Hogstad (1978) examined the foraging behavior of *montanus*, *ater*, and *cristatus* in spruce forests along an elevational gradient in Norway. Crested Tits were present only at the lower elevations; where they were absent, Willow Tits foraged more frequently in the outermost branches and less often on branches close to the trunk. There was no change in the foraging position of *ater*, however, nor did *montanus* evidence any shift in association with the absence of *ater* from the high-elevation site. The sites Hogstad studied differed somewhat in vegetation, in the complex of other non-*Parus* foliage-feeders present, and in elevation, so the comparisons are not so clear as one might wish.

In these studies, different mainland areas were compared with one another. In other situations, island–mainland comparisons have been made, as for the Corsican *Parus* species discussed above (Fig. 11.3). Alatalo *et al.* (1985) compared central Sweden, where Willow Tits occupy coniferous forests and Marsh Tits occur in deciduous woods, with the Åland Islands, where only the Willow Tit is present. On Åland, *montanus* has expanded to occupy deciduous as well as coniferous habitats, although none of the other six species in the pariform guild exhibits any sort of niche shift. Densities of Willow Tits are equivalent in the two areas, so the habitat expansion on Åland cannot be attributed simply to a density increase. On the island of Gotland, three of the 10 woodland species present in winter on the adjacent Swedish mainland are missing (Alerstam *et al.* 1974, Alatalo *et al.* 1986). The three missing species are all parids (*montanus*, *palustris*, and *cristatus*) and, of the seven species on Gotland, only *Parus ater* exhibits any noticeable changes in ecology. On the basis of a 3-month study during an exceptionally mild winter, Alerstam *et al.* (1974) reported that densities of Coal Tits were 2–3 times greater in deciduous woodland and 5–15 times greater in coniferous habitats on the island than on the mainland. The tits on Gotland also shifted their foraging niches, using interior positions much more frequently than on the mainland (Fig. 11.6). The generality of these patterns might be questioned on the basis of the brevity of the study, but Alatalo *et al.* (1986) have gathered additional information from Gotland that confirms the patterns of density compensation and niche shifts. Moreover, the shift on Gotland is not evidenced by all individuals. Older, dominant individuals have shifted their foraging behavior to use the interior foraging positions not used on the mainland, whereas younger, subordinate birds continue to forage in the same peripheral locations as used by

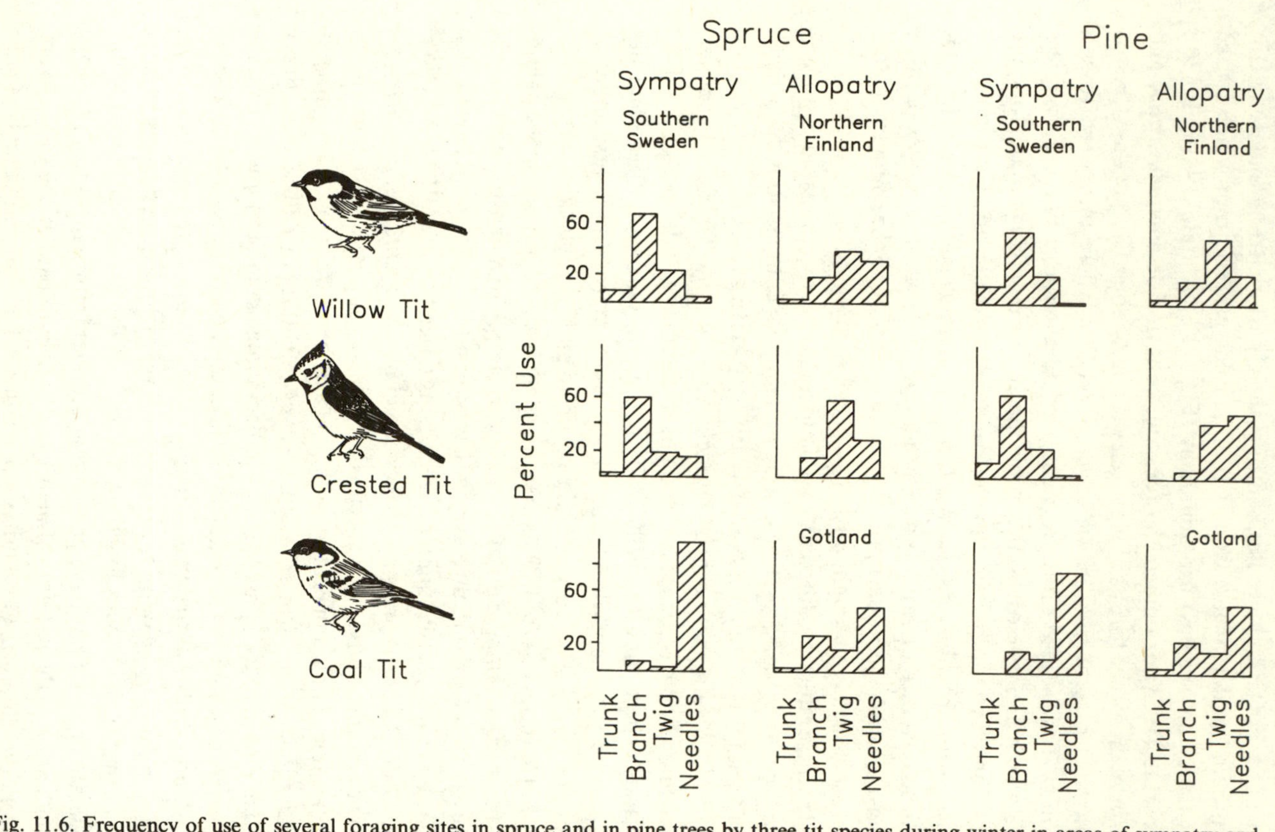

Fig. 11.6. Frequency of use of several foraging sites in spruce and in pine trees by three tit species during winter in areas of sympatry and allopatry in Scandinavia. After Alatalo (1982b).

ater in southern Sweden. If one assumes that increased intraspecific competition associated with the greater densities of *ater* on Gotland forces subordinates into marginal foraging locations, one must conclude that all *ater* individuals are constrained to use ecologically marginal peripheral locations on the mainland.

All of these comparisons suffer from the criticism that the areas compared are geographically separated, and thus environmental features other than the presence or absence of putative competitors must surely differ as well (Alatalo 1982b). One way to reduce this problem is to compare the niche patterns of species in feeding flocks of differing composition in the same area, as Hogstad (1978) and Alatalo (1981b, 1982b) have done. Hogstad found that *montanus* exhibited the same shifts in foraging positions in mixed flocks in which *cristatus* was absent as it did in the broader elevational comparisons. In northern Finland, Alatalo studied foraging shifts in flocks consisting of *cristatus*, *montanus*, *major*, *ater* (in some years), and *Regulus regulus*. Overall, all of the species tended to avoid one another's foraging positions when moving in the same flock. When occurring in flocks with *cristatus* (which used the upper, outer portions of trees), *montanus* shifted to the lower, inner portions; it used the upper, outer locations when foraging in flocks containing *major* (which foraged in the lower, inner areas) but not *cristatus* (Fig. 11.7). *Parus montanus* is a subordinate in social interactions with these species and thus may be more likely to shift foraging positions than social dominants such as *major* or *cristatus*.

In central Sweden, *cristatus* and *montanus* forage in the inner portions of trees, while the smaller, subordinate species (*ater* and *Regulus regulus*) forage in the outer branches. When Alatalo *et al.* (1987) experimentally removed Coal Tits and Goldcrests from mixed-species feeding flocks, *cristatus* feeding in spruce trees expanded to forage on the outer branches, but *montanus* did not. In pines, however, *montanus* did respond to the absence of the Coal Tit and Goldcrest in the expected manner. Because the removed species are normally subordinate, Alatalo and his colleagues surmised that in unmanipulated flocks they must restrict the activities of *cristatus* and *montanus* by means of food exploitation and depletion. The difference in the response of *montanus* in the different conifers was attributed to the scarcity in the pines of *cristatus*, which may inhibit the foraging patterns of *montanus*.

In some cases, niche shifts in *Parus* are accompanied by morphological changes. On Gotland, for example, Coal Tits are larger than on the Swedish mainland, approaching the missing Crested and Willow tits in morphology (Alatalo *et al.* 1986, Norberg and Norberg unpublished). In several other

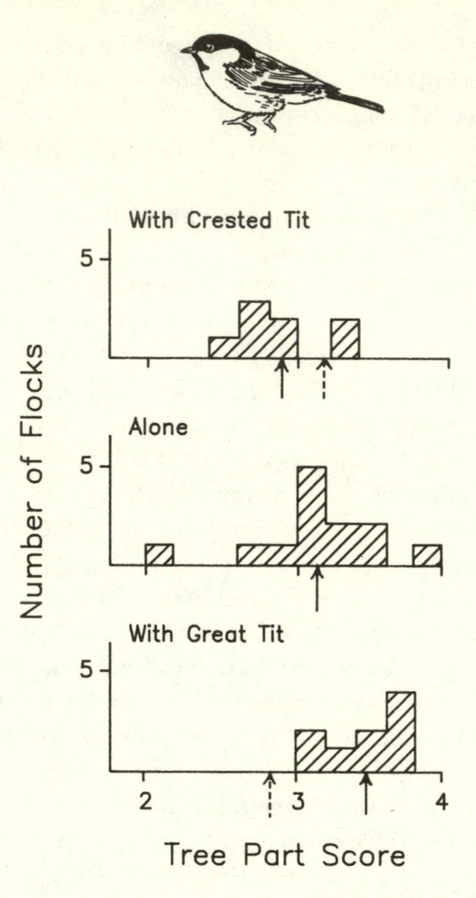

Fig. 11.7. Use of portions of spruce trees during winter by individuals in *Parus montanus* flocks in northern Finland in the presence of *Parus cristatus*, in the absence of *cristatus* and *Parus major*, and in the presence of *major*. Tree parts are scored from larger to smaller diameter (1 = trunk, 2 = branch diameter > 8 mm, 3 = twig, 4 = needled twig). Solid arrows indicate the average (over flocks) tree part score for *montanus* in each type of flock; dashed arrows give the average tree part score of the other species. After Alatalo (1982b).

situations, populations differ in morphology in the absence of a putative competitor. Even if no morphological shift occurs, it appears that the species most likely to shift in ecology in the absence of a putative competitor is the one most similar to it in morphology (Fig. 11.8; Alatalo *et al.* 1986).

The most intensively studied case of character release in *Parus* involves the populations of *caeruleus* occurring on the Canary Islands. Lack and Southern (1949) originally drew attention to the fact that these birds have longer, narrower bills than do populations on the mainland of Europe and North Africa. They also have shorter wings and longer tarsi and are less

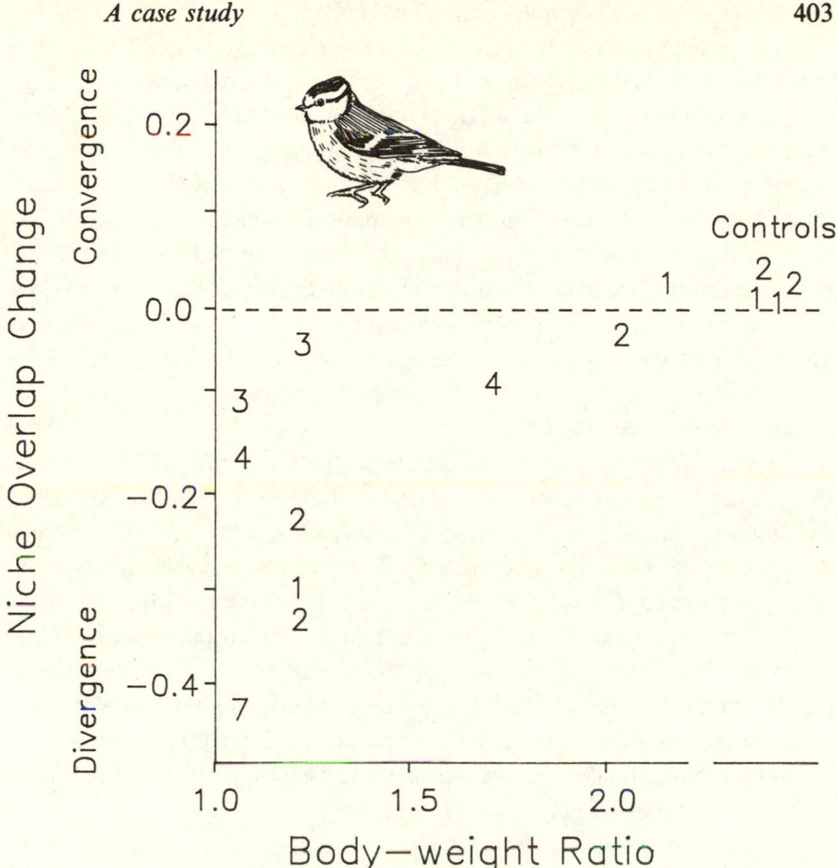

Fig. 11.8. The relationship between body–weight ratios of a given species of *Parus* and a congener that is absent from an area to the change in foraging-site niche overlap between the species produced by the niche shift (of the species present in both areas) where the congener is absent. Location codes: 1 = *cristatus*, *montanus*, and *ater* on Gotland; 2 = *cristatus*, *montanus*, *ater*, *major*, and *Regulus regulus* in northern Finland, Sweden, and Åland; 3 = *montanus*, *cristatus*, and *ater* in Norway; 4 = *cristatus*, *caeruleus*, and *major* in Spain; 7 = *montanus* and *cinctus* in Swedish Lapland. 'Controls' are cases in which interspecific competition should not affect foraging sites because of great morphological differences (treecreeper compared with tits) or because they are based on observations of a few individuals of the 'missing' species in allopatry. After Alatalo *et al.* (1986).

variable than mainland forms (Grant 1979b). These are traits often associated with foraging on pine foliage, and Lack and Southern suggested that the absence of the pine-dwelling Coal Tit on the Canaries promoted this morphological shift. Their observations of the foraging behavior of the Canary Island birds were anecdotal, but Grant (1979b) gathered limited quantitative data that supported the conclusion that Blue Tits spend more time foraging in pine foliage in the Canaries than in North Africa. Partridge and Pring-Mill (1977) made field observations and conducted laboratory trials to test the notion that the long-billed Canary Island Blue Tits should be more similar to mainland Coal Tits (long, thin bill) than to mainland Blue Tits (short, stout bills) in their foraging behavior. They found that birds on the Canaries indeed spent most of their foraging time in the pines, although they tended to concentrate their activities about the trunk and less on the needles. Their methods of prey capture more closely approximated those of mainland (English) Blue Tits than mainland Coal Tits, however. In the laboratory, Canary Island Blue Tits performed in a manner generally intermediate between the English Blue Tits and Coal Tits but did not differ significantly from their conspecifics. Partridge and Pring-Mill suggested that Canary Island *caeruleus* showed little behavioral difference from their mainland counterparts and that the established morphological differences were possibly associated with climatic differences on the islands, an hypothesis also favored by Snow (1954, 1955). Grant (1979b) rejected this idea in its pure form, suggesting instead that the morphological pattern was influenced by a combination of the past history of the island populations, climatic factors, and selection for foraging efficiency in pine foliage. In any case, the *pattern* of morphological divergence of Canary Island *caeruleus* from their mainland conspecifics toward the phenotype of the absent *ater* seems well established.

Conclusions

That bird populations occur at greater densities, occupy different niche positions and greater niche space, and exhibit altered morphology on islands in relation to populations of the same species in similar mainland sites can be accepted as a 'proven fact' (Blondel 1985a), at least in some situations. The expected patterns are also absent in many seemingly appropriate situations, however. Density compensation, niche shifts, and character release are not ubiquitous features of bird communities, but they occur frequently enough to be of considerable interest, especially inasmuch as these patterns may shed light on the possible role of competition or other factors in structuring communities.

Establishing the reality of these patterns is hindered by the same sorts of methodological difficulties noted in previous chapters, plus some additional ones. Most studies suffer from a short-term bias. Population census values from mainland locations, for example, characteristically exhibit substantial seasonal and annual variation, and the discovery that mainland densities of a species differ from those on islands is therefore not very surprising (Williamson 1981). Densities of some species on islands *do* seem to be unusually large and appear to remain consistently so (e.g. Black-throated Sparrows on Baja California islands). There is a tendency among ecologists, however, to emphasize such examples and ignore the many situations that do not fit their expectations. When Alerstam *et al.* (1974) compared bird populations on Gotland with those in southern Sweden, for example, they focused on the one species in coniferous forests most likely to exhibit niche shifts and density compensation and gave scant attention to the other guild members present. Because the Marsh Tit is also absent from deciduous woodlands on Gotland, one might expect niche shifts there as well, but this possibility has not been investigated (Alatalo personal communication).

Island–mainland comparisons have also been hampered by small samples, especially of mainland locations. Typically, several islands are compared to one or two mainland survey plots. This often precludes statistical tests of the population differences, especially if sampling is restricted to a single time; if different factors influence densities or niche characteristics in different areas at different times, limited sampling may produce quite misleading indications of patterns. Wright (1980) suggested that the solution to this problem may be to compare measures among a large series of ecologically similar sites that differ in species number, but any insights derived in this way are gained at the expense of an understanding of specific local patterns and their probable causation.

The emphasis on islands in the determination of these patterns also has benefits and costs. Island populations are usually relatively closed, and they are therefore free to develop responses to local environmental conditions without the erosive influences of dispersal, immigration, migration, or gene flow. As a result, patterns may be more clearly expressed on islands. On the other hand, there are often consistent climatic differences between island and mainland areas that are not considered in comparisons. Also, island populations are often established by small founding groups; combined with the effects of isolation, this may foster genetic divergences of the populations that are not always in directions consistent with ecological expectations.

Even when patterns of ecological shifts are apparent, interpreting them is not always straightforward. When one compares different areas it is unlikely that the composition of a species' guild or community will remain unchanged. It is therefore always possible to attribute the shifts to the absence of some species that is a 'presumed' competitor or to changes in an undefined set of diffuse competitors. Rarely is the designation of the supposed competitors based on anything more substantive than the investigator's intuitions. When these are colored by preconceptions about patterns and their causes, almost any species can become a potential competitor or contribute to a network of diffuse competitors. Congeners are often targeted as the most likely competitors (as in the *Parus* studies), but in their absence other species may be assigned that role. The approach followed by Alatalo *et al.* (1986; Fig. 11.8) reduces these problems somewhat.

Studies of ecological release also rely heavily on the *ceteris paribus* assumption. In comparing island and mainland situations or different mainland settings, one assumes that the sites are equivalent in all important respects except species number or the presence or absence of the presumed competitors. The status of food resource levels, habitat structure or composition, climate or microclimate, predation levels, intraspecific competitive effects, and the like are rarely assessed in any detail. Such measures are necessary, however, if one wishes to verify the *ceteris paribus* assumption or to disentangle the effects of species composition from those of other environmental factors (Schoener 1975, Blondel 1985a). Differences in the densities of a species in the compared areas, for example, may produce density-dependent expansions or contractions in its habitat occupancy or use of microhabitats, quite independently of the effects of other species. It is therefore necessary to consider alternative hypotheses to account for the observed patterns of niche changes, a point I develop further in Volume 2.

12

Convergence of species and communities

When Europeans first explored regions of the New World and Australia, they named many of the new species they encountered according to the European counterparts of these species with which they were familiar. The New World tyrannids were called flycatchers because of their behavioral similarities with the Old World muscicapids, and several were initially assigned to the genus *Muscicapa*. Some of the South American furnariids were likewise considered to be sylviids because of their behavioral and morphological similarities to European warblers. In Australia, one has 'wrens' that belong to the Maluridae rather than the Troglodytidae, 'warblers' that belong to the Acanthizidae rather than the Sylviidae (or Parulidae), 'treecreepers' that are climacterids rather than certhiids, 'chats' that are ephthianurids rather than turdids, and 'magpies' that belong to the Cracticidae, not the Corvidae. Darwin (1897) recognized that many such similarities involved members of taxonomically distinct families, and he drew specific attention to the Chilean diving petrel *Pelecanoides*, which 'offers an example of those extraordinary cases, of a bird evidently belonging to one well-marked family, yet in both its habits and structure allied to a very distinct tribe. The form of its beak and nostrils . . . show that this bird is a petrel: on the other hand, its short wings . . . its form of body . . . its habit of diving . . . make it at first doubtful whether its relationship is not equally close with the auks' (Darwin 1897: 277; Cody and Mooney 1978).

The notion that species of different phylogenies that occupy similar niches in similar environmental settings on different continents may be convergent in features of morphology, behavior, and ecology is thus an old one. In many cases, the extent of convergence between the species is only superficial, but some examples of intercontinental convergences are quite impressive. The African pipit *Macronyx croceus* is remarkably similar to the North American meadowlark *Sturnella neglecta* in plumage pattern and coloration, habitat, nest construction, and foraging behavior (Fig. 12.1).

Fig. 12.1. Patterns of ecological convergence between New World icterids (left) and Africal ploceids and a motacillid (right). (a) The monogamous insectivorous oriole *Icterus galbula* and *Malimbus scutatus*; (b) the polygynous marsh-nesting blackbird *Agelaius phoeniceus* and *Euplectes orix*, both of which nest in grouped territories; (c) the extremely colonial marsh-nesting blackbird

Several species of ploceid finches breeding in marsh and savannah habitats in Africa are closely paralleled in ecology and breeding habits (if not in morphology) by New World icterids (Fig. 12.1; Lack 1968). Both North American *Toxostoma* thrashers and South American *Upucerthia* ovenbirds are terrestrial, open-country birds with decurved bills, and forage extensively in litter and soil (Cody 1973a).

The premise of community convergence

If species converge in morphology, behavior, and ecology in similar environments, then it follows that there might be convergence in guild or community patterns as well. The basic premise of such convergences has been stated clearly by Cody and Diamond (1975: 7):

> If the observed patterns in community structure are products of natural selection, then similar selection by similar environments should produce similar optimal solutions to community structure. In particular, if species are assembled nonrandomly into communities and if the fine structure of such assemblages is determined by the physical and biological environment, then patterns of community structure should be reproducible, independent of the species pool from which the component species (and the biological components of the selective background) are drawn.

More specifically, similarities in the physical environment on different continents may produce vegetation of similar structure and resources of similar types and abundances and place similar stresses and constraints on individuals (Fig. 12.2). The resource opportunities are therefore similar, resulting in convergent adaptations among unrelated species. The similarities in physical environment and resources may also determine the patterns of diversification that are possible among species, producing convergences in guild structuring (Karr 1980) and in community attributes such as diversity or total biomass (Giller 1984, Emlen 1986). These arguments seem compelling enough that, bolstered by a few examples, the idea of community convergence has become widely accepted (Giller 1984, Blondel *et al.* 1984).

Fig. 12.1. (*cont.*)
Agelaius tricolor and *Quelea quelea*; (d) the colonial oropendola *Zarhynchus wagleri* and *Ploceus cucullatus*; (e) the grassland meadowlark *Sturnella neglecta* and the pipit *Macronyx croceus*; (f) the parasitic cowbird *Molothrus ater* and the cuckoo-weaver *Anomalospiza imberis*. From Lack (1968).

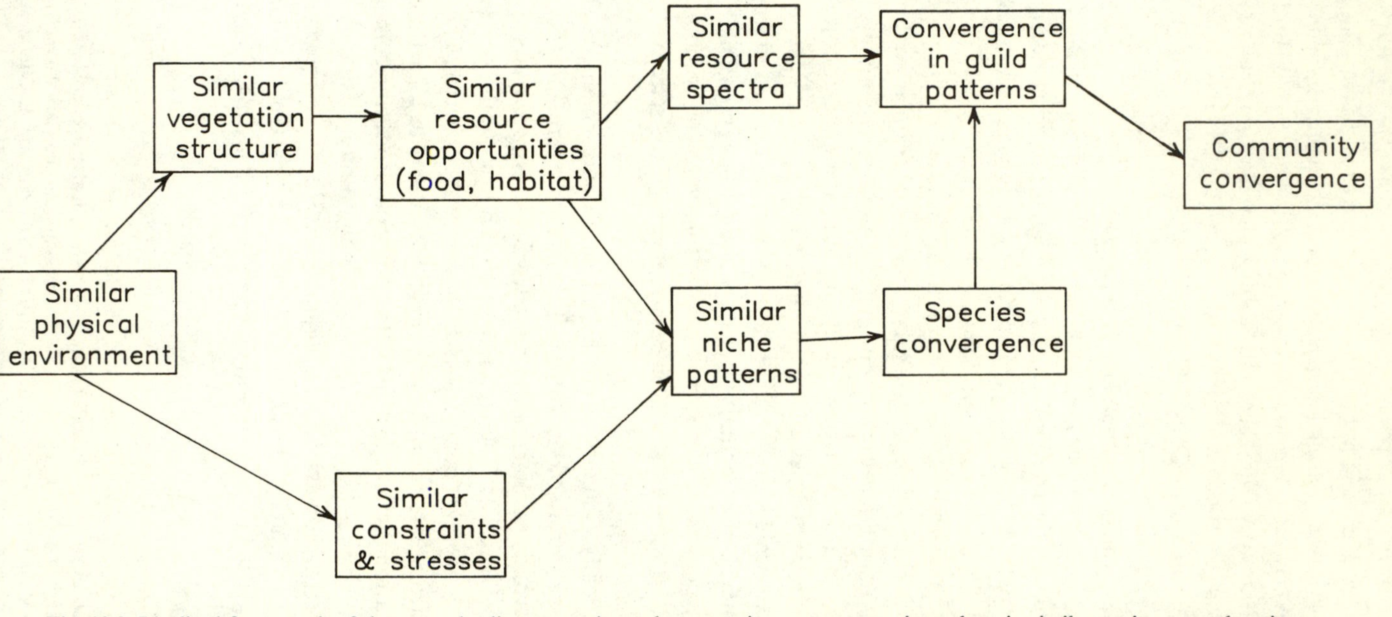

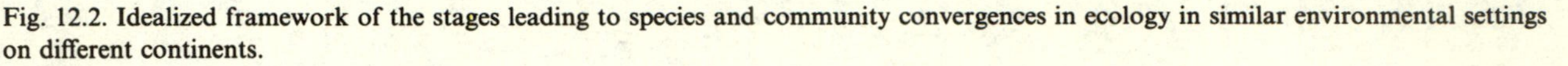

Fig. 12.2. Idealized framework of the stages leading to species and community convergences in ecology in similar environmental settings on different continents.

Evidence of convergence among species

What is the evidence of convergence, either between species occuping similar settings in different continents or between guilds or communities? It turns out that detailed, quantitative evidence is rather scanty and produces an inconsistent picture. The evidence is perhaps most compelling at the species level. In addition to the anecdotal examples given above, Pearson (1982) noted that the Paradise Kingfisher (*Tanysiptera galatea*) of lowland tropical forests in New Guinea forages in the same manner as and is similar in shape and color to the Blue-crowned Motmot (*Momotus momota*) of similar habits in western Amazonia, and the White-tailed Ant-thrush (*Neocrossyphus poensis*) of West Africa is quite similar to the Black-spotted Bare-eye (*Phlegopsis nigromaculata*) of western Amazonia in size, call-notes, and ant-following behavior. On the basis of comparisons of the environment and biota of sites in the Sonoran Desert of Arizona and in the Argentinian Monte, Orians and Solbrig (1977) listed several species pairs that are taxonomically unrelated but at least generally convergent in foraging behavior and diet (Table 12.1). In several of these pairs, however, the convergences in ecology are not accompanied by obvious parallels in morphology. Karr and James (1975) noted several examples of one-to-one species convergences in the avifaunas of tropical lowland forests in Panama and Liberia, defining these on the basis of both ecology and morphology (Table 12.2, Fig. 12.3).

It is apparent from these examples that convergence between species may be evident in some attributes but not in others. Terborgh and Robinson (1986) argued that, because the morphology of species is constrained by their past history and phylogeny, species that are widely divergent taxonomically may not be able to attain close morphological similarity. We should expect the least amount of convergence between species in morphological traits. On the other hand, aspects of behavior are more flexible. Because sites are selected for comparisons so as to match habitats, the structure of the habitats and the spatial distribution of resources should be similar, and aspects of foraging behavior are therefore most likely to evidence convergence. Terborgh and Robinson used the results of several studies of convergence in tropical forest birds to examine this notion and found general agreement with their expectations. Their analysis of behavioral convergence, however, was based on relatively coarse rank-order distributions of general foraging tactics among individuals, so the pattern is more qualitative than quantitative.

There is little detailed information available to test Terborgh and Robin-

Table 12.1. *Ecologically convergent species pairs from different families in Sonoran Desert habitats of Arizona and Monte habitats of Argentina.*

Convergence is judged on the basis of similarity of diet, foraging behavior, and foraging substrates used.

Foraging mode	Monte	Sonoran Desert
Snatching and gleaning insects from foliage of shrubs	Lesser Wagtail-tyrant (*Stigmatura budytoides*) (Tyrannidae)	Black-tailed Gnatcatcher (*Polioptila melanura*) (Sylviidae)
Gleaning and probing for insects on foliage	Tufted Tit-spinetail (*Leptasthenura platensis*) (Furnariidae)	Verdin (*Auriparus flaviceps*) (Paridae)
Digging and probing on ground; gleaning from lower branches	Sandy Gallito (*Teledromus fuscus*) (Rhinocryptidae)	Thrasher (*Toxostoma* spp.) (Mimidae)
Gleaning and probing on and in crevices of bark	Short-billed Canastero (*Asthenes baeri*) (Furnariidae)	Cactus Wren (*Campylorhynchus brunneicapillus*) (Troglodytidae)
Pouncing on large insects and small vertebrates from elevated perches	Spot-winged Falconet (*Spiziapteryx circumcinctus*) (Falconidae)	Loggerhead Shrike (*Lanius ludovicianus*) (Laniidae)
Gleaning seeds and fruit from ground	Elegant Crested Tinamou (*Eudromia elegans*) (Tinamidae)	Gambel's Quail (*Lophortyx gambelii*) (Phasianidae)
General carrion feeding	Crested Caracara (*Polyborus plancus*) (Falconidae)	Raven (*Corvus corax*) (Corvidae)
	Chimango Caracara (*Milvago chimango*) (Falconidae)	White-necked Raven (*Corvus cryptoleucus*) (Corvidae)

Source: From Orians and Solbrig (1977).

Table 12.2. *Pairs of ecologically convergent species in lowland tropical forests of Panama and Liberia.*

All species are insectivorous passerines. Morphological similarities between these species are shown in Fig. 12.3.

	Species		
Pair code	Panama	Liberia	Ecology
A	*Tyrannus melancholicus*	*Meleanornis annamarulae*	High stratum, sallying, forest border
B	*Onychorhychus mexicanus*	*Terpsiphone rufiventer*	Low-to-medium stratum, sallying, forest
C	*Rhynchocyclus olivaceus*	*Fraseria ocreata*	Medium stratum, sallying forest
D	*Myiobius sulphureipygius*	*Trochocerus nitens*	Low-to-medium stratum, sallying, forest
E	*Polioptila plumbea*	*Apalis nigriceps*	High stratum, foliage-gleaning, forest
F	*Henicorhina leucosticta*	*Stiphrornis erythrothorax*	Ground-gleaning, forest
G	*Microcerculus philomela*	*Sheppardia cyornithopsis*	Ground-gleaning, forest
H	*Cyphorhinus phaeocephalus*	*Trichastoma cleaveri*	Ground-gleaning, forest
I	*Grallaria perspicillata*	*Trichastoma puveli*	Ground-gleaning, forest

Source: From Karr and James (1975).

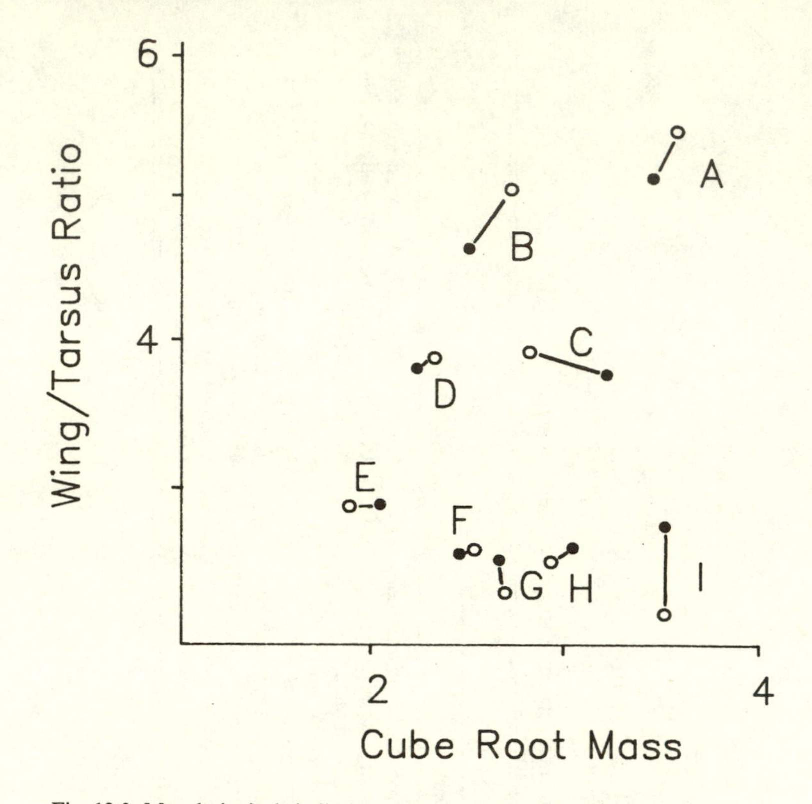

Fig. 12.3. Morphological similarities of ecologically equivalent species in Panama (open circles) and Liberia (closed circles). See Table 12.2 for species names. From Karr and James (1975).

son's prediction, and what evidence is available is inconclusive. In some situations, behavioral convergence is more apparent than morphological convergence (e.g. Orians and Solbrig (1977); Table 12.1). Holmes and Recher (1986a,b) found that birds of northern hardwood forests in North America foraged in generally similar ways to the birds of eucalyptus woodlands in Australia, despite their morphological differences. The species differed in the details of these behaviors, however, presumably because of differences in foliage structure and food resource types available in the two regions. In other instances, morphological convergence is evident (e.g. Fig. 12.3).

Because the degree of morphological similarity among species is constrained by their phylogeny, however, it may be more difficult to detect morphological than behavioral convergence. After all, convergence refers to the tendency for species occupying similar but separate environments to

Table 12.3. *Tests for convergence in body size and the structure of the bill, hind limb, and flight apparatus (indexed by PCA components) for five sets of warbler and sparrow congeners occupying shrub versus coniferous forest peatland habitats in North America and Finland.*

For each genus, the morphology of the shrub-dwelling species was compared with that of the forest-dwelling species; the + or − signs indicate its value relative to that of the forest species. Convergence is indicated by the same sign for a trait for all five genera.

			Shape (PCA)				
			Bill (relative)		Hind limb (relative)	Flight (relative)	
Genus[a]	Species compared	Size (absolute)	Thicker (PC1)	Longer (PC2)	Longer (PC1)	Flatter sternum (PC1)	Longer humerus (PC2)
Dendroica	*petechia* / *coronata*	−	+	+	+	+	+
Geothlypis	*trichas* / *agilis*	−	+	+	+	+	+
Zonotrichia	*georgiana* / *albicollis*	−	−	+	+	+	+
Spizella	*pallida* / *passerina*	−	−	+	+	+	0
Emberiza	*schoeniclus* / *rustica*	−	−	−	+	+	+
Exact probability[b]		$P=0.03$	$P=0.50$	$P=0.19$	$P=0.03$	$P=0.03$	$P=0.19$

Notes:
[a] *Empidonax* was excluded because its body form and behavior are different from those of warblers and sparrows.
[b] From sign test applied to the five within-genus comparisons.
Source: From Niemi (1985).

be more similar to one another than were their ancestors. Although this is most obvious when the result is a close similarity of the species, it is evidenced as well if the species have diverged from their ancestors (or congeners) in similar directions, even if this does not produce close matching.

Niemi (1985) followed this theme to examine the morphological convergence among species occupying peatland habitats in Finland and North America. Using quantitative habitat measurements, he first determined that shrub and forest peatlands in the two regions were occupied by the birds in similar ways – this established that the habitats were indeed similar enough for convergence to occur. Niemi then examined the patterns of morphological variation within sets of congeneric species that were distributed over the habitat gradient in each region, to determine whether or not species occupying the shrub habitat differed from their forest-dwelling congeners in similar fashions for the different genera. They did. In all five genera, the species in shrub vegetation were consistently smaller, had relatively larger pelves and legs, and had wider and shallower sterna than did their forest congeners (Table 12.3). The most abundant species in shrub peatlands (*Cistothorus platensis* in Minnesota, *Acrocephalus schoenobaenus* in Finland) were among the most extreme species in these characteristics. The species therefore exhibited convergence in morphological features related to their movement through the vegetation (hind-limb structure, size, sternum shape) but not in bill features, which are dominated by phylogeny (finches vs warblers). In his analysis, Niemi assumed that he could assess the degree of evolutionary convergence of species in different groups by focusing on patterns of radiation within genera. It would be worthwhile to extend this approach by charting the relationships of species within compared lineages through cladistic analysis (Eldredge and Cracraft 1980, Brooks 1985).

Although these examples provide some compelling evidence of convergences between species occupying similar environments in widely separated areas, close convergence is apparently atypical. Pearson (1982) noted that examples of species convergences were rare in the six tropical forest systems he compared, and the listings of Tables 12.1 and 12.2 represent a small fraction of the avifaunas of those locations. Cody (1973a, 1974a) diagrammed several examples of broad ecological and behavioral convergences between species occupying mediterranean environments in California and Chile, but he was unable to document many species–level convergences in morphology in a broader comparison of mediterranean systems (Cody and Mooney 1978). Cody's (1973b) description of apparent convergences in the use of offshore foraging zones by alcid species of the North Pacific and those of the North Atlantic is flawed by the problems noted in Chapter 10.

Guild and community convergence

Communities in Mediterranean climates

Extensive studies of community-level convergence have been carried out in several mediterranean-climate systems. Cody (1973a, 1974a, 1975, 1983b, Cody *et al.* 1977, Cody and Mooney 1978) conducted surveys on scrub-woodland habitat gradients in California, Chile, Sardinia, and South Africa. The species–area curves for these areas are generally similar, although the spatial patterns of diversity vary considerably. The communities exhibit similar relationships between alpha diversity and changes in vegetational complexity on a local gradient, but point diversities (individual spatial overlap among species), beta diversity, and gamma diversity follow different patterns in the different continents (see Fig. 5.12). Cody found little evidence of species-by-species matching of positions in a multivariate morphological space, but the communities did have similar overall mean values and ranges. To Cody, this indicated a general convergence in the array of permissible phenotypes. Multivariate analyses of foraging heights and behaviors produced similar patterns of community-level convergence. Although Cody suggested that there was general similarity among the four scrub communities in patterns of resource partitioning and in the distribution of species among diet-foraging behavior guilds, an inspection of his data shows that these similarities are not very close.

These results illustrate the difficulty of documenting ecological convergence, especially at the community level. It is unrealistic to expect guilds or communities to be structured in an identical way in different areas, but what degree of similarity then indicates convergence? What level of similarity might be expected to occur by chance alone, simply because there are general constraints on the design of birds and the assembly of communities? One way to approach these questions is by comparing the attributes of the supposedly convergent communities with those of communities in different habitat types in the same or different areas, in a manner analogous to outgroup comparisons in cladistics (Eldredge and Cracraft 1980). To determine whether the patterns of apparent convergence that he found in the mediterranean communities might also typify a nonmediterranean community, Cody conducted the multivariate analyses for oak woodlands in California and England. Both exhibited different distributions of species in the multivariate morphological or ecological spaces. The mediterranean

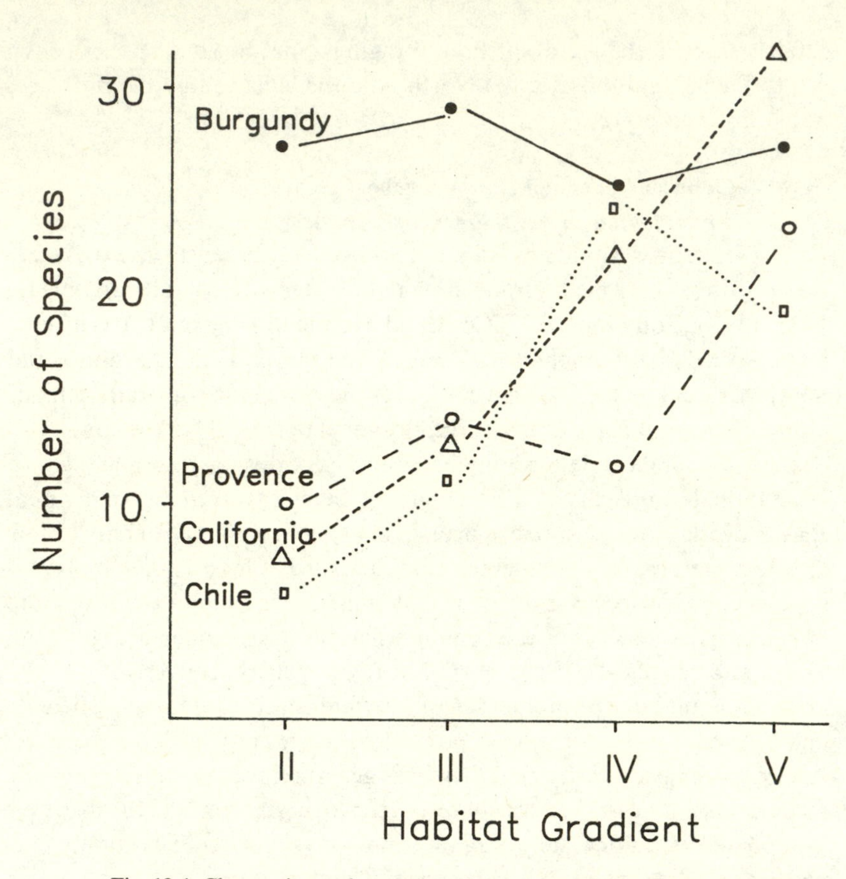

Fig. 12.4. Changes in species number along a standardized habitat gradient in scrub habitats in three mediterranean-climate regions, (Chile, California, and Provence, France) and in a 'control' temperate-climate location (Burgundy, France). From Blondel *et al.* (1984).

communities therefore differ consistently from oak woodland communities, providing additional, indirect evidence of convergence among the mediterranean sites.

The results of a somewhat similar analysis of ecomorphological convergences between mediterranean bird communities in California, Chile, and France (Blondel *et al.* 1984) only partially agree with Cody's conclusions. Blondel *et al.* used data from Cody's (1975) surveys of habitat gradients in California and Chile and from Blondel's (1979, 1981) censuses in southern France for the mediterranean comparisons; surveys from a similar scrub habitat gradient in temperate France (Burgundy; Ferry and Frochot 1970) were used as a nonmediterranean 'control'. The comparisons are thus compromised from the outset because data gathered by different observers

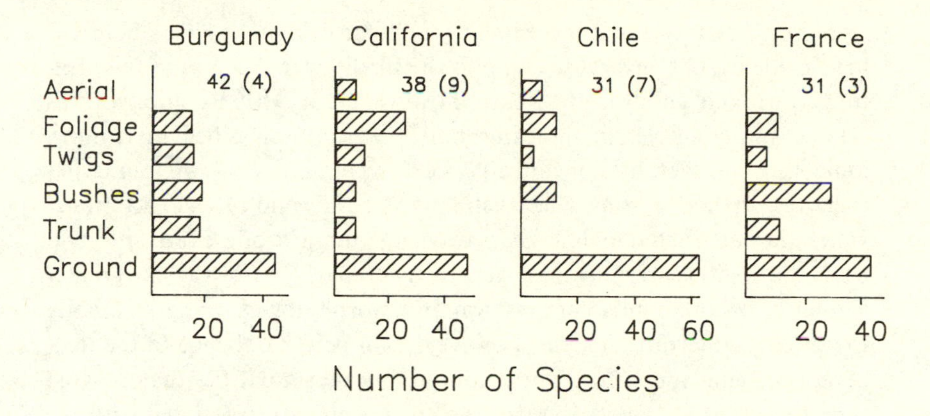

Fig. 12.5. Distributions of species among foraging-substrate guilds in mediterranean-climate sites in California, Chile, and France and in a temperate-climate location in France (Burgundy). The total number of species present on the habitat sequence in each region is given, with the number of species following mixed foraging-substrate patterns (not included in the histograms) in parentheses. After Blondel *et al.* (1984).

using somewhat different methods in areas of different sizes are used. In addition, Cody's species lists included species that were not actually recorded during his surveys but that he expected to find. Because the habitat gradients were originally defined in rather different ways, Blondel *et al.* also had some difficulty in matching habitat samples from the four regions.

The methodological problems preclude any interpretations of small-scale variations among the four areas, but broad-scale similarities and differences are probably valid. Thus, although the details of increases in species number with increasing vegetational complexity on the habitat gradient differed among the three mediterranean locations, the generally similar patterns of increase contrasted sharply with the roughly constant number of species found across the habitat gradient in Burgundy (Fig. 12.4). There were considerable differences among the mediterranean sites, however, in the allocation of species to foraging guilds (Fig. 12.5) or dietary groups. In particular, the feeding habits of ground-foragers in California were quite different from those in France, due largely to taxonomic differences in guild composition (insectivorous turdids in France, granivorous finches in California). In Chile, on the other hand, levels of insectivory among the ground-foraging species were quite similar to those in France (especially Burgundy), although these species belonged to several taxonomic groups (Picidae, Furnariidae, Rhinocryptidae, Tyrannidae, Mimidae, Turdidae, and Icteridae).

The distribution of species in multivariate morphological space showed recognizable clusters that correspond roughly to foraging guilds. There was little evidence of convergence among the mediterranean sites in the locations of similar guilds in this space, however. In a DFA using regions as classes, many species in the same guilds were misclassified to regions, indicating considerable nonconvergence. Overall, Blondel *et al.* found that each region had a somewhat distinctive ecomorphological patterning, which they attributed to differences in the taxonomies of the birds present. Still, the regions were at least generally similar. There was no greater similarity in community composition and morphological patterns among the three mediterranean areas, however, than between them and the temperate Burgundy 'control'. The community-level similarities that do exist (and they are only very general) are thus not directly associated with the mediterranean climate and its effects.

Shrub deserts in North America and Australia

As part of our studies of bird communities in shrubsteppe, I surveyed populations at several shrub-desert sites in western New South Wales, Australia during 1984–85 (Wiens unpublished). Because the observations were gathered by the same individual using the same procedures, the results can be compared directly with those of our North American studies, which I have described previously. These comparisons provide a basis for determining the extent of convergence in attributes of species and communities occupying similar semiarid shrublands in the two continents.

To establish the basis for such comparisons, it is first necessary to determine the degree of similarity in the habitats in the two regions. Using quantitative information on vegetation structure and coverage, I constructed a similarity matrix comparing the four Australian sites with each of 68 North American samples. From this matrix I selected for detailed comparison seven North American sites that showed high similarities to the Australian plots. By placing the Australian sites in the multivariate PCA space determined by the vegetation attributes of the North American sites, I confirmed that the sites were indeed remarkably similar, although the vegetation in the Australian shrublands was generally somewhat shorter than that in the North American shrubsteppe.

The breeding bird communities in the two regions contained similar numbers of species (Table 12.4). 'Peripheral' species that are characteristic of other habitats but occupy the shrublands (such as Galahs and Little Corellas in Australia and Brown-headed Cowbirds (*Molothrus ater*) and Loggerhead Shrikes in North America) contributed equally to the avifaunas of the two regions. Total breeding densities, however, were nearly

Table 12.4. *Mean values of various community attributes and ecological/life-history traits of the bird species breeding in shrub desert habitats in Australia (*N=*4 sites) and North America (*N=*7).*

The values were obtained by multiplying the index value for each species by its relative density in the community; the ecological/life-history trait means are therefore weighted on the basis of relative abundances of the species.

Trait	Australia	North America	P^a
Community			
Species number	5.5	6.3	NS
Total density (indiv km^{-2})	157.5	301.0	*
Peripheral species	1.3	1.1	NS
Species			
Duration of nesting (days)	27.6	22.8	**
Breeding period (months)	5.9	3.1	***
Clutch size	3.4	3.3	NS
Broods per year	2.1	1.1	**
Mating system index[b]	1.1	1.0	NS
Communal breeding index[c]	1.5	1.0	***
Nest type index[d]	2.8	2.0	***
Nest height index[e]	2.0	1.9	NS
Breeding aggregation index[f]	2.6	1.0	***
Territory size index[g]	3.4	3.9	*
Diet index[h]	2.5	2.9	NS
Flock feeding index[i]	2.2	1.1	**
Migration index[j]	1.2	3.0	***

Notes:

[a] NS = not significant, * = $P<0.05$, ** = $P<0.01$, *** = $P<0.001$

[b] 1 = monogamous, 2 = polygynous

[c] 1 = no, 2 = yes

[d] 1 = platform, 2 = cup, 3 = domed cup, 4 = cavity

[e] 1 = ground, 2 = shrub, 3 = tree

[f] 1 = pairs, 2 = group, 3 = small colonies (2–10 pairs), 4 = large colonies (> 10 pairs)

[g] 1 = none, 2 = nest-centered, 3 = small, 4 = medium, 5 = large

[h] 1 = granivorous, 2 = omnivorous, 3 = insectivorous (during breeding season)

[i] 1 = never, 2 = small flocks, 3 = medium flocks, 4 = large flocks (during breeding season)

[j] 1 = sedentary, 2 = nomadic, 3 = migratory

Source: From Wiens (unpublished).

twice as great in North America as in Australia (Table 12.4). This depression in densities of individuals relative to North America appears to characterize other terrestrial vertebrates in Australian deserts as well – densities of small mammals are very low in most arid and semiarid habitats in Australia (Morton 1982, A. Cockburn, personal communication), and densities of lizards are also relatively low in many areas (D. Wiens, unpublished observation, R. Shine, personal communication), despite the greater number of species in the Australian deserts (Pianka 1986).

There is thus some degree of matching in community-level attributes between the two systems. Can we also detect species pairs that are unusually similar to each other? To make this comparison, I tallied information on several ecological and life-history traits of each species and then subjected this information to a cluster analysis based on the similarity matrix of the species. If convergence in suites of attributes occurs, some North America species should be clustered with some Australian species in the analysis. This did not happen. The clustering procedure instead defined sets of North American species and separate sets of Australian species, these sets combining only at relatively low similarity levels (Fig. 12.6). Only one matching of a North American with an Australian species emerged (the Western Meadowlark with the Brown Songlark (*Cinclorhamphus curalis*)); this pairing appears to represent a true ecological convergence.

If no close matching of species occurs between these avifaunas, might there nonetheless be convergence in the frequency of occurrence of the life-history traits in the entire communities? I examined this possibility in two ways. First, I calculated the mean value of each trait for each site and then tested the association of these values with the variation in habitat structure depicted by the PCA components. None of the life-history variables correlated with variation in PC I, which indexed a gradient in coverage of shrubs and standing dead vegetation. The second PCA component represented a gradient from areas with considerable grass cover to areas with substantial vertical stratification of the vegetation. Several life-history features varied significantly on this gradient: areas with greater grass cover supported more species (including more peripheral species) that had larger clutch sizes and were more frequently polygynous than did the vertically heterogeneous sites. Because the Australian and North American sites were distributed in the same way on this gradient, these correlations indicate similar responses to the environmental variation on both continents. These aspects of the communities, therefore, appear to be convergent.

Most of the ecological and life-history traits, however, differed significantly in their occurrence in the Australian and North American assemblages (Table 12.4). The Australian species, on average, have a much longer breeding period (Wyndham 1986) and more broods per year than the North American species, even though clutch sizes are quite similar. More of the Australian species breed in communal groups and/or form breeding aggregations (small colonies). The Australian species also feed in flocks during

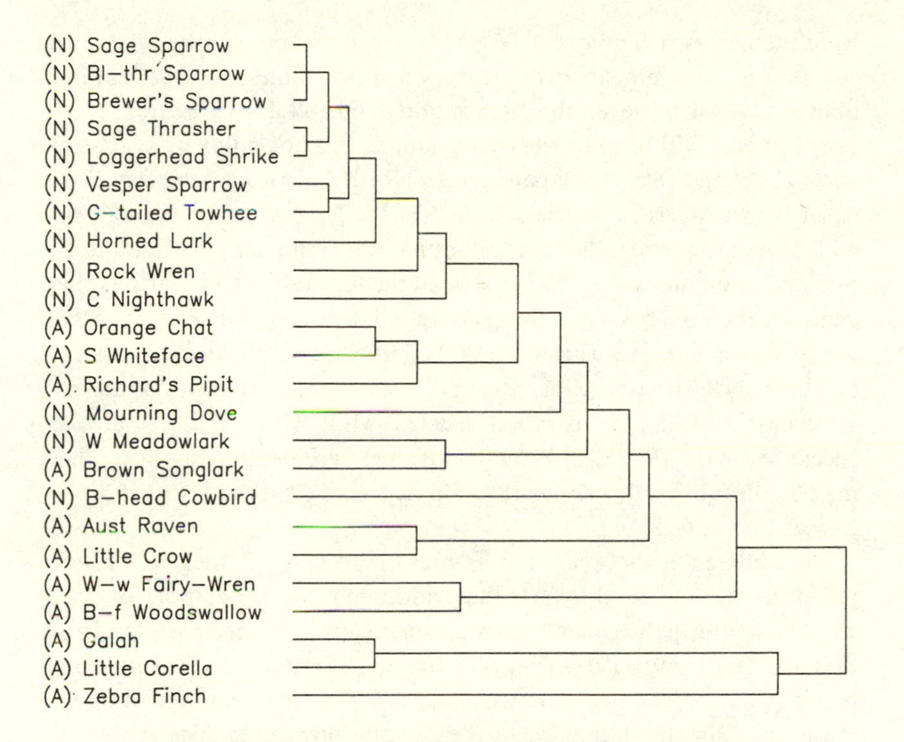

Fig. 12.6. Dendrogram of similarities among Australian (A) and North American (N) shrub desert bird species derived from a cluster analysis based on several ecological and life-history traits of the species. Note that, with one exception (Western Meadowlark/Brown Songlark), the primary clusterings group together sets of North American or sets of Australian species. From Wiens (unpublished).

the breeding period more frequently than do the North American species, although there is no overall difference in diets. Most impressive, however, is the difference in their residency status: all of the North American species are typical migrants, whereas the Australian species are either sedentary or nomadic – none is a true migrant.

I also calculated the degree of morphological similarity of the Australian and North American species by deriving Euclidean distances for species pairs based on measures of eight features. If convergence is common, we should expect the nearest neighbor of a species in this morphological space to be a member of the other continental avifauna. In fact, several such pairings occurred. Brewer's Sparrows and Southern Whitefaces (*Aphelocephala leucopsis*) and Horned Larks and Richard's Pipits (*Anthus novaeseelandiae*) were each other's nearest neighbors, and Brown Songlarks were the nearest neighbors of Western Meadowlarks (strengthening the suggestion of convergence in life-history features of these two species noted above). Overall, however, the distribution of morphological nearest-neighbor distances within versus between faunas did not differ from that expected by chance: 8 of the 20 pairings for North American species were with other North American species, while 5 of the 12 Australian pairings were with other members of the same avifauna. A discriminant analysis yielded one significant dimension that separated the Australian and North American species on the basis of differences in bill depth and tail length (both of which were smaller in Australia) (Fig. 12.7). Of the North American species, however, 20% (Brewer's Sparrow and Rock Wren (*Salpinctes obsoletus*)) were misclassified as Australian species, while 33% of the Australian species (Richard's Pipit and Brown Songlark) were also misclassified. These misclassifications indicate possible convergences or at least close morphological similarities.

The differences between the faunas, especially in their life-history attributes, point up the fact that, although the vegetation of these shrublands on the two continents is quite similar in structure (and to a degree in taxonomy – *Atriplex* species are often the dominant shrub in both regions), the environments are vastly different in other ways that influence the birds. Climatic differences are especially noticeable. The Australian shrublands are warmer than the North American counterparts, especially during winter. Average annual precipitation is nearly the same in the North American and Australian locations, but the predictability of rainfall is much lower in Australia. Droughts are not only more frequent there but they are more widespread and last longer. The absence of mountains in the landscape of the Australian arid zone reinforces the climatic stress accompanying droughts, as there is no spring runoff to provide water and no nearby refuge of more mesic habitats like that characteristic of western North America. As a consequence, the Australian birds exhibit a variety of adaptations that are not seen in most North American species. As Ricklefs (1987) and I (Wiens 1986) have noted, it is necessary to consider factors at a regional scale in order to interpret local-scale patterns.

Other studies

Other examinations of convergence at the guild or community level have also produced mixed results. In comparisons of six lowland tropical forest communities in different regions, Pearson (1982) found no indications of convergences in species number or in guild composition and size. Karr (1980) likewise documented substantial differences in species

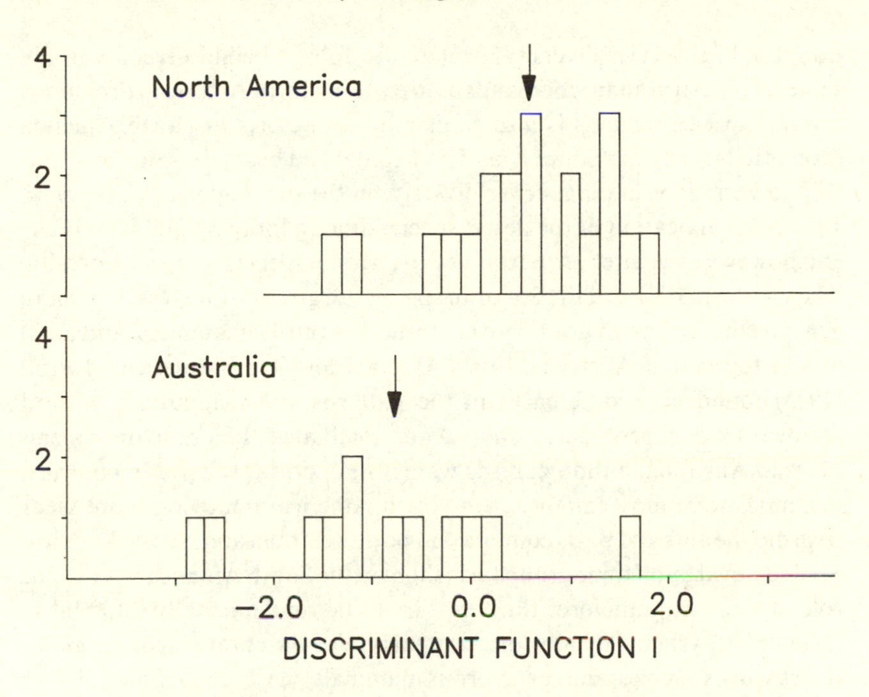

Fig. 12.7. Distribution of North American and Australian shrub desert bird species on the first axis of a discriminant function based on several morphological size variables. Vertical arrows indicate mean positions for the North American and Australian faunas. From Wiens (unpublished).

richness and guild composition among tropical forest understory avifaunas of Central America, Africa, and Malaysia (see Chapter 6). Australian eucalyptus woodlands contain nine recognizable guilds in comparison with the four found in North American hardwoods forests (Holmes and Recher 1986a). In these avifaunas, the guilds are defined primarily by differences in foraging height in both regions, but use of different foraging zones in trees is more important in grouping the North American species whereas details of foraging methods and food substrates used are more important in Australia. These differences may reflect the use by several North American species of a resource category (bark-burrowing insects) that is not present in the Australian woodlands.

Cody (1966, 1968) compared breeding bird communities in grasslands in North America and Chile and suggested that the grassland species converged in foraging behaviors, although the matches were not precise. The number of species present was associated with vegetation height in the same way in the two areas, however. This parallels Recher's (1969) demonstration that bird species diversity is related to foliage height diversity in the same way in Australian woodlands as in eastern North American deciduous forests. Species richness is also similar in hot deserts in North America (Sonoran Desert) and South America (Monte) and bears the same relationship to variations in plant cover diversity in the two regions (Mares *et al.* 1977). The allocation of the desert species among foraging guilds is different, however: a greater proportion of the Monte species forage arboreally, whereas the majority of the Sonoran species are ground-foragers. In a more general comparison of granivore communities (birds, mammals, and ants) in desert systems in Australia, North America, and South America, Morton (1985) found large differences in the patterns and magnitudes of seed removal from experimental trays. Ants dominated the granivore assemblage in Australia, although birds were also important there. In contrast, mammals were most important in North America, removing more seeds than did the ants and birds combined in both Australia and North America. Seed removal by all three groups was slight in the South America desert. The role of birds as granivores thus appears to be fundamentally different in these desert systems. Finally, in a comparison of vertebrate predator guilds (hawks, owls, snakes, and carnivorous mammals) in Chile and Spain, Jaksić and Delibes (1987) found the Spanish assemblages to contain roughly twice as many species as those in Chile. Some taxonomically unrelated species appeared to be close ecological counterparts of one another, whereas other closely related species differed markedly in ecology. Some ecological roles were represented in only one of the areas, presumably because of differences in the resources available to the predators. The studies of Morton and of Jaksić and Delibes illustrate the importance of adopting a taxonomically unrestricted view of guild membership.

Part of the difficulty in assessing whether or not convergence has occurred involves determining how the species or assemblages being compared have diverged from some ancestral condition. Schluter (1986) has approached this problem in an indirect way. If one compares the patterns of variation in attributes of species or communities over a broad habitat gradient in one region with those in communities from similar gradients in widely separated regions, convergence should be indicated by a strong effect of habitat and a minor effect of region on the patterns. Species occupying grassland habitats in different regions, for example, should differ in the same ways from other species occurring in forest habitats in the same regions, even though the grassland species in the different regions may not be particularly similar to one another. Schluter used ANOVA procedures to separate the effects of habitat on patterns of variations in wintering assemblages of granivorous finches occurring over habitat gradients from tropical lowland wet forest to cold temperate deserts in several regions. He found common responses to the habitat gradient (convergence) in several features. Species number varied in the same way on the gradient in the different regions, and mean body size also exhibited habitat-dependent variation (large in forests, cold temperate desert, and mediterranean scrub, small in tropical savanna woodlands, tropical thorn steppe, and warm temperate grasslands). There was some evidence of convergence in measures of bill shape, although there was a strong regional component to shape as well (a reflection of the biogeographic distributions of finch familes that differ in morphology).

Schluter's analysis was based on published and unpublished species lists from different areas that were obtained by different observers using different techniques in areas of different sizes. The heterogeneity of the data sets could lead to errors in the analysis, although one could justly claim that because these errors are not habitat dependent they only contribute to the error term of the ANOVA and the procedure is thus a conservative one. More important, perhaps, are the complications produced by variations in species lists within a general habitat type in a given region. Schluter did not use any North American cold temperate desert sites in his analysis (one from Patagonia contained three species, another from Kazakhstan two species). Depending on where one surveys wintering bird communities in North American cold temperate deserts, one may record from 0 to 6 finch species (unpublished observations). Ideally, surveys should be replicated *within* habitat types in a given region in such an analysis. The procedure would also seem to work best when the range of habitats considered is quite large, as in Schluter's analysis. Habitat effects on community attributes or species morphologies may be apparent when one compares faunas over gradients from tropical forests to deserts, but they might be more easily swamped by other sources of variation in a comparison of gradients, say, within forest types. This only means that the patterns of convergence that are demonstrated using this procedure are as general or coarse as the habitat gradients that define the analysis.

Conclusions

The evidence that convergence between pairs of ecologically similar species and among guilds and communities in similar environmental settings is a common and widespread phenomenon is equivocal. There are some impressive instances of close similarities in attributes of species and of communities, some suggestions of more general similarities, and clear indications of nonconvergences among many species and communities. Convergences seem more apparent in attributes of species that are strongly affected by the physical environment than in those influenced by more complex biotic interactions (Orians and Solbrig 1977) and in low-diversity communities than in species-rich assemblages (Cody *et al.* 1977).

There are many reasons to expect close convergence to be relatively infrequent. Convergence requires that the selective regimes in the different areas be similar, that the populations have the genetic potential to respond in similar ways, that there be sufficient time for the responses to evolve, and that confusing historical and chance events be absent (Cody and Mooney 1978, Orians and Solbrig 1977, Orians and Paine 1983, Ricklefs 1987). In other words, the optimal phenotypes and community structuring envisioned by Cody and Diamond (1975) must be attainable, and the community must be saturated.

Violations of any of these requirements may produce nonconvergent patterns. Thus, although close similarity of environments in different regions is necessary for close convergence, the degree of environmental similarity among compared areas is often not quantified and may be less than imagined. Blondel *et al.* (1984) used arbitrary rather than quantitative criteria to position habitat samples along environmental gradients in the mediterranean areas they compared, and, although they were conscientious in determining sample relationships, inaccuracies may have contributed to some of the nonconvergence in patterns that they found. Cody (1983b) also noted the importance of differences in factors such as habitat area, resource spectra, and past history in contributing to nonconvergences between South African mediterranean avifaunas and those elsewhere. Milewski (1982) has shown that scrub vegetation in similar mediterranean climates in southwestern Australia and southern Africa is remarkably similar in overall structure, but the underlying soils in Australia are quite nutrient-poor. As a consequence, the plants there produce tiny seeds that are consumed chiefly by ants, whereas the richer soils of southern Africa support plants that produce fleshy fruits that are often taken by birds. There is thus a fundamental dissimilarity in resource spectra that is not apparent from climate and vegetation structure alone. Morton (1985) documented other differences between the superficially similar hot desert environments of Australia, North America, and South America, and Shmida and Whittaker (1979) drew attention to a wide variety of contrasts in desert areas of the world and the imperfect nature of convergence in plant growth-forms and in community patterns in these environments.

Taxonomic differences between the avifaunas of different regions may impose fundamental constraints on morphology and behavior that also preclude close convergence. A morphological comparison of nectarivore guilds on different continents that includes tarsal measurements, for example, is bound to demonstrate nonconvergence, simply because the extremely short tarsi of hummingbirds are not matched by other nectarivore families and the evolution of longer tarsi in hummingbirds is constrained by features of their overall design (Karr and James 1975). The assumption that the different areas have experienced similar histories and that sufficient time has been available for the evolution of convergence may also be invalid in many situations. Differences in Pleistocene history, for example, have been used to explain some of the nonconvergences of community patterns among tropical locations (Pearson 1982, Karr 1980) or mediterranean-climate situations (Cody 1983b). The prospects for precise species or community convergence are also diminished if local communities are not in equilibrium and are not saturated with species (Wiens 1974a) or if selection does not produce optimal phenotypes (Wiens 1984b).

Analyses of convergence are also complicated by the difficulty of assessing 'similarity', either of environmental settings or of species or community attributes. How similar is 'similar', and what level of similarity constitutes evidence for convergence (Mooney *et al.* 1977, Terborgh and Robinson 1986)? Ecologists have tended to emphasize convergences to the exclusion of considerations of nonconvergent patterns (Karr 1980), and the convergences have often been documented anecdotally and qualitatively. Attempts to subject presumed patterns of convergence to quantitative analysis and to assess similarity levels with reference to some suitable 'control' or 'outgroup' samples from a different climatic regime (e.g. Blondel *et al.* 1984) or to determine parallels in the direction of change from ancestral states (e.g. Niemi 1985, Schluter 1986) are clearly steps in the right direction. Equal quantitative attention must be given, however, to documenting the extent of similarity in environments and habitats and to determining the precise relationships between control and sample areas.

There are some compelling similarities between a few selected pairs of species on different continents and some intriguing suggestions of convergences in some community patterns. It is premature, however, to conclude that 'there is a lot of evidence to suggest that convergence of bird assemblages from different continents does occur' (Giller 1984: 129).

13

Bioenergetic approaches to communities

Community patterns are usually expressed in terms of numbers of species, densities, or derivatives of these values such as diversity. These are not the only criteria for documenting patterns, however. Odum (1953), Brown (1981), and others have advocated basing community analyses on energetics. Biomass and energy flows then become important criteria for defining community patterns. A focus on the energy dynamics of communities may be especially appropriate for several reasons. First, the availability of energy may limit individuals and populations, and, if community patterns are expressed in terms of energy, they may be related more directly to a presumed limiting factor. Second, the demands placed on resources by community members may be estimated quantitatively, which allows an assessment of the potential impact of populations on resources and of their role in the trophic dynamics of ecosystems. Finally, using biomass or energy flow measures permits the calculation of additional community parameters, such as total energy flow, allocation of this flow to respiration or production, or standing crop biomass, and it provides an alternative way of expressing community patterns such as guild structure or diversity.

Patterns of biomass or energy distribution are functions of the sizes of the organisms in a community. Many features of life history, behavior, physiology, or ecology vary as a function of body mass (Peters 1983, Calder 1984, Schmidt-Nielsen 1984). Larger animals within a general trophic category generally occupy larger territories or home ranges, for example, and they are usually less abundant than smaller animals, although this relationship does not hold particularly well for birds (Peters and Wassenberg 1983). Body mass may be constrained within groups using certain categories of resources, and this may affect community patterns in biomass and energy patterns. Nectarivorous birds, for example, may be limited to a minimum mass of *ca* 2 g because any smaller individuals would be subjected to intense competition with arthropod nectarivores and would not be able to acquire and store sufficient energy reserves to support the demands of fasting, temperature regulation, and reproduction. An upper limit of *ca* 80 g may be set by limitations on the amount of nectar that plants can produce to foster reliable long-distance pollination (Brown *et al.* 1978). Patterns of nectarivore biomass distributions and energetics must fall within these limits.

Perhaps the most important ecological correlate of body mass is its close association with metabolism. Metabolic rates of various sorts – basal, standard, existence – increase as a fractional power of body mass. Thus, although larger animals require more energy to support their overall metabolism, their demand per gram mass is less than that of small animals. Physiologists have now derived mass–metabolism relationships for a wide assortment of birds and have expressed these in several general equations (Aschoff and Pohl 1970, Calder 1974, Kendeigh 1970, Kendeigh *et al.* 1977, Walsberg 1983). The metabolism–mass function for nonpasserines is generally lower than that for passerines, and these functions are also different at different temperatures or photoperiod lengths (Fig. 13.1; Kendeigh *et al.* 1977). Among nonpasserines, seabirds generally have slightly greater basal metabolic rates than do other species of equivalent weight (Ellis 1984). Among seabirds, however, the slope of the basal metabolism–body mass relationship is significantly lower among procellariids than in other seabird orders or birds in general, whereas the regression slope for charadriids is greater than that of other birds (Rahn and Whittow 1984). This is apparently because many small charadriids are tropical whereas the larger species have high-latitude distributions; tropical species generally have lower metabolic rates for a given body mass (Rahn and Whittow 1984, Ellis 1984). Among passerines, the metabolic rates of nectarivorous Australian honeyeaters is about two-thirds of that expected on the basis of general mass–metabolism relationships and more closely approximates the nonpasserine relationship (MacMillen 1985).

There are thus important taxonomic, distributional, and environmental components to the mass–metabolism relationship. Faaborg (1977) employed some of these factors to explain (1) the greater abundance of nonpasserines in tropical terrestrial communities, (2) the tendency for nonpasserine species in tropical forests to be larger than passerines of similar feeding status, and (3) the greater number of nonpasserines on tropical islands (especially small ones) in relation to the mainland. Because passerines have higher metabolic rates and higher activity levels, Faaborg argued, they may have a competitive edge over nonpasserines in feeding on small, abundant resources. The metabolically more conservative non-

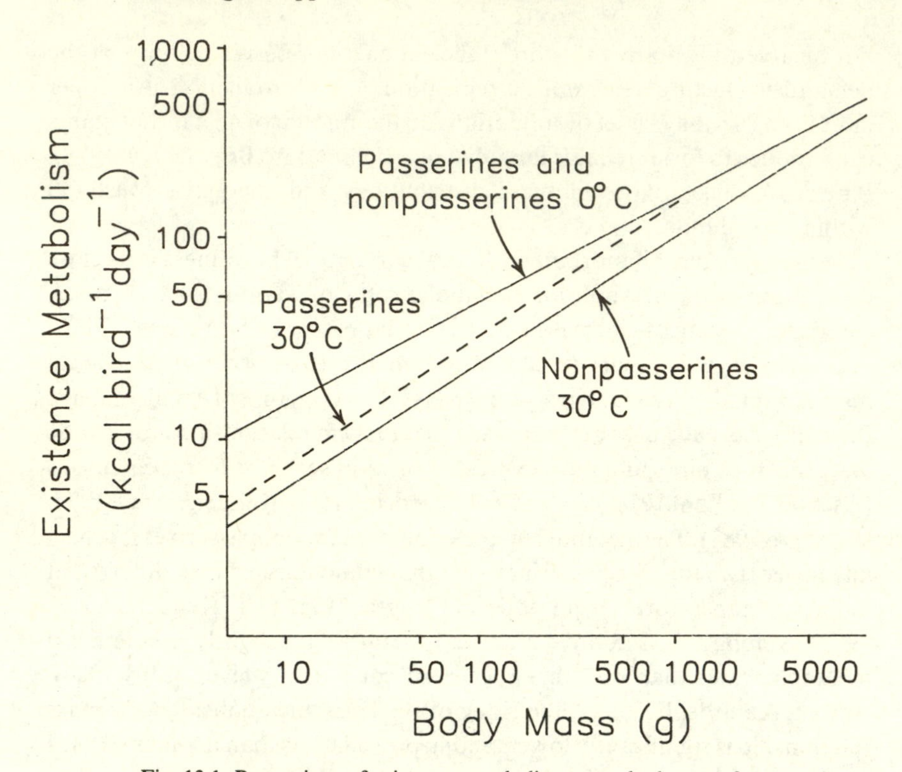

Fig. 13.1. Regressions of existence metabolic rate on body mass for passerines and nonpasserines. From Kendeigh *et al.* (1977).

passerines, however, may be able to support larger populations on a given amount of rarer, larger resources and expend less energy searching for them. These types of resources are more diverse and plentiful in the tropics than in temperate locations, contributing to patterns (1) and (2). Faaborg also suggested that the conservative metabolism of nonpasserines enables them to attain higher population densities and thereby persist through periods of environmental stress. If tropical species are subjected to increased stress on islands, this might contribute to pattern (3). Faaborg's documentations of the patterns are not entirely convincing, and his arguments are general and somewhat speculative. The application of a metabolic perspective to the interpretation of biogeographic patterns in communities is innovative, however, and it merits closer study.

The use of biomass measures

Of the several levels at which bioenergetic approaches may be applied to community analysis, the expression of patterns in terms of biomass is the simplest. The standing crop biomass of a species' population is simply its density multiplied by the average body mass of individuals, and the total community biomass is thus the sum of the values of the individual species. Standing crop biomasses of species' populations and of entire communities have been calculated in a number of community studies (e.g. Raitt and Pimm 1976, Wiens 1973b, 1974b, Väisänen and Järvinen 1977, Głowacinski 1979, Folse 1982). Grant (1986b) used measures of available seed biomass to explain variations in the abundance and biomass of finches on the Galápagos Islands. Folse (1982) found that sites in the Serengeti with low vegetation biomass and low vegetation height (but well-developed vertical vegetation structure) supported the greatest avian biomass. Among ponderosa pine stands in Arizona subjected to various management practices, avian biomass was generally greatest in strip-cut or silviculturally-cut stands that in uncut controls, lower in severely thinned stands, and quite low in clear-cut plots (Szaro and Balda 1979). In a successional sequence of managed pine forests in southern Poland, standing crop biomass increased dramatically over the decade following clear-cutting, was somewhat reduced in a 35-yr-old stand, and then increased more than twofold in an 80-yr stand (Głowacinski 1979).

The patterns that emerge when communities are considered in terms of biomass may differ from those determined by density calculations. Because a few individuals of large species may contribute considerably more standing crop biomass than many individuals of small species, the patterns of species dominance based on biomass may be quite different from those based on density. In a South Dakota mixed-grass prairie, for example, Long-billed Curlews (*Numenius americanus*) occurred at low densities (15.8 individuals km^{-2}) and accounted for only 5.1% of the total avian density. They are large birds (592 g), however, and they contributed 40.9% of the total community biomass (Wiens 1973b). Not surprisingly, species-abundance distributions calculated from biomass values differed considerably from the more conventional patterns derived from densities.

Biomass measures may also produce patterns that are not at all apparent when density values are used. Over a series of grassland locations in North America, for example, I found that vegetational heterogeneity increased with decreasing grass coverage and increasing coverage of shrubs and bare ground, a sequence that follows the tallgrass–shortgrass–shrubsteppe gradient associated with increasing aridity (Wiens 1974b). Total breeding bird densities showed no relationship to this heterogeneity gradient, but community biomass decreased significantly as heterogeneity increased (and, presumably, primary production decreased) (Fig. 13.2).

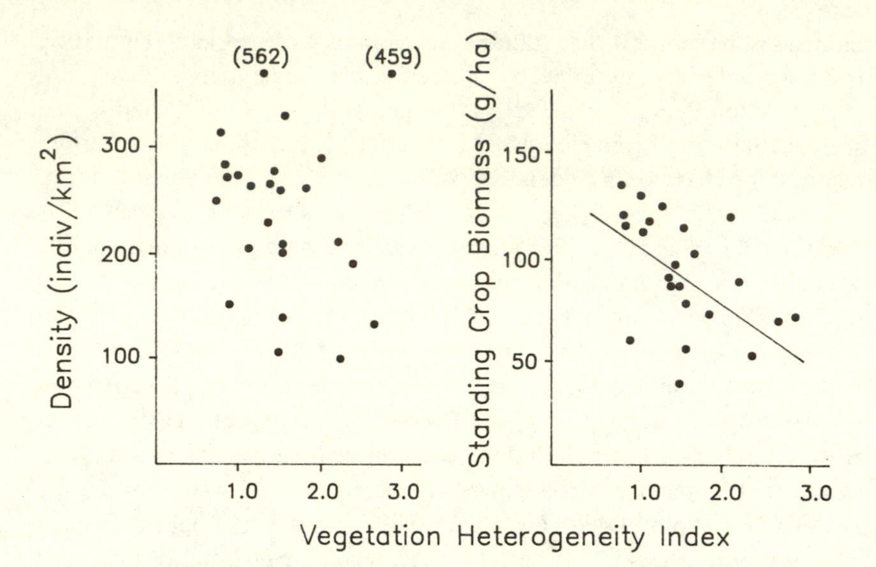

Fig. 13.2. Relationships between the vegetational heterogeneity of a series of grassland and shrubsteppe census plots in North America and plot values of total breeding bird density (left) and total standing crop biomass (right). After Wiens (1974b).

In some regards, standing crop biomass is simply another way to express the abundance of birds. It may also bear a closer relationship to the demands placed by bird populations on the energy resources of an environment than does density. Large birds certainly consume more food than do small birds, but measures of mass alone overestimate this difference. Salt (1957) first noted this problem in a community context and suggested calculating 'consuming biomass' by multiplying density by body mass raised to the 0.7 power ($M_B{}^{0.7}$), in recognition of the relatively lower metabolic rate per gram mass of larger species. Salt also proposed that the ratio of consuming biomass to standing crop biomass of a community (CB/SCB) might be used as a measure of community efficiency, because communities dominated by large species that require less energy per gram body mass exhibit a greater discrepancy between CB and SCB than do communities of smaller species. Salt analyzed breeding avifaunas in Wyoming coniferous forests and concluded that there was an increase in the energetic efficiency of communities (as measured by this ratio) as succession proceeds toward the climax vegetation. Karr (1968) noted similar relationships among successional stages of Illinois strip-mine habitats, as did Kilgore (1968) in undisturbed sequoia forests in the Sierra Nevada Mountains of California. The CB/SCB values decreased in several experimentally manipulated plots in Kilgore's study, suggesting to him a slight reversal of the successional process. On the other hand, Bock and Lynch (1970) found that this measure of efficiency of energy processing was greater in a burned than in an unburned forest in the same region. They suggested that foraging opportunities favored larger ground- or brush-foraging species in the burned area and smaller foliage insectivores in the unburned plots. The trends in consuming biomass over the series of managed pine stands studied by Szaro and Balda (1979) followed the same pattern as those of standing crop biomass, and there were therefore no systematic changes in the CB/SCB ratio over this sequence.

There is little doubt that consuming biomass values are more indicative of the potential energy flow through bird populations than are biomass values alone. In view of recent work, however, CB should be calculated using a somewhat lower exponent of M_B; Szaro and Balda (1979) used $M_B^{0.633}$, and Walsberg (1983) reported that the total energy demand of a wide range of birds scales with $M_B^{0.61}$. Whether the ratio of CB to SCB measures anything resembling 'community efficiency' or simply indexes the relative size distribution of species in a somewhat cryptic fashion, however, is unclear.

Approaches based on energetics

If one is interested in expressing community patterns in terms of measures related to potential energy flow through the populations, it makes sense to estimate energy flow directly rather than to rely on indirect measures such as biomass. Several approaches have been followed, all based in one way or another on physiological functions relating metabolic rate to body mass (see Gessaman 1973, King 1974, Kendeigh *et al.* 1977, Walsberg 1983 for general background). Puttick (1980), McLachlan and Wooldridge (1981), and Schneider and Hunt (1982), for example, used equations relating basal metabolic rate (BMR) to body mass, adjusted in a simple fashion for the energetic costs of activity and for assimilation efficiency, to estimate energy budgets of Curlew Sandpipers (*Calidris ferruginea*), South African shorebirds, and Bering Sea seabirds, respectively. Drent and Daan (1980) assembled data from several studies that indicate that the daily metabolizable energy demands of adult birds feeding nestlings cluster between 3.5 and 4.2 times BMR. Pienkowski *et al.* (1984) surveyed studies of shorebird energetics and found that the daily energy intake of individuals varied from 2.5 to 6.8 times BMR.

Other investigators have used standard (resting) metabolic rate (SMR) or existence metabolism (EMR) as the basis for calculations. Holmes and Sturges (1973) multiplied SMR (obtained from the Lasiewski–Dawson

(1967) equation) by 2.5 to obtain daily energy demands for individuals of each species in a hardwoods forest in northeastern USA. Multiplied by population densities, this yielded estimates of overall community energetics. Schneider *et al.* (1987) and Duffy and Siegfried (1987) used a similar approach to estimate the energy demands of seabirds. Väisänen and Järvinen (1977) multiplied the value of EMR obtained from the equations of Kendeigh *et al.* (1977) by 3.0 to obtain an estimate of overall community energy demand, and Szaro and Balda (1979) used EMR without any further adjustments to express energy flows in the communities they studied. In other studies (e.g. West and DeWolfe 1974, Holmes *et al.* 1979c), metabolic rates of species have been determined directly in the laboratory and then adjusted for the additional costs of free-living activities.

Most attempts to estimate community energy demands have been based on EMR. Existence metabolism is the rate of energy use by caged birds that are not engaged in reproduction, molt, premigratory fat accumulation, or growth and that are provided with food and water *ad libitum*. It integrates BMR, temperature regulation, the heat increment of feeding (SDA), and the energy expended in activity while in the cage (Fig. 13.3). Kendeigh (1970) developed equations estimating EMR as a function of body mass and ambient air temperature. These equations were based on a limited number of studies, however, and their use in models led to confidence intervals on estimates of population energy demands of as much as ± 50% (Furness 1978). Pimm (1976) reanalyzed Kendeigh's data and derived an equation for EMR by stepwise multiple regression analysis that explained more of the variance than did the original equations. The relationships derived by Kendeigh *et al.* (1977), however, were based on substantially larger sample sizes, and analyses made using these equations have much tighter confidence intervals (Furness 1981, 1982). In these equations, EMR is scaled separately for passerines and nonpasserines as a function of body mass, ambient temperature, and photoperiod. The inclusion of air temperature as a variable in these equations is convenient, but it may produce some inaccuracy in the energetic estimates because the complicating effects of other avenues of heat exchange, such as radiation or convection, are ignored (King 1974, Lustick 1984). Weathers *et al.* (1984) suggested that models based on heat-transfer theory may provide more accurate estimates of daily energy demands than those based on general existence metabolism regressions.

Calculations of daily EMR for multi-species assemblages become complex if they are extended over a season or year, during which temperatures and photoperiod change, population densities undergo flux, and body

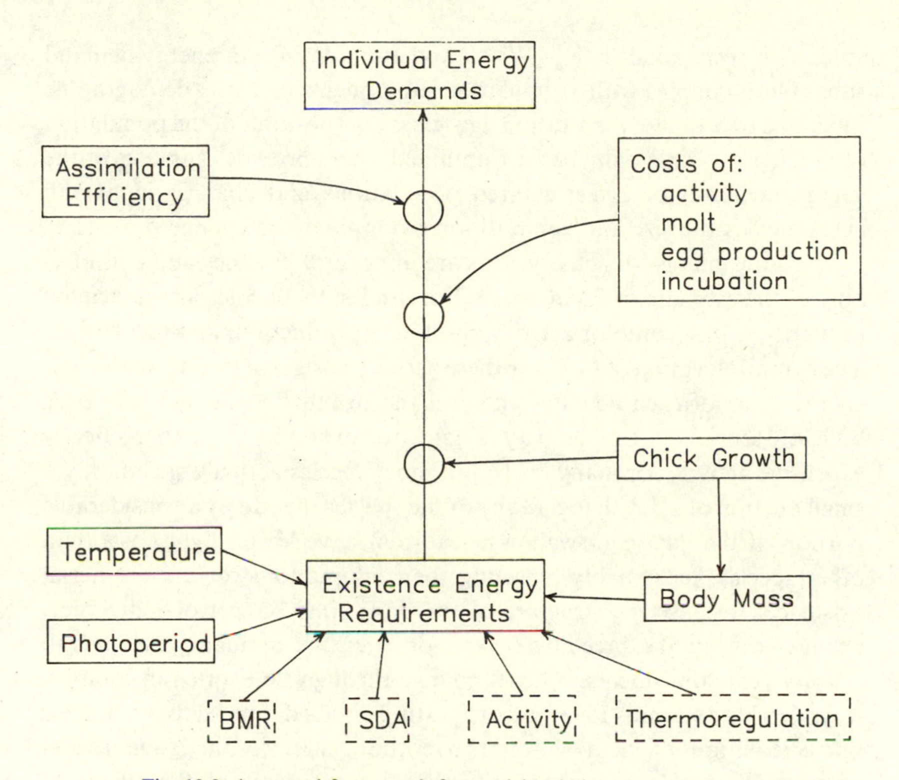

Fig. 13.3. A general framework for models relating components of individual metabolism to the energy demands placed on environmental resources. From Wiens (1984c).

masses of individuals may vary. Computer simuation models simplify such analyses and, more important, permit one to vary population or environmental parameters in a simulation 'experiment' to explore the effects of such changes on population or community energetics. The several simulation models that have been developed (Wiens and Innis 1974, Furness 1978, 1981, Weiner and Głowacinski 1975, Głowacinski and Weiner 1980, 1983) are similar in that calculations are based on EMR equations that are dependent on body mass and temperature (and, in some, photoperiod) and are then adjusted for the additional costs of various activities; the resulting values are adjusted in turn for assimilation efficiency to obtain an estimate of the energy demands that individuals place on the environment (Fig. 13.3). The energy demands of chicks (especially those associated with growth) may be derived in a similar fashion. These individual energy demands are then multiplied by population density to obtain population energy demands, which are then summed across species to give community values. In some models (e.g. Wiens and Innis 1974), an energy-demand submodel is coupled with a population submodel, which uses demographic functions to estimate the size and age–class composition of the population of each species on a daily basis. Combined, these provide estimates of the energy demands of age-structured populations and changes in overall energetics as the size and composition of populations change over time.

The adjustments for activity costs are important, for they add considerably to the estimate of EMR alone. For adults, these costs are associated primarily with locomotor activity, molt, egg production, and incubation. The overall elevation of EMR by these activity costs is a product of the unit energy costs of each activity type and the proportion of the daily time budget occupied in that activity. Flight, for example, is an energetically expensive activity for many birds, but most species actually spend only a small portion of a day flying. Many of the species that do fly a considerable portion of the day (e.g. swallows, seabirds) have lower flight costs than other species, presumably reflecting their adaptations for a more aerial existence. Because the structure of the daily time budget of individuals changes during the breeding season or over the annual cycle, overall activity-cost adjustments to EMR change through time, often in complex ways (Wiens and Innis 1973). Incorporating detailed time–activity budgets into calculations of energy flow improves the accuracy of the estimates and permits an examination through simulation exercises of the effects of varying the time budget on energy demands. To do this, however, requires an accurate determination of time budgets in the field, which is often difficult (if not biased; see Chapter 3). It also requires a knowledge of the metabolic costs of various activities. These have been determined for some species, but general functions must be used for most. Walsberg (1983) and Wiens (1984c) have reviewed much of the available information on these activity costs.

By the time one has gone through such a series of adjustments to a basic metabolic function such as EMR or BMR, one may wonder whether the estimates produced have much substance in reality. One way to determine the accuracy of energy estimates based on time–energy budget models is to compare them with direct measurements of the overall metabolism of free-living birds, which can be made using doubly-labeled water ($D_2{}^{18}O$). The $D_2{}^{18}O$ method may overestimate or underestimate actual metabolism slightly (3–13%; Bryant *et al.* 1985), but, as the mean error rarely exceeds 10% (Nagy 1980), the method provides a reasonable standard for comparisons of model-derived estimates. Measurements of the daily energy expenditure of free-living individuals made using $D_2{}^{18}O$ have been compared with

Table 13.1. *Comparisons of estimates of daily individual energy demands (DEE) obtained using various metabolic/time-energy budget models with those measured using doubly-labelled water ($D_2{}^{18}O$) for Savannah Sparrows (*Passerculus sandwichensis*) (Williams and Nagy 1984) and Dippers (*Cinclus cinclus*) (Bryant* et al. *1985).*

Method[a]	Reference	Predicted DEE (kJd^{-1})	% difference from $D_2{}^{18}O$
Savannah Sparrow			
$D_2{}^{18}O$	Williams and Nagy 1984	80.3	standard
TEB	Schartz and Zimmerman 1971	92.6	+19.1
TEB	Utter and LeFebvre 1973	40.5	−48.3
TEB	Walsberg 1977	73.6	−5.6
Regression	Kendeigh *et al.* 1977	80.9	+4.9
TEB	Holmes *et al.* 1979c	122.6	+57.0
Regression	Walsberg 1980	71.3	−8.3
TEB	Mugaas and King 1981	82.5	+6.0
Dipper			
$D_2{}^{18}O$	Bryant *et al.* 1985	206	standard
TEB	Bryant *et al.* 1984	194	−5.8
Regression	King 1974	176	−14.6
Regression	Bryant *et al.* 1985	175	−15.1
Regression	Kendeigh *et al.* 1977	156	−24.3
Regression	Bryant *et al.* 1985	155	−24.8
Regression	Dolnik & Kinzhewskaja 1980	154	−25.2
TEB	Dolnik 1982	153	−25.7
Regression	Walsberg 1983	151	−26.7
TEB	Koplin *et al.* 1980	146	−29.1
TEB	Walsberg 1977	144	−30.1
TEB	Yom-Tov and Hilborn 1981	130	−36.9
M multiple	Drent and Daan 1980	114	−44.7

Notes:

[a] TEB = time-activity budgets combined with laboratory measurements of activity costs
Regression = DEE determined from regression equations based on metabolic measurements
M multiple = DEE measured as a direct multiple of BMR with no specific activity adjustments

model estimates for Savannah Sparrows and Dippers (*Cinclus cinclus*) (Williams and Nagy 1984, Bryant *et al.* 1985). The model estimates for both species vary considerably in relation to the $D_2{}^{18}O$ standard (Table 13.1), but some models produce estimates that are statistically indistinguishable from the standard values. In particular, the models of Kendeigh *et al.* (1977) and Walsberg (1977) estimate sparrow energetics quite closely, although the performance of these models for Dippers is less satisfactory (Table 13.1).

The $D_2{}^{18}O$ value for Dippers, however, is higher than any of the model estimates, suggesting either errors in the $D_2{}^{18}O$ calculations or (more likely) the extraordinary activity costs associated with underwater 'diving' in this species. Most of the variation in model estimates is associated with differences in the energy costs assigned to activities such as flight.

These comparisons indicate that model-based estimations of avian energetics may be reasonably accurate, especially if appropriate models are used. Both regression equations relating daily expenditure to body mass and more detailed time–energy budgets used in conjunction with metabolic estimates may be useful in obtaining general estimates of population and community energetics, especially where the differences among species or habitats reduce the importance of intraspecific variations in energy costs (Bryant and Westerterp 1980). A more accurate procedure is required if one wishes to examine the energetics of a single species in a given environment (Bryant *et al.* 1985).

Some patterns of energetics in bird communities

Models of energy flow have not been used to estimate community energetics in very many studies, but patterns of energy demands have been explored on habitat gradients in forests, grasslands, and agricultural areas, and energy demands have also been estimated for seabirds and shorebirds in several areas. These studies focus on patterns of overall energy consumption and how it is partitioned among respiration, production, and excretion, on the energetic dominance of particular species or guilds in communities, and on the extension of the energetic values to estimate magnitudes of food consumption. The latter is especially important, for in combination with some measure of resource availability it offers a way of assessing food limitation quantitatively rather than subjectively. Although most studies have used the models of Weiner and Głowacinski (1975), Wiens and Innis (1974), or Furness (1978) or variations on these themes, the models are not exactly equivalent and the results of the studies are thus comparable only in a general way.

Forests

Community energy flows have been estimated along successional sequences following cutting or for various management treatments in Poland and the USA and on elevational gradients in western USA (Table 13.2). Not surprisingly, the magnitude of energy flow increases with increasing stand age and vegetational development, although the increase is

Table 13.2. *Model-derived estimates of energy flow through bird communities in different environments.*

Habitat	Number of species	Time period	Total energy flow[a] (kJ km^{-2})	Mean daily energy flow (kJ ha^{-1})	Estimated food consumption[a]	Energy estimation method[b]	Reference
Forests							
Deciduous, Poland:							
pioneer stage		15 Mar–15 Aug (153 days)	58.7×10^5	0.38×10^3		W & G model	Głowacinski & Weiner 1983
15-yr		15 Mar–15 Aug (153 days)	925.1×10^5	6.05×10^3		W & G model	Głowacinski & Weiner 1983
20–30yr		15 Mar–15 Aug (153 days)	536.6×10^5	3.51×10^3		W & G model	Głowacinski & Weiner 1983
climax		15 Mar–15 Aug (153 days)	981.0×10^5	6.41×10^3		W & G model	Głowacinski & Weiner 1983
mixed alder-oak		15 Mar–15 Aug (153 days)	113.6×10^6	7.43×10^3		W & G model	Głowacinski & Weiner 1983
oak-hornbeam	40[c]	year	632.4×10^5	1.73×10^3	665 g m^{-2} yr^{-1}	W & G model	Weiner & Głowacinski 1975
Deciduous, northeastern USA	31	year	413.3×10^5	1.13×10^3		2.5 × SMR	Holmes & Sturges 1975
Montane gradient, western USA:							
alpine meadow	3[d]	year	$17.6–25.6 \times 10^5$	$0.48–0.70 \times 10^2$		W & I model	Smith & MacMahon 1981
aspen woods	12–24[d]	year	$83.8 - 288.7 \times 10^5$	$2.30–7.91 \times 10^2$		W & I model	Smith & MacMahon 1981
fir forests	20–24[d]	year	$273.2–370.4 \times 10^5$	$7.48–10.14 \times 10^2$		W & I model	Smith & MacMahon 1981
spruce forest	20–22[d]	year	$250.1–357.4 \times 10^5$	$6.85–9.79 \times 10^2$		W & I model	Smith & MacMahon 1981
Managed pine forests, Poland:							
1-yr clear-cut	2	15 Mar–15 Aug (153 days)	15.2×10^5	0.99×10^2		W & G model	Głowacinski & Weiner 1980
4-yr culture	6	15 Mar–15 Aug (153 days)	98.8×10^5	6.45×10^2		W & G model	Głowacinski & Weiner 1980
10-yr thicket	9	15 Mar–15 Aug (153 days)	292.2×10^5	1.91×10^3		W & G model	Głowacinski & Weiner 1980
35-yr pole pine	11	15 Mar–15 Aug (153 days)	285.0×10^5	1.86×10^3		W & G model	Głowacinski & Weiner 1980
80-yr forest	30	15 Mar–15 Aug (153 days)	568.6×10^5	3.72×10^3		W & G model	Głowacinski & Weiner 1980

Table 13.2. (*cont.*)

Habitat	Number of species	Time period	Total energy flow[a] (kJ km^{-2})	Mean daily energy flow (kJ ha^{-1})	Estimated food consumption[a]	Energy estimation method[b]	Reference
Managed pine, southwestern USA:							
clear cut	6–10	15 Mar–15 Aug[e] (153 days)	8.1–12.8 × 10^5	0.53 – 0.84 × 10^2		W & G model	Szaro & Balda 1979
selectively thinned	20–26	15 Mar–15 Aug[e] (153 days)	39.1–63.2 × 10^5	2.55–4.12 × 10^2		W & G model	Szaro & Balda 1979
strip cut	23–29	15 Mar–15 Aug[e] (153 days)	45.1–96.4 × 10^5	2.95–6.29 × 10^2		W & G model	Szaro & Balda 1979
silviculturally cut	23–28	15 Mar–15 Aug[e] (153 days)	42.8–85.5 × 10^5	2.79–5.59 × 10^2		W & G model	Szaro & Balda 1979
uncut control	19–27	15 Mar–15 Aug[e] (153 days)	34.0–78.9 × 10^5	2.21–5.16 × 10^2		W & G model	Szaro & Balda 1979
Coniferous, northwestern USA:							
low-elevation dry douglas-fir	12	1 Apr–7 Oct (190 days)	499.6 × 10^5	2.37 × 10^3		W & I model	Wiens & Nussbaum 1975
low-elevation mesic hemlock	12	1 Apr–7 Oct (190 days)	439.5 × 10^5	2.31 × 10^3		W & I model	Wiens & Nussbaum 1975
low-elevation moist hemlock	12	1 Apr–7 Oct (190 days)	697.6 × 10^5	3.67 × 10^3		W & I model	Wiens & Nussbaum 1975
mid-elevation transitional	15	1 Apr–7 Oct (190 days)	869.8 × 10^5	4.58 × 10^3		W & I model	Wiens & Nussbaum 1975
high-elevation dry fir hemlock	13	1 Apr–7 Oct (190 days)	509.1 × 10^5	2.68 × 10^3		W & I model	Wiens & Nussbaum 1975
high-elevation moist fir	7	1 Apr–7 Oct (190 days)	513.7 × 10^5	2.70 × 10^3		W & I model	Wiens & Nussbaum 1975
Grasslands							
Tallgrass prairie, Oklahoma	4	1 Apr–31 Aug (153 days)	72.1 × 10^5	4.71 × 10^2		W & I model	Wiens 1977c
Mixed-grass prairie, S. Dakota	5	1 Apr–31 Aug (153 days)	62.9 × 10^5	4.10 × 10^2		W & I model	Wiens 1977c
Shortgrass prairie, Texas	3	1 Apr–31 Aug (153 days)	49.4 × 10^5	3.23 × 10^2		W & I model	Wiens 1977c

Palouse Shrubsteppe, Washington	3	1 Apr–31 Aug (153 days)	48.6×10^5	3.18×10^2		W & I model	Wiens 1977c
Palouse Shrubsteppe, Washington	5	year	121.9×10^5	3.34×10^2	0.21 g m^{-2} yr^{-1}	W & I model	Rotenberry 1980b
Marine seabirds							
Finland:							
boulder skerries	13	1 May–30 Sept[e] (153 days)	125.8×10^7	82.20×10^3		3.0 × EMR	Väisänen & Järvinen 1977
grassy skerries	26	1 May–30 Sept[e] (153 days)	221.9×10^7	145.06×10^3		3.0 × EMR	Väisänen & Järvinen 1977
grassy islands	29	1 May–30 Sept[e] (153 days)	42.3×10^7	27.62×10^3		3.0 × EMR	Väisänen & Järvinen 1977
wooded islands	28	1 May–30 Sept[e] (153 days)	59.6×10^5	0.39×10^3		3.0 × EMR	Väisänen & Järvinen 1977
Oregon coast	4	year	322.4×10^9 kJ yr^{-1}	8.67×10^2	62,562 tonnes yr^{-1}	W & I model	Wiens & Scott 1975
Foula, Shetland Islands[f]							
total	9	year	503.0×10^8 kJ yr^{-1}	1.38×10^8 kJ day^{-1}	1.9 g m^{-2} yr^{-1}	F model	Furness 1978, 1984
terns and skuas	2	year	17.8×10^8 kJ yr^{-1}	4.88×10^6 kJ day^{-1}		F model	Furness 1982
Southern Benguela, S. Africa[f]	3	year	1.3×10^8 kJ yr^{-1}(?)	3.6×10^5 kJ day^{-1}(?)	16,500 tonnes yr^{-1}	F model	Furness & Cooper 1982
South Georgia[f]	12	(breeding)	510.0×10^{10} kJ $season^{-1}$			C & P model	Croxall & Prince 1982
	22	(breeding)			7.8×10^6 tonnes yr^{-1}	C & P model	Croxall *et al.* 1984, Croxall & Prince 1987
Norway[f]	9	year	366.8×10^8 kJ $year^{-1}$	1.00×10^8 kJ day^{-1}	9000 tonnes yr^{-1}	F model	Furness & Barrett 1985
Kodiak Island, Alaska	3	1 May–30 Sept[g]	437.1×10^8 kJ $season^{-1}$	2.85×10^8 kJ day^{-1}	8100 tonnes $season^{-1}$	W & I model	Wiens 1984c
Pribilof Islands, Bering Sea	11	1 May–30 Sept[g]	29.1×10^{10} kJ $season^{-1}$	19.02×10^8 kJ day^{-1}	53600 tonnes $season^{-1}$	W & I model	Wiens 1984c
Bristol Bay, Bering Sea	11+	year	11.7×10^5–22.8×10^5 kJ yr^{-1}			2.8 × SMR + assimilation	Schneider *et al.* 1987
Georges Bank, NW Atlantic	7+	year	4.7×10^5 kJ yr^{-1}			2.8 × SMR + assimilation	Schneider *et al.* 1987

Notes:

[a] Total consumption; original values adjusted using 75% assimilation efficiency when necessary.

[b] Simulation models: W & G = Weiner and Głowacinski 1975; W & I = Wiens and Innis 1974; F = Furness 1978; C & P = Croxall and Prince 1982.

[c] Total species; 31 breeding, 19 wintering.

[d] Breeding species; range during 3 yr.

[e] Values originally expressed as daily estimates; converted to this time period (153 days) to facilitate comparisons.

[f] No area specified.

[g] Values originally calculated for 1 May–20 Sept; adjusted for comparability with other values.

generally not uniform. In deciduous forests in southern Poland, community energy flow was nearly as great in a 15-yr-old stand as in a mature climax forest; in a successional sequence of managed pine stands in the same region, energy flow plateaued in the 10–35-yr stands before doubling in the 80-yr forest (Głowacinski and Weiner 1980, 1983). In pine stands in Arizona subjected to different management treatments, on the other hand, energy flow was greater in several manicured stands than in the uncut control (Szaro and Balda 1979).

These studies were all conducted during the breeding season, when bird densities and biomass are greatest, young are produced, and energy flow is greatest. Magnitudes of energy flow in the managed pine forests in Poland were generally somewhat lower than in the deciduous forests, which occur on more fertile soils. In both sequences, however, the community energy consumption was substantially greater than in the Arizona stands, despite the greater number of species in the latter (Table 13.2). The studies used the same model to estimate energy demands, and the reason for this difference is not apparent.

Smith and MacMahon (1981) used the Wiens–Innis model to estimate annual energy consumption by the bird communities occurring over an elevational gradient in Utah. There, energy flow was consistently low in species-poor alpine meadows, intermediate in aspen woodlands, and greatest in fir forests (Table 13.2). This study was conducted over a 3-yr period that included a severe drought, and the effects of the drought were most evident in the aspen stands. In the year before the drought, energy flow in the aspen was 78% of that in fir, 97% of that in spruce. In the drought year, energy consumption in the aspen woods dropped to 30% of both fir and spruce, and it increased to only 40% of both in the following post-drought year. Daily energy demands, averaged over the entire year, were an order of magnitude lower in these Utah stands than in the Polish forests, although the spruce and fir values were twice those of the Arizona pine stands.

In a series of coniferous stands in the Cascade Mountains of Oregon, breeding-season energy demands were similar to those in some of the Polish deciduous woods and mature pine forests and were substantially greater than in the more speciose Utah coniferous stands (Wiens and Nussbaum 1975; Table 13.2). In Oregon, total energy flow peaked in a mid-elevation transitional stand and was lower in both dry low-elevation stands and in dry or moist high-elevation stands. The proportion of total breeding-season energy intake consumed by species that are seasonal rather than year-around occupants of the forests decreased steadily from the drier, low-elevation stands to the more mesic, higher-elevation stands. This stand sequence reflects a shortening of the growing season, later persistence of low

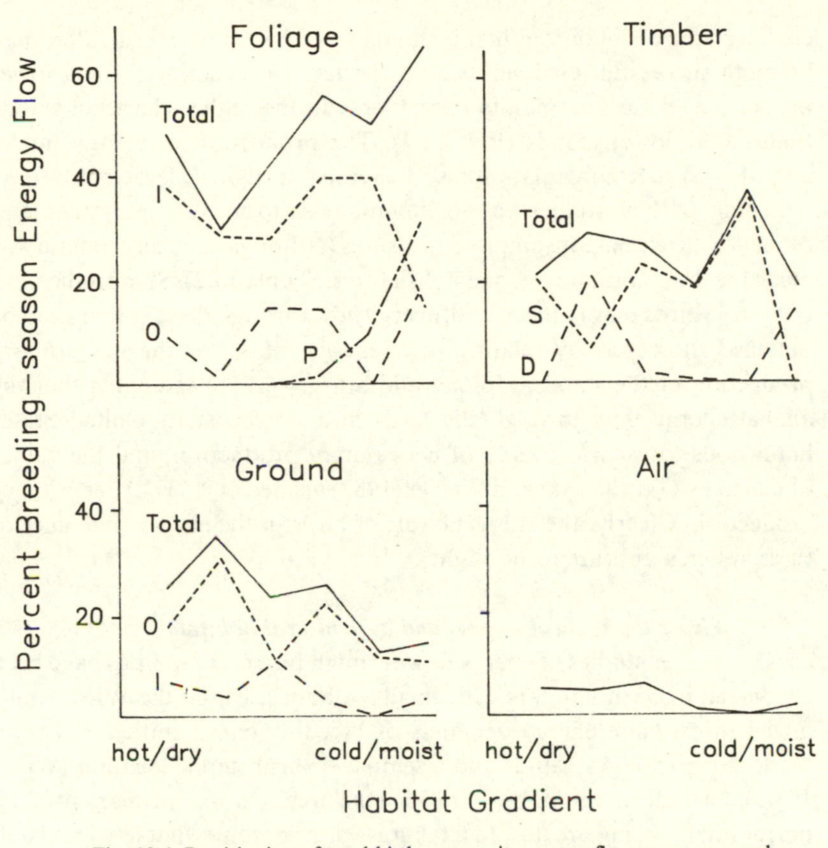

Fig. 13.4. Partitioning of total bird community energy flow among several guilds in coniferous forest stands on an environmental gradient in the Oregon Cascade Mountains. For foliage-feeders, P = plant (seed) feeders, O = omnivores, I = insectivores; for timber-feeders, S = searchers, D = drillers; for ground-feeders, O = omnivores, I = insectivores. Modified from Wiens and Nussbaum (1975).

temperatures (and retardation of insect emergence), and generally more severe environmental conditions persisting later into the breeding season. Under such conditions, the seasonal flush of insect abundance that is exploited by migrants may be increasingly constrained in time, favoring the more flexible resident species and enhancing total community energy flow. In the dry high-elevation stand, however, seasonal residents accounted for a greater proportion of the total energy flow than in any other stand. This was accompanied by a change in dominance from insectivorous species to seed- and fruit-eating species in the foliage-feeding guild (Fig. 13.4). Several of these species were highly irruptive species that were generally absent from the other stands, suggesting that resident species may find it difficult to exploit resource conditions in this stand and opportunistic seed-eaters may be more successful. Ground-feeding species also accounted for a lower proportion of the community energy flow in the higher-elevation stands than in the lower stands (Fig. 13.4). The proportion of energy intake allocated to thermoregulation also increased at higher elevations.

In none of these studies was an attempt made to convert energy demand estimates into food-consumption measures (although such an estimate was made for the community in one Polish forest; Table 13.2). Some indication of the role birds play in the overall energy flow through these systems can be obtained, however, by relating their energy intake to the net primary production of the stands. Holmes and Sturges (1975) estimated that the annual energy flow through the birds in a northeastern United States hardwoods forest was 0.17% of net primary production, and the values obtained by Głowacinski and Weiner (1983) peaked at 0.5% of net primary production. Clearly, the energetic role of birds in the energy dynamics of these systems appears to be slight.

Grasslands, shrubsteppe, and agricultural habitats

Fewer studies of energy flow through bird communities have been conducted in open habitats with small avifaunas. I used the Wiens–Innis model to estimate energy demands of breeding communities in several North American grasslands and a semi-arid shrubsteppe location (Wiens 1977c). Overall energy intake paralleled the precipitation–primary production gradient, being greatest in a tallgrass prairie, somewhat less in mixed-grass prairie, and least in shortgrass prairie and shrubsteppe (Table 13.2). In shortgrass prairie locations, energy flow was somewhat greater in heavily grazed than in lightly grazed or ungrazed plots, but this pattern was reversed at the mixed-grass location. There was substantial annual variation in these estimates, however, reflecting annual changes in population densities and in the relative densities of species of different biomass, and few of the differences between sites or grazing treatments were statistically significant. Average daily energy intake by birds in these grasslands was an order of magnitude lower than that estimated for the forests in Poland and Oregon, although the grassland birds consumed a similarly low proportion of the net primary production (0.02–0.11%).

These estimates of energy demands were combined with information on the dietary composition of the species and the caloric content of the prey to derive estimates of the consumption of various prey categories by the birds. These estimates also varied considerably between years and sites (Table 13.3), not only because of variations in bird abundances but also because

Table 13.3. *Estimated consumption of prey groups ($g\ m^{-2} \times 10^{-2}$) by birds during the 153-day breeding season at a grazed tallgrass and a grazed shortgrass prairie location during 3 yr.*

		Seeds			Arthropods						
							Phytophagous				
Site	Year	Grass	Forb	Other	Predaceous	Omnivorous	Chewing	Sucking	Flower-feeding	Scavenging	Total
Tallgrass	1970	2.67	—	—	2.26	4.52	18.28	0.52	0.02	1.09	29.36
	1971	3.79	0.02	—	6.40	3.40	14.53	0.13	0.02	0.21	28.50
	1972	1.73	0.02	—	7.01	1.22	8.59	0.14	—	1.19	19.90
Shortgrass	1970	1.84	6.05	4.77	1.34	1.24	11.68	2.03	0.21	0.68	29.84
	1971	4.66	9.10	1.51	13.89	3.42	9.60	0.16	0.12	0.26	42.72
	1972	1.19	1.14	—	3.80	3.58	23.19	0.95	0.05	0.99	34.89

Source: From Wiens (1977c).

their diets underwent annual changes (Wiens and Rotenberry 1979). Rotenberry's (1980b) analysis of the annual bioenergetics of the bird community at the shrubsteppe location indicated substantial seasonal variation in overall energy flow and prey-consumption magnitudes and patterns. Rotenberry compared the model estimates of arthropod consumption with independent measures of arthropod standing crops taken at the same times. The daily demands of the birds never exceeded 0.7% of the arthropod standing crop. Using less accurate standing-crop estimates, I calculated that the breeding birds at the grassland sites might consume 0.4–7.4% of the arthropod standing crop per day. These values hint at the potential for the birds to be food-limited or to have an impact on the arthropods. In the absence of information on the actual *availability* of the prey to the birds or on recruitment or turnover rates of the arthropods, however, such estimates are of limited value. Still, the use of bioenergetic models to estimate prey consumption magnitudes brings one closer to assessing food limitation than does information on diet composition alone.

In another exercise using the Wiens–Innis model, Dyer and I examined patterns of energy flow and food consumption for 24 predominantly granivorous species in different sections of Illinois (Wiens and Dyer 1977). We used data assembled by Graber and Graber (1963), who compared census results for several agricultural habitats obtained in the 1950s with those from similar surveys in these areas taken half a century before. The surveys thus provide a view of the population changes of these species over this time period, as well as an indication of the differences between the agricultural habitats. The 24 species constituted in excess of 60% of the individuals present in most habitats. Total density was greatest in the southern third of the state, which receive heavy use by wintering individuals. Densities increased markedly in different habitat types in several sections of the state over the 50-yr period, largely due to increases in the abundance of Red-winged Blackbirds.

Estimates of total energy flow through these granivore populations ranged from 40.2×10^5 to 637.3×10^5 kJ km^{-2} yr^{-1}, values generally similar to those recorded in shorter time periods in the forest studies. Energy intake was greatest in hayfields and fallow fields undergoing secondary succession in 1958, and was lowest in corn croplands (Fig. 13.5). Energy intake was greater in 1958 than in 1909 in most habitats, especially in the southern sector of the state, and increases were especially apparent in the hayfield habitat.

By coupling these estimates of energy demands with general information on the proportions of seed and animal prey in the diets of the species, food

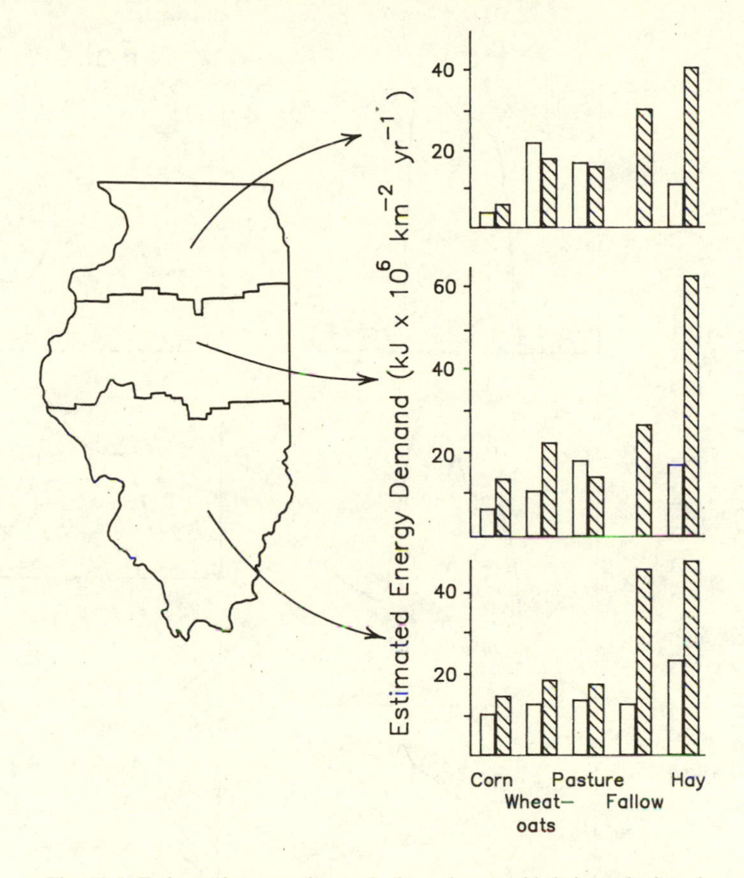

Fig. 13.5. Estimated energy demand of granivorous birds in agricultural habitats in three regions of Illinois in 1909 (open) and in 1958 (hatched). From Wiens and Dyer (1977) and Graber and Graber (1963).

consumption rates could be estimated as well. In hayfields in 1958, for example, consumption of both seed and animal prey was greatest during summer in the northern and central sections of the state but was more evenly spread over the year in the south (Fig. 13.6). Prey-consumption patterns in fallow fields followed generally similar patterns in the northern and central sections (although granivory was more predominant), but seed consumption in the southern region was substantially greater during the winter than in summer (Fig. 13.6). This reflects the concentration of wintering granivores (especially blackbirds) in that region.

Finally, estimates of seed consumption on a statewide basis were derived by adjusting the consumption rates per km^2 to reflect the total area of each habitat type in the state at the two times. In 1909, projected seed consump-

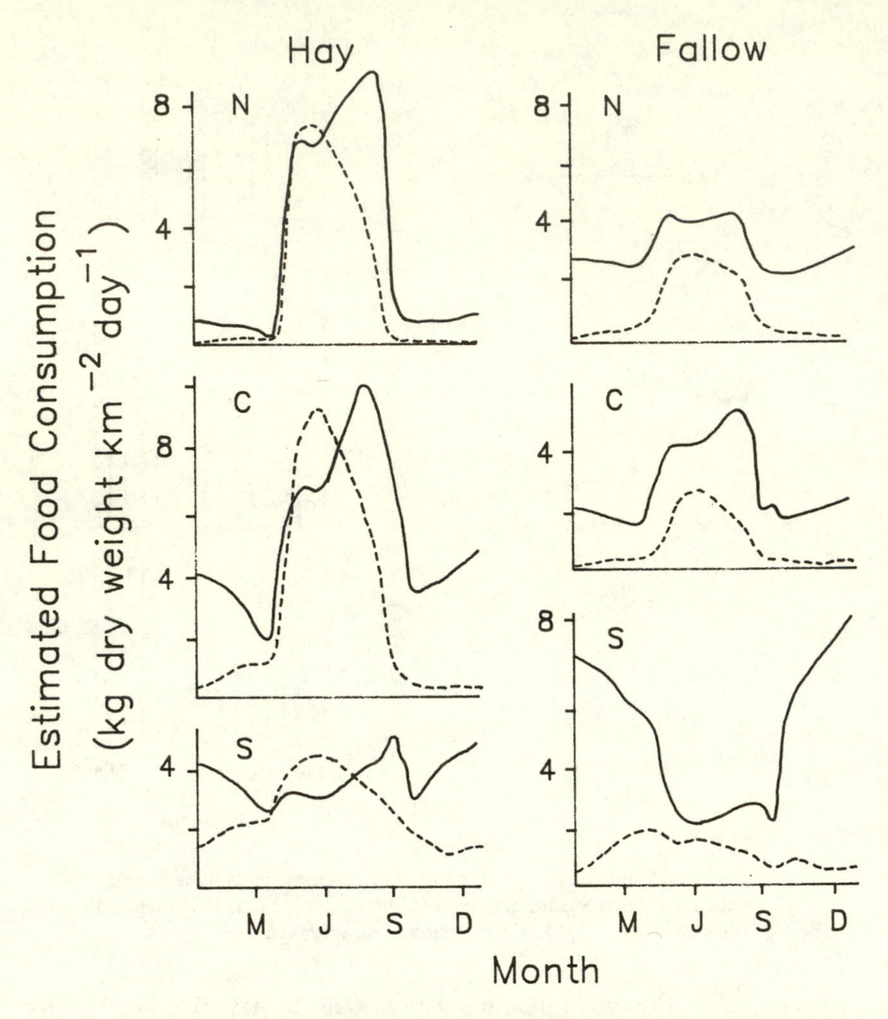

Fig. 13.6. Estimated daily food consumption through the year by granivorous birds in hayfields and fallow fields in three regions of Illinois (N = north, C = central, S = south) in 1958. Solid line = seed consumption, dashed line = consumption of animal prey. From Wiens and Dyer (1977).

tion was high and relatively similar (*ca* 10000–12500 tonnes yr^{-1}) in pasture, wheat-oat fields, and corn, and relatively low in hayfields and fallow fields. Fifty years later, overall seed consumption was relatively unchanged in cornfields (despite the considerable reduction in area under cultivation), lower in wheat–oat fields and pastures, somewhat higher in fallow fields, and had increased over 250% in hayfields (despite the lack of change in the area of hayfields in the state). In all five agricultural types combined, total annual statewide seed consumption by these birds increased slightly over the 50-yr interval, despite the substantial decrease in area devoted to these crops as the cultivation of soybeans increased. In the absence of direct information, these model-derived estimates provide some indication of the potential impact of this avian assemblage on their food resources, which in this case may be economically important (see also Wiens and Dyer 1975).

Marine birds and shorebirds

Modeling approaches have been applied more broadly to seabirds and have been extended more frequently to estimate food consumption than in other assemblages, perhaps because of the potential interactions of seabirds with marine fisheries (Furness 1982). Because it is difficult to determine the area over which seabirds breeding at a colony are distributed, density estimates are not often available, and energy flow into seabird populations is more frequently expressed as total colony values than as measures per unit area. The area-based estimates that are available suggest that magnitudes of energy flow of widely dispersed Oregon seabird populations may be similar to those of landbirds in some forested habitats (Table 13.2). Seabird energy demands in the vicinity of dense breeding colonies in Finland, however, are substantially greater than those of birds in terrestrial communities.

Averaging energy demands per unit area may not be very realistic for birds such as seabirds, whose distributions at sea are quite uneven. Estimates of the total energy intake by seabird assemblages, unadjusted for area, vary over two orders of magnitude, from the relatively low value for a 3-species group in the Southern Benguela current region of South Africa (Furness and Cooper 1982) to high values for a 4-species assemblage over the entire Oregon coast (Wiens and Scott 1975) and for an 11-species community breeding on the Pribilof Islands in the Bering Sea (Wiens 1984c). Some indication of the degree to which these demands are spatially concentrated is evident from an analysis of the distribution of energy intake rates with increasing distance from the breeding colonies on the Pribilofs (Fig. 13.7). There, most of the total community energy demand occurred within 40 km of the breeding island. Much of this demand was by murres (*Uria* spp.), which consumed 77% of their energy within this zone. Northern Fulmars (*Fulmaris glacialis*), on the other hand, accounted for only a moderate proportion of the total community energy demand but assumed greatest importance in areas beyond 120 km from the islands (Fig. 13.7). This analysis indicates clearly how the energetic dominance of various

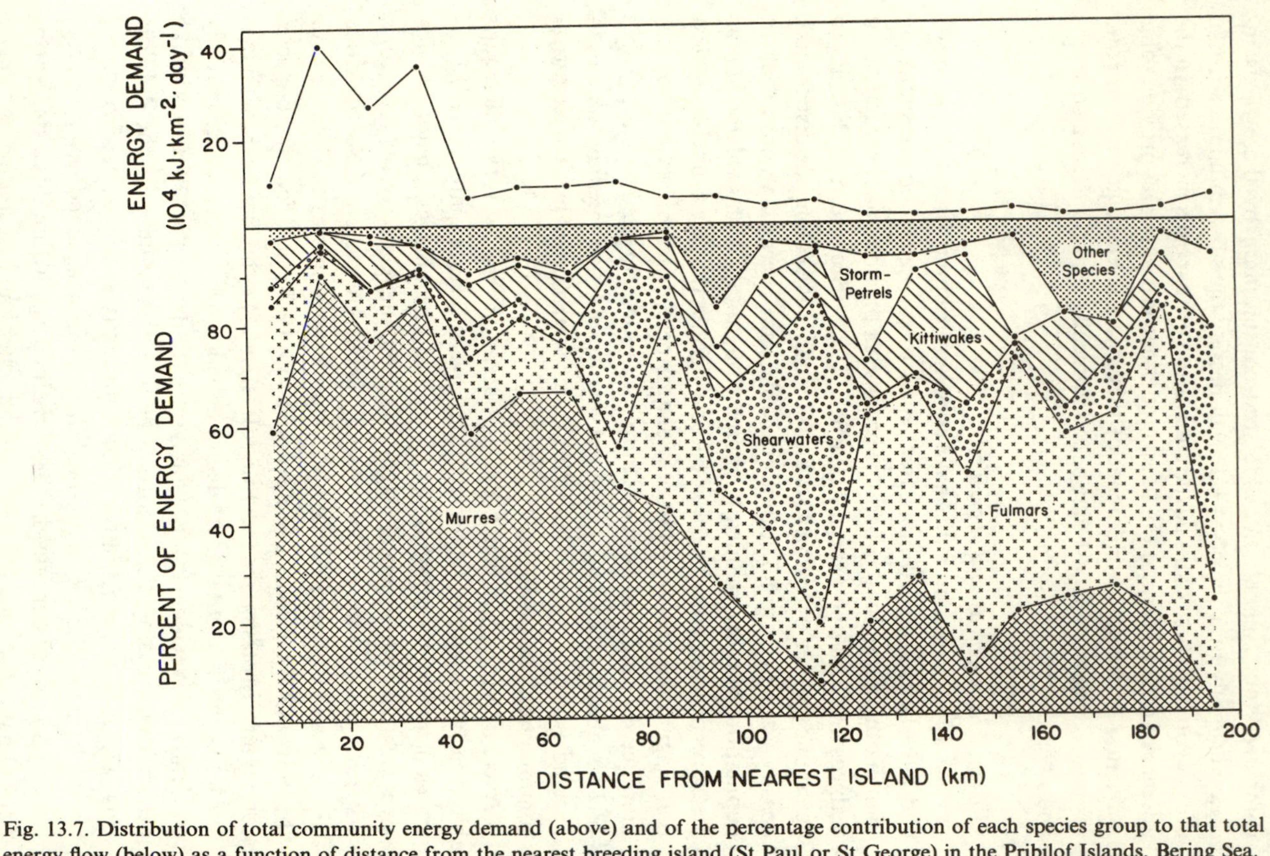

Fig. 13.7. Distribution of total community energy demand (above) and of the percentage contribution of each species group to that total energy flow (below) as a function of distance from the nearest breeding island (St Paul or St George) in the Pribilof Islands, Bering Sea. From Wiens (1984c).

species in the community may change over space, especially for species that forage radially over different distances from a central place (see also Fig. 10.10).

When these estimates of overall energy flow into seabird communities are converted into estimates of food consumption, the magnitude of food intake by these birds and their potential role in marine systems become apparent. Various procedures have been used to estimate total food consumption. By assuming that individuals consumed 20–40% of their body weight in prey per day, Hunt *et al.* (1981) conservatively estimated that the 53 million seabirds occupying the shelf waters of the eastern Bering Sea consumed $0.58–1.15 \times 10^8$ tonnes yr^{-1}. More detailed simulation-model analyses indicated that the breeding seabirds on the Pribilof Islands in this same region consumed 5.36×10^4 tonnes during the 4-month breeding season. By way of contrast, a smaller assemblage of seabirds breeding on Kodiak Island in the Gulf of Alaska was estimated to consume only 8.1×10^3 tonnes during the same period (Table 13.2) (Wiens 1984c). Croxall *et al.* (1984) used a different but equally detailed model to estimate that annual food consumption by the much larger South Georgia seabird community (21 species, roughly 36 million individuals) was approximately 7.8×10^9 tonnes (73% of it krill). Macaroni Penguins (*Eudyptes chrysolophus*) accounted for over half of the annual consumption. The several species of penguins breeding on South Georgia comprised only 13% of the total seabird density but accounted for 76% of the biomass and consumed 53% of the food. Scott and I (Wiens and Scott 1975) estimated that the four species we considered on coastal Oregon consumed 62.5×10^3 tonnes of prey per year. Of this total, 43% was anchovies, and 86% of the consumption of anchovies was by Sooty Shearwaters (*Puffinus griseus*) during their brief spring and fall movements along the coast (Fig. 13.8). Briggs and Chu (1987) estimated that seabirds off the California coast consumed about 44×10^3 tonnes of plankton and 149×10^3 tonnes of fish and squid in a normal year (37% less in years of warm-water currents); most of the consumption was by Sooty Shearwaters and Common Murres (*Uria aalge*). In the northwestern Hawaiian Islands, seabirds have been estimated to consume roughly 4.1×10^5 tonnes of prey (54% squid) per year (Harrison and Seki 1987). Here, albatrosses accounted for 64% of the overall consumption, birds feeding in association with predatory fish another 29%. Furness and Barrett (1985) estimated that the seabirds breeding at two colonies on the northern Norwegian coast consumed 9.0×10^3 tonnes yr^{-1}, 83% of it capelin. Four of the nine species present accounted for virtually all of the prey consumption, and Herring Gulls (*Larus argentatus*) alone accounted for 75% of the total community food consumption.

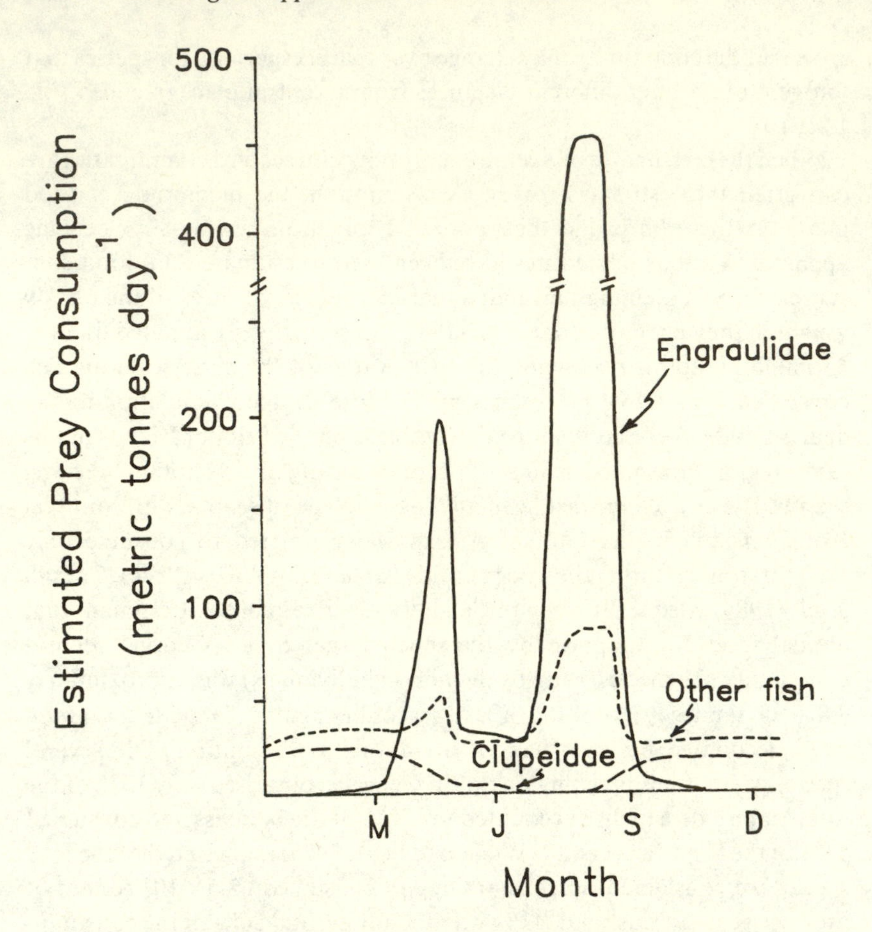

Fig. 13.8. Model-derived estimates of seasonal variations in the consumption of herring, anchovies, and other fish taxa by a four-species community of seabirds occurring on the Oregon coast. From Wiens and Scott (1975).

Seabird food consumption has been related to standing stocks or production rates of the prey in several studies. For the Oregon coastal seabirds, consumption of small pelagic fish may represent as much as 22% of the annual production of these fish within 185 km of the coast (Wiens and Scott 1975). Furness (1978) used a similar modelling approach to estimate that seabirds breeding on one of the Shetland Islands consumed 29% of the annual fish production within a 45-km radius of the colony, and Furness and Cooper (1982) used the same approach to calculate that the seabirds in a South African colony might consume about 20% of the herring and anchovy biomass each year. They also reanalyzed data from Peru and found that seabirds there were responsible for roughly 20% of the fish mortality in that system. The degree of concordance among these estimates is encouraging, but Bourne (1983) criticized them as being unrealistically high. He suggested that seabird consumption of fish in the North Sea might be as low as 0.25% or as high as 48% of the annual fish production, depending on the method of calculation. Furness (1984), however, noted several errors in Bourne's calculations, corrected these, and came up with a value closer to 20%. Duffy and Siegfried (1987) estimated that consumption of anchovetta by seabirds in the Humboldt current region off Peru (1.73×10^9 tonnes yr^{-1}) comprised 7–8% of the fish stock at peak seabird abundances, reaching perhaps 11% in some years. Here, seabird consumption rates fluctuated dramatically (Fig. 13.9) in response to variations in both bird abundances and fish stocks associated with El Niño events and overexploitation of the fishery by humans.

In any case, it appears that seabirds may consume a sizeable proportion of the annual production of their prey in at least some situations. Whether this may lead to resource limitation, however, remains uncertain. In other situations, seabird consumption appears to represent a small fraction of the standing stock of prey. In northern Norway, for example, the commercial catch of capelin exceeds the estimated consumption of the birds by more than two orders of magnitude (Furness and Barrett 1985).

Studies of the energy demands of shorebirds in estuarine and intertidal environments have been less detailed than the seabird studies. Baird *et al.* (1985) reviewed these studies and attempted to relate estimates of the energy demands of the birds to estimates of invertebrate production. These comparisons suggest that the birds may consume anywhere from 6% to 44% of the annual invertebrate production (Table 13.4). The potential impact of this consumption may vary considerably among prey species, however. In a mudflat area in northeastern England, for example, the total estimated consumption of two prey species by three shorebird species amounted to 4.5–9% of their estimated annual production, whereas consumption of a third prey species was calculated to be 43% of its production (Baird *et al.* 1985). The potential for shorebirds to have a major impact on their prey populations appears to be great, at least in some situations (see also Fig. 10.3).

Conclusions

Bioenergetic analyses have not been a prominent feature of community studies, perhaps because they involve aspects of the physiology of individuals that many community ecologists have considered to be unnecessary autecological detail. The results reviewed in this chapter, however, hint

Table 13.4. *Estimated energy consumption by shorebirds and annual production of their invertebrate prey in several European intertidal and estuarine environments.*

Consumption efficiency indicates the proportion of the invertebrate production estimated to be consumed by the birds.

Locality	Energy equivalent ($kJ\ m^{-2}\ yr^{-1}$) of: Consumption by birds	Invertebrate production	Consumption efficiency, invertebrates to birds (%)
Dutch Wadden Sea (intertidal zone)	103.6	619.2	17
Grevelingen estuary	71.5	1201.4	6
Ythan estuary	873.6	2448.1	36
Tees estuary	367.0	851.0	44
Langebaan Lagoon	141.6	705.0	20

Source: From Baird *et al.* (1985).

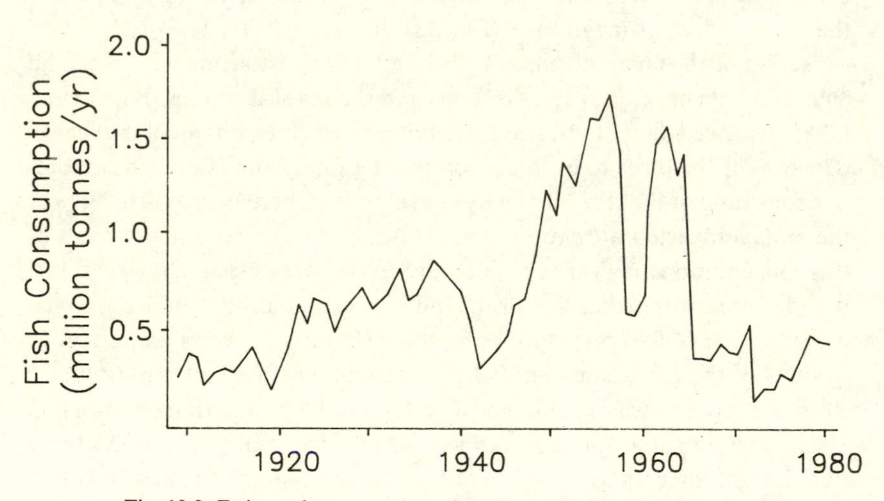

Fig. 13.9. Estimated consumption of anchoveta by breeding Peruvian seabirds, 1909–1980. After Duffy and Siegfried (1987).

at some of the different views of patterns and ecological relationships in communities that may emerge from such a perspective. Expressing community patterns in terms of biomass or energy flows brings one that much closer to a part of the resource base that may influence community structure. If estimates of energy flow are coupled to information on dietary composition, the projections of food consumption provide a much more realistic assessment of dietary overlap among species than does a matrix of diet taxa recorded. If such estimates can be coupled with measures of food abundance or availability, then they may also permit a quantitative assessment of how food limits populations. By relating energy flows or food-consumption rates to production values, the role of birds as consumers in the energy dynamics of ecosystems may also be evaluated.

This is not to say that bioenergetic analyses, especially those based upon models, are free of problems or constraints. Most models require information on a large number of parameters in order to generate realistic estimates. Some of the metabolic and population parameters are straightforward and easily measured, but many are not. Not all parameters need be specified with great accuracy in order to generate reasonably robust model estimates, but these estimates may be quite sensitive to errors in the measurement or estimation of some other parameters. Variations in values for population densities, assimilation efficiency, population phenology, adult activity budgets, and existence metabolism may be especially critical (Wiens and Innis 1974, Wiens *et al.* 1984, Furness 1982). Modeling approaches also rely on assumptions, many of which relate to aspects of environmental physiology about which our knowledge is meager. Although several validation studies using $D_2{}^{18}O$ measurements indicate that the estimates obtained with at least some models may be reasonably accurate, model projections of population or total community energetics should still be regarded with some caution, as general rather than precise estimates.

14

A re-examination of the recent history of avian community ecology

In the previous chapters I examined the patterns of bird communities, largely in the context of what I have called the MacArthurian paradigm. I suggested that our approach to studies of avian communities – the questions we have asked and the ways we have gone about answering them – underwent a major transformation during the late 1950s and early 1960s, followed and expanded this new theme during the later 1960s and early 1970s, and began to change again during the later 1970s and early 1980s.

Perhaps I have painted a false picture of the conceptual development of the discipline, however. Perhaps the 'MacArthurian paradigm' was never as dominant as I have implied, the search for patterns not as focused, and the explanations of the patterns not so closely linked to competition. In this chapter I re-examine these recent historical trends in a more objective manner. I then conclude by emphasizing several points that, whatever the past or current status of the paradigm, must be part of our continuing efforts to document and understand the patterns of bird communities.

Trends in the development of avian community ecology, 1950–84

In order to chart the historical progression of thinking and practice in avian community ecology over the past several decades, I surveyed the contents of four ornithological journals (*The Auk*, *The Ibis*, *The Condor*, and *The Wilson Bulletin*) and five ecological journals (*Ecology*, *Ecological Monographs*, *Oikos*, *Journal of Animal Ecology*, and *The Americal Naturalist*) by 5-yr periods from 1950 through 1984. I tallied the papers dealing with bird communities and characterized each paper by criteria relating to the questions on which they were focused, the interpretations that the authors offered for their findings, and the procedures used.

The initial portion of this period has been characterized as one of explosive growth in ecology (Schoener 1974a), and a charting of the total number of papers published in ecology journals shows that this trend continued unabated into the mid-1980s. The number of papers published in ornithological journals also increased over this period, although far less dramatically (Fig. 14.1A). Although only a small proportion of the papers published in ecology journals dealt with bird communities, the number increased nearly twofold between the early 1960s and the later 1970s. Reports on bird community studies constituted a larger fraction of the ornithological literature, and here as well the proportion increased between 1965 and 1979 (Fig. 14.1B). Clearly, activity in this discipline increased dramatically during the 1960s.

Papers published during the 1950s were largely descriptive. During the 1960s, however, the proportion of studies specifically designed to address conceptual questions (if not to test hypotheses) increased rapidly, then stabilized during the 1970s and early 1980s (Fig. 14.1C–D). Thus, the shift from descriptive to question-oriented studies began around 1960 and was well established before studies of bird communities became popular later in the decade, so most of these late studies were also question-oriented. I contend that this pattern is a clear manifestation of the impact of MacArthur and his colleagues on avian community ecology.

Many of the early, descriptive studies were focused on documenting the species composition or distributional patterns of assemblages (Fig. 14.1E). The decrease in the frequency of such studies was rapid in the ecology journals but lagged by a decade in the ornithology journals. In each case, this decrease was accompanied by a new emphasis on species-diversity patterns, which was particularly pronounced in the ecological literature during the 5-yr interval in which MacArthur and MacArthur published their paper on bird species diversity (1961) (Fig. 14.1F). The differences in the timing of these shifts between the ecology and ornithology journals probably reflect a variety of factors – editorial personalities and practices, journal orientations, perceptions among authors about what is of interest to ecologists versus ornithologists, and readiness to accept mathematically based arguments about diversity.

I have suggested that a major focus of the MacArthurian paradigm was on niche differences among co-existing species. In fact, many of the papers published during the 1950s emphasized habitat differences, often in the context of pre-Hutchinsonian niche theory (Fig. 14.2A), so this focus had a clear historical antecedent. The emphasis on habitat differences diminished during the late 1950s and 1960s as avian ecologists turned their attention more to the forms of niche separation operating within local habitats. In the ornithological journals, a shift toward consideration of dietary differences was especially pronounced, whereas behavioral means of niche partitioning

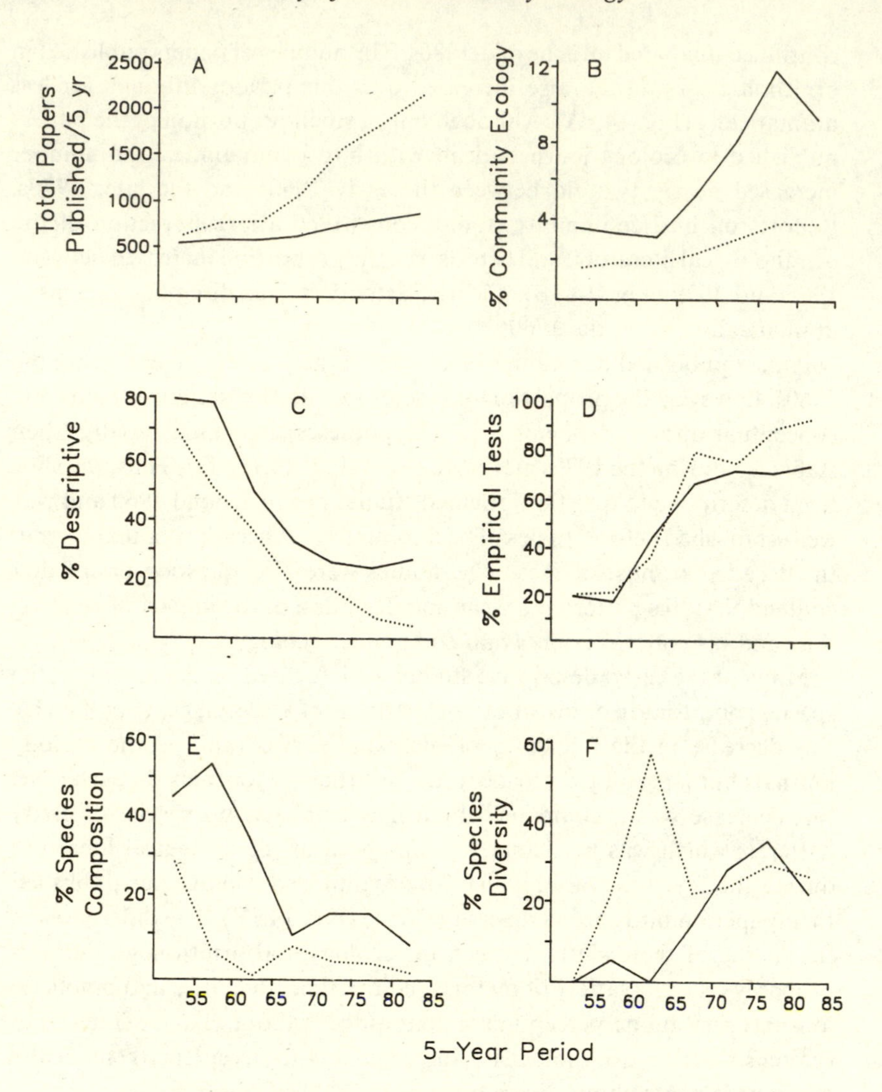

Fig. 14.1. Patterns of publication in ornithological (solid line) and ecological (dotted line) journals, 1950–84 by 5-yr intervals. A. Total papers published; B. Proportion of those papers dealing with avian community ecology; C. Proportion of the avian community ecology papers reporting descriptions only; D. Proportions of the avian community papers addressing questions or concepts or testing hypotheses empirically; E. Proportions of avian community papers dealing with the species composition of the community; F. Proportion of avian community papers focussing on species diversity.

suddenly became a popular topic in the ecological papers (Fig. 14.2B–C). The proportion of papers focused on behavioral or habitat niche differences has remained relatively stable in both types of journals since the late 1960s or early 1970s. During the 1960–74 period, studies of niche shifts and density compensation were relatively frequent, especially in the ecological literature, but this emphasis diminished in the late 1970s (Fig. 14.2D). This change might signal a growing dissatisfaction with the *ceteris paribus* assumption that is so central to the documentation and interpretation of these patterns.

Root's paper drawing attention to the utility of the guild concept in community studies was published in 1964, and the idea rapidly caught on, especially in the ecological literature (Fig. 14.2E). In ornithological publications, there was nearly a three-fold increase in the proportion of papers focusing on guilds between the late 1970s and the early 1980s. Once again, there were antecedents: aspects of the guild structure of communities received some attention in the ornithological journals during the 1950s, although different terms were used to describe the patterns. This chronology provides an especially clear example of the way in which a single paper that poses a new question and attaches a simple term to a concept may prompt a flurry of activity in a discipline.

In Chapter 13 I drew attention to the bioenergetic approach to communities. Figure 14.2F shows that interest in this approach has been relatively short-lived, not developing until the early 1970s, peaking in the late 1970s, and then nearly disappearing in the early 1980s. Energetic perspectives were not part of the MacArthurian conceptual framework and did not contribute to the major shift in the focus of community studies that occurred in the 1960s. The emergence of this perspective in the 1970s may have resulted at least in part from the ecosystem-oriented studies conducted as part of the International Biological Program. Rather than indicating the inapplicability of the approach to community question, its apparent demise in the early 1980s may reflect a degree of stagnation in the methodology used (mainly computer modelling) and the tendency of physiological ecologists to focus on single species.

As the focus on patterns has changed over this 35-yr period, so also have the interpretations of these patterns. The authors of many ornithological papers of the 1950s offered no causal interpretations of their findings at all. As these investigations became more question-oriented during the 1960s, however, explanations of one sort or another were offered with increasing frequency (Fig. 14.3A). Ecological papers, on the other hand, generally included some form of interpretation throughout this period, except for a short period in the early 1960s.

I have proposed that an important feature of the MacArthurian paradigm was its emphasis on interspecific competition as the primary cause of community patterns, and many of the examples I have discussed exemplify this emphasis. In fact, competition was a prevalent explanation of the observations in papers published in the ecological journals during the 1950s, prior to the emergence of the MacArthurian views (Fig. 14.3B). Of course, competition was the major feature of the Lotka–Volterra theory on which Hutchinson and MacArthur built their concepts (Kingsland 1985), and the idea of competitive exclusion was central to Lack's thinking about communities (Lack 1945b, 1947). This early interest in competition was less evident in the ornithological publications. There, the impact of the emerging MacArthurian view was more evident; nearly two-thirds of the papers appearing in 1965–69 attributed their findings to competition. Recourse to this explanation began to diminish thereafter, although it continued to be frequent into the 1980s.

What are the chronologies of other interpretations of community patterns? The notion that community patterns were determined by differences in the habitat responses of species (whether based on competition or on other selective forces) was frequent in the ecological literature in the early 1950s but then dropped, only to increase again during the 1970s, when it also became more fashionable in the ornithological literature (Fig. 14.3C). The frequency of references to the effects of past history or disturbance remained relatively unchanged (and relatively low) in the ecological papers over the entire time period; it was never a prevalent interpretation in the ornithological papers (Fig. 14.3D). Explanations based on the characteristics of individual species, independent of any interspecific interactions, increased during the late 1960s in the ecological journals and a decade later in the ornithological journals, reaching relatively high frequencies in the early 1980s (Fig. 14.3E). Such interpretations were also frequent in the ecological papers of the early 1950s, although not in the ornithological literature. This pattern may be a manifestation of the resurgence of an 'individualistic', Gleasonian view of communities among ecologists at that time, a movement that apparently did not influence those publishing in ornithological journals. Throughout the 35-yr period there was a slow but steady decrease in the tendency of authors to consider only a single interpretation of their observations, although consideration of alternative explanations still is not widespread (Fig. 14.3F).

These changes in the focus and interpretations of avian community

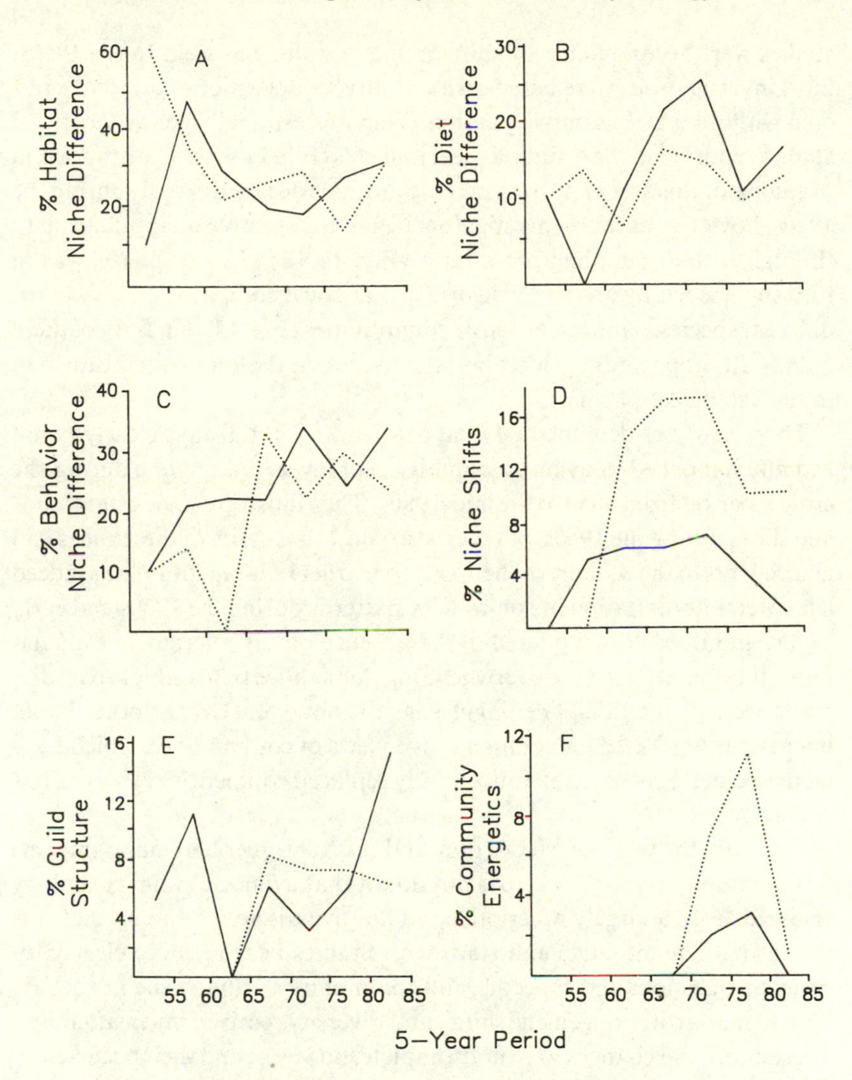

Fig. 14.2. Proportions of papers on avian community ecology in ornithological (solid line) and ecological (dotted line) journals focused on (A) habitat niche differences among species, (B) diet niche differences, (C) behavioral niche differences, (D) patterns of niche shifts or density compensation, (E) guild structure, and (F) community energetics, by 5-yr intervals, 1950–84.

studies were accompanied by shifts in methodology as well. In the 1950s, most investigations were based on qualitative observations that were gathered without a clear sampling design. This was especially characteristic of studies reported in the ornithological journals (Fig. 14.4A). The proportion of studies of this sort appearing in these journals declined rapidly during the 1960s, however, as more investigators began to assemble quantitative data through formal sampling procedures (Fig. 14.4B). This emphasis was in turn superseded by an increasing use of statistical comparisons of data for different species, habitats, areas, or communities (Fig. 14.4C). Both of these latter shifts appeared considerably later in the ornithological literature than in the ecological papers.

These analyses demonstrate quantitatively (if not always clearly) what actually happened in avian community ecology as a discipline during the critical period from 1950 to the mid-1980s. They illustrate some clear trends and shifts during the 1960s, but they also show that many of these shifts had antecedents in the ecology of the 1950s (or earlier). Competition was indeed a favored interpretation of community patterns during the 1960s and early 1970s, and it was often offered in the absence of any alternative explanations. It never attained the overwhelming dominance that a simple paradigmatic view of the discipline might suggest, however. On the other hand, interpretations based on noninteractive views of communities, which have increased in frequency, have not entirely replaced competitive explanations either.

Certainly the work of MacArthur and his colleagues had major impacts on community ecology. There is not doubt that avian community ecology changed from a largely descriptive, qualitative discipline to one that was increasingly quantitative and statistical. Studies became more clearly focused on questions and concepts, and community attributes such as behavioral niche partitioning, niche shifts, and diversity received much attention. These shifts and changes were not complete, however, and their occurrence, chronology, and magnitude frequently differed between papers appearing in ornithological journals and those published in ecology journals.

Concluding comments

Whether or not the body of theory and ideas that crystallized into a coherent research tradition in the 1960s merits designation as a formal paradigm, there is little doubt that it generated an outburst of scientific activity and excitement that has grown and diversified. Our search for community patterns has been dominated by expectations derived from this approach. That these patterns have not always been found does not dimin-

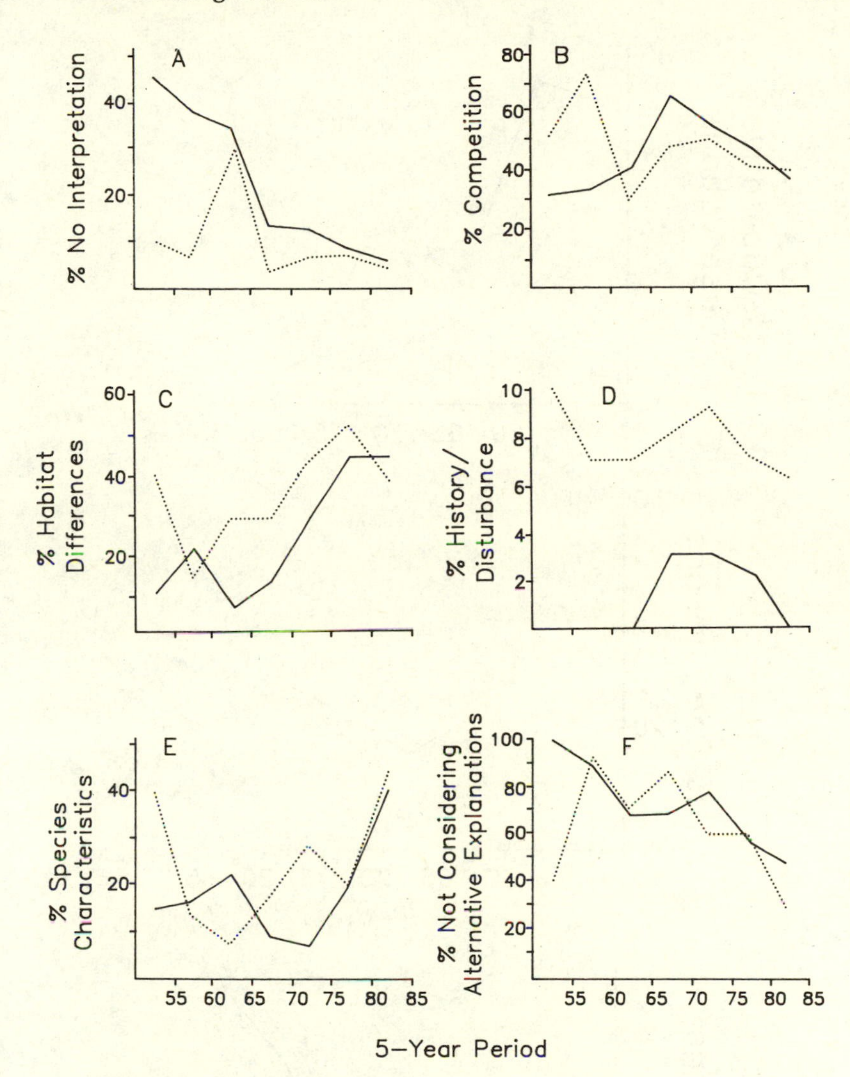

Fig. 14.3. Proportions of papers on avian community ecology in ornithological (solid line) and ecological (dotted line) journals (A) offering no causal interpretation of the observations, or interpreting the results in terms of (B) interspecific competition, (C) habitat differences among species, (D) effects of history of disturbance, or (E) species-specific attributes; panel (F) shows the proportions of papers offering no alternative explanations or hypotheses to the one given. Data are grouped by 5-yr periods, 1950–84.

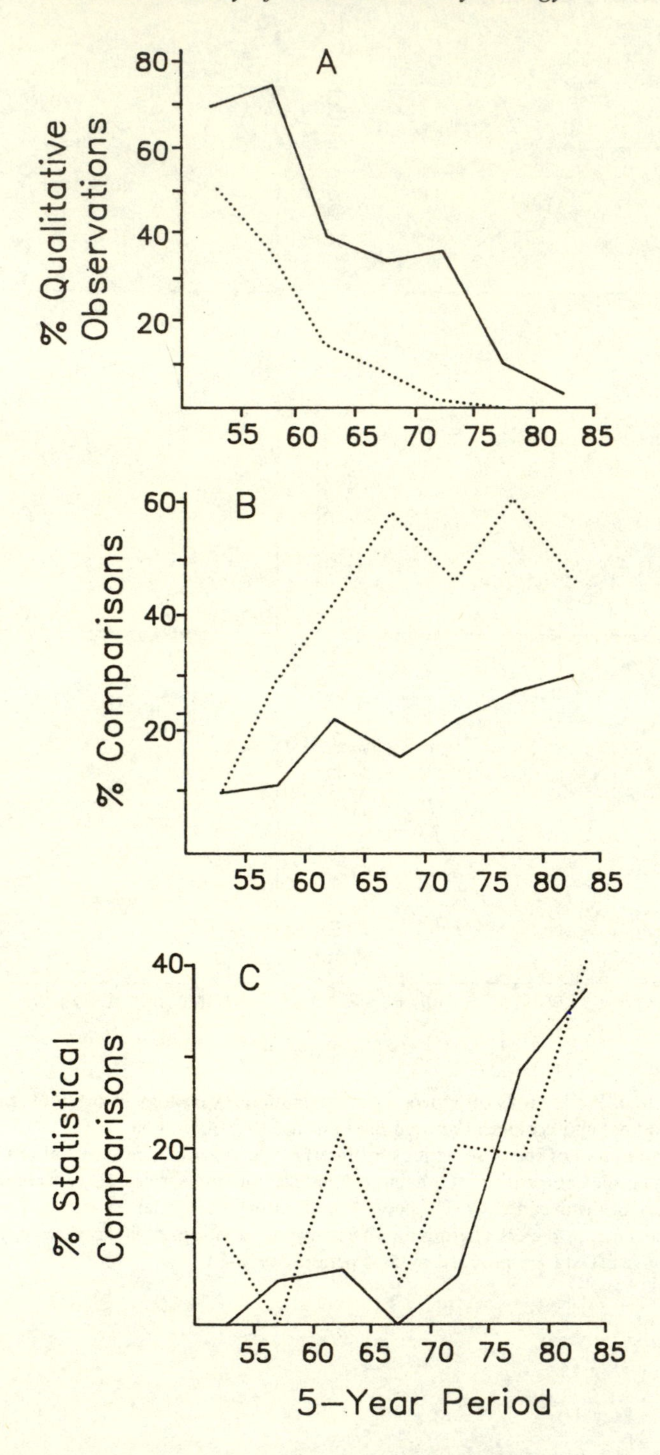

A
% Qualitative Observations
80
60
40
20
55 60 65 70 75 80 85
B
% Comparisons
60
40
20
55 60 65 70 75 80 85
C
% Statistical Comparisons
40
20
55 60 65 70 75 80 85
5-Year Period

ish the importance of the concepts or the value of the studies. MacArthur's hope that the variations in communities could be distilled into a few simple, general patterns has not yet been realized. Instead, the variety of patterns and their possible causes had led us to a new awareness of the complexity of natural communities, providing new avenues for ecological investigations. Several of the points I developed in previous chapters are worth re-emphasizing here, for they bear on how these investigations should be done.

First, the patterns of communities reflect what goes on during community assembly. Traditionally, ecologists have centered their attention on what they believe to be the end products of this process. Assembly, however, is likely to be an ongoing process. It is ongoing because each of the components in Fig. 4.1 is dynamic – each is influenced by a variety of factors that exhibit varying degrees of independence, interdependence, and chance. This is why the expectation of equilibrium is unlikely to be realized very often. To understand why community patterns are as they are, more attention must be devoted to the components influencing the assembly process.

Second, the degree to which competition may determine community patterns is a function of how species' demands for resources relate to resource availability. If explanations of community patterns are to be based on resource-related processes, then the resources must be identified and measured correctly. In so doing, it is essential that the components of resource systems (Fig. 10.2) be considered and that one knows which components one is measuring.

Third, it is short-sighted to believe that all important patterns in communities are consequences of resource-based interactions, that the only niche dimensions of importance are resource dimensions. A wide variety of other factors may influence the distribution and abundance of populations (Grinnell 1917, Andrewartha and Birch 1954, 1984) and thus contribute to community patterns. We need to conduct our studies within a broader conceptualization of the niche, one that includes both resource and nonresource dimensions.

Fourth, whether hypotheses about community patterns are tested following a formal Popperian protocol or some other approach, it is important

Fig. 14.4. Proportions of papers on avian community ecology in ornithological (solid line) and ecological (dotted line) journals in which the methodology involved primarily (A) qualitative observations obtained following no particular sampling design, (B) comparisons of either qualitative descriptions or quantitative data obtained without sampling, or (C) statistical comparisons of observations; the data are grouped by 5-yr intervals, 1950–84.

that the hypothesis, the tests, and the interpretations be logically sound. The observations and comparisons that constitute the tests must be made with appropriate attention to methodology and the possible biases introduced by various methods. Because comparisons are especially dependent on the *ceteris paribus* assumption, that assumption should be validated whenever possible. It is particularly important that the pattern and process components of hypotheses be clearly differentiated and, to the extent possible, tested separately.

Finally (and perhaps most importantly), it is obvious that there is tremendous variation in community patterns in both time and space. Some of this variation is orderly and predictable (seasonal changes, for example), but much of it is not. The variations reflect the influences of a variety of factors – differences in causal processes, intrinsic dynamics of the systems, shifts in the scale of operation of processes or in the scales on which we study the communities, the effects of history or disturbance, and chance. Rather than attempt to collapse this variation into some simple, general model of communities, we must focus our attention on the variation as something of interest in its own right. Understanding variation in community patterns holds the key to understanding the patterns themselves. This, in one way or another, is my theme in the second volume of this set.

References

Abbott, I. (1977) The role of competition in determining the differences between Victorian and Tasmanian passerine birds. *The Australian Journal of Ecology*, **25**, 429–47.

(1980) Theories dealing with the ecology of landbirds on islands. *Advances in Ecological Research*, **11**, 329–71.

(1981) The composition of landbird faunas of islands round southwestern Australia: is there evidence for competitive exclusion? *Journal of Biogeography*, **8**, 135–44.

(1983) The meaning of z in species/area regressions and the study of species turnover in island biogeography. *Oikos*, **41**, 385–90.

Abbott, I. & Grant, P.R. (1976) Nonequilibrial bird faunas on islands. *The American Naturalist*, **110**, 507–28.

Abbott, I. & Van Heurck, P. (1985) Tree species preferences of foraging birds in Jarrah Forest in Western Australia. *Australian Wildlife Research*, **12**, 461–66.

Abbott, I., Abbott, L.K. & Grant, P.R. (1977) Comparative ecology of Galápagos ground finches (*Geospiza* Gould): Evaluation of the importance of floristic diversity and interspecific competition. *Ecological Monographs*, **49**, 151–84.

Able, K.P. & Noon, B.R. (1976) Avian community structure along elevational gradients in the northeastern United States. *Oecologia*, **26**, 275–94.

Abraham, D.M. & Ankney, C.D. (1984) Partitioning of foraging habitat by breeding Sabine's Gulls and Arctic Terns. *The Wilson Bulletin*, **96**, 161–72.

Abrams, P. (1975) Limiting similarity and the form of the competition coefficient. *Theoretical Population Biology*, **8**, 356–75.

(1976) Niche overlap and environmental variability. *Mathematical Biosciences*, **28**, 357–75.

(1980) Some comments on measuring niche overlap. *Ecology*, **61**, 44–9.

(1983) The theory of limiting similarity. *Annual Review of Ecology and Systematics*, **14**, 359–76.

(1986) Character displacement and niche shift analyzed using consumer-resource models of competition. *Theoretical Population Biology*, **29**, 107–60.

(1988) How should resources be counted? *Theoretical Population Biology*, **33**, 226–42

Abrams, R.W. & Griffiths, A.M. (1981) Ecological structure of the pelagic seabird community in the Benguela Current region. *Marine Ecology – Progress Series*, **5**, 269–77

Adams, J. (1985) The definition and interpretation of guild structure in ecological communities. *Journal of Animal Ecology*, **54**, 43–59.

Airola, D.A. & Barrett, R.H. (1985) Foraging and habitat relationships of insect-gleaning birds in a Sierra Nevada mixed-conifer forest. *The Condor*, **87**, 205–16.

Alatalo, R.V. (1980) Seasonal dynamics of resource partitioning among foliage-gleaning passerines in Northern Finland. *Oecologia*, **45**, 190–6.
(1981a) Habitat selection of forest birds in the seasonal environment of Finland. *Annales Zoologici Fennici*, **18**, 103–14.
(1981b) Interspecific competition in tits *Parus* spp. and the goldcrest *Regulus regulus*: foraging shifts in multispecific flocks. *Oikos*, **37**, 335–44.
(1982a) Bird species distributions in the Galápagos and other archipelagos: competition or chance? *Ecology*, **63**, 881–7.
(1982b) Evidence for interspecific competition among European tits *Parus* spp.: a review. *Annales Zoologici Fennici*, **19**, 309–18.
(1982c) Multidimensional foraging niche organization of foliage–gleaning birds in northern Finland. *Ornis Scandinavica*, **13**, 56–71.
Alatalo, R.V. & Alatalo, R.H. (1980) Seasonal variation in evenness of forest bird communities. *Ornis Scandinavica*, **11**, 217–22.
Alatalo, R.V., Eriksson, D., Gustafsson, L. & Larsson, K. (1987) Exploitation competition influences the use of foraging sites by tits: experimental evidence. *Ecology*, **68**, 284–90.
Alatalo, R.V., Gustafsson, L. & Lundberg, A. (1986) Interspecific competition and niche changes in tits (*Parus* spp.): evaluation of nonexperimental data. *The American Naturalist*, **127**, 819–34.
Alatalo, R.V., Gustafsson, L., Lundberg, A. & Ulfstrand, S. (1985) Habitat shift of the Willow Tit *Parus montanus* in the absence of the Marsh Tit *Parus palustris*. *Ornis Scandinavica*, **16**, 121–8.
Alerstam, T. (1985) Fågelsamhället i Borgens lövskogsomåde. *Anser*, **24**, 213–34.
Alerstam, T. & Ulfstrand, S. (1977) Niches of tits *Parus* spp. in two types of African woodland. *Ibis*, **119**, 521–24.
Alerstam, T., Nilsson, S.G. & Ulfstrand, S. (1974) Niche differentiation during winter in woodland birds in southern Sweden and the Island of Gotland. *Oikos*, **25**, 321–30.
Alexander, R.D. (1979) *Darwinism and human affairs*. Seattle: University of Washington Press.
Alldredge, J.R. & Ratti, J.T. (1986) Comparison of some statistical techniques for analysis of resource selection. *Journal of Wildlife Management*, **50**, 157–65.
Allee, W.C., Emerson, A.E., Park, O., Park, T. & Schmidt, K.P. (1949) *Principles of Animal Ecology*. Philadelphia: W.B. Saunders Company.
Altmann, J. (1974) Observational study of behaviour: sampling methods. *Behaviour*, **49**, 227–67.
(1984) Observational sampling methods for insect behavioral ecology. *The Florida Entomologist*, **67**, 50–6.
Amat, J.A. (1984) Ecological segregation between Red-crested Pochard *Netta rufina* and Pochard *Aythya ferina* in a fluctuating environment. *Ardea*, **72**, 229–33.
Ambuel, B. & Temple, S.A. (1983) Area-dependent changes in the bird communities and vegetation of southern Wisconsin forests. *Ecology*, **64**, 1057–68.
Anderson, B.W., Ohmart, R.D. & Rice, J. (1983) Avian and vegetation community structure and their seasonal relationships in the lower Colorado River valley. *The Condor*, **85**, 392–405.
Anderson, D.J. & Horowitz, R.J. (1979) Competitive interactions among vultures and their avian competitors. *Ibis*, **121**, 505–9.
Anderson, S.H. (1981) Correlating habitat variables and birds. *Studies in Avian Biology*, **6**, 538–42.
Anderson, S.H. & Shugart, H.H. Jr (1974) Habitat selection of breeding birds in an east Tennessee deciduous forest. *Ecology*, **55**, 828–37.

Andersson, M & Norberg, R.Å. (1981) Evolution of reversed sexual size dimorphism and role partitioning among predatory birds, with a size scaling of flight performance. *Biological Journal of the Linnean Society*, **15**, 105–30.

Andrewartha, H.G. & Birch, L.C. (1954) *The distribution and abundance of animals.* Chicago: University of Chicago Press.

(1984) *The ecological web. More on the distribution and abundance of animals.* Chicago: University of Chicago Press.

Armstrong, R.A. & McGehee, R. (1980) Competitive exclusion. *The American Naturalist*, **115**, 151–70.

Arthur, W. (1982) The evolutionary consequences of interspecific competition. *Advances in Ecological Research*, **12**, 127–87.

Aschoff, J. & Pohl, H. (1970) Der Ruheumsatz von Vögeln als Function der Tageszeit und der Körpergrösse. *Journal für Ornithologie*, **111**, 38–47.

Ashmole, N.P. (1968) Body size, prey size, and ecological segregation in five sympatric tropical terns (Aves: Laridae). *Systematic Zoology*, **17**, 292–304.

(1971) Seabird ecology and the marine environment. In *Avian Biology*, ed. Farner, D.S. & King, J.R., pp. 223–86. New York: Academic Press.

Askenmo, C., von Bromssen, A., Ekman, J. & Jansson, C. (1977) Impact of some wintering birds on spider abundance in spruce. *Oikos*, **28**, 90–4.

Austin, G.T. (1970) Breeding birds of desert riparian habitat in southern Nevada. *The Condor*, **72**, 431–6.

Austin, G.T. & Tomoff, C.S. (1978) Relative abundance in bird populations. *The American Naturalist*, **112**, 695–9.

Austin, G.T., Blake, E.R., Brodkorb, P., Browning, M.R., Godfrey, W.E., Hubbard, J.P., McCaskie, G., Marshall, J.T. Jr, Monson, G., Olson, S.L., Ouellet, H., Palmer, R.S., Phillips, A.R., Pulich, W.M., Ramos, M.A., Rea, A.M. & Zimmerman, D.A. (1981) Ornithology as science. *The Auk*, **98**, 636–7.

Austin, M.P. (1979) Current approaches to the non-linearity problem in vegetation analysis. In *Contemporary quantitative ecology and related ecometrics*, ed Patil, E.P. & Rosenzweig, M., pp. 197–210. Fairland, Maryland: International Co-operative Publishers.

(1985) Continuum concept, ordination methods, and niche theory. *Annual Review of Ecology and Systematics*, **16**, 39–61.

Austin, M.P., Cunningham, R.B. & Fleming, P.M. (1984) New approaches to direct gradient analysis using environmental scalers and statistical curve-fitting procedures. *Vegetatio*, **55**, 11–27.

Baird, D., Evans, P.R., Milne, H. & Pienkowski, M.W. (1985) Utilization by shorebirds of benthic invertebrate production in intertidal areas. *Oceanography and Marine Biology Annual Review*, **23**, 573–97.

Bakeman, R. & Gottman, J.M. (1986) *Observing interaction. An introduction to sequential analysis.* Cambridge: Cambridge University Press.

Baker, M.C. (1977) Shorebird food habits in the eastern Canadian arctic. *The Condor*, **79**, 56–62.

(1979) Morphological correlates of habitat selection in a community of shorebirds (Charadriiformes). *Oikos*, **33**, 121–6.

Baker, M.C. & Baker, A.E.M. (1973) Niche relationships among six species of shorebirds on their wintering and breeding ranges. *Ecological Monographs*, **43**, 193–212.

Balda, R.P. (1969) Foliage used by birds of the oak-juniper woodland and ponderosa pine forest in southeastern Arizona. *The Condor*, **71**, 399–412.

(1975) Vegetation structure and breeding bird diversity. In *USDA Forest Service General Technical Report WO–1*, pp. 59–80. Washington, D.C.: USDA Forest Service.

Balen, J.H. van, Booy, C.J.H., Franeker, J.A. van & Osieck, E.R. (1982) Studies on hole-nesting birds in natural nest sites. I. Availability and occupation of natural nest sites. *Ardea*, **70**, 1–24.

Baltz, D.M. & Morejohn, G.V. (1977) Food habits and niche overlap of seabirds wintering on Monterey Bay, California. *The Auk*, **94**, 526–43.

Bart, J. & Schoultz, J.D. (1984) Reliability of singing bird surveys: changes in observer efficiency with avian density. *The Auk*, **101**, 307–18.

Bartholomew, G.A. (1958) The role of physiology in the distribution of terrestrial vertebrates. In *Zoogeography*, ed. Hubbs, C.L., pp. 81–95. Washington, D.C.: American Association for the Advancement of Science.

Beals, E. (1960) Forest bird communities in the Apostle Islands of Wisconsin. *The Wilson Bulletin*, **72**, 156–81.

Bédard, J. (1976) Coexistence, coevolution and convergent evolution in seabird communities: a comment. *Ecology*, **57**, 177–84.

Beecher, W.J. (1962) The biomechanics of the bird skull. *Bulletin of the Chicago Academy of Sciences*, **11**, 10–33.

Beedy E.C. (1981) Bird communities and forest structure in the Sierra Nevada of California. *The Condor*, **83**, 97–105.

Beever, J.W. (1979) The niche-variation hypothesis: an examination of assumptions and organisms. *Evolutionary Theory*, **4**, 181–91.

Bell, H.L. (1984) A bird community of lowland rainforest in New Guinea. 6. Foraging ecology and community structure of the avifauna. *The Emu*, **84**, 142–58.

(1985) The social organization and foraging behaviour of three syntopic thornbills *Acanthiza* spp. In *Birds of Eucalypt Forests and Woodlands: Ecology, Conservation, Management*, ed. Keast, A., Recher, H.F., Ford, H. & Saunders D., pp. 151–63. Chipping Norton, NSW: Surrey Beatty & Sons.

Belopol'skii, L.O. (1957) *Ecology of sea colony birds of the Barents Sea*. Jerusalem: Israel Program for Scientific Translations.

Bender, E.A., Case, T.J. & Gilpin, M.E. (1984) Perturbation experiments in community ecology: theory and practice. *Ecology*, **65**, 1–13.

Bengtson, S-A. & Bloch, D. (1983) Island land bird population densities in relation to island size and habitat quality on the Faroe Islands. *Oikos*, **41**, 507–22.

Berger, A.J. (1981) *Hawaiian birdlife*. Honolulu: University of Hawaii Press.

Berndt, R. & von Henss, M. (1967) Die kohlmeise, *Parus major*, als invasionsvögel. *Die Vögelwart*, **24**, 17–37.

Berthold, P. (1976) Methoden der Bestandserfassung in der Ornithologie: Ubersicht und Kritische Betrachtung. *Journal für Ornithologie*, **117**, 1–69.

Best, L.B. (1975) Interpretational errors in the 'mapping method' as a census technique. *The Auk*, **92**, 452–60.

(1981) Seasonal changes in detection of individual bird species. *Studies in Avian Biology*, **6**, 252–61.

Best, L.B. & Stauffer, D.F. (1986) Factors confounding evaluation of bird-habitat relationships. In *Wildlife 2000. Modeling habitat relationships of terrestrial vertebrates*, ed. Verner, J., Morrison, M.L. & Ralph, C.J., pp. 209–16. Madison: The University of Wisconsin Press.

Betts, M.M. (1955) The food of titmice in oak woodland. *Journal of Animal Ecology*, **24**, 282–323.

Bilcke, G. (1982) Breeding songbird community structure: influences of plot size and vegetation. *Acta Œcologica Œcologia Generalis*, **3**, 511–21.

(1984) Seasonal changes in habitat use of resident passerines. *Ardea*, **72**, 95–9.

Birks, H.J.B. (1987) Recent methodological developments in quantitative descriptive biogeography. *Annales Zoologici Fennici*, **24**, 165–78.

Blake, J.G. (1983) Trophic structure of bird communities in forest patches in east-central Illinois. *The Wilson Bulletin*, **83**, 416–30.

(1986) Species–area relationship of migrants in isolated woodlots in east-central Illinois. *The Wilson Bulletin*, **98**, 291–96.

Blakers, M., Davies, S.J.J.F. & Reilly, P.N. (1984) *The atlas of Australian birds*. Carlton, Victoria: Melbourne University Press.

Block, W.M., With, K.A. & Morrison, M.L. (1987) On measuring bird habitat: influence of observer variability and sample size. *The Condor*, **89**, 241–51.

Blondel, J. (1979) *Biogéographie et écologie*, Paris: Masson.

(1981) Structure and dynamics of bird communities in Mediterranean habitats. In *Mediterranean-Type Shrublands*, ed. di Castri, F., Goodall, D.W. & Specht, R.L., pp. 361–85. Amsterdam: Elsevier.

(1985a) Habitat selection in island versus mainland birds. In *Habitat Selection in Birds*, ed. Cody, M.L., pp. 477–516. New York: Academic Press.

(1985b) Breeding strategies of the blue tit and coal tit (*Parus*) in mainland and island mediterranean habitats: a comparison. *Journal of Animal Ecology*, **54**, 531–56.

Blondel, J., Vuilleumier, F., Marcus, L.F. & Terouanne, E. (1984) Is there ecomorphological convergence among Mediterranean bird communities of Chile, California, and France? *Evolutionary Biology*, **18**, 141–213.

Boag, P.T. (1983) The heritability of external morphology in Darwin's ground finches (*Geospiza*) on Isla Daphne Major, Galápagos. *Evolution*, **37**, 877–94.

Boag, P.T. & Grant, P.R. (1978) Heritability of external morphology in Darwin's finches. *Nature*, **274**, 793–4.

(1981) Intense natural selection in a population of Darwin's finches (Geospizinae) in the Galápagos. *Science*, **214**, 82–5.

(1984a) The classical case of character release. Darwin's finches (*Geospiza*) on Isla Daphne Major, Galápagos. *Biological Journal of the Linnean Society*, **22**, 243–87.

(1984b) Darwin's finches (*Geospiza*) on Isla Daphne Major, Galápagos: breeding and feeding ecology in climatically variable environment. *Ecological Monographs*, **54**, 463–89.

Bock, C.E. (1984) Geographical correlates of abundance vs. rarity in some North America winter landbirds. *The Auk*, **101**, 266–73.

(1987) Distribution–abundance relationships of some Arizona landbirds: a matter of scale. *Ecology*, **68**, 124–9.

Bock, C.E. & Lepthien, L.W. (1976) Synchronous eruption of boreal seed-eating birds. *The American Naturalist*, **110**, 559–71.

Bock, C.E. & Lynch, J.F. (1970) Breeding bird populations of burned and unburned conifer forest in the Sierra Nevada. *The Condor*, **72**, 182–9.

Bock, C.E. & Ricklefs, R.E. (1983) Range size and local abundance of some North American songbirds: a positive correlation. *The American Naturalist*, **122**, 295–9.

Bock, W.J. (1964) Kinetics of the avian skull. *Journal of Morphology*, **144**, 1–42.

Boecklen, W.J. & Gotelli, N.J. (1984) Island biogeographic theory and conservation practice: species–area or specious–area relationships? *Biological Conservation*, **29**, 63–80.

Boecklen, W.J. & Simberloff, D. (1986) Area-based extinction models in conservation. In *Dynamics of Extinction*, ed. Elliott, D.K., pp. 247–76. New York: John Wiley & Sons.

Bond, J. (1971) *Birds of the West Indies*, 2nd edn. Boston: Houghton-Mifflin Company.

Bond, R.R. (1957) Ecological distribution of breeding birds in the upland forests of southern Wisconsin. *Ecological Monographs*, **27**, 351–84.

Bookstein, F.L., Chernoff, B., Elder, R.L., Humphries, J.M. Jr, Smith, G.R. & Strauss, R.E. (1985) *Morphometrics in Evolutionary Biology*. Philadelphia: The Academy of Natural Sciences of Philadelphia.

Boström, U. & Nilsson, S.G. (1983) Latitudinal gradients and local variations in species richness and structure of bird communities on raised peat-bogs in Sweden. *Ornis Scandinavica*, **14**, 213–26.

Bourne, W.R.P. (1981) Some factors underlying the distribution of seabirds. In *Proceedings of the Symposium on birds of the sea and shore*, ed. Cooper, J., pp. 119–34. Cape Town: African Seabird Group.

(1983) Reappraisal of threats to seabirds. *Marine Pollution Bulletin*, **14**, 1.

Bowers, M.A. & Brown, J.H. (1982) Body size and coexistence in desert rodents: chance or community structure. *Ecology*, **63**, 391–400.

Bowman, R.I. (1961) Morphological differentiation and adaptation in the Galapagos finches. *University of California Publications in Zoology*, **58**, 1–302.

Bradley, D.W. (1985) The effects of visibility bias on time-budget estimates of niche breadth and overlap. *The Auk*, **102**, 493–9.

Bradley, R.A. & Bradley, D.W. (1983) Co-occurring groups of wintering birds in the lowlands of southern California. *The Auk*, **100**, 491–3.

(1985) Do non-random patterns of species in niche space imply competition? *Oikos*, **45**, 443–6.

Brady, R.H. (1982) Dogma and doubt. *Biological Journal of the Linnean Society*, **17**, 79–96.

Brandl, R & Utschick, H. (1985) Size, ecology and wading birds: a nonparsimonious view. *Naturwissenschaften*, **72**.

Bridges, L. (1980) 'Some examples of the Behaviour and Feeding Ecology of the Rufous (*Pachycephala rufiventris*) and Golden (*Pachycephala pectoralis*) Whistler'. B.Sc. (Honours) Thesis, Armidale, NSW: University of New England.

Briggs, K.T. & Chu, E.W. (1987) Trophic relationships and food requirements of California seabirds: updating models of trophic impact. In *Seabirds: Feeding Ecology and Role in Marine Ecosystems*, ed. Croxall, J.P., pp. 279–304. Cambridge: Cambridge University Press.

Bronowski, J. (1977) *A sense of the future*. Cambridge, Massachusetts: MIT Press.

Brooks, D.R. (1985) Historical ecology: a new approach to studying the evolution of ecological associations. *Annals of the Missouri Botanical Garden*, **72**, 660–80.

Broome, L.S. (1985) Sightability as a factor in aerial survey of bird species and communities. *Australian Wildlife Research*, **12**, 57–67.

Brown, J.H. (1981) Two decades of homage to Santa Rosalia; toward a general theory of diversity. *American Zoologist*, **21**, 877–88.

(1984) On the relationship between abundance and distribution of species. *The American Naturalist*, **124**, 255–79.

Brown, J.H. & Bowers, M.A. (1984) Patterns and processes in three guilds of terrestrial vertebrates. In *Ecological communities. Conceptual issues and the evidence*, ed. Strong, D.R. Jr, Simberloff, D., Abele, L.G. & Thistle, A.B., pp. 282–96. Princeton: Princeton University Press.

(1985) Community organization in hummingbirds: relationships between morphology and ecology. *The Auk*, **102**, 251–69.

Brown, J.H. & Gibson, A.C. (1983) *Biogeography*. St Louis: Mosby.

Brown, J.H. & Kodric-Brown, A. (1977) Turnover rates in insular biogeography: effect of immigration on extinction. *Ecology*, **58**, 445–9.

(1979) Convergence, competition, and mimicry in a temperate community of hummingbird-pollinated flowers. *Ecology*, **60**, 1022–35.

Brown, J.H. & Maurer, B.A. (1986) Body size, ecological dominance and Cope's Rule. *Nature*, **324**, 248–50.

(1987) Evolution of species assemblages: effects of energetic constraints and species dynamics on the diversification of the North American avifauna. *The American Naturalist*, **130**, 1–17.

Brown, J.H., Calder, W.A. III & Kodric-Brown, A. (1978) Correlates and consequences of body size in nectar-feeding birds. *The American Zoologist*, **68**, 687–700.

Brown, J.H., Davidson, D.W., Munger, J.C. & Inouye, R.S. (1986) Experimental community ecology: the desert granivore system. In *Community ecology*, ed. Diamond, J. & Case, T.J., pp. 41–61. New York: Harper & Row.

Brown, J.L. (1969) Territorial behavior and population regulation in birds. A review and re-evaluation. *The Wilson Bulletin*, **81**, 293–329.

Brown, R.G.B. (1980) Seabirds as marine animals. In *Behavior of Marine Animals. Volume 4: Marine Birds*, ed. Burger, J., Olla, B.L. & Winn, H.E., pp. 1–39. New York: Plenum Press.

Brown, R.N. (1977) Character convergence in bird song. *Canadian Journal of Zoology*, **55**, 1523–9.

Brown, R.N. & Lemon, R.E. (1979) Structure and evolution of song form in the wrens *Thyrothorus sinaloa* and *T. felix*. *Behavioral Ecology and Sociobiology*, **5**, 111–31.

Brown, W.L. Jr & Wilson, E.O. (1956) Character displacement. *Systematic Zoology*, **7**, 49–64.

Brush, T. & Stiles, E.W. (1986) Using food abundance to predict habitat use by birds. In *Wildlife 2000. Modeling habitat relationships of terrestrial vertebrates*, ed. Verner, J., Morrison, M.L. & Ralph, C.J., pp. 57–63. Madison: The University of Wisconsin Press.

Bryant, D.M. & Westerterp, K.R. (1980) The energy budget of the House Martin (*Delichon urbica*). *Ardea*, **68**, 91–102.

Bryant, D.M., Hails, C.J. & Prys-Jones, R. (1985) Energy expenditure by free-living Dippers (*Cinclus cinclus*) in winter. *The Condor*, **87**, 177–86.

Bryant, D.M., Hails, C.J. & Tatner, P. (1984) Reproductive energetics of two tropical bird species. *The Auk*, **101**, 25–37.

Burdon-Sanderson, J.S. (1893) Inaugural address. *Nature*, **48**, 464–72.

Burger, J., Howe, M.A., Hahn, D.C. & Chase, J. (1977) Effects of tide cycles on habitat selection and habitat partitioning by migrating shorebirds. *The Auk*, **94**, 743–58.

Burnham, K.P. & Anderson, D.R. (1984) The need for distance data in transect counts. *Journal of Wildlife Management*, **48**, 1248–54.

Burnham, K.P., Anderson, D.R. & Laake, J.L. (1980) Estimation of density from line transect sampling of biological populations. *Wildlife Monographs*, **72**, 1–202.

(1981) Line transect estimation of bird population density using a fourier series. *Studies in Avian Biology*, **6**, 466–82.

(1985) Efficiency and bias in strip and line transect sampling. *Journal of Wildlife Management*, **49**, 1012–18.

Calder, W.A. III. (1974) The consequences of body size for avian energetics. In *Avian Energetics*, ed. Paynter, R.A. Jr, pp. 86–157. Cambridge, Massachusetts: Nuttall Ornithological Club.

(1984) *Size, Function, and Life History*. Cambridge: Harvard University Press.

Calver, M.C. & Wooller, R.D. (1981) Seasonal differences in the diets of small birds in the Karri forest understorey. *Australian Wildlife Research*, **8**, 653–7.

Cameron, E. (1985) Habitat usage and foraging behaviour of three fantails (Rhipidura: Pachycephalidae). In *Birds of Eucalypt Forests and Woodlands: Ecology, Conservation, Management*, ed. Keast, A., Recher, H.F., Ford, H., & Saunders, D., pp. 177–91. Chipping Norton, NSW: Surrey Beatty & Sons.

Capen, D.E., ed. (1981) *The Use of Multivariate Statistics in Studies of Wildlife Habitat (General Technical Report RM-87)*. Fort Collins, Colorado: Rocky Mountain Forest and Range Experiment Station, USDA Forest Service.

Capen, D.E., Fenwick, J.W., Inkley, D.B. & Boynton, A.C. (1986) Multivariate models of songbird habitat in New England forests. In *Wildlife 2000. Modeling Habitat Relationships of Terrestrial Vertebrates*, ed. Verner, J., Morrison, M.L., & Ralph, C.J., pp. 171–5. Madison: University of Wisconsin Press.

Carnap, R. (1923) Über die Aufgabe der Physik und die Andwendung des Grundsatze der Einfachstheit. *Kant-Studien*, **28**, 90–107.

Carnes, B.A. & Slade, N.A. (1982) Some comments on niche analysis in canonical space. *Ecology*, **63**, 888–93.

Carothers, J.H. (1982) Effects of trophic morphology and behavior on foraging rates of three Hawaiian honeycreepers. *Oecologia*, **55**, 157–9.

(1986) Homage to Huxley: on the conceptual origin of minimum size ratios among competing species. *The American Naturalist*, **128**, 440–2.

Carpenter, F.L. (1978) A spectrum of nectar-eater communities. *The American Zoologist*, **18**, 809–19.

Case, T.J. (1981) Niche separation and resource scaling. *The American Naturalist*, **118**, 554–60.

(1982) Coevolution in resource-limited competition communities. *Theoretical Population Biology*, **21**, 69–91.

(1984) Niche overlap and resource weighting terms. *The American Naturalist*, **124**, 604–8.

Case, T.J. & Cody, M.L., eds. (1983) *Island Biogeography in the Sea of Cortez*. Berkeley: University of California Press.

Case, T.J. & Sidell, R. (1983) Pattern and chance in the structure of model and natural communities. *Evolution*, **37**, 832–49.

Case, T.J., Faaborg, J. & Sidell, R. (1983) The role of body size in the assembly of West Indian bird communities. *Evolution*, **37**, 1062–74.

Case, T.J., Gilpin, M.E. & Diamond, J.M. (1979) Over exploitation, interference competition, and excess density compensation in insular faunas. *The American Naturalist*, **113**, 843–54.

Caswell, H. (1976) Community structure: a neutral model analysis. *Ecological Monographs*, **46**, 327–54.

Catchpole, C.K. (1973) Conditions of co-existence in sympatric breeding populations of *Acrocephalus* warblers. *Journal of Animal Ecology*, **42**, 623–35.

(1978) Interspecific territorialism and competition in *Acrocephalus* warblers as revealed by playback experiments in areas of sympatry and allopatry. *Animal Behaviour*, **26**, 1072–80.

Caughley, G. (1977) *Analysis of Vertebrate Populations*. New York: John Wiley & Sons.

Chamberlin, T.C. (1965) The method of multiple working hypotheses. *Science*, **148**, 754–9.

Charnov, E.L. (1976) Optimal foraging: the marginal value theorem. *Theoretical Population Biology*, **9**, 129–36.

Clark, L.R., Geier, P.W., Hughes, R.D. & Morris, R.F. (1967) *The Ecology of Insect Populations in Theory and Practice*. London: Methuen and Company.

Clements, F.E. (1916) Plant succession: an analysis of the development of vegetation. *Publication no. 242*, Washington, D.C.: Carnegie Institute.

Clout, M.N. & Gaze, P.D. (1984) Effects of plantation forestry on birds in New Zealand. *Journal of Applied Ecology*, **21**, 795–815.

Clutton-Brock, T.H. & Harvey, P.H. (1979) Comparison and adaptation. *Proceedings of the Royal Society of London*, **205**, 547–65.

Cody, M.L. (1966) The consistency of intra- and inter-continental grassland bird species counts. *The American Naturalist*, **100**, 371–6.

(1968) On the methods of resource division in grassland bird communities. *The American Naturalist*, **102**, 107–47.

(1969) Convergent characteristics in sympatric species: a possible relation to interspecific competition and agression. *The Condor*, **71**, 222–39.

(1973a) Character convergence. *Annual Review of Ecology and Systematics*, **4**, 189–211.

(1973b) Coexistence, coevolution and convergent evolution in seabird communities. *Ecology*, **56**, 31–44.

(1974a) *Competition and the structure of bird communities*. Princeton: Princeton University Press.

(1974b) Optimization in ecology. *Science*, **183**, 1156–64.

(1975) Towards a theory of continental species diversities: bird distributions over mediterranean habitat gradients. In *Ecology and Evolution of Communities*, ed. Cody, M.L. & Diamond, J. M., pp. 214–57. Cambridge: Harvard University Press.

(1978) Habitat selection and interspecific territoriality among the sylviid warblers of England and Sweden. *Ecological Monographs*, **48**, 351–96.

(1979) Resource allocation patterns in Palaearctic warblers (Sylviidae). *Fortschritte der Zoologie*, **25**, 223–34.

(1980) Species-packing in insectivorous bird communities: density, diversity, and productivity. In *Acta XVII Congressus Internationalis Ornithologici*, ed. Nohring, R., pp. 1071–77. Berlin: Deutsche Ornithologen–Gesellschaft.

(1981) Habitat selection in birds: the roles of vegetation structure, competitors, and productivity. *BioScience*, **31**, 107–11.

(1983a) The land birds. In *Island Biogeography in the Sea of Cortez*, ed. Case, T.J. & Cody, M.L., pp. 210–45. Berkeley: University of California Press.

(1983b) Continental diversity patterns and convergent evolution in bird communities. In *Mediterranean-Type Ecosystems*, ed. Kruger, F.J., Mitchell, D.T., & Jarvis, J.U.M., pp. 357–402. Berlin: Springer-Verlag.

(1983c) Bird diversity and density in south African forests. *Oecologia*, **59**, 201–15.

ed. (1985) *Habitat Selection in Birds*. New York: Academic Press.

Cody, M.L. & Brown, J.H. (1970) Character convergence in Mexican finches. *Evolution*, **24**, 304–10.

Cody, M.L. & Diamond, J.M., eds. (1975) *Ecology and Evolution of Communities*. Cambridge: Harvard University Press.

Cody, M.L. & Mooney, H.A. (1978) Convergence versus nonconvergence in Mediterranean-climate ecosystems. *Annual Review of Ecology and Systematics*, **9**, 265–321.

Cody, M.L. & Walter, H. (1976) Habitat selection and interspecific interactions among Mediterranean sylviid warblers. *Oikos*, **27**, 210–38.

Cody, M.L., Fuentes, E.R., Glanz, W., Hunt, J.H. & Moldenke, A.R. (1977) Convergent evolution in the consumer organisms of mediterranean Chile and California. In *Convergent Evolution In Chile and California*, ed. Mooney, H.A., pp. 144–92. Stroudsberg, Pennsylvania: Dowden, Hutchinson & Ross.

Cody, M.L., Moran, R. & Thompson, H. (1983) The plants. In *Island Biogeography in the Sea of Cortez*, ed. Case, T.J. & Cody, M.L., pp. 49–97. Berkeley: University of California Press.

Coleman, B.D. (1981) On random placement and species–area relations. *Mathematical Biosciences*, **54**, 191–215.

Coleman, B.D., Mares, M.A., Willig, M.R. & Hsieh, Y-H. (1982) Randomness, area and species richness. *Ecology*, **63**, 1121–33.

Collias, N.E. & Collias, E.C. (1984) *Nest Building and Bird Behavior*. Princeton: Princeton University Press.

Collins, S.L. (1983a) Geographic variation in habitat structure of the Black-throated Green Warbler (*Dendroica virens*). *The Auk*, **100**, 382–89.

(1983b) Geographic variation in habitat structure for the wood warblers in Maine and Minnesota. *Oecologia*, **59**, 246–52.

Collins, S.L., James, F.C. & Risser, P.G. (1982) Habitat relationship of wood warblers (Parulidae) in northern central Minnesota. *Oikos*, **39**, 50–8.

Colwell, R.K. & Futuyma, D.J. (1971) On the measurement of niche breadth and overlap. *Ecology*, **52**, 567–76.

Colwell, R.K. & Winkler, D.W. (1984) A null model for null models in biogeography. In *Ecological Communities. Conceptual Issues and the Evidence*, ed. Strong, D.R. Jr, Simberloff, D., Abele, L.G. & Thistle, A.B., pp. 344–59. Princeton: Princeton University Press.

Connell, J.H. (1975) Some mechanisms producing structure in natural communities: a model and evidence from field experiments. In *Ecology and Evolution of Communities*, ed. Cody, M.L. & Diamond, J.M., pp. 460–90. Cambridge: Harvard University Press.

(1980) Diversity and coevolution of competitors, or the ghost of competition past. *Oikos*, **35**, 131–8.

(1983) On the prevalence and relative importance of interspecific competition: evidence from field experiments. *The American Naturalist*, **122**, 661–96.

Connor, E.F. & McCoy, E.D. (1979) The statistics and biology of the species–area relationship. *The American Naturalist*, **113**, 791–833.

Connor, E.F. & Simberloff, D. (1978) Species number and compositional similarity of the Galápagos flora and avifauna. *Ecological Monographs*, **48**, 219–48.

(1979) The assembly of species communities: chance or competition? *Ecology*, **60**, 1132–40.

(1983) Interspecific competition and species co-occurrence patterns on islands: null models and the evaluation of evidence. *Oikos*, **41**, 455–65.

(1984a) Neutral models of species' co-occurrence patterns. In *Ecological Communities. Conceptual Issues and the Evidence*, ed. Strong, D.R. Jr, Simberloff, D., Abele, L.G. & Thistle, A.B., pp. 316–31. Princeton: Princeton University Press.

(1984b) Rejoinder. In *Ecological Communities. Conceptual Issues and the Evidence*, ed. Strong, D.R. Jr, Simberloff, D., Abele, L.G. & Thistle, A.B., pp. 341–3. Princeton: Princeton University Press.

(1986) Competition, scientific method, and null models in ecology. *American Scientist*, **74**, 155–62.

Cook, R.E. (1969) Variation in species density in North American birds. *Systematic Zoology*, **18**, 63–84.

Cox, G.W. & Ricklefs, R.E. (1977) Species diversity and ecological release in Caribbean land bird faunas. *Oikos*, **28**, 113–22.

Cracraft, J. (1985) Biological diversification and its causes. *Annals of the Missouri Botanical Garden*, **72**, 794–822.

Craig, R.J. (1984) Comparative foraging ecology of Louisiana and Northern waterthrushes. *The Wilson Bulletin*, **96**, 173–83.

Crawford, H.S., Hooper, R.G. & Titterington, R.W. (1981) Songbird population response to silvicultural practices in central Appalachian hardwoods. *Journal of Wildlife Management*, **45**, 680–92.

Crawford, R.J. & Shelton, P.A. (1978) Pelagic fish and seabird interrelationships off the coast of South West and South Africa. *Biological Conservation*, **14**, 85–109.

Crome, F.H.J. (1978) Foraging ecology of an assemblage of birds in lowland rainforest in northern Queensland. *Australian Journal of Ecology*, **3**, 195–212.

Crowe, T.M. & Crowe, A.A. (1982) Patterns of distribution, diversity and endemism in Afrotropical birds. *Journal of Zoology (London)*, **198**, 417–42.

Crowell, K.L. (1962) Reduced interspecific competition among the birds of Bermuda. *Ecology*, **43**, 75–88.

Croxall, J.P., ed. (1987). *Seabirds: feeding ecology and role in marine ecosystems.* Cambridge: Cambridge University Press.

Croxall, J.P. & Prince, P.A. (1980) Food, feeding ecology and ecological segregation of seabirds at South Georgia. *Biological Journal of the Linnean Society*, **14**, 103–31.

(1982) A preliminary assessment of the impact of seabirds on marine resources at South Georgia. *Com. Nat. Franc. Recherch. Antarct.*, **51**, 501–9.

(1987) Seabirds as predators on marine resources, especially krill, at South Georgia. In *Seabirds: Feeding Ecology and Role in Marine Ecosystems*, ed. Croxall, J.P.. pp. 347–68. Cambridge: Cambridge University Press.

Croxall, J.P., Ricketts, C. & Prince, P.A. (1984) Impact of seabirds on marine resources, especially krill, of South Georgia waters. In *Seabird Energetics*, ed. Whittow, G.C. & Rahn, H., pp. 285–317. New York: Plenum Press.

Curtis, J.T. (1959) *The Vegetation of Wisconsin*. Madison: The University of Wisconsin Press.

Custer, T.W. & Pitelka, F.A. (1975) Correction factors for digestion rates for prey taken by Snow Buntings (*Plectrophenax nivalis*). *The Condor*, **77**, 210–12.

Dansereau, P., Buell, P.F. & Dagon, R. (1966) A universal system for recording vegetation. II. A methodological critique and an experiment. *Sarracenia*, **10**, 1–64.

Darwin, C. (1859) *The origin of species by means of natural selection*. London: Murray.

(1897) *Voyage of a Naturalist*. London: Murray.

Davies, N.B. (1982) Behaviour and competition for scarce resources. In *Current Problems in Sociobiology*, ed. King's College Sociobiology Group, pp. 363–80. Cambridge: Cambridge University Press.

Dawson, W.R. (1981) Adjustments of Australian birds to thermal conditions and water scarcity in arid zones. In *Ecological Biogeography of Australia*, ed. Keast, A., pp. 1651–74. The Hague: W. Junk.

(1984) Physiological studies of desert birds: present and future considerations. *Journal of Arid Environments*, **7**, 133–55.

Dayton, P.K. (1979) Ecology: a science and a religion. In *Ecological Processes in Coastal and Marine Systems*, ed. Livingston, R.J., pp. 3–18. New York: Plenum Press.

Dayton, P.K. & Oliver, J.S. (1980) An evaluation of experimental analyses of population and community patterns in benthic marine environments. In *Marine Benthic Dynamics*, ed. Tenore, K.R. & Coull, B.C., pp. 93–120. Columbia, South Carolina: University of South Carolina Press.

DeJong, M.J. & Emlen, J.T. (1985) The shape of the auditory detection function and its implications for songbird censusing. *Journal of Field Ornithology*, **56**, 213–23.

den Boer, P.J. (1981) On the survival of populations in a heterogeneous and variable environment. *Oecologia*, **50**, 39–53.

DeSante, D.F. (1981) A field test of the variable circular-plot censusing technique in a California coastal scrub breeding bird community. *Studies in Avian Biology*, **6**, 177–85.

(1986) A field test of the variable circular-plot censusing method in a Sierran subalpine forest habitat. *The Condor*, **88**, 129–42.

De Vita, J. (1979) The broken-stick model: a response to Pielou's critique and some general comments. *The American Naturalist*, **119**, 576–78.

(1982) Niche separation and the broken-stick model. *The American Naturalist*, **114**, 171–8.

Diamond, A.W. (1984) Feeding overlap in some tropical and temperate seabird communities. *Studies in Avian Biology*, **8**, 24–46.

Diamond, J.M. (1970) Ecological consequences of island colonization by southwest Pacific birds. I. Types of niche shifts. *Proceedings of the National Academy of Sciences USA*, **67**, 529–36.

(1972) Biogeographic kinetics: estimation of relaxation times for avifaunas of southwest Pacific islands. *Proceedings of the National Academy of Sciences, USA*, **69**, 3199–203.

(1973) Distributional ecology of New Guinea birds. *Science*, **179**, 759–69.

(1975a) Assembly of species communities. In *Ecology and evolution of communities*, ed. Cody, M.L. & Diamond, J.M., pp. 342–444. Cambridge, Massachusetts: Harvard University Press.

(1975b) The island dilemma: lessons of modern biogeographic studies for the design of natural reserves. *Biological Conservation*, **7**, 129–45.

(1978) Niche shifts and the rediscovery of interspecific competition. *The American Scientist*, **66**, 322–31.

(1979) Population dynamics and interspecific competition in bird communities. *Fortschritte der Zoologie*, **25**, 389–402.

(1982) Effect of species pool size on species occurrence frequencies: musical chairs on islands. *Proceedings of the National Academy of Sciences, USA*, **79**, 2420–24.

(1983) Laboratory, field and natural experiments. *Nature*, **304**, 586–87.

(1984) Distributions of New Zealand birds on real and virtual islands. *New Zealand Journal of Ecology*, **7**, 37–55.

(1986a) Overview: Laboratory experiments, field experiments, and natural experiments. In *Community ecology*, ed. Diamond, J. & Case, T.J., pp. 3–22. New York: Harper & Row.

(1986b) Evolution of ecological segregation in the New Guinea montane avifauna. In *Community ecology*, ed. Diamond, J. & Case, T.J., pp. 98–125. New York: Harper & Row.

Diamond, J.M. & Gilpin, M.E. (1982) Examination of the 'null' model of Connor and Simberloff for species co-occurences on islands. *Oecologia*, **52**, 64–74.

Diamond, J.M. & Jones, H.L. (1980) Breeding land birds of the Channel Islands. In *The California Islands: Proceedings of the Multidisciplinary Symposium*, ed. Power, D.M., pp. 597–612. Santa Barbara, California: Santa Barbara Museum of Natural History.

Diamond, J.M. & Marshall, A.G. (1977) Niche shifts in New Hebridean birds. *Emu*, **77**, 61–72.

Diamond, J.M. & Mayr. E. (1976) Species–area relations for birds of the Solomon archipelago. *Proceedings of the National Academy of Sciences, USA*, **73**, 262–6.

Diamond, J.M. & Veitch, C.R. (1981) Extinctions and introductions in the New Zealand avifauna: cause and effect? *Science*, **211**, 499–501.

Diehl, B. (1986) Factors confounding predictions of bird abundance from habitat data. In *Wildlife 2000. Modeling habitat relationships of terrestrial vertebrates*, ed. Verner, J., Morrison, M.L. & Ralph, C.J., pp. 229–33. Madison: The University of Wisconsin Press.

Dilger, W.C. (1956) Adaptive modifications and ecological isolating mechanisms in the thrush genera *Catharus* and *Hylocichla*. *The Wilson Bulletin*, **68**, 171–99.

Dolnik, V.R. (1982) (Time and energy budgets in free-living birds) (in Russian). *Zool. Zh.*, **61**, 1009–22.

Dolnik, V.R. & Kinzhewskaja, L.I. (1980) (Time and energy budgets during the nest period in the Swift (*Apus apus*) and swallows (*Delichon urbica*, *Hirundo rustica*)) (in Russian). *Zool. Zh.*, **59**, 1841–51.

Downhower, J.F. (1976) Darwin's finches and the evolution of sexual dimorphism in body size. *Nature*, **263**, 558–63.

Drent, R.H. & Daan, S. (1980) The prudent parent: energetic adjustments in avian breeding. *Ardea*, **68**, 225–52.

Dueser, R.D. & Shugart, H.H. Jr. (1978) Microhabitats in a forest-floor small mammal fauna. *Ecology*, **59**, 89–98.

(1979) Niche pattern in a forest floor small mammal fauna. Ecology, **60**, 108–18.

Duffy, D.C. (1983) Competition for nesting space among Peruvian guano birds. *The Auk*, **100**, 680–8.

Duffy, D.C. & La Cock, G.D. (1985) Partitioning of nesting space among seabirds of the Benguela upwelling region. *Ostrich*, **56**, 186–201.

Duffy, D.C. & Siegfried, W.R. (1987) Historical variations in food consumption by breeding seabirds of the Humboldt and Benguela upwelling regions. In *Seabirds: Feeding Ecology and Role in Marine Ecosystems*, ed. Croxall, J.P., pp. 327–46. Cambridge: Cambridge University Press.

Dunbar, R.I.M. (1982) Adaptation, fitness and the evolutionary tautology. In *Current problems in sociobiology*, ed. King's College Sociobiology Group, pp. 9–28. Cambridge: Cambridge University Press.

Dunning, J.B. Jr. (1986) Shrub-steppe bird assemblages revisited: implications for community theory. *The American Naturalist*, **128**, 82–98.

Eadie, J.M., Broekhoven, L. & Cogan, P. (1987) Size ratios and artifacts: Hutchinson's Rule revisited. *The American Naturalist*, **129**, 1–17.

East, R. (1981) Species–area curves and populations of large mammals in African savanna reserves. *Biological Conservation*, **21**, 111–26.

East, R. & Williams, G.R. (1984) Island biogeography and the conservation of New Zealand's indigenous forest-dwelling avifauna. *New Zealand Journal of Ecology*, **7**, 27–35.

Edington, J.M. & Edington, M.A. (1972) Spatial patterns and habitat partition in the breeding birds of an upland wood. *Journal of Animal Ecology*, **41**, 331–57.

(1983) Habitat partitioning and antagonistic behaviour amongst the birds of a West African scrub and plantation plot. *Ibis*, **125**, 74–89.

Ekman, J. (1979) Coherence, composition and territories of winter social groups of the Willow Tit *Parus montanus* and the Crested Tit *P. cristatus*. *Ornis Scandinavica*, **10**, 56–68.

Eldredge, N. & Cracraft, J. (1980) *Phylogenetic Patterns and the Evolutionary Process.* New York: Columbia University Press.

Ellis, H.I. (1984) Energetics of free-ranging seabirds. In *Seabird Energetics*, ed. Whittow, G.C. & Rahn, H., pp. 203–34. New York: Plenum Press.

Ellis, J.E., Wiens, J.A., Rødell, C.F. & Anway, J.C. (1976) A conceptual model of diet selection as an ecosystem process. *Journal of Theoretical Biology*, **60**, 93–108.

Elton, C. (1927) *Animal Ecology*. London: Sidgwick and Jackson.

(1950) *The Ecology of Animals*. London: Methuen and Company.

Emlen, J.M. (1968) Optimal choice in animals. *The American Naturalist*, **102**, 385–9.

Emlen, J.T. (1956) A method for describing and comparing avian habitats. *Ibis*, **98**, 565–76.

(1971) Population densities of birds derived from transect counts. *The Auk*, **88**, 323–42.

(1972) Size and structure of a wintering avian community in southern Texas. *Ecology*, **53**, 317–29.

(1974) An urban bird community in Tucson, Arizona: derivation, structure, regulation. *The Condor*, **76**, 184–97.

(1977a) Estimating breeding season bird densities from transect counts. *The Auk*, **94**, 455–68.

(1977b) Land bird communities of Grand Bahama Island: the structure and dynamics of an avifauna. *Ornithological Monographs*, **24**, 1–129.

(1978) Density anomalies and regulatory mechanisms in land bird populations on the Florida peninsula. *The American Naturalist*, **112**, 265–86.

(1979) Land bird densities on Baja California islands. *The Auk*, **96**, 152–67.

(1981) Divergence in the foraging responses of birds on two Bahama Islands. *Ecology*, **62**, 289–95.

(1986) Land-bird densities in matched habitats on six Hawaiian islands: a test of resource-regulation theory. *The American Naturalist*, **127**, 125–39.

Emlen, J.T. & DeJong, M.J. (1981) Intrinsic factors in the selection of foraging substrates by Pine Warblers: a test of an hypothesis. *The Auk*, **98**, 294–8.

Emlen, J.T., DeJong, M.J., Jaeger, M.J., Moermond. T.C., Rusterholz, K.A. & White, R.P. (1986) Density trends and range boundary constraints of forests birds along a latitudinal gradient. *The Auk*, **103**, 791–803.

Endler, J.A. (1977) *Geographic Variation, Speciation, and Clines*. Princeton: Princeton University Press.

Engstrom, T. (1981) The species–area relationship in sport-map censusing. *Studies in Avian Biology*, **6**, 421–5.

Erdelen, M. (1984) Bird communities and vegetation structure: I. Correlations and comparisons of simple and diversity indices. *Oecologia*, **61**, 277–84.

Ericksson, M.O.G. (1985) Prey detectability for fish-eating birds in relation to fish density and water transparency. *Ornis Scandinavica*, **16**, 1–7.

Faaborg, J. (1977) Metabolic rates, resources, and the occurrence of nonpasserines in terrestrial avian communities. *The American Naturalist*, **111**, 903–16.

(1982) Trophic and size structure of West Indian bird communities. *Proceedings of the National Academy of Sciences USA*, **79**, 1563–67.

(1985) Ecological constraints on West Indian bird distributions. *Ornithological Monographs*, **36**, 621–53.

Faanes, C.A. & Bystrak, D. (1981) The role of observer bias in the North American Breeding Bird Survey. *Studies in Avian Biology*, **6**, 353–59.

Faeth, S.H. (1984) Density compensation in vertebrates and invertebrates: a review and an experiment. In *Ecological Communities. Conceptual Issues and the Evidence*, ed. Strong, D.R. Jr, Simberloff, D., Abele, L.G. & Thistle, A.B., pp. 491–509. Princeton: Princeton University Press.

Fawver, B.J. (1947) Bird population of an Illinois floodplain forest. *Transactions of the Illinois Academy of Sciences*, **40**, 178–89.

Feinsinger, P. (1976) Organization of a tropical guild of nectarivorous birds. *Ecological Monographs*, **46**, 257–91.

Feinsinger, P. & Colwell, R.K. (1978) Community organization among neotropical nectar-feeding birds. *The American Zoologist*, **18**, 779–95.

Feinsinger, P. & Swarm, L.A. (1982) "Ecological release," seasonal variation in food supply, and the hummingbird *Amazilia tobaci* on Trinidad and Tobago. *Ecology*, **63**, 1574–87.

Fenton, M.B. & Fleming, T.H. (1976) Ecological interactions between bats and nocturnal birds. *Biotropica*, **8**, 104–10.

Ferry, C. & Frochot, B. (1970) L'avifaune nidificatrice d'une forêt de Chênes pédonculés en Bourgogne: étude de deux successions écologiques. *Terre Vie*, **24**, 153–250.

Ferry, C., Frochot, B. & Leruth, Y. (1981) Territory and home range of the Blackcap

(*Sylvia atricapilla*) and some other passerines, assessed and compared by mapping and capture–recapture. *Studies in Avian Biology*, **6**, 119–20.

Ferson, S., Downey. P., Klerks, P., Weissburg, M., Kroot, I., Stewart, S., Jacquez, G., Ssemakula, J., Malenky, R. & Anderson, K. (1986) Competing reviews, or Why do Connell and Schoener disagree? *The American Naturalist*, **127**, 571–76.

Feyerabend, P. (1975) *Against Method*, London: Verso.

Ficken, R.W., Ficken, M.S. & Morse, D.H. (1968) Competition and character displacement in two sympatric pine-dwelling warblers (Dendroica, Parulidae). *Evolution*, **22**, 307–14.

Findley, J.S. & Wilson, D.E. (1982) Ecological significance of chiropteran morphology. In *Ecology of Bats*, ed. Kunz, T.H., pp. 243–60. New York: Plenum Publishing Company.

Fischer, D.H. (1981) Wintering ecology of thrashers in southern Texas. *The Condor*, **83**, 340–46.

Fitzpatrick, J.W. (1985) Form, foraging behavior, and adaptive radiation in the Tyrannidae. *Ornithological Monographs*, **36**, 447–70.

Fjeldså, J. (1983) Ecological character displacement and character release in grebes Podicipedidae. *Ibis*, **125**, 463–81.

Flack, J.A.D. (1976) Bird populations of aspen forests in western North America. *Ornithological Monographs*, **19**, 1–97.

Fleming, P. (1980) 'The Comparative Ecology of Four Sympatric Robins.' B.Sc. (Honours) Thesis, Armidale, NSW: University of New England.

Folse, L.J. Jr (1979) Analysis of community census data: a multivariate approach. In *The Role of Insectivorous Birds in Forest Ecosystems*, ed. Dickson, J.G., Conner, R.N. Fleet, R.R., Kroll, J.C. & Jackson, J.A., pp. 9–22. New York: Academic Press.

(1981) Ecological relationships of grassland birds to habitat and food supply in East Africa. In *The Use of Multivariate Statistics in Studies of Wildlife Management. USDA Forest Service General Technical Report RM-87*, ed. Capen, D.E., pp. 160–66. Fort Collins, Colorado: Rocky Mountain Forest and Range Experiment Station.

(1982) An analysis of avifauna–resource relationships on the Serengeti Plains. *Ecological Monographs*, **52**, 111–27.

Fonstad, T. (1984) Reduced territorial overlap between the Willow Warbler *Phylloscopus trochilus* and the Brambling *Fringilla montifringilla* in heath birch forest: competition or different habitat preferences? *Oikos*, **42**, 314–22.

Ford, H.A. & Paton, D.C. (1976) Resource partitioning and competition in honeyeaters of the genus Meliphaga. *Australian Journal of Ecology*, **1**, 281–7.

(1977) The comparative ecology of ten species of honeyeaters in South Australia. *Australian Journal of Ecology*, **2**, 399–407.

Ford, R.G., Wiens, J.A., Heinemann, D. & Hunt, G.L. (1982) Modelling the sensitivity of colonially breeding marine birds to oil spills: guillemot and kittiwake populations on the Pribilof Islands, Bering Sea. *Journal of Applied Ecology*, **19**, 1–31.

Franzreb, K.E. (1981a) A comparative analysis of territorial mapping and variable-strip transect censusing methods. *Studies in Avian Biology*, **6**, 164–9.

(1981b) The determination of avian densities using the variable-strip and fixed-width transect surveying methods. *Studies in Avian Biology*, **6**, 139–45.

(1983) Intersexual habitat partitioning in Yellow-rumped Warblers during the breeding season. *The Wilson Bulletin*, **95**, 581–90.

Fretwell, S.D. (1972) *Populations in a Seasonal Environment*. Princeton: Princeton University Press.

(1975) The impact of Robert MacArthur on ecology. *Annual Review of Ecology and Systematics*, **6**, 1–13.

(1978) Competition for discrete versus continuous resources: tests for predictions from the MacArthur–Levins model. *The American Naturalist*, **112**, 73–81.

Fretwell, S.D. & Lucas, H.L. (1969) On territorial behavior and other factors influencing habitat distribution in birds. I. Theoretical development. *Acta Biotheoretica*, **19**, 16–36.

Fritz, R.S. (1980) Consequences of insular population structure: distribution and extinction of Spruce Grouse populations in the Adirondack Mountains. In *Acta XVII Congressus Internationalis Ornithologici*, ed. Nohring, R., pp. 757–63. Berlin: Verlag der Deutschen Ornithologen-Gesellschaft.

Fuller, R.J. (1982) *Bird Habitats in Britain*. Calton: T & A D Poyser.

Fuller, R.J. & Langslow, D.R. (1984) Estimating numbers of birds by point counts: how long should counts last? *Bird Study*, **31**, 195–202.

Furness, R.W. (1978) Energy requirements of seabird communities: a bioenergetics model. *Journal of Animal Ecology*, **47**, 39–53.

(1981) Estimating the food requirements of seabird and seal populations and their interactions with commercial fisheries and fish stocks. In *Proceedings of the Symposium on Birds of the Sea and Shore*, ed. Cooper, J., pp. 1–13. Cape Town: African Seabird Group.

(1982) Modelling relationships among fisheries, seabirds, and marine mammals. In *Marine Birds: Their Feeding Ecology and Commercial Fisheries Relationships*, ed. Nettleship, D.N., Sanger, G.A. & Springer, P.F., pp. 117–26. Ottawa: Canadian Wildlife Service Special Publication.

(1984) Seabird biomass and food consumption in the North Sea. *Marine Pollution Bulletin*, **15**, 244–8.

Furness, R.W. & Barrett, R.T. (1985) The food requirements and ecological relationships of a seabird community in North Norway. *Ornis Scandinavica*, **16**, 305–13.

Furness, R.W. & Cooper, J. (1982) Interactions between breeding seabirds and pelagic fish populations in the southern Benguela region. *Marine Ecology Progress Series*, **8**, 243–50.

Garnett, M.C. (1981) Body size, its heritability and influence on juvenile survival among Great Tits, *Parus major*. *Ibis*, **123**, 31–41.

Gaubert, H. (1985) Étude comparée de la croissance pondérale des jeunes de deux populations de mésanges bleues, *Parus caeruleus* L., en Corse et an Provence: augmentation expérimentale de la taille des nichées corses. *Acta Oecologica Oecologica Generalis*, **6**, 305–16.

Gauch, H.G. Jr (1982) *Multivariate Analysis in Community Ecology*. Cambridge: Cambridge University Press.

Gauch, H.G., Chase, G.B. & Whittaker, R.H. (1974) Ordination of vegetation samples by Gaussian species distributions. *Ecology*, **55**, 1382–90.

Gause, G.F. (1934) *The Struggle for Existence*. Baltimore: Williams & Wilkins.

George, T.L. (1987a) Greater land bird densities on island vs. mainland: relation to nest predation level. *Ecology*, **68**, 1393–400.

(1987b) 'Factors influencing the abundance of land birds on Baja California islands: tests of alternative hypotheses.' Ph.D. dissertation. Albuquerque: University of New Mexico.

Gessaman, J.A. (1973) *Ecological Energetics of Homeotherms*. Logan: Utah State University Press.

Gibb, J. (1954) Feeding ecology of tits, with notes on treecreeper and goldcrest. *Ibis*, **96**, 513–43.

Giller, P.S. (1984) *Community Structure and the Niche*. London: Chapman and Hall.

Gilpin, M.E. & Diamond, J.M. (1976) Calculation of immigration and extinction curves from the species–area–distance relation. *Proceedings of the National Academy of*

Sciences, USA, **73**, 4130–4.
(1982) Factors contributing to non-randomness in species co-occurrences on islands. *Oecologia*, **52**, 75–84.
(1984a) Are species co-occurrences on islands non-random, and are null hypotheses useful in community ecology? In *Ecological Communities. Conceptual Issues and the Evidence*, ed. Strong, D.R. Jr, Simberloff, D., Abele, L.G., & Thistle, A.B., pp. 297–315. Princeton: Princeton University Press.
1984b) Rejoinder. In *Ecological Communities. Conceptual Issues and the Evidence*, ed. Strong, D.R. Jr, Simberloff, D., Abele, L.G. & Thistle, A.B., pp. 332–41. Princeton: Princeton University Press.
Gilpin, M.E., Carpenter, M.P. & Pomerantz, M.J. (1986) The assembly of a laboratory community: multispecies competition in Drosophila. In *Community Ecology*, ed. Diamond, J. & Case, T.J., pp. 23–40. New York: Harper & Row.
Gleason, H.A. (1917) The structure and development of the plant association. *Bulletin of the Torrey Botanical Club*, **43**, 463–81.
(1926) The individualistic concept of the plant association. *Bulletin of the Torrey Botanical Club*, **53**, 1–20.
Głowacinski, Z. (1979) Some ecological parameters of avian communities in the successional series of cultivated pine forest. *Bulletin De L'Academie Polonaise Des Sciences Serie des sciences biologiques*, **29**, 169–77.
Głowacinski, Z. & Järvinen, O. (1975) Rate of secondary succession in forest bird communities. *Ornis Scandinavica*, **6**, 33–40.
Głowacinski, Z. & Weiner, J. (1980) Energetics of bird fauna in consecutive stages of semi-natural pine forest. *Ekologia Polska*, **28**, 71–94.
(1983) Successional trends in the energetics of forest bird communities. *Holarctic Ecology*, **6**, 305–14.
Gochfeld, M. (1979) Interspecific territoriality in Red-breasted Meadowlarks and a method for estimating the mutuality of their participation. *Behavioral Ecology and Sociobiology*, **5**, 159–70.
Gotelli, N.J. & Abele, L.G. (1982) Statistical distributions of West Indian land bird families. *Journal of Biogeography*, **9**, 421–35.
Gotelli, N.J. & Simberloff, D. (1987) The distribution and abundance of tall grass prairie plants: a test of the core–satellite hypothesis. *American Naturalist*, **130**, 18–35.
Gotfryd, A. & Hansell, R.I.C. (1985) The impact of observer bias on multivariate analyses of vegetation structure. *Oikos*, **45**, 223–34.
Gould, S.J. & Lewontin, R.C. (1979) The spandrels of San Marco and the Panglossian paradigm: a critique of the adaptationist programme. *Proceedings of the Royal Society of London*, **205**, 581–98.
Graber, R.R. & Graber, J.W. (1963) A comparative study of bird populations in Illinois, 1906–1909 and 1956–1958. *Illinois Natural History Survey Bulletin*, **28**, 383–528.
Granholm, S.L. (1983) Bias in density estimates due to movement of birds. *The Condor*, **85**, 243–8.
Grant, B.R. (1985) Selection on bill characters in a population of Darwin's finches: *Geospiza conirostris* on Isla Genovesa, Galápagos. *Evolution*, **39**, 523–32.
Grant, B.R. & Grant, P.R. (1981) Exploitation of *Opuntia* cactus by birds on the Galápagos. *Oecologia*, **49**, 179–87.
(1982) Niche shifts and competition in Darwin's finches: *Geospiza conirostris* and congeners. *Evolution*, **36**, 637–57.
(1983) Fission and fusion in a population of Darwin's finches: an example of the value of studying individuals in ecology. *Oikos*, **41**, 530–47.
Grant, P.R. (1966) Ecological compatibility of bird species on islands. *The American*

Naturalist, **100**, 451–62.
(1967) Bill length variability in birds of the Tres Marias Island, Mexico. *Canadian Journal of Zoology*, **45**, 805–15.
(1968) Bill size, body size, and the ecological adaptations of bird species to competitive situations on islands. *Systematic Zoology*, **17**, 319–33.
(1972) Convergent and divergent character displacement. *Biological Journal of the Linnean Society*, **4**, 39–68.
(1975) The classical case of character displacement. *Evolutionary Biology*, **8**, 237–337.
(1977) Island Biology, Illustrated by the land birds of Jamaica [book review]. *Bird-Banding*, **48**, 296–300.
(1979a) Evolution of the Chaffinch, *Fringilla coelebs*, on the Atlantic Islands. *Biological Journal of the Linnean Society*, **11**, 301–32.
(1979b) Ecological and morphological variation of Canary Island Blue Tits, *Parus caeruleus* (Aves: Paridae). *Biological Journal of the Linnean Society*, **11**, 103–29.
(1981) Speciation and the adaptive radiation of Darwin's finches. *American Scientist*, **69**, 653–63.
(1983a) Inheritance of size and shape in a population of Darwin's finches, *Geospiza conirostris*. *Proceedings of the Royal Society of London*, **220**, 219–36.
(1983b) The role of interspecific competition in the adaptive radiation of Darwin's finches. In *Patterns of Evolution in Galápagos Organisms*, ed. Bowman, R.I., Berson, M., & Leviton. A.E., pp. 187–99. San Francisco, California: Pacific Division, AAAS.
(1986a) *Ecology and Evolution of Darwin's Finches*. Princeton: Princeton University Press.
(1986b) Interspecific competition in fluctuating environments. In *Community Ecology*, ed. Diamond, J. & Case, T.J., pp. 173–91. New York: Harper & Row.
Grant, P.R. & Abbott, I. (1980) Interspecific competition, island biogeography and null hypotheses. *Evolution*, **34**, 332–41.
Grant, P.R. & Grant, B.R. (1980) Annual variation in finch numbers, foraging and food supply on Isla Daphne Major, Galápagos. *Oecologia*, **46**, 55–62.
Grant, P.R. & Price. T.D. (1981) Population variation in continuously varying traits as an ecological genetics problem. *The American Zoologist*, **21**, 795–811.
Grant, P.R. & Schluter, D. (1984) Interspecific competition inferred from patterns of guild structure. In *Ecological Communities. Conceptual Issues and the Evidence*, ed. Strong, D.R. Jr, Simberloff, D., Abele, L.G. & Thistle, A.B., pp. 201–33. Princeton: Princeton University Press.
Grant, P.R., Abbott, I., Schluter, D., Curry, R.L. & Abbott, L.K. (1985) Variation in size and shape of Darwin's finches. *Biological Journal of the Linnean Society*, **25**, 1–39.
Grant, P.R., Grant, B.R., Smith, J.N.M., Abbott, I.J. & Abbott, L.K. (1976) Darwin's finches: population variation and natural selection. *Proceedings of the National Academy of Sciences, USA*, **73**, 257–61.
Graves, G.R. (1985) Elevation correlates of speciation and intraspecific geographic variation in plumage in Andean forest birds. *The Auk*, **102**, 556–79.
Graves, G.R. & Gotelli, N.J. (1983) Neotropical land-bridge avifaunas: new approaches to null hypotheses in biogeography. *Oikos*, **41**, 322–33.
Gray, L. & King, J.A. (1986) The use of multidimensional scaling to determine principal resource axes. *The American Naturalist*, **127**, 577–92.
Green, R.H. (1979) *Sampling Design and Statistical Methods for Environmental Biologists*. New York: John Wiley & Sons.
Greene, H.W. & Jaksić, F.M. (1983) Food–niche relationships among sympatric predators: effects of level of prey identification. *Oikos*, **40**, 151–4.
Greene, J.C. (1981) *Science, Ideology, and World View*. Berkeley: University of California Press.

Greenslade, P.J.M. (1968) Island patterns in the Solomon Islands bird fauna. *Evolution*, **22**, 751–61.

Greenwood, P.J., Harvey, P.H. & Perrins, C.M. (1979) The role of dispersal in the Great Tit (*Parus major*): the causes, consequences and heritability of natal dispersal. *Journal of Animal Ecology*, **48**, 123–42.

Grice, D., Caughley, G. & Short, J. (1985) Density and distribution of Emus. *Australian Wildlife Research*, **12**, 69–73.

(1986) Density and distribution of the Australian Bustard *Ardeotis australis*. *Biological Conservation*, **35**, 259–67.

Griffiths, D. (1986) Size–abundance relations in communities. *The American Naturalist*, **127**, 140–66.

Grinnell, J. (1917) The niche-relationships of the California Thrasher. *The Auk*, **34**, 427–33.

(1922) The trend of avian populations in California. *Science*, **56**, 671–6.

(1924) Geography and evolution. *Ecology*, **5**, 225–9.

(1928) Presence and absence of animals. *University of California Chronicle*, **30**, 429–50.

Grzybowski, J.A. (1982) Population structure in grassland bird communities during winter. *Condor*, **84**, 137–52.

Gutting, G., ed. (1980) *Paradigms and Revolutions. Appraisals and Applications of Thomas Kuhn's Philosophy of Science*. Notre Dame: University of Notre Dame Press.

Haefner, J.W. (1981) Avian community assembly rules: the foliage-gleaning guild. *Oecologia*, **50**, 131–42.

Haffer, J. (1974) Avian speciation in tropical South America: with a systematic survey of the toucans (Ramphastidae) and jacamars (Galbulidae). *Publications of the Nuttall Ornithological Club*, **14**.

Haftorn, S. (1956) Contribution to the food biology of tits especially about storing of surplus food. IV. A comparative analysis of *Parus atricapillus* L., *P. cristatus* L. and *P. ater* L. *K. Norske Vidensk. Selsk. Skr.*, 1956 (4), 1–54.

(1978) Energetics of incubation by the Goldcrest *Regulus regulus* in relation to ambient air temperatures and the geographical distribution of the species. *Ornis Scandinavica*, **9**, 22–30.

Haila, Y. (1981) Winter bird communities in the Åland archipelago: an island biogeographic point of view. *Holarctic Ecology*, **4**, 174–83.

(1982) Hypothetico-deductivism and the competition controversy in ecology. *Annales Zoologica Fennici*, **19**, 255–63.

(1983) Colonization of islands in a north-boreal Finnish lake by land birds. *Annales Zoologici Fennici*, **20**, 179–97.

(1986) On the semiotic dimension of ecological theory: the case of island biogeography. *Biology and Philosophy*, **1**, 377–87.

Haila, Y. & Järvinen, O. (1981) The underexploited potential of bird censuses in insular ecology. *Studies in Avian Biology*, **6**, 559–65.

(1982) The role of theoretical concepts in understanding the ecological theater: a case study on island biogeography. In *Conceptual Issues in Ecology*, ed. Saarinen, E., pp. 261–78. Boston: D. Reidel Publishing Company.

(1983) Land bird communities on a Finnish island: species impoverishment and abundance patterns. *Oikos*, **41**, 255–73.

Haila, Y., Hanski, I.K. & Ravio, S. (1987) Breeding bird distribution in fragmented coniferous taiga in southern Finland. *Ornis Fennica*, **64**, 90–106.

Haila, Y., Järvinen, O. & Kuusela, S. (1983) Colonization of islands by land birds: prevalence functions in a Finnish archipelago. *Journal of Biogeography*, **10**, 499–531.

Haila, Y., Järvinen, O., & Väisänen, R.A. (1980) Habitat distribution and species associations of land bird populations on the Åland Islands, SW Finland. *Annales*

Zoologici Fennici, **17**, 87–106.
Hailman, J.P. (1986) The heritability concept applied to wild birds. *Current Ornithology*, **4**, 71–95.
Hairston, N.G. (1984) Inferences and experimental results in guild structure. In *Ecological Communities: Conceptual Issues and the Evidence*, ed. Strong, D.R. Jr, Simberloff, D., Abele, L.G. & Thistle, A.B., pp. 19–27. Princeton, NJ: Princeton University Press.
(1985) The interpretation of experiments on interspecific competition. *The American Naturalist*, **125**, 321–5.
Hall, B.P. & Moreau, R.E. (1962) A study of the rare birds of Africa. *Bulletin of the British Museum of Natural History, Zoology*, **8**, 316–78.
Hamel, P.B. (1984) Comparison of variable circular-plot and spot-map censusing methods in temperate deciduous forest. *Ornis Scandinavica*, **15**, 266–74.
Haney, J.C. (1986) Seabird segregation at Gulf Stream frontal eddies. *Marine Ecology – Progress Series*, **28**, 279–85.
Hanski, I. (1978) Some comments on the measurement of niche metrics. *Ecology*, **59**, 168–74.
(1982) Dynamics of regional distribution: the core and satellite species hypothesis. *Oikos*, **38**, 210–21.
Harner, E.J. & Whitmore, R.C. (1977) Multivariate measures of niche overlap using discriminant analysis. *Theoretical Population Biology*, **12**, 21–36.
Harris, M.P. (1973) The Galapagos avifauna. *The Condor*, **75**, 265–78.
Harrison, C.S. and Seki, M.P. (1987) Trophic relationships among tropical seabirds at the Hawaiian Islands. In *Seabirds: Feeding Ecology and Role in Marine Ecosystems*, ed. Croxall, J.P., pp. 305–26. Cambridge: Cambridge University Press.
Harrison, C.S., Hida, T.S. & Seki, M.P. (1983) Hawaiian seabird feeding ecology. *Wildlife Monographs*, **47**, 1–70.
Harvey, P.H., Colwell, R.K. & Silverton, J.W. (1983) Null models in ecology. *Annual Review of Ecology and Systematics*, **14**, 189–211.
Heinemann, D. (1984) 'Interactions among Rufous Hummingbirds, Hymenopterans, and a Shared Resource: Exploitative Exclusion of a Vertebrate from a Nectar Source, *Scrophularia montana*, by Insects.' Ph.D. Dissertation. Albuquerque: University of New Mexico.
Helle, E. & Helle, P. (1982) Edge effect on forest bird densities on offshore islands in the northern Gulf of Bothnia. *Annales Zoologici Fennici*, **19**, 165–9.
Helle, P. (1984) Effects of habitat area on breeding bird communities in northeastern Finland. *Annales Zoologici Fennici*, **21**, 421–5.
Heller, H.C. (1971) Altitudinal zonation of chipmunks (*Eutamias*): interspecific aggression. *Ecology*, **52**, 312–9.
Hendrickson, J.A. Jr. (1981) Community-wide character displacement reexamined. *Evolution*, **35**, 794–810.
Hengeveld, R. & Haeck, J. (1982) The distribution of abundance. I. Measurements. *Journal of Biogeography*, **9**, 303–16.
Hennemann, W.W. III. (1985) Energetics, behavior and the zoogeography of Anhingas and Double-crested Cormorants. *Ornis Scandinavica*, **16**, 319–23.
Henry, C. (1979) Le concept de niche ecologique illustre par le cas de populations congeneriques sympatriques du genre acrocephalus. *Terre Vie, Rev. Ecology*, **33**, 458–92.
Herrera, C.M. (1977) 'Composicion y estructura de los comunidades mediterraneas de passeriformes en el sur de Espana.' Doctoral Thesis. Sevilla: Universidad de Sevilla.
(1978a) On the breeding distribution pattern of European migrant birds: MacArthur's theme reexamined. *The Auk*, **95**, 496–509.
(1978b) Niche-shift in the genus *Parus* in southern Spain. *Ibis*, **120**, 235–40.

(1981) Combination rules among western European *Parus* species. *Ornis Scandinavica*, **12**, 140–47.

Herrera, C.M. & Hiraldo, F. (1976) Food-niche and trophic relationships among European owls. *Ornis Scandinavica*, **7**, 29–41.

Hertz, P.E., Remsen, J.V. Jr & Jones, S.I. (1976) Ecological complementarity of three sympatric parids in a California oak woodland. *The Condor*, **78**, 307–16.

Hespenheide, H.A. (1975) Prey characteristics and predator niche width. In *Ecology and Evolution of Communities*, ed. Cody, M.L. & Diamond, J.M., pp. 158–80. Cambridge: Harvard University Press.

Higuchi, H. (1980) Colonization and coexistence of woodpeckers in the Japanese islands. *Miscellaneous Reports Yamashina Institute for Ornithology*, **12**, 139–56.

Hilborn, R. & Stearns, S.C. (1982) On inference in ecology and evolutionary biology: the problem of multiple causes. *Acta Biotheoretica*, **31**, 145–64.

Hildén, O. (1965) Habitat selection in birds. *Annales Zoologici Fennici*, **2**, 53–75.

(1982) Winter ecology and partial migration of the Goldcrest *Regulus regulus*, in Finland. *Ornis Fennica*, **59**, 99–122.

Hill, M. O. (1973) Diversity and evenness: a unifying notation and its consequences. *Ecology*, **54**, 427–32.

Hoffman, W., Heinemann, D. & Wiens, J.A. (1981) The ecology of seabird feeding flocks in Alaska. *The Auk*, **98**, 437–56.

Hogstad, O. (1975) Interspecific relations between Willow Warbler (*Phylloscopus trochilus*) and Brambling (*Fringilla montifringilla*) in subalpine forests. *Norwegian Journal of Zoology*, **23**, 223–34.

(1978) Differentiation of foraging niche among tits, *Parus* spp., in Norway during winter. *The Ibis*, **120**, 139–46.

Holmes, R.T. (1981) Theoretical aspects of habitat use by birds. In *The Use of Multivariate Statistics in Studies of Wildlife Habitat. USDA Forest Service General Technical Report RM 87*, ed. Capen, D.E., pp. 33–7. Fort Collins, CO: Rocky Mountain Forest and Range Experiment Station.

(1986) Foraging patterns of forest birds: male–female differences. *The Wilson Bulletin*, **98**, 196–213.

Holmes, R.T. & Pitelka, F.A. (1968) Food overlap among coexisting sandpipers on northern Alaskan tundra. *Systematic Zoology*, **17**, 305–18.

Holmes, R.T. & Recher, H.F. (1986a) Determinants of guild structure in forest bird communities: an intercontinental comparison. *The Condor*, **88**, 427–39.

(1986b) Search tactics of insectivorous birds foraging in an Australian eucalypt forest. *The Auk*, **103**, 515–30.

Holmes, R.T. & Robinson, S.K. (1981) Tree species preferences of foraging insectivorous birds in a northern hardwoods forest. *Oecologia*, **48**, 31–5.

Holmes, R.T. & Sturges, F.W. (1973) Annual energy expenditures by the avifauna of a northern hardwoods ecosystem. *Oikos*, **24**, 24–9.

(1975) Bird community dynamics and energetics in a northern hardwoods ecosystem. *Journal of Animal Ecology*, **44**, 175–200.

Holmes, R.T., Black, C.P. & Sherry, T.W. (1979c) Comparative population bioenergetics of three insectivorous passerines in a deciduous forest. *The Condor*, **81**, 9–20.

Holmes, R.T., Bonney, R.E. Jr & Pacala, S.W. (1979a) Guild structure of the Hubbard Brook bird community: a multivariate approach. *Ecology*, **60**, 512–20.

Holmes, R.T., Schultz, J.C. & Nothnagle, P. (1979b) Bird predation on forest insects: an exclosure experiment. *Science*, **206**, 462–3.

Holt, R.D. (1987) On the relation between niche overlap and competition: the effect of incommensurable niche dimensions. *Oikos*, **48**, 110–4.

Horn, H.S. & MacArthur, R.H. (1972) Competition among fugitive species in a harlequin environment. *Ecology*, **53**, 749–52.

Horn, H.S. & May, R.M. (1977) Limits to similarity among coexisting competitors. *Nature*, **270**, 660–1.

Howe, R.W. (1984) Local dynamics of bird assemblages in small forest habitat islands in Australia and North America. *Ecology*, **56**, 1585–601.

Howell, T.R. (1971) An ecological study of the birds of the lowland pine savanna and the adjacent rain forest in northeastern Nicaragua. *The Living Bird*, **10**, 185–242.

Hull, D.L. (1974) *Philosophy of biological science*. Englewood Cliffs, NJ: Prentice-Hall.

Hulsman, K. (1981) Width of gape as a determinant of prey size eaten by terns. *The Emu*, **81**, 29–32.

Hunt, G.L. Jr, Burgeson, B. & Sanger, G.A. (1981) Feeding ecology of seabirds of the Eastern Bering Sea. In *The Eastern Bering Shelf: Oceanography and Resources*, ed. Hood, D.W. & Calder, J.A., Rockville, Maryland: US Department of Commerce, National Oceanic and Atmospheric Administration, Office of Marine Pollution Assessment.

Hurlbert, S.H. (1971) The nonconcept of species diversity: a critique and alternative parameters. *Ecology*, **52**, 577–586.

(1978) The measurement of niche overlap and some relatives. *Ecology*, **59**, 67–77.

(1984) Pseudoreplication and the design of ecological field experiments. *Ecological Monographs*, **54**, 187–211.

Hutchinson, G.E. (1951) Copepodology for the ornithologist. *Ecology*, **32**, 571–7.

(1953) The concept of pattern in ecology. *Proceedings of the Academy of Natural Sciences of Philadelphia*, **105**, 1–12.

(1957) Concluding Remarks. *Cold Spring Harbor Symposia on Quantitative Biology*, **22**, 415–27.

(1959) Homage to Santa Rosalia, or Why are there so many kinds of animals? *The American Naturalist*, **93**, 145–59.

(1978) *An Introduction to Population Ecology*. New Haven, Connecticut: Yale University Press.

Hutchinson, G.E. & MacArthur, R.H. (1959) A theoretical ecological model of size distribution among species of animals. *The American Naturalist*, **93**, 117–25.

Hutto, R.L. (1981) Temporal patterns of foraging activity in some wood warblers in relation to the availability of insect prey. *Behavioral Ecology and Sociobiology*, **9**, 195–98.

Huxley, J.S. (1942) *Evolution. The Modern Synthesis*. London: Allen & Unwin.

Inouye, R.S. (1981) Interactions among unrelated species: granivorous rodents, a parasitic fungus, and a shared prey species. *Oecologia*, **49**, 425–27.

Jackson, J.B.C. (1981) Interspecific competition and species distributions: the ghosts of theories and data past. *The American Zoologist*, **21**, 889–901.

Jaksić, F.M. (1981) Abuse and misuse of the term 'guild' in ecological studies. *Oikos*, **37**, 397–400.

Jaksić, F.M. & Braker, H.E. (1983) Food–niche relationships and guild structure of diurnal birds of prey: competition versus opportunism. *Canadian Journal of Zoology*, **61**, 2230–41.

Jaksić, F.M. & Delibes, M. (1987) A comparative analysis of food-niche relationships and trophic guild structure in two assemblages of vertebrate predators differing in species richness: causes, correlations, and consequences. *Oecologia*, **71**, 461–72.

James, F.C. (1970) Geographic size variation in birds and its relationship to climate. *Ecology*, **51**, 365–90.

(1971) Ordinations of habitat relationships among breeding birds. *The Wilson Bulletin*, **83**, 215–36.

(1982) The ecological morphology of birds: a review. *Annales Zoologici Fennici*, **19**, 265–72.

(1983) Environmental component of morphologic differentiation in birds. *Science*, **221**, 184–6.

James, F.C. & Boecklen, W.J. (1984) Interspecific morphological relationships and the densities of birds. In *Ecological Communities. Conceptual Issues and the Evidence*, ed. Strong, D.R. Jr, Simberloff, D., Abele, L.G. & Thistle, A.B., pp. 458–77. Princeton: Princeton University Press.

James, F.C. & McCulloch, C.E. (1985) Data analysis and the design of experiments in ornithology. *Current Ornithology*, **2**, 1–63.

James, F.C. & Rathbun, S. (1981) Rarefaction, relative abundance, and diversity of avian communities. *The Auk*, **98**, 785–800.

James, F.C. & Shugart, H.H. Jr (1970) A quantitative method of habitat description. *Audubon Field Notes*, **24**, 727–36.

James, F.C. & Wamer, N.O. (1982) Relationships between temperate forest bird communities and vegetation structure. *Ecology*, **63**, 159–71.

James, F.C., Johnston, R.F., Wamer, N.O., Niemi, G.J. & Boecklen, W.J. (1984) The Grinnellian niche of the Wood Thrush. *The American Naturalist*, **124**, 17–30.

Jansson, C., Ekman, J. & von Bromssen, A. (1981) Winter mortality and food supply in tits *Parus* spp. *Oikos*, **37**, 313–22.

Jarman, P. (1982) Prospects for interspecific comparison in sociobiology. In *Current Problems in Sociobiology*, ed. King's College Sociobiology Group, pp. 323–42. Cambridge: Cambridge University Press.

Järvinen, A. (1983) Breeding strategies of hole-nesting passerines in northern Lapland. *Annales Zoologici Fennici*, **20**, 129–49.

Järvinen, O. (1978) Are northern bird communities saturated? *Anser*, **3**, 112–16.

(1979) Geographical gradients of stability in European land bird communities. *Oecologia*, **38**, 51–69.

Järvinen, O. & Haila, Y. (1984). Assembly of land bird communities on northern islands: a quantitative analysis of insular impoverishment. In *Ecological Communities. Conceptual Issues and the Evidence*, ed. Strong, D.R. Jr, Simberlof, D., Abele, L.G. & Thistle, A.B., pp. 138–47. Princeton: Princeton University Press.

Järvinen, O. & Ranta, E. (1987) Patterns and processes in species assemblages on northern Baltic islands. *Annales Zoologici Fennici*, **24**, 249–66.

Järvinen, O. & Sammalisto, L. (1976) Regional trends in the avifauna of Finnish peatland bogs. *Annales Zoologici Fennici*, **13**, 31–43.

Järvinen, O. & Ulfstrand, S. (1980) Species turnover of a continental bird fauna: Northern Europe, 1850–1970. *Oecologia*, **46**, 186–95.

Järvinen, O. & Väisänen, R.A. (1975) Estimating relative densities of breeding birds by the line transect method. *Oikos*, **26**, 316–22.

(1976) Between-year component of diversity in communities of breeding land birds. *Oikos*, **27**, 34–9.

(1977) Recent quantitative changes in the populations of Finnish land birds. *Polish Ecological Studies*, **3**, 177–88.

(1979) Climatic changes, habitat changes, and competition: dynamics of geographical overlap in two pairs of congeneric bird species in Finland. *Oikos*, **33**, 261–71.

(1981) Methodology for censusing land bird faunas in large regions. *Studies in Avian Biology*, **6**, 146–51.

Joern, A. & Lawlor, L.R. (1980) Food and microhabitat utlization by grasshoppers from arid grasslands: comparisons with neutral models. *Ecology*, **61**, 591–9.

Johnson, D.H. (1981) The use of misuse of statistics in wildlife habitat studies. In *The Use of Multivariate Statistics in Studies of Wildlife Habitat. USDA Forest Service General Technical Report RM-87*, ed. Capen, D. E., pp. 11–9. Fort Collins, Colorado: Rocky Mountain Forest and Range Experiment Station, USDA Forest Service.

Johnson, N.K. (1966) Bill size and the question of competition in allopatric and sympatric populations of Dusky and Gray flycatchers. *Systematic Zoology*, **15**, 70–87.

(1975) Controls of number of bird species on montane islands in the Great Basin. *Evolution*, **29**, 545–67.

(1978) Patterns of avian geography and speciation in the intermountain region. In *Great Basin Naturalist Memoirs*, pp. 137–59.

Johnston, D.W. & Odum, E.P. (1956) Breeding bird populations in relation to plant succession on the Piedmont of Georgia. *Ecology*, **37**, 50–62.

Jones, H.L. & Diamond, J.M. (1976) Short-time-base studies of turnover in breeding bird populations on the California Channel Islands. *The Condor*, **78**, 526–49.

Jones, S.E. (1977) Coexistence in mixed species antwren flocks. *Oikos*, **29**, 366–75.

Juanes, F. (1986) Population density and body size in birds. *The American Naturalist*, **128**, 921–9.

Juvik, J.O. & Austring, A.P. (1979) The Hawaiian avifauna: biogeographic theory in evolutionary time. *Journal of Biogeography*, **6**, 205–24.

Kamil, A.C. & Sargent, T.D., eds. (1981) *Foraging behaviour. Ecological, Ethological, and Psychological Approaches*. New York: Garland STPM Press.

Kangas, P. (1987) On the use of species area curves to predict extinctions. *Bulletin of the Ecological Society of America*, **68**, 158–62.

Karr, J.R. (1968) Habitat and avian diversity on strip-mined land in east-central Illinois. *The Condor*, **70**, 348–57.

(1971) Structure of avian communities in selected Panama and Illinois habitats. *Ecological Monographs*, **41**, 207–33.

(1976) Within- and between-habitat avian diversity in African and neotropical lowland habitats. *Ecological Monographs*, **46**, 457–81.

(1980) Geographical variation in the avifaunas of tropical forest undergrowth. *The Auk*, **97**, 283–98.

(1981) Surveying birds with mist nets. *Studies in Avian Biology*, **6**, 62–7.

(1982a) Avian extinction on Barro Colorado Island, Panama: a reassessment. *The American Naturalist*, **119**, 220–39.

(1982b) Population variability and extinction in the avifauna of a tropical land bridge island. *Ecology*, **63**, 1975–8.

(1983) Commentary. In *Perspectives in Ornithology*, ed. Brush, A.H. & Clark, G.A. Jr, pp. 403–10. Cambridge: Cambridge University Press.

Karr, J.R. & Freemark, K.E. (1983) Habitat selection and environmental gradients: dynamics in the "stable" tropics. *Ecology*, **64**, 1481–94.

Karr, J.R. & James, F.C. (1975) Eco-morphological configurations and convergent evolution in species and communities. In *Ecology and Evolution of Communities*, ed. Cody, M.L. & Diamond, J.M., pp. 258–91. Cambridge: Harvard University Press.

Karr, J.R. & Martin, T.E. (1981) Random numbers and principal components: further searches for the unicorn? In *The Use of Multivariate Statistics in Studies of Wildlife Habitat. USDA Forest Service General Technical Report RM-87*, ed. Capen, D.E., pp. 20–4. Fort Collins, Colorado: Rocky Mountain Forest and Range Experiment Station, US Forest Service.

Karr, J.R. & Roth, R.R. (1971) Vegetation structure and avian diversity in several New World areas. *American Naturalist*, **105**, 423–35.

Keast, A. (1961) Bird speciation on the Australian continent. *Bulletin of the Museum of Comparative Zoology*, **123**, 305–495.

(1968) Competitive interactions and the evolution of ecological niches as illustrated by the Australian honeyeater genus *Melithreptus* (Meliphagidae). *Evolution*, **22**, 762–84.

(1970) Adaptive evolution and the shifts in niche occupation in island birds. *Biotropica*, **2**, 61–75.

(1976) Ecological opportunities and adaptive evolution on islands, with special reference to evolution in the isolated forest outliers of Southern Australia. *Proceedings of the 16th International Ornithological Congress, Canberra*, 573–83.

(1981) Distributional patterns, regional biotas, and adaptations in the Australian biota: a synthesis. In *Ecological Biogeography of Australia*, ed. Keast, A., pp. 1891–997. The Hague: Dr W. Junk bv Publishers.

(1985) Bird community structure in southern forests and northern woodlands: a comparison. In *Birds of Eucalpyt Forests and Woodlands* ed. Keast, A., Recher, H.F., Ford, H. & Saunders, D., pp. 97–116. Chipping Norton, NSW: Surrey Beatty & Sons.

Keast, A. & Morton, E.S. eds. (1980) *Migrant Birds in the Neotropics. Ecology, Behavior, Distribution, and Conservation*. Washington, DC: Smithsonian Institution Press.

Kempton, R.A. & Taylor, L.R. (1976) Models and statistics for species diversity. *Nature*, **262**, 818–9.

Kendeigh, S.C. (1934) The role of environment in the life of birds. *Ecological Monographs*, **4**, 229–417.

(1944) Measurement of bird populations. *Ecological Monographs*, **14**, 67–106.

(1945) Community selection by birds on the Helderberg Plateau of New York. *The Auk*, **62**, 418–36.

(1948) Bird populations and biotic communities in northern Lower Michigan. *Ecology*, **29**, 101–14.

(1970) Energy requirements for existence in relation to size of bird. *The Condor*, **72**, 60–5.

Kendeigh, S.C., Dol'nik, V.R. & Gavrilov, V.M. (1977) Avian energetics. In *Granivorous Birds in Ecosystems*, ed. Pinowski, J. & Kendeigh, S.C., pp. 127–204. Cambridge: Cambridge University Press.

Kenkel, N.C. & Orlóci, L. (1986) Applying metric and nonmetric multidimensional scaling to ecological studies: some new results. *Ecology*, **67**, 919–28.

Kepler, C.B. & Kepler, A.K. (1970) Preliminary comparison of bird species diversity and density in Luquillo and Guanica forests. In *A Tropical Rain Forest: A Study of Irradiation and Ecology at El Verde, Puerto Rico*, ed. Odum, H.T., pp. E183–E186. Oak Ridge, Tennessee: US Atomic Energy Commission.

Kepler, C.B. & Scott, J.M. (1981) Reducing bird count variability by training observers. *Studies in Avian Biology*, **6**, 366–71.

Kikkawa, J. (1982) Ecological association of birds and vegetation structure in wet tropical forests of Australia. *Australian Journal of Ecology*, **7**, 325–45.

Kilgore, B.M. (1968) 'Breeding bird populations in managed and unmanaged stands of *Sequoia gigantea*.' Ph.D. dissertation. Berkeley, California: University of California.

King, J.R. (1974) Seasonal allocation of time and energy resources in birds. In *Avian Energetics*, ed. Paynter, R.A. Jr, pp. 4–85. Cambridge, Massachusetts: Nuttall Ornithological Club.

Kingsland, S.E. (1985) *Modeling Nature. Episodes in the History of Population Ecology*. Chicago: University of Chicago Press.

Kitchener, D.J., Chapman, A., Dell, J. & Muir, B.G. (1980) Lizard assemblage and reserve size and structure in the Western Australian wheatbelt – some implications for conservation. *Biological Conservation*, **17**, 25–62.

Kitchener, D.J., Dell, J. & Muir, B.G. (1982) Birds in western Australian wheatbelt reserves – implications for conservation. *Biological Conservation*, **22**, 127–63.

Klopfer, P.H. & MacArthur, R.H. (1961) On the causes of tropical species diversity: niche overlap. *The American Naturalist*, **95**, 223–6.

Kobayashi, S. (1982) The rarefaction diversity measurement and the spatial distribution of individuals. *Japanese Journal of Ecology*, **32**, 255–8.

(1983) Another calculation for the rarefaction diversity measurement for different spatial distributions. *Japanese Journal of Ecology*, **33**, 101–2.

Kodric-Brown, A., Brown, J.H., Byers, G.S. & Gori, D.F. (1984) Organization of a tropical island community of hummingbirds and flowers. *Ecology*, **65**, 1358–68.

Koplin, J.R., Collopy, M.W., Bannman, A.R. & Levenson, H. (1980) Energetics of two wintering raptors. *The Auk*, **97**, 795–806.

Korpimäki, E. (1986) Niche relationships and life-history tactics of three sympatric *Strix* owl species in Finland. *Ornis Scandinavica*, **17**, 126–32.

Krebs, J.R., Stephens, D.W. & Sutherland, W.J. (1983) Perspectives in optimal foraging. In *Perspectives in ornithology*, ed. Brush, A.H. & Clark, G.A. Jr, pp. 165–216. Cambridge: Cambridge University Press.

Kricher, J.C. (1972) Bird species diversity: the effect of species richness and equitability on the diversity index. *Ecology*, **53**, 278–82.

Kuhn, T.S. (1970a) *The Structure of Scientific Revolutions*, 2nd edn. Chicago: University of Chicago Press.

(1970b) Reflections on my critics. In *Criticism and the Growth of Knowledge*, ed. Lakatos, I. & Musgrave, A., pp. 231–78. Cambridge: Cambridge University Press.

(1970c) Logic of discovery or psychology of research? In *Criticism and the Growth of Knowledge*, ed. Lakatos, I. & Musgrave, A., pp. 1–23. Cambridge: Cambridge University Press.

Lack, D. (1933) Habitat selection in birds, with special reference to the effects of afforestation on the Breckland avifauna. *Journal of Animal Ecology*, **2**, 239–62.

(1937) The psychological factor in bird distribution. *British Birds*, **31**, 130–6.

(1944) Correlation between beak and food in the crossbill *Loxia curvirostra*. *The Ibis*, **86**, 552–3.

(1945a) The Galapagos finches (Geospizinae): a study in variation. *Occasional Papers of the California Academy of Sciences*, **21**, 1–159.

(1945b) The ecology of closely related species with special reference to cormorant (*Phalacrocorax carbo*) and Shag (*P. aristotelis*). *Journal of Animal Ecology*, **14**, 12–6.

(1946) Competition for food by birds of prey. *Journal of Animal Ecology*, **15**, 123–9.

(1947) *Darwin's Finches*. Cambridge: Cambridge University Press.

(1968) *Ecological Adaptations for Breeding in Birds*. London: Methuen & Co.

(1969a) The numbers of bird species on islands. *Bird Study*, **16**, 193–209.

(1969b) Tit niches in two worlds: or Homage to Evelyn Hutchinson. *The American Naturalist*, **103**, 43–9.

(1971) *Ecological Isolation in Birds*. Cambridge: Harvard University Press.

(1976) *Island Biology Illustrated by the Land Birds of Jamaica*. Oxford: Blackwell Scientific Publications.

Lack, D. & Southern, H.N. (1949) Birds on Tenerife. *The Ibis*, **91**, 607–26.

Lack, P.C. (1981) Some results from different methods of censusing birds in winter. In *Bird Census and Mediterranean Landscape*, ed. Purroy, F.J., pp. 5–12. Leon: Depto. de Zoologia, Fac. de Biologia, University de Leon.

Lacy, R.C. & Bock, C.E. (1986) The correlation between range size and local abundance of some North American birds. *Ecology*, **67**, 258–60.

Lakatos, I. (1978) *The Methodology of Scientific Research Programmes*. Cambridge: Cambridge University Press.

Lakatos, I. & Musgrave, A., eds. (1970) *Criticism and the Growth of Knowledge*. Cambridge: Cambridge University Press.

Lande, R. (1979) Quantitative genetic analysis of multivariate evolution, applied to brain:body size allometry. *Evolution*, **33**, 402–16.
Landres, P.B. (1983) Use of the guild concept in environmental impact assessment. *Environmental Management*, **7**, 393–8.
Landres, P.B. & MacMahon, J.A. (1980) Guilds and community organization: analysis of an oak woodland avifauna in Sonora, Mexico. *The Auk*, **97**, 351–65.
(1983) Community organization of arboreal birds in some oak woodlands of western North America. *Ecological Monographs*, **53**, 183–208.
Lanyon, S.M. & Thompson, C.F. (1986) Site fidelity and habitat quality as determinants of settlement pattern in male Painted Buntings. *The Condor*, **88**, 206–10.
Lanyon, W.E. (1981) Breeding birds and old field succession on fallow Long Island farmland. *Bulletin of the American Museum of Natural History*, **168**, 5–57.
Larson, D.L. & Bock, C.E. (1986) Determining avian habitat preference by bird-centred vegetation sampling. In *Wildlife 2000. Modeling Habitat Relationships of Terrestrial Vertebrates*, ed. Verner, J., Morrison, M.L. & Ralph, C.J., pp. 37–43. Madison: University of Wisconsin Press.
Lasiewski, R.C. & Dawson, W.R. (1967) A re-examination of the relation between standard metabolic rate and body weight in birds. *The Condor*, **69**, 13–23.
Laudan, L. (1977) *Progress and its Problems. Towards a Theory of Scientific Growth.* Berkeley: University of California Press.
Lawlor, L.R. (1980) Overlap, similarity, and competition coefficients. *Ecology*, **61**, 245–51.
Lederer, R.J. (1975) Bill size, food size, and jaw forces of insectivorous birds. *The Auk*, **92**, 385–7.
(1984) A view of avian ecomorphological hypotheses. *Okologie der Vogel*, **6**, 119–26.
Leger, D.W., Owings, D.H. & Coss, R.G. (1983) Behavioral ecology of time allocation in California ground squirrels (*Spermophilus beecheyi*): microhabitat effects. *Journal of Comparative Psychology*, **97**, 283–91.
Legrende, L. & Legrende, P. (1983) *Numerical Ecology*. Amsterdam: Elsevier Scientific Publishing Company.
Lein, M.R. (1972) A trophic comparison of avifaunas. *Systematic Zoology*, **21**, 135–50.
Leisler, B. & Thaler, E. (1982) Differences in morphology and foraging behavior in the goldcrest *Regulus regulus* and firecrest *R. ignicapillus*. *Annales Zoologici Fennici*, **19**, 277–84.
Leisler, B. & Winkler, H. (1985) Ecomorphology. *Current Ornithology*, **2**, 155–86.
Lewin, R. (1983a) Santa Rosalia was a goat. *Science*, **221**, 636–9.
(1983b) Predators and hurricanes change ecology. *Science*, **221**, 737–40.
Lewontin, R.C. (1972) Testing the theory of natural selection. *Nature*, **236**, 181–2.
Lidicker, W.Z. Jr (1975) The role of dispersal in the demography of small mammals. In *Small Mammals: Their Productivity and Population Dynamics*, ed. Golley, F.B., Petrusewicz, K. & Ryszkowski, L., pp. 103–28. Cambridge: Cambridge University Press.
Lifjeld, J.T. (1984) Prey selection in relation to body size and bill length of five species of waders feeding in the same habitat. *Ornis Scandinavica*, **15**, 217–26.
Lima, S.L., Wiebe, K.L. & Dill, L.M. (1987) Protective cover and the use of space by finches: is closer better? *Oikos*, **50**, 225–30.
Linton, L.R., Davies, R.W. & Wrona, F.J. (1981) Resource utilization indices: an assessment. *Journal of Animal Ecology*, **50**, 283–92.
Loman, J. (1986) Use of overlap indices as competition coefficients: tests with field data. *Ecological Modelling*, **34**, 231–43.
Lovejoy, T.E. (1975) Bird diversity and abundance in Amazon rain forest communities. *The Living Bird*, **13**, 127–91.

Lundberg, A., Alatalo, R.V., Carlson, A. & Ulfstrand, S. (1981) Biometry, habitat distribution and breeding success in the Pied Flycatcher *Ficedula hypoleuca. Ornis Scandinavica*, **12**, 68–79.

Lustick, S. (1984) Thermoregulation in adult seabirds. In *Seabird Energetics*, ed. Whittow, G.C. & Rahn, H., pp. 183–201. New York: Plenum Press.

Lynch, J.F. & Johnson, N.K. (1974) Turnover and equilibria in insular avifaunas, with special reference to the California Channel Islands. *The Condor*, **76**, 370–84.

MacArthur, R.H. (1957) On the relative abundance of bird species. *Proceedings of the National Academy of Sciences USA*, **43**, 293–5.

(1958) Population ecology of some warblers in northeastern coniferous forests. *Ecology*, **39**, 599–619.

(1965) Patterns of species diversity. *Biological Reviews*, **40**, 510–33.

(1966) Note on Mrs Pielou's comments. *Ecology*, **47**, 1074.

(1968) The theory of the niche. In *Population Biology and Evolution*, ed. Lewontin, R.C., pp. 159–76. Syracuse: Syracuse University Press.

(1969) Patterns of communities in the tropics. *Biological Journal of the Linnean Society*, **1**, 19–30.

(1970) Species packing and competitive equilibrium for many species. *Theoretical Population Biology*, **1**, 1–11.

(1971) Patterns of terrestrial bird communities. In *Avian Biology*, ed. Farner, D.S. & King, J.R., pp. 189–221. New York: Academic Press.

(1972) *Geographical Ecology*. New York: Harper & Row.

MacArthur, R.H. & Connell, J.H. (1966) *The Biology of Populations*. New York: John Wiley & Sons.

MacArthur, R.H. & Levins, R. (1967) The limiting similarity, convergence, and divergence of coexisting species. *The American Naturalist*, **101**, 377–85.

MacArthur, R.H. & MacArthur, J.W. (1961) On bird species diversity. *Ecology*, **42**, 594–8.

MacArthur, R.H. & Pianka, E.R. (1966) On optimal use of a patchy environment. *The American Naturalist*, **100**, 603–9.

MacArthur, R.H. & Wilson, E.O. (1967) *The Theory of Island Biogeography*. Princeton: Princeton University Press.

MacArthur, R.H., Diamond, J.M. & Karr, J.R. (1972) Density compensation in island faunas. *Ecology*, **53**, 330–42.

MacArthur, R.H., Recher, H. & Cody, M. (1966) On the relation between habitat selection and species diversity. *The American Naturalist*, **100**, 319–32.

Macdonald, J.D. (1973) *Birds of Australia. A Summary of Information*. London: H.F. & G. Witherby.

Machlis, L., Dodd, P.W.D. & Fentress, J.C. (1985) The pooling fallacy: problems arising when individuals contribute more than one observation to the data set. *Zeitschrift fur Tierpsychologie*, **68**, 201–14.

MacMahon, J.A., Schimpf, D.J., Andersen, D.C., Smith, K.G. & Bayn, R.L. Jr (1981) An organism-centered approach to some community and ecosystem concepts. *Journal of Theoretical Biology*, **88**, 287–307.

MacMillen, R.E. (1985) Energetic pattern and lifestyle in the Meliphagidae. *New Zealand Journal of Zoology*, **12**, 623–9.

Maiorana, V.C. (1978) An explanation of ecological and developmental constants. *Nature*, **273**, 375–7.

Manuwal, D.A. (1983) Feeding locations of coniferous forest birds during a spruce budworm outbreak in western Montana. *The Murrelet*, **64**, 12–17.

Mares, M.A. & Rosenzweig, M.L. (1978) Granivory in North and South American deserts: rodents, birds, and ants. *Ecology*, **59**, 235–41.

Mares, M.A., Blair, W.F., Enders, F.A., Greegor, D., Hulse, A.C., Hunt, J.H., Otte, D., Sage, R.D. & Tomoff, C.S. (1977) The strategies and community patterns of desert animals. In *Convergent Evolution in Warm Deserts*, ed. Orians, G.H. & Solbrig, O.T., pp. 107–63. Stroudsburg, Pennysylvania: Dowden, Hutchinson & Ross.

Martin, J-L. (1983a) Le diagnostic de la compensation de densité dans les peuplements insulaires d'oiseaux par la méthode des échantillonnages fréquentiels progressifs (EPF). *Acta Oecologica Oecologica Generalis*, **4**, 167–79.

(1983b) Impoverishment of island bird communities in a Finnish archipelago. *Ornis Scandinavica*, **14**, 66–77.

(1984) Island biogeography of Corsican birds: some trends. *Holarctic Ecology*, **7**, 211–7.

Maurer, B.A. (1986) Predicting habitat quality for grassland birds using density-habitat correlations. *Journal of Wildlife Management*, **50**, 556–66.

Maurer, B.A. & Whitmore, R.C. (1981) Foraging of five bird species in two forests with different vegetation structure. *The Wilson Bulletin*, **93**, 478–90.

May, P.G. (1982) Secondary succession and breeding bird community structure: patterns of resource utilization. *Oecologia*, **55**, 208–16.

May, R.M. (1973) *Stability and Complexity of Model Ecosystems*. Princeton: Princeton University Press.

(1974) On the theory of niche overlap. *Theoretical Population Biology*, **5**, 297–332.

(1975) Patterns of species abundance and diversity. In *Ecology and Evolution of Communities*, ed. Cody, M.L. & Diamond, J.M., pp. 81–120. Cambridge: Harvard University Press.

ed. (1976) *Theoretical Ecology. Principles and Applications*. Philadelphia: W.B. Saunders Company.

(1978) The dynamics and diversity of insect faunas. In *Diversity of Insect Faunas. Symposium of the Royal Entomological Society No. 9*, ed. Mound, L.A. & Waloff, N., pp. 188–204. Oxford: Blackwell Scientific Publications.

(1981) Models for two interacting populations. In *Theoretical Ecology. Principles and Applications*, ed. May, R.M., pp. 78–104. Sunderland, Massachusetts: Sinauer Associates.

(1984) An overview: real and apparent patterns in community structure. In *Ecological Communities. Conceptual Issues and the Evidence*, ed. Strong, D.R. Jr, Simberloff, D., Abele, L.G. & Thistle, A.B., pp. 3–16. Princeton: Princeton University Press.

May, R.M. & MacArthur, R.H. (1972) Niche overlap as a function of environmental variability. *Proceedings of the National Academy of Sciences, USA*, **69**, 1109–13.

Mayfield, H.F. (1960) *The Kirtland's Warbler*. Bloomfield Hills, Michigan: Cranbrook Institute of Science.

(1978) Brood parasitism: reducing interactions between Kirtland's Warblers and Brown-headed Cowbirds. In *Endangered Birds. Management Techniques for Preserving Threatened Species*, ed. Temple, S.A., pp. 85–91. Madison: University of Wisconsin Press.

(1981) Problems in estimating population size through counts of singing males. *Studies in Avian Biology*, **6**, 220–24.

Maynard Smith, J. (1974) *Models in ecology*. Cambridge: Cambridge University Press.

Mayr, E. (1942) *Systematics and the Origin of Species*. New York: Columbia University Press.

(1982) *The Growth of Biological Thought*. Cambridge: Harvard University Press.

McGuinness, K.A. (1984a) Equations and explanations in the study of species–area curves. *Biological Reviews*, **59**, 423–40.

(1984b) Species–area relations of communities on intertidal boulders: testing the null

hypothesis. *Journal of Biogeography*, **11**, 439–56.
McIntosh, R.P. (1967) The continuum concept of vegetation. *Botanical Review*, **33**, 130–87.
(1970) Community, competition, and adaptation. *Quarterly Review of Biology*, **45**, 259–80.
(1975) H.A. Gleason, 'individualistic ecologist,' 1882–1975: His contributions to ecological theory. *Bulletin of the Torrey Botanical Club*, **102**, 253–73.
(1982) The background and some current problems of theoretical ecology. In *Conceptual Issues in Ecology*, ed. Saarinen, E., pp. 1–61. Boston: D. Reidel Publishing Company.
(1985) *The Background of Ecology. Concept and Theory*. Cambridge: Cambridge University Press.
McLachlan, A. & Wooldridge, T. (1981) The role of birds in the ecology of eastern Cape sandy beaches, South Africa. In *Proceedings of the Symposium on Birds of the Sea and Shore*, ed. Cooper, J., pp. 117–8. Cape Town: African Seabird Group.
McMurtrie, R. (1976) On the limit to niche overlap for nonuniform niches. *Theoretical Population Biology*, **10**, 96–107.
McNaughton, S.J. & Wolf, L.L. (1970) Dominance and the niche in ecological systems. *Science*, **167**, 131–9.
Meents, J.K., Rice, J., Anderson, B.W. & Ohmart, R.D. (1983) Nonlinear relationships between birds and vegetation. *Ecology*, **64**, 1022–7.
Mehlhop, P. & Lynch, J.F. (1986) Bird/habitat relationships along a successional gradient in the Maryland coastal plain. *The American Midland Naturalist*, **116**, 225–39.
Menge, B.A. (1976) Organization of the New England rocky intertidal community: role of predation, competition and environmental heterogeneity. *Ecological Monographs*, **46**, 355–93.
Mengel, R.M. & Jackson, J.A. (1977) Geographic variation of the Red-cockaded Woodpecker. *Condor*, **79**, 349–55.
Merikallio, E. (1951) On the numbers of land-birds in Finland. *Acta Zoologica Fennica*, **65**, 1–16.
Miles, D.B. & Ricklefs, R.E. (1984) The correlation between ecology and morphology in deciduous forest passerine birds. *Ecology*, **65**, 1629–40.
Miles, D.B., Ricklefs, R.E. & Travis, J. (1987) Concordance of ecomorphological relationships in three assemblages of passerine birds. *The American Naturalist*, **129**, 347–64.
Milewski, A.V. (1982) The occurrence of seed and fruits taken by ants versus birds in mediterranean Australia and southern Africa, in relation to the availability of soil potassium. *Journal of Biogeography*, **9**, 505–16.
Miller, A.H. (1951) An analysis of the distribution of the birds of California. *University of California Publications in Zoology*, **42**, 1–80.
Miller, R.S. (1968) Conditions of competition between redwings and yellowheaded blackbirds. *Journal of Animal Ecology*, **37**, 43–61.
Minot, E.O. (1981) Effects of interspecific competition for food in breeding Blue and Great tits. *Journal of Animal Ecology*, **50**, 375–85.
Moldenhauer, R.R. & Wiens, J.A. (1970) The water economy of the Sage Sparrow, *Amphispiza belli nevadensis*. *The Condor*, **72**, 265–75.
Monroe, B.L. Jr (1968) A distributional survey of the birds of Honduras. *Ornithological Monographs*, **7**, 1–458.
Mooney, H.A., Solbrig, O.T. & Cody, M.L. (1977) Introduction. In *Convergent Evolution in Chile and California*, ed. Mooney, H.A., pp. 1–12. Stroudsburg, Pennsylvania: Dowden, Hutchinson & Ross.

Moore, J. (1983) Responses of an avian predator and its isopod prey to an acanthocephalan parasite. *Ecology*, **64**, 1000–15.
(1984) Parasites and altered host behavior. *Scientific American*, **250**, 108–15.
Moreau, R.E. (1966) *The Bird Faunas of Africa and its Islands*. New York: Academic Press.
Moreno, J. (1981) Feeding niches of woodland birds in a montane coniferous forest in central Spain during winter. *Ornis Scandinavica*, **12**, 148–59.
Morrison, M.L. (1984) Influence of sample size and sampling design on analysis of avian foraging behavior. *The Condor*, **86**, 146–50.
Morrison, M.L. & Meslow, E.C. (1983) Bird community structure on early-growth clearcuts in western Oregon. *The American Midland Naturalist*, **110**, 129–37.
Morrison, M.L., Timossi, I.C., With, K.A. & Manley, P.N. (1985) Use of tree species by forest birds during winter and summer. *Journal of Wildlife Management*, **49**, 1098–102.
Morrison, M.L., With, K.A. & Timossi, I.C. (1986) The structure of a forest bird community during winter and summer. *The Wilson Bulletin*, **98**, 214–30.
Morse, D.H. (1967) The contexts of songs in the Black-throated Green and Blackburnian warblers. *The Wilson Bulletin*, **79**, 62–72.
(1968) A quantitative study of foraging of male and female spruce-woods warblers. *Ecology*, **49**, 779–84.
(1971a) The foraging of warblers isolated on small islands. *Ecology*, **52**, 216–28.
(1971b) Effects of the arrival of a new species upon habitat utilization by two forest thrushes in Maine. *The Wilson Bulletin*, **83**, 57–65.
(1977) The occupation of small islands by passerine birds. *The Condor*, **79**, 399–412.
(1978) Structure and foraging patterns of tit flocks in an English woodland. *Ibis*, **120**, 298–312.
(1980) Foraging and coexistence of spruce-woods warblers. *The Living Bird*, **18**, 7–25.
Morton, E.S. (1975) Ecological sources of selection on avian sound. *The American Naturalist*, **109**, 17–34.
Morton, S.R. (1982) Dasyurid marsupials of the Australian arid zone: an ecological review. In *Carnivorous marsupials*, ed. Archer, M., pp. 117–30. Sydney: Royal Zoological Society of New South Wales.
(1985) Granivory in arid regions: comparison of Australia with North and South America. *Ecology*, **66**, 1859–66.
Mosimann, J.E. & James, F.C. (1979) New statistical methods for allometry with application to Florida Red–winged Blackbirds. *Evolution*, **33**, 444–59.
Mosimann, J.E. & Malley, J.D. (1979) Size and shape variables. In *Statistical Ecology. Multivariate Methods in Ecological Work*, ed. Orlóci, L., Rao, C.R. & Stiteler, W.M., pp. 175–89. Fairland, Maryland: International Co-operative Publishing House.
Moss, D. (1978) Diversity of woodland song-bird populations. *Journal of Animal Ecology*, **47**, 521–7.
Moss, D., Taylor, P.N. & Easterbee, N. (1979) The effects on song-bird populations of upland afforestation with spruce. *Forestry*, **52**, 129–50.
Moulton, M.P. (1985) Morphological similarity and coexistence of congeners: an experimental test with introduced Hawaiian birds. *Oikos*, **44**, 301–5.
Moulton, M.P. & Pimm, S.L. (1983) The introduced Hawaiian avifauna: biogeographic evidence for competition. *The American Naturalist*, **121**, 669–90.
(1986) The extent of competition in shaping an introduced avifauna. In *Community Ecology*, ed. Diamond, J. & Case, T.J., pp. 80–97. New York: Harper & Row.
(1987) Morphological assortment in introduced Hawaiian passerines. *Evolutionary Ecology*, **1**, 113–24.

Mountainspring, S. & Scott, J.M. (1985) Interspecific competition among Hawaiian forest birds. *Ecological Monographs*, **55**, 219–39.

Moynihan, M. (1968) Social mimicry: character convergence versus character displacement. *Evolution*, **22**, 315–31.

Mueller, L.D. & Altenberg, L. (1985) Statistical inference on measures of niche overlap. *Ecology*, **66**, 1204–10.

Mugaas, J. & King, J.R. (1981) Annual variation of daily energy expenditure by the Black-billed Magpie: a study of thermal and behavioral energetics. *Studies in Avian Biology*, **5**.

Murphy, E.C. (1985) Bergmann's rule, seasonality, and geographic variation in body size of House Sparrows. *Evolution*, **39**, 1327–34.

Murray, B.G. Jr (1971) The ecological consequences of interspecific territorial behavior in birds. *Ecology*, **52**, 414–23.

(1976) A critique of interspecific territoriality and character convergence. *The Condor*, **78**, 518–25.

(1987) Comments on interspecific territoriality in *Luscinia*. *Ornis Scandinavica*, **18**, 64–5.

Murray, B.G. Jr & Hardy, J.W. (1981) Behavior and ecology in four syntopic species of finches in Mexico. *Zeitschrift fur Tierpsychologie*, **57**, 51–72.

Myers, R.H. (1971) *Response Surface Methodology*. Boston: Allyn and Bacon.

Nagy, K.A. (1980) CO_2 production in animals: analysis of potential errors in the doubly-labelled water method. *American Journal of Physiology*, **238**, 466–73.

Newton, I. (1967) The adaptive radiation and feeding ecology of some British finches. *The Ibis*, **109**, 33–98.

Niemi, G.J. (1985) Patterns of morphological evolution in bird genera of New World and Old World peatlands. *Ecology*, **66**, 1215–28.

Nilsson, S.G. (1977) Density compensation and competition among birds breeding on small islands in a South Swedish lake. *Oikos*, **28**, 170–6.

(1979a) Density and species richness of some forest bird communities in south Sweden. *Oikos*, **33**, 392–401.

(1979b) Seed density, cover, predation and the distribution of birds in a beech wood in southern Sweden. *Ibis*, **121**, 177–85.

Nilsson, S.G. & Alerstam, T. (1976) Resource division among birds in North Finnish coniferous forest in autumn. *Ornis Fennica*, **53**, 16–27.

Nilsson, S.G. & Ebenman, B. (1981) Density changes and niche differences in island and mainland Willow Warblers *Phylloscopus trochilus* at a lake in southern Sweden. *Ornis Scandinavica*, **12**, 62–7.

Noon, B.R, (1981) The distribution of an avian guild along a temperate elevational gradient: the importance and expression of competition. *Ecological Monographs*, **51**, 105–24.

Noon, B.R., Dawson, D.K., Inkley, D.B., Robbins, C.S. & Anderson, S.H. (1980) Consistency in habitat preference of forest bird species. In *Transactions of the 45th North American Wildlife and Natural Resources Conference*, pp. 226–44.

Noordwijk, A.J. van, Balen, J.H. van & Scharloo, W. (1980) Heritability of ecologically important traits in the Great Tit. *Ardea*, **68**, 193–203.

Norberg, U.M. (1979) Morphology of the wings, legs and tail of three coniferous forest tits, the goldcrest, and the treecreeper in relation to locomotor pattern and feeding station selection. *Philosophical Transactions of the Royal Society of London (B)*, **287**, 131–65.

(1981) Flight, morphology and the ecological niche in some birds and bats. *Symposia of the Zoological Society of London*, **48**, 173–97.

Noske, R.A. (1979) Co-existence of three species of treecreepers in north-eastern New South Wales. *The Emu*, **79**, 120–8.

(1985) Habitat use by three bark-foragers of eucalypt forests. In *Birds of Eucalypt Forests and Woodlands*, ed. Keast, A., Recher, H.F., Ford, H. & Saunders, D., pp. 193–204. Chipping Norton, NSW: Surrey Beatty & Sons.

(1986) Intersexual niche segregation among three bark-foraging birds of eucalypt forests. *Australian Journal of Ecology*, **11**, 255–67.

Nudds, T.D. (1983) Niche dynamics and organization of waterfowl guilds in variable environments. *Ecology*, **64**, 319–30.

Nudds, T.D., Abraham, K.F., Ankney, C.D. & Tebbel, P.D. (1981) Are size gaps in dabbling- wading-bird arrays real? *The American Naturalist*, **118**, 549–53.

O'Connor, R.J. (1980) Population regulation in the Yellowhammer, *Emberiza citrinella*. In *Bird Census Work and Nature Conservation. Proceedings of the VI International Conference on Bird Census Work*, ed. Oelke, H., pp. 190–200. Gottingen, West Germany: University of Gottingen.

(1981) The influence of observer and analyst efficiency in mapping method censuses. *Studies in Avian Biology*, **6**, 372–6.

(1985) Behavioral regulation of bird populations: a review of habitat use in relation to migration and residency. In *Behavioral Ecology: Ecological Consequences of Adaptive Behavior*, ed. Sibley, R.M. & Smith, R.H., pp. 105–42. Oxford: Blackwell Scientific Publications.

(1986) Dynamical aspects of avian habitat use. In *Wildlife 2000. Modeling Habitat Relationships of Terrestrial Vertebrates*, ed. Verner, J., Morrison, M.L. & Ralph, C.J., pp. 235–40. Madison: University of Wisconsin Press.

O'Connor, R.J. & Fuller, R.J. (1985) Bird population responses to habitat. In *Bird Census and Atlas Studies: Proceedings of the VII International Conference on Bird Census Work*, ed. Taylor, K., Fuller, R.J. & Lack, P.C., pp. 197–211. Tring, England: British Trust for Ornithology.

Odum, E.P. (1950) Bird populations of the Highlands (North Carolina) Plateau in relation to plant succession and avian invasion. *Ecology*, **31**, 587–605.

(1953) *Fundamentals of Ecology*, 1st edn Philadelphia: W.B. Saunders Company.

(1969) The strategy of ecosystem development. *Science*, **164**, 262–70.

Oelke, H. (1981) Limitations of estimating bird populations because of vegetation structure and composition. *Studies in Avian Biology*, **6**, 316–21.

Oksanen, L., Fretwell, S.D. & Järvinen, O. (1979) Interspecific aggression and the limiting similarity of close competitors: the problem of size gaps in some community arrays. *The American Naturalist*, **114**, 117–29.

Olson, S.L. & James, H.F. (1982) Fossil birds from the Hawaiian Islands: evidence for wholesale extinction by man before western contact. *Science*, **217**, 633–5.

O'Neil, L.J. & Carey, A.B. (1986) Introduction: When habitats fail as predictors. In *Wildlife 2000. Modeling Habitat Relationships of Terrestrial Vertebrates*, ed. Verner, J., Morrison, M.L. & Ralph, C.J., pp. 207–8. Madison: University of Wisconsin Press.

Opdam, P. (1975) Inter- and intraspecific differentiation with respect to feeding ecology in two sympatric species of the genus *Accipiter*. *Ardea*, **63**, 30–54.

Orians, G.H. (1969) The number of bird species in some tropical forests. *Ecology*, **50**, 783–801.

Orians, G.H. & Collier, G. (1963) Competition and blackbird social systems. *Evolution*, **17**, 449–59.

Orians, G.H. & Horn, H.S. (1969) Overlap in foods of four species of blackbirds in the potholes of central Washington. *Ecology*, **50**, 930–8.

Orians, G.H. & Paine, R.T. (1983) Convergent evolution at the community level. In *Coevolution*, ed. Futuyma, D.J. & Slatkin, M., pp. 431–58. Sunderland, Massachusetts: Sinauer Associates.

Orians, G.H. & Solbrig, O.T. (1977) Degree of convergence of ecosystem characteristics. In *Convergent Evolution in Warm Deserts*, ed. Orians, G.H. & Solbrig, O.T., pp. 225–55. Stroudsburg, Pennsylvania: Dowden, Hutchinson & Ross.

Orians, G.H. & Willson, M.F. (1964) Interspecific territories of birds. *Ecology*, **45**, 736–45.

Orlóci, L. & Kenkel, N.C. (1985) *Introduction to Data Analysis with examples from Population and Community Ecology*. Fairland, Maryland: International Co-operative Publishing House.

Pacala, S. & Roughgarden, J. (1982) Resource partitioning and interspecific competition in two two-species insular *Anolis* lizard communities. *Science*, **217**, 444–46.

Palmgren, A. (1915–1917) Studier öfver löfängsomradena på Åland. Ett bidrag till kännedomen om vegetationen och floran på torr och på frisk Kalkhaltig grund. I–III. *Acta Societie Fauna Flora Fennica*, **42**, 1–634.

(1922) Über Artenzahl und Areal sowie über die Konstitution der Vegetation. Eine vegetations-statistische Untersuchung. *Acta Forest Fennica*, **22**, 1–136.

Palmgren, P. (1930) Quantitative Untersuchungen über die Vogelfauna in den Wäldern Südfinnlands, mit besonderer Berücksichtingung Ålands. *Acta Zoologica Fennica*, **7**, 1–218.

Partridge, L. (1976a) Field and laboratory observations on the foraging and feeding techniques of Blue Tits (*Parus caeruleus*) and Coal Tits (*P. ater*) in relation to their habitats. *Animal Behaviour*, **24**, 534–44.

(1976b) Some aspects of the morphology of Blue Tits (*Parus caeruleus*) and Coal Tits (*Parus ater*) in relation to their behaviour. *Journal of Zoology, London*, **179**, 121–33.

(1978) Habitat selection. In *Behavioural Ecology. An Evolutionary Approach*, ed. Krebs, J.R. & Davies, N.B., pp. 351–76. Sunderland, Massachusetts: Sinauer Associates.

Partridge, L. & Pring-Mill, F. (1977) Canary Island Blue Tits and English Coal Tits: convergent evolution? *Evolution*, **31**, 657–65.

Pearson, D.L. (1982) Historical factors and bird species richness. In *Biological Diversification in the Tropics*, ed. Prance, G.T., pp. 441–52. New York: Columbia University Press.

Pearson, O.P. & Ralph, C.P. (1978) The diversity and abundance of vertebrates along an altitudinal gradient in Peru. *Memorias del Museo de Historia Natural 'Javier Prado'*, **18**, 1–97.

Pearson, T.H. (1968) The feeding biology of sea-bird species breeding on the Farne Islands, Northumberland. *Journal of Animal Ecology*, **37**, 521–52.

Peters, R.H. (1983) *The Ecological Implications of Body Size*. Cambridge: Cambridge University Press.

Peters, R.H. & Wassenberg, K. (1983) The effect of body size on animal abundance. *Oecologia*, **60**, 89–96.

Pianka, E.R. (1966) Latitudinal gradients in species diversity: a review of concepts. *The American Naturalist*, **100**, 33–46.

(1980) Guild structure in desert lizards. *Oikos*, **35**, 194–201.

(1981) Competition and niche theory. In *Theoretical Ecology. Principles and Applications*, ed. May, R.M., pp. 167–96. Sunderland, Massachusetts: Sinauer Associates.

(1986) *Ecology and Natural History of Desert Lizards*. Princeton: Princeton University Press.

Pielou, E.C. (1974) *Population and Community Ecology: Principles and Methods*. New York: Gordon and Breach Science Publishers.

(1975) *Ecological Diversity*. New York: John Wiley & Sons.
(1977) *Mathematical Ecology*. New York: John Wiley & Sons.
(1979) *Biogeography*. New York: John Wiley & Sons.
(1981) The broken-stick model: a common misunderstanding. *The American Naturalist*, **117**, 609–10.
Pienkowski, M.W., Ferns, P.N., Davidson, N.C. & Worrall, D.H. (1984) Balancing the budget: measuring the energy intake and requirements of shorebirds in the field. In *Coastal Waders and Wildfowl in Winter*, ed. Evans, P.R., Goss-Custard, J.D., & Hale, W.G., pp. 29–56. Cambridge: Cambridge University Press.
Pimm, S.L. (1976) Existence metabolism. *The Condor*, **78**, 121–4.
Pirsig, R.M. (1974) *Zen and the Art of Motorcycle Maintenance*. New York: William Morrow and Company.
Pitelka, F.A. (1941) Distribution of birds in relation to major biotic communities. *The American Midland Naturalist*, **25**, 113–37.
Platt, J.R. (1964) Strong interference. *Science*, **146**, 347–53.
Poole, R.W. (1974) *An Introduction to Quantitative Ecology*. New York: McGraw-Hill Book Company.
Popper, K.R. (1959) *The Logic of Scientific Discovery*. London: Hutchinson.
(1962) *Conjectures and Refutations. The Growth of Scientific Knowledge*. New York: Basic Books.
(1970) Normal science and its dangers. In *Criticism and the Growth of Knowledge*, ed. Lakatos, I. & Musgrave, A., pp. 51–8. Cambridge; Cambridge University Press.
(1983) *Realism and the Aim of Science*. London: Hutchinson.
Popper, K.R. & Miller, D. (1983) A proof of the impossibility of inductive probability. *Nature*, **302**, 687–8.
Power, D.M. (1972) Numbers of bird species on the California islands. *Evolution*, **26**, 451–63.
(1976) Avifauna richness on the California Channel Islands. *The Condor*, **78**, 394–8.
Pöysä, H. (1983) Morphology-mediated niche organization in a guild of dabbling ducks. *Ornis Scandinavica*, **14**, 317–26.
(1986) Foraging niche shifts in multispecies dabbling duck (*Anas* spp.) feeding groups: harmful and beneficial interactions between species. *Ornis Scandinavica*, **17**, 333–46.
Pregill, G.K. & Olson, S.L. (1981) Zoogeography of West Indian vertebrates in relation to Pleistocene climatic cycles. *Annual Review of Ecology and Systematics*, **12**, 75–98.
Preston, F.W. (1948) The commonness, and rarity, of species. *Ecology*, **29**, 254–83.
(1960) Time and space and the variation of species. *Ecology*, **41**, 611–27.
(1962) The canonical distribution of commonness and rarity. *Ecology*, **43**, 185–215, 410–32.
Price, P.W. (1975) *Insect Ecology*. New York: John Wiley & Sons.
(1980) *Evolutionary Biology of Parasites*. Princeton: Princeton University Press.
(1984) Alternative paradigms in community ecology. In *A New Ecology. Novel Approaches to Interactive Systems*, ed. Price, P.W., Slobodchikoff, C.N. & Gaud, W.S., pp. 353–83. New York: John Wiley & Sons.
Price, T.D. (1984a) Sexual selection and body size, plumage and territory variables in a population of Darwin's finches. *Evolution*, **38**, 327–41.
(1984b) The evolution of sexual size dimorphism in a population of Darwin's finches. *The American Naturalist*, **123**, 500–18.
(1987) Diet variation in a population of Darwin's finches. *Ecology*, **68**, 1015–28.
Price, T.D., Grant, P.R. & Boag, P.T. (1984a) Genetic changes in the morphological differentiation of Darwin's ground finches. In *Population Biology and Evolution*, ed. Wohrmann, K. & Loeschcke, V., pp. 49–66. Berlin: Springer-Verlag.
Price, T.D., Grant, P.R., Gibbs, H.L. & Boag, P.T. (1984b) Recurrent patterns of natural

selection in a population of Darwin's finches. *Nature*, **309**, 787–9.

Prodon, R. & Lebreton, J-D. (1981) Breeding avifauna of a Mediterranean succession: the holm oak and cork oak series in the eastern Pyrenees, 1. Analysis and modelling of the structure gradient. *Oikos*, **37**, 21–38.

Pulliam, H.R. (1975) Coexistence of sparrows: a test of community theory. *Science*, **184**, 474–6.

(1983) Ecological community theory and the coexistence of sparrows. *Ecology*, **64**, 45–52.

(1986) Niche expansion and contraction in a variable environment. *The American Zoologist*, **26**, 71–9.

Pulliam, H.R. & Enders, F. (1971) The feeding ecology of five sympatric finch species. *Ecology*, **52**, 557–66.

Pulliam, H.R. & Mills, G.S. (1977) The use of space by wintering sparrows. *Ecology*, **58**, 1393–9.

Puttick, G.M. (1980) Energy budgets of Curlew Sandpipers at Langebaan Lagoon, South Africa. *Estuarine and Coastal Marine Science*, **11**, 207–15.

Quinn, J.F. & Dunham, A.E. (1983) On hypothesis testing in ecology and evolution. *The American Naturalist*, **122**, 602–17.

Rabenold, K.N. (1978) Foraging strategies, diversity, and seasonality in bird communities of Appalachian spruce-fir forests. *Ecological Monographs*, **48**, 397–424.

(1979) A reversed latitudinal diversity gradient in avian communities of eastern deciduous forests. *The American Naturalist*, **114**, 275–86.

Rabinovich, J.E. & Rapoport, E.H. (1975) Geographical variation of diversity in Argentine passerine birds. *Journal of Biogeography*, **2**, 141–57.

Rabøl, J. (1987) Coexistence and competition between overwintering Willow Warblers *Phylloscopus trochilus* and local warblers at Lake Naivasha, Kenya. *Ornis Scandinavica*, **18**, 101–21.

Rafe, R.W., Usher, M.B. & Jefferson, R.G. (1985) Birds on reserves: the influence of area and habitat on species richness. *Journal of Applied Ecology*, **22**, 327–35.

Rahn, H & Whittow, G.C. (1984) Introduction. In *Seabird Energetics*, ed. Whittow, G.C. & Rahn, H., pp. 1–32. New York: Plenum Press.

Raitt, R.J. & Pimm, S.L. (1976) Dynamics of bird communities in the Chihauhuan Desert, New Mexico. *The Condor*, **78**, 427–42.

Ralph, C.J. & Scott, J.M., eds. (1981) *Estimating numbers of terrestrial birds. Studies in Avian Biology*, vol. 6.

Ranta, E. (1982) Animal communities in rock pools. *Annales Zoologici Fennici*, **19**, 337–47.

Ranta, E. & Järvinen, O. (1987) Ecological biogeography: its history in Finland and recent trends. *Annales Zoologici Fennici*, **24**, 157–63.

Ratcliffe, L.M. & Boag, P.T. (1983) Introduction and notes. In *Darwin's Finches*, Lack, D., pp. xv–liii. Cambridge: Cambridge University Press.

Ratcliffe, L.M. & Grant, P.R. (1983a) Species recognition in Darwin's finches (*Geospiza*, Gould). I. Discrimination by morphological cues. *Animal Behaviour*, **31**, 1139–53.

(1983b) Species recognition in Darwin's finches (*Geospiza*, Gould). II. Geographic variation in mate preference. *Animal Behaviour*, **31**, 1154–65.

Recher, H.F. (1969) Bird species diversity and habitat diversity in Australia and North America. *The American Naturalist*, **103**, 75–80.

(1977) Ecology of co-existing White-cheeked and New Holland honeyeaters. *The Emu*, **77**, 136–42.

(1981) Introductory remarks: environmental influences. *Studies in Avian Biology*, **6**, 251.

Recher, H.F. & Holmes, R.T. (1985) Foraging ecology and seasonal patterns of abundance in a forest avifauna. In *Birds of Eucalypt Forests and Woodlands: Ecology,*

Conservation, Management, ed. Keast, A., Recher, H.F., Ford, H. & Saunders, D., pp. 79–96. Chipping Norton, NSW: Surrey Beatty & Sons.

Recher, H.F., Gowing, G., Kavanagh, R., Shields, J. & Rohan-Jones, W. (1983) Birds, resources and time in a tablelands forest. *Proceedings of the Ecological Society of Australia*, **12**, 101–23.

Reed, J.M. (1985) A comparison of the 'flush' and spot-map methods for estimating the size of Vesper Sparrow territories. *Journal of Field Ornithology*, **56**, 131–7.

Reed, T.M. (1984) The numbers of landbird species on the Isles of Scilly. *Biological Journal of the Linnean Society*, **21**, 431–7.

Reichman, O.J. (1981) Factors influencing foraging in desert rodents. In *Foraging Behavior. Ecological, Ethological, and Psychological Approaches*, ed. Kamil, A.C. & Sargent, T.D., pp. 195–213. New York: Garland STPM Press.

(1984) Spatial and temporal variation of seed distributions in Sonoran Desert soils. *Journal of Biogeography*, **11**, 1–11.

Remsen, J.V. Jr (1985) Community organization and ecology of birds of high elevation humid forest of the Bolivian Andes. *Ornithological Monographs*, **36**, 733–56.

Remsen, J.V. & Parker, T.A. III (1983) Contribution of river-created habitats to bird species richness in Amazonia. *Biotropica*, **15**, 223–31.

Reynolds, R.T. & Meslow, E.C. (1984) Partitioning of food and niche characteristics of coexisting *Accipiter* during breeding. *The Auk*, **101**, 761–79.

Reynolds, R.T., Scott, J.M. & Nussbaum, R.A. (1980) A variable circular-plot method for estimating bird numbers. *The Condor*, **82**, 309–13.

Rice, J. (1978a) Ecological relationships of two interspecifically territorial vireos. *Ecology*, **59**, 526–38.

(1978b) Behavioural interactions of interspecifically territorial vireos. I. Song discrimination and natural interactions. *Animal Behaviour*, **26**, 527–49.

(1978c) Behavioural interactions of interspecifically territorial vireos. II. Seasonal variation in response intensity. *Animal Behaviour*, **26**, 550–61.

Rice, J., Anderson, B.W. & Ohmart, R.D. (1980) Seasonal habitat selection by birds in the lower Colorado River Valley. *Ecology*, **61**, 1402–11.

(1984) Comparison of the importance of different habitat attributes to avian community organization. *Journal of Wildlife Management*, **48**, 895–911.

Rice, J., Ohmart, R.D. & Anderson, D.W. (1983a) Habitat selection attributes of an avian community: a discriminant analysis investigation. *Ecological Monographs*, **53**, 263–90.

(1983b) Turnovers in species composition of avian communities in contiguous riparian habitats. *Ecology*, **64**, 1444–55.

(1986) Limits in a data-rich model: modeling experience with habitat management on the Colorado River. In *Wildlife 2000. Modeling Habitat Relationships of Terrestrial Vertebrates*, ed. Verner, J., Morrison, M.L. & Ralph, C.J., pp. 79–86. Madison: University of Wisconsin Press.

Richards, D.G. (1981) Environmental acoustics and censuses of singing birds. *Studies in Avian Biology*, **6**, 297–300.

Richardson, F. (1942) Adaptive modifications for treetrunk foraging in birds. *University of California Publications in Zoology*, **46**, 317–68.

Richardson, J.L. (1980) The organismic community: resilience of an embattled ecological concept. *BioScience*, **30**, 465–71.

Ricklefs, R.E. (1970) Stage of taxon cycle and distribution of birds on Jamaica, Greater Antilles. *Evolution*, **24**, 475–7.

(1975) Competition and the structure of bird communities (review). *Evolution*, **29**, 581–5.

(1977) A discriminant function analysis of assemblages of fruit-eating birds in Central

America. *The Condor*, **79**, 228–31.
(1987) Community diversity: relative roles of local and regional processes. *Science*, **235**, 167–71.
Ricklefs, R.E. & Cox, G.W. (1972) Taxon cycles in the West Indian avifauna. *The American Naturalist*, **106**, 195–219.
(1977) Morphological similarity and ecological overlap among passerine birds on St Kitts, British West Indies. *Oikos*, **29**, 60–6.
(1978) Stage of taxon cycle, habitat distribution, and population density in the avifauna of the West Indies. *The American Naturalist*, **112**, 875–95.
Ricklefs, R.E. & Travis, J. (1980) A morphological approach to the study of avian community organization. *The Auk*, **97**, 321–38.
Ricklefs, R.E., Cochran, D., & Pianka, E.R. (1981) A morphological analysis of the structure of communities of lizards in desert habitats. *Ecology*, **62**, 1474–83.
Ridgway, R. (1901–18) The birds of North and Middle America. *United States National Museum Bulletin*, **50**.
Robbins, C.S. (1981a) Effect of time of day on bird activity. *Studies in Avian Biology*, **6**, 275–86.
(1981b) Bird activity levels related to weather. *Studies in Avian Biology*, **6**, 301–10.
Robinson, S.K. & Holmes, R.T. (1984) Effects of plant species and foliage structure on the foraging behavior of forest birds. *The Auk*, **101**, 672–84.
Root, R.B. (1967) The niche exploitation pattern of the Blue-gray Gnatcatcher. *Ecological Monographs*, **37**, 317–50.
Root, T. (1988a) Environmental factors associated with avian distributional boundaries. *Journal of Biogeography*, **15**, 489–505.
(1988b) Energy constraints on avian distributions and abundances. *Ecology*, **62**, 330–9.
(1988c) Factors influencing the continent-wide distribution and abundance patterns of birds wintering in North America.
(1988d) *Atlas of Wintering North American Birds*. Chicago: University of Chicago Press.
Rosenberg, A. (1985) *The Structure of Biological Science*, Cambridge University Press: Cambridge.
Rosenzweig, M.L. (1975) On continental steady states of species diversity. In *Ecology and Evolution of Communities*, ed. Cody, M.L. & Diamond, J.M., pp. 121–40. Cambridge: Harvard University Press.
(1979) Optimal habitat selection in two-species competitive systems. *Fortschritte fur Zoologie*, **25**, 283–93.
(1981) A theory of habitat selection. *Ecology*, **62**, 327–35.
Rotenberry, J.T. (1978) Components of avian diversity along a multifactorial climatic gradient. *Ecology*, **59**, 693–9.
(1980a) Dietary relationships among shrubsteppe passerine birds: competition or opportunism in a variable environment? *Ecological Monographs*, **50**, 93–110.
(1980b) Bioenergetics and diet in a simple community of shrubsteppe birds. *Oecologia*, **46**, 7–12.
(1985) The role of habitat in avian community composition: physiognomy or floristics? *Oecologia*, **67**, 213–7.
(1986) Habitat relationships of shrubsteppe birds: even 'good' models cannot predict the future. In *Wildlife 2000: Modeling Habitat Relationships of Terrestrial Vertebrates*, ed. Verner, J., Morrison, M.L. & Ralph, C.J., pp. 217–21. Madison: University of Wisconsin Press.
Rotenberry, J.T. & Wiens, J.A. (1978) Nongame bird communities in northwestern rangelands. In *Proceedings of the Workshop of Nongame Bird Habitat Management in Coniferous Forests of the Western United States. USDA Forest Service General*

Technical Report PNW-64, ed. DeGraaf, R.M., pp. 32–46. Portland, Oregon: Pacific Northwest Forest and Range Experiment Station.

(1980) Habitat structure, patchiness, and avian communities in North American steppe vegetation: a multivariate analysis. *Ecology*, **61**, 1228–50.

(1981) A synthetic approach to principal component analysis of bird/habitat relationships. In *The use of multivariate statistics in studies of wildlife habitat. USDA Forest Service General Technical Report RM-87*, ed. Capen, D.E., pp. 197–208. Fort Collins, Colorado: USDA Forest Service Rocky Mountain Forest and Range Experiment Station.

(1985) Statistical power analysis and community-wide patterns. *The American Naturalist*, **125**, 164–8.

Rotenberry, J.T., Fitzner, R.E. & Rickard, W.H. (1979) Seasonal variation in avian community structure: differences in mechanisms regulating diversity. *The Auk*, **96**, 499–505.

Roth, R.R. (1976) Spatial heterogeneity and bird species diversity. *Ecology*, **57**, 773–82.

Roth, V.L. (1981) Constancy in the size ratios of sympatric species. *The American Naturalist*, **118**, 394–404.

Rothstein, S.I. (1973) Relative variation of avian morphological features: relation to the niche. *The American Naturalist*, **107**, 796–9.

Roughgarden, J. (1972) The evolution of niche width. *The American Naturalist*, **106**, 683–718.

(1974a) Species packing and the competition function with illustrations from coral reef fish. *Theoretical Population Biology*, **5**, 163–86.

(1974b) Niche width: biogeographic patterns among *Anolis* lizard populations. *The American Naturalist*, **108**, 429–42.

(1976) Resource partitioning among competing species: a coevolutionary approach. *Theoretical Population Biology*, **9**, 388–424.

(1979) *Theory of Population Genetics and Evolutionary Ecology: an Introduction*. New York: Macmillan.

(1983) Competition and theory in community ecology. *The American Naturalist*, **122**, 583–601.

(1986) A comparison of food-limited and space-limited competition communities. In *Community Ecology*, ed. Diamond, J. & Case, T.J., pp. 492–516. New York: Harper & Row.

Routledge, R.D. (1979) Diversity indices: Which ones are admissible? *Journal of Theoretical Biology*, **76**, 503–15.

(1980) Bias in estimating the diversity of large, uncensused communities. *Ecology*, **61**, 276–81.

Røv, N. (1975) Breeding bird community structure and species diversity along an ecological gradient in deciduous forest in western Norway. *Ornis Scandinavica*, **6**, 1–14.

Rowley, I. (1975) *Bird Life*. Sydney: Collins.

Rummel, J.D. & Roughgarden, J. (1983) Some differences between invasion-structured and coevolution-structured competitive communities: a preliminary theoretical analysis. *Oikos*, **41**, 477–86.

(1985) A theory of faunal buildup for competition communities. *Evolution*, **39**, 1009–33.

Rusterholz, K.A. & Howe, R.W. (1979) Species–area relation of birds on small islands in a Minnesota lake. *Evolution*, **33**, 468–77.

Sabo, S.R. (1980) Niche and habitat relations in subalpine bird communities of the White Mountains of New Hampshire. *Ecological Monographs*, **50**, 241–59.

Sabo, S.R. & Holmes, R.T. (1983) Foraging niches and the structure of forest bird communities in contrasting montane habitats. *The Condor*, **85**, 121–38.

Sabo, S.R. & Whittaker, R.H. (1979) Bird niches in a subalpine forest: an indirect ordination. *Proceedings of the National Academy of Sciences, USA*, **76**, 1338–42.

Sæther, B-E (1982) Foraging niches in a passerine bird community in a grey alder forest in central Norway. *Ornis Scandinavica*, **13**, 149–63.

(1983) Habitat selection, foraging niches and horizontal spacing of Willow Warbler *Phylloscopus trochilus* and Chiffchaff *P. collybita* in an area of sympatry. *Ibis*, **125**, 24–32.

Sale, P.F. (1978) Coexistence of coral reef fishes – a lottery for living space. *Environmental Biology of Fishes*, **3**, 85–102.

(1984) The structure of communities of fish on coral reefs and the merit of a hypothesis-testing, manipulative approach to ecology. In *Ecological Communities. Conceptual Issues and the Evidence*, ed. Strong, D.R. Jr, Simberloff, D., Abele, L.G. & Thistle, A.B., pp. 478–90. Princeton: Princeton University Press.

Sale, P.F. & Dybdahl, R. (1975) Determinants of community structure for coral reef fishes in an experimental habitat. *Ecology*, **56**, 1343–55.

(1978) Determinants of community structure for coral reef fishes in isolated coral heads at lagoonal and reef slope sites. *Oecologia*, **34**, 57–74.

Salt, G.W. (1953) An ecologic analysis of three California avifaunas. *The Condor*, **55**, 258–73.

(1957) An analysis of avifaunas in the Teton Mountains and Jackson Hole, Wyoming. *The Condor*, **59**, 373–93.

Sanders, H.L. (1968) Marine benthic diversity: a comparative study. *The American Naturalist*, **102**, 243–82.

Sarkar, H. (1983) *A Theory of Method*. Berkeley: University of California Press.

Saunders, D.A., Rowley, I. & Smith, G.T. (1985) The effects of clearing for agriculture on the distribution of cockatoos in the southwest of Western Australia. In *Birds of Eucalypt Forests and Woodlands: Ecology, Conservation, Management*, ed. Keast, A., Recher, H.F., Ford, H. & Saunders, D., pp. 309–21. Chipping Norton, NSW: Surrey Beatty & Sons.

Saunders, D.A., Smith, G.T. & Rowley, I. (1982) The availability and dimensions of tree hollows that provide nest sites for cockatoos (Psittaciformes) in Western Australia. *Australian Wildlife Research*, **9**, 541–56.

Savile, D.B.O. (1957) Adaptive evolution in the avian wing. *Evolution*, **11**, 212–24.

Schaffer, W.M. (1981) Ecological abstraction: the consequences of reduced dimensionality in ecological models. *Ecological Monographs*, **51**, 383–401.

Schall, J.J. & Pianka, E.R. (1978) Geographical trends in numbers of species. *Science*, **201**, 679–86.

Schamberger, M.L. & O'Neil, L.J. (1986) Concepts and constraints of habitat-model testing. In *Wildlife 2000. Modeling Habitat Relationships of Terrestrial Vertebrates*, ed. Verner, J., Morrison, M.L. & Ralph, C.J., pp. 5–10. Madison: University of Wisconsin Press.

Schartz, R.L. & Zimmerman, J.L. (1971) The time and energy budget of the male Dickcissel (*Spiza americana*). *The Condor*, **73**, 65–76.

Schluter, D. (1982a) Seed and patch selection by Galápagos ground finches: relation to foraging efficiency and food supply. *Ecology*, **63**, 1106–20.

(1982b) Distributions of Galápagos ground finches along an altitudinal gradient: the importance of food supply. *Ecology*, **63**, 1504–17.

(1984) A variance test for detecting species associations with some example applications. *Ecology*, **65**, 998–1005.

(1986) Tests for similarity and convergence of finch communities. *Ecology*, **67**, 1073–85.

Schluter, D. & Grant, P.R. (1982) The distribution of *Geospiza difficilis* in relation to *G.*

fuliginosa in the Galápagos Islands: tests of three hypotheses. *Evolution*, **36**, 1213–26.
(1984) Determinants of morphological patterns in communities of Darwin's finches. *The American Naturalist*, **123**, 175–96.
Schluter, D., Price, T.D. & Grant, P.R. (1985) Ecological character displacement in Darwin's finches. *Science*, **227**, 1056–9.
Schmidt-Nielsen, K. (1984) *Scaling. Why is Animal Size so Important?* Cambridge: Cambridge University Press.
Schneider, D.C. & Hunt, G.L. Jr. (1982) Carbon flux to seabirds in waters with different mixing regimes in the southeastern Bering Sea. *Marine Biology*, **67**, 337–44.
Schneider, D.C., Hunt, G.L. Jr & Powers, K.D. (1987) Energy flux to pelagic birds: a comparison of Bristol Bay (Bering Sea) and Georges Bank (Northwest Atlantic). In *Seabirds: Feeding Ecology and Role in Marine Ecosystems*, ed. Croxall, J.P., pp. 259–77. Cambridge: Cambridge University Press.
Schodde, R. (1982) Origin, adaptation and evolution of birds in arid Australia. In *Evolution of the Flora and Fauna of Arid Australia*, ed. Barker, W.R. & Greenslade, P.J.M., pp. 191–224. Frewville, Australia: Peacocke.
Schoener, T.W. (1965) The evolution of bill size differences among symaptric congeneric species of birds. *Evolution*, **19**, 189–213.
(1970) Size patterns in West Indian *Anolis* lizards. II. Correlations with the sizes of particular sympatric species – displacement and convergence. *The American Naturalist*, **104**, 155–74.
(1971a) Large-billed insectivorous birds: a precipitous diversity gradient. *The Condor*, **73**, 154–61.
(1971b) Theory of feeding strategies. *Annual Review of Ecology and Systematics*, **2**, 369–404.
(1974a) Resource partitioning in ecological communities. *Science*, **185**, 27–39.
(1974b) Competition and the form of habitat shift. *Theoretical Population Biology*, **6**, 265–307.
(1975) Presence and absence of habitat shift in some widespread lizard species. *Ecological Monographs*, **45**, 233–58.
(1976) The species–area relation within archipelagos: models and evidence from island land birds. *Proceedings 16th International Ornithological Congress (Canberra)*, 629–42.
(1982) The controversy over interspecific competition. *The American Scientist*, **70**, 586–95.
(1983a) Reply to John Wiens (Letter to the Editor). *The American Scientist*, **71**, 235.
(1983b) Field experiments on interspecific competition. *The American Naturalist*, **122**, 240–85.
(1984) Size differences among sympatric, bird-eating hawks: a worldwide survey. In *Ecological Communities. Conceptual Issues and the Evidence*, ed. Strong, D.R. Jr, Simberloff, D., Abele, L.G. & Thistle, A.B., pp. 254–81. Princeton: Princeton University Press.
(1985a) Some comments on Connell's and my reviews of field experiments on interspecific competition. *The American Naturalist*, **125**, 730–40.
(1985b) On the degree of consistency expected when different methods are used to estimate competition coefficients from census data. *Oecologia*, **67**, 591–92.
(1986a) Mechanistic approaches to community ecology: a new reductionism? *The American Zoologist*, **26**, 81–106.
(1986b) Patterns in terrestrial vertebrate versus arthropod communities: Do systematic differences in regularity exist? In *Community Ecology*, ed. Diamond, J & Case, T.J., pp. 556–86. New York: Harper & Row.

(1986c) Resource partitioning. In *Community Ecology. Pattern and Process*, ed. Kikkawa, J. & Anderson, D.J., pp. 91–126. Oxford: Blackwell Scientific Publications.

Schoener, T.W. & Schoener, A. (1983a) Distribution of vertebrates on some very small islands. I. Occurrence sequences of individual species. *Journal of Animal Ecology*, **52**, 209–35.

(1983b) Distribution of vertebrates on some very small islands. II. Patterns in species number. *Journal of Animal Ecology*, **52**, 237–62.

Schoener, T.W. & Toft, C.A. (1983) Spider populations: extraordinarily high densities on islands without top predators. *Science*, **219**, 1353–55.

Schulenberg, T.S. (1983) Foraging behavior, eco-morphology, and systematics of some antshrikes (Formicariidae: *Thamnomanes*). *The Wilson Bulletin*, **95**, 505–21.

Scott, J.M. & Ramsey, F.L. (1981) Length of count period as a possible source of bias in estimating bird densities. *Studies in Avian Biology*, **6**, 409–13.

Scott, J.M., Mountainspring, S., Ramsey, F.L. & Kepler, C.B. (1986) Forest bird communities of the Hawaiian Islands: their dynamics, ecology, and conservation. *Studies in Avian Biology*, **9**, 1–431.

Seagle, S.W. & McCracken, G.F. (1986) Species abundance, niche position, and niche breadth for five terrestrial animal assemblages. *Ecology*, **67**, 816–18.

Selven, H.C. & Stuart, A. (1966) Data-dredging procedures in survey analysis. *American Statistician*, **20**, 20–3.

Sherry, T.W. (1984) Comparative dietary ecology of sympatric, insectivorous neotropical flycatchers (Tyrannidae). *Ecological Monographs*, **54**, 313–38.

Sherry, T.W. & Holmes, R.T. (1985) Dispersion patterns and habitat responses of birds in northern hardwoods forests. In *Habitat Selection in Birds*, ed. Cody, M.L., pp. 283–309. New York: Academic Press.

Sherry, T.W. & McDade, L.A. (1982) Prey selection and handling in two neotropical hover–gleaning birds. *Ecology*, **63**, 1016–28.

Shmida, A. & Whittaker, R.H. (1979) Convergent evolution of deserts in the Old and New Worlds. In *Werden und Vergehen von Pflanzengesellschaften*, ed. Wilmanns, O. & Tuxen, R., pp. 437–50. Braunschweig: J. Cramer.

Shmida, A. & Wilson, M.V. (1985) Biological determinants of species diversity. *Journal of Biogeography*, **12**, 1–20.

Short, J.J. (1979) Patterns of alpha-diversity and abundance in breeding bird communities across North America. *The Condor*, **81**, 21–7.

Short, L.L. (1978) Sympatry in woodpeckers of lowland Malayan forest. *Biotropica*, **10**, 122–33.

(1979) Burdens of the picid hole-excavating habit. *The Wilson Bulletin*, **91**, 16–28.

Shugart, H.H. Jr & James, D. (1973) Ecological succession of breeding bird populations in northwestern Arkansas. *The Auk*, **90**, 62–77.

Shy, E. (1984) Habitat shift and geographical variation in North American tanagers (Thraupinae: *Piranga*). *Oecologia*, **63**, 281–85.

Siegfried, W.R. & Crowe, T.M. (1983) Distribution and species diversity of birds and plants in fynbos vegetation of Mediterranean-climate zone, South Africa. In *Mediterranean-Type Ecosystems*, ed. Kruger, F.J., Mitchell, D.T. & Jarvis, J.U.M., pp. 404–16. Berlin: Springer-Verlag.

Simberloff, D. (1970) Taxonomic diversity of island biotas. *Evolution*, **24**, 23–47.

(1976) Species turnover and equilibrium island biogeography. *Science*, **194**, 572–78.

(1978) Using island biogeographic distributions to determine if colonization is stochastic. *The American Naturalist*, **112**, 713–26.

(1982) A succession of paradigms in ecology: essentialism to materialism and probabilism. In *Conceptual Issues in Ecology*, ed. Saarinen, E., pp. 63–99. Boston: D. Reidel Publishing Company.

(1983a) Competition theory, hypothesis-testing, and other community ecological buzzwords. *The American Naturalist*, **122**, 626–35.
(1983b) Sizes of coexisting species. In *Coevolution*, ed. Futuyma, D.J. & Slatkin, M., pp. 404–30. Sunderland, Massachusetts: Sinauer Associates.
(1984) Properties of coexisting bird species in two archipelagoes. In *Ecological Communities. Conceptual Issues and the Evidence*, ed. Strong, D.R. Jr, Simberloff, D., Abele, L.G. & Thistle, A.B., pp. 234–53. Princeton: Princeton University Press.
Simberloff, D. & Boecklen, W. (1981) Santa Rosalia reconsidered: size ratios and competition. *Evolution*, **35**, 1206–28.
Simberloff, D. & Connor, E.F. (1981) Missing species combinations. *The American Naturalist*, **118**, 215–39.
Simpson, B.B. (1974) Glacial migrations of plants: island biogeographical evidence. *Science*, **185**, 698–700.
Slagsvold, T. (1975) Competition between the Great Tit *Parus major* and the Pied Flycatcher *Ficedula hypoleuca* in the breeding season. *Ornis Scandinavica*, **6**, 179–90.
(1980) Habitat selection in birds: On the presence of other bird species with special regard to *Turdus pilaris*. *Journal of Animal Ecology*, **49**, 523–36.
Slatkin, M. (1980) Ecological character displacement. *Ecology*, **61**, 163–77.
Slobodkin, L.B. & Saunders, H.L. (1969) On the contribution of environmental predictability to species diversity. *Brookhaven Symposia in Biology*, **22**, 82–95.
Slud, P. (1976) Geographic and climatic relationships of avifaunas with special reference to comparative distribution in the Neotropics. *Smithsonian Contributions in Zoology*, **212**, 1–149.
Smith, C.C. & Balda, R.P. (1979) Competition among insects, birds and mammals for conifer seeds. *The American Zoologist*, **19**, 1065–83.
Smith, J.N.M. & Zach, R. (1979) Heritability of some morphological characters in a Song Sparrow population. *Evolution*, **33**, 460–67.
Smith, J.N.M., Grant, P.R., Grant, B.R., Abbot, I.J. & Abbot, L.K. (1978) Seasonal variation in feeding habits of Darwin's ground finches. *Ecology*, **59**, 1137–50.
Smith, K.G. (1977) Distribution of summer birds along a forest moisture gradient in an Ozark watershed. *Ecology*, **58**, 810–19.
Smith, K.G. & MacMahon, J.A. (1981) Bird communities along a montane sere: community structure and energetics. *The Auk*, **98**, 8–28.
Smyth, M. & Bartholomew, G.A. (1966) The water economy of the Black-throated Sparrow and the Rock Wren. *The Condor*, **68**, 447–58.
Snow, D.W. (1954) The habitats of Eurasian tits (*Parus* spp.). *Ibis*, **96**, 565–85.
(1955) Geographical variation of the Coal Tit *Parus ater* L. *Ardea*, **43**, 195–226.
Sober, E. (1984) *The Nature of Selection*. Cambridge, Massachusetts: The MIT Press.
Sorjonen, J. (1986) Mixed singing and interspecific territoriality – consequences of secondary contact of two ecologically and morphologically similar nightingale species in Europe. *Ornis Scandinavica*, **17**, 53–67.
(1987) Interspecific territoriality in *Luscinia*; an example of interspecific competition for space. *Ornis Scandinavica*, **18**, 65.
Southwood, T.R.E. (1978) *Ecological Methods. With Particular Reference to the Study of Insect Populations*, 2nd edn. London: Chapman and Hall.
Sowles, A.L., Hatch, S.A. & Lensink, C.J. (1978) *Catalog of Alaskan Seabird Colonies*. Washington, DC: Fish and Wildlife Service, US Department of the Interior.
Squibb, R.C. & Hunt, G.L. Jr (1983) A comparison of nesting-ledges used by seabirds on St George Island. *Ecology*, **64**, 727–34.
Stauffer, D.F., Garton, E.O. & Steinhorst, R.K. (1985) A comparison of principal components from real and random data. *Ecology*, **66**, 1693–98.
Steadman, D.W. & Olson, S.L. (1985) Bird remains from an archeological site on

Henderson Island, South Pacific: Man–caused extinctions on an 'uninhabited' island. *Proceedings of the National Academy of Sciences, USA*, **82**, 6191.

Stiles, E.W. (1980) Bird community structure in alder forests in Washington. *The Condor*, **82**, 20–30.

Stjernberg, T. (1979) Breeding biology and population dynamics of the Scarlet Rosefinch *Carpodacus erythrinus*. *Acta Zoologica Fennica*, **157**, 1–88.

Storer, R.W. (1966) Sexual dimorphism and food habits in three North American accipiters. *The Auk*, **83**, 423–36.

Strauss, R.E. (1982) Statistical significance of species clusters in association analysis. *Ecology*, **63**, 634–9.

Strong, D.R. Jr (1982) Null hypotheses in ecology. In *Conceptual Issues in Ecology*, ed. Saarinen, E., pp. 245–59. Boston: D. Reidel Publishing Company.

(1983) Natural variability and the manifold mechanisms of ecological communities. *The American Naturalist*, **122**, 636–60.

Strong, D.R. Jr & Simberloff, D.S. (1981) Straining at gnats and swallowing ratios: character displacement. *Evolution*, **35**, 810–12.

Strong, D.R. Jr, Szyska, L.A. & Simberloff, D. (1979) Tests of community-wide character displacement against null hypotheses. *Evolution*, **33**, 897–913.

Sugihara, G. (1981) $S=CA^z$, $z=\frac{1}{4}$: a reply to Connor and McCoy. *The American Naturalist*, **117**, 790–93.

Suppe, F., ed. (1977) *The Structure of Scientific Theories*, 2nd edn. Urbana: University of Illinois Press.

Svärdson, G. (1949) Competition and habitat selection in birds. *Oikos*, **1**, 157–74.

Svensson, S. & Williamson, K. (1970) Recommendations for an international standard for a mapping method in bird census work. In *Bird Census Work and Environmental Monitoring. Bulletin of the Ecological Research Committee No. 9*, ed. Svensson, S., pp. 49–52. Stockholm: Swedish Natural Science Research Council.

Swennen, C., De Bruijn, P. & Duiven, M.F. (1983) Differences in bill form of the Oystercatcher *Haematopus ostralegus*, a dynamic adaptation to specific foraging techniques. *Netherlands Journal of Sea Research*, **17**, 57–83.

Szaro, R.C. & Balda, R.P. (1979) Bird community dynamics in a ponderosa pine forest. *Studies in Avian Biology*, **3**, 1–66.

Szaro, R.C. & Jakle, M.D. (1982) Comparisons of variable circular-plot and spot-map methods in desert riparian and scrub habitats. *The Wilson Bulletin*, **94**, 546–50.

Tacha, T.C., Vohs, P.A. & Iverson, G.C. (1985) A comparison of interval and continuous sampling methods for behavioral observation. *Journal of Field Ornithology*, **56**, 258–64.

Talbot, F.H., Russell, B.C. & Anderson, G.R.V. (1978) Coral reef fish communities: unstable high-diversity systems? *Ecological Monographs*, **49**, 425–40.

Tanner, J.T. (1978) *A Guide to the Study of Animal Populations*. Knoxville: The University of Tennessee Press.

Taylor, J.A., Friend, G.R. & Dudzinski, M.L. (1984) Influence of sampling strategy on the relationships between fauna and vegetation structure, plant lifeform and floristics. *Australian Journal of Ecology*, **9**, 281–7.

Telleria, J.L. & Garza, V. (1981) Methodological features in the study of a mediterranean forest bird community. In *Bird Census and Mediterranean Landscape. Proceedings VII International Congress Bird Census IBCC/V Meeting EOAC*, ed. Purroy, F.J., pp. 89–92. Leon, Spain: Universidad de Leon.

Terborgh, J. (1971) Distribution on environmental gradients: theory and a preliminary interpretation of distributional patterns in the avifauna of the Cordillera Vilcabamba, Peru. *Ecology*, **52**, 23–40.

(1973) Chance, habitat, and dispersal in the distribution of birds in the West Indies. *Evolution*, **27**, 338–49.

(1974) Preservation of natural diversity: the problem of extinction prone species. *BioScience*, **24**, 715–22.

(1977) Bird species diversity on an Andean elevational gradient. *Ecology*, **58**, 1007–19.

(1980a) Causes of tropical species diversity. In *Acta XVII Congressus Internationalis Ornithologici*, ed. Nohring. R., pp. 955–61. Berlin: Deutschen Ornithologen-Gesellschaft.

(1980b) The conservation status of neotropical migrants: present and future. In *Migrant Birds in the Neotropics. Ecology, Behavior, Distribution, and Conservation*, ed. Keast, A. & Morton, E.S., pp. 21–30. Washington, DC: Smithsonian Institution Press.

(1985a) The role of ecotones in the distribution of Andean birds. *Ecology*, **66**, 1237–46.

(1985b) Habitat selection in Amazonian birds. In *Habitat Selection in Birds*, ed. Cody, M.L., pp. 311–38. New York: Academic Press.

Terborgh, J. & Faaborg, J. (1980a) Saturation of bird communities in the West Indies. *The American Naturalist*, **116**, 178–95.

(1980b) Factors affecting the distribution and abundance of North American migrants in the eastern Caribbean region. In *Migrant Birds in the Neotropics. Ecology, Behavior, Distribution, and Conservation*, ed. Keast, A. & Morton, E.S., pp. 145–55. Washington, DC: Smithsonian Institution Press.

Terborgh, J. & Robinson, S. (1986) Guilds and their utility in ecology. In *Community Ecology. Pattern and Process*, ed. Kikkawa, J. & Anderson, D.J., pp. 65–90. Oxford: Blackwell Scientific Publications.

Terborgh, J. & Weske, J.S. (1975) The role of competition in the distribution of Andean birds. *Ecology*, **56**, 562–76.

Terborgh, J. & Winter, B. (1982) Evolutionary circumstances of species with small ranges. In *Biological Diversification in the Tropics*, ed. Prance, G.T., pp. 587–600. New York: Columbia University Press.

Terborgh, J., Faaborg, J.W. & Brockmann, H.J. (1978) Island colonization by Lesser Antillean birds. *The Auk*, **95**, 59–72.

Ter Braak, C.J.F. (1985) Correspondence analysis of incidence and abundance data: properties in terms of a unimodal response model. *Biometrics*, **41**, 859–73.

(1986) Canonical correspondence analysis: a new eigenvector technique for multivariate direct gradient analysis. *Ecology*, **67**, 1167–79.

Thiollay, J.M. (1986) Structure comparee du peuplement avien dans trois sites de foret primaire en Guyane. *Rev. Ecol. (Terre Vie)*, **41**, 59–105.

Thompson, P.M. & Lawton, J.H. (1983) Seed size diversity, bird species diversity and interspecific competition. *Ornis Scandinavica*, **14**, 327–36.

Thomson, J.D. (1980) Implications of different sorts of evidence for competition. *The American Naturalist*, **116**, 719–26.

Thornhill, R. (1984) Scientific methodology in entomology. *The Florida Entomologist*, **67**, 74–96.

Thornhill, R. & Alcock, J. (1983) *The Evolution of Insect Mating Systems*. Cambridge: Harvard University Press.

Tiainen, J. (1980) Regional trends in bird communities of mature pine forests between Finland and Poland. *Ornis Scandinavica*, **11**, 85–91.

(1982) Ecological significance of morphometric variation in three sympatric *Phylloscopus* warblers. *Annales Zoologici Fennici*, **19**, 285–95.

Tiainen, J., Vickholm, M., Pakkala, T., Piiroinen, J. & Virolainen, E. (1983) The habitat and spatial relations of breeding *Phylloscopus* warblers and the Goldcrest *Regulus regulus* in southern Finland. *Annales Zoologici Fennici*, **20**, 1–12.

Tilman, D. (1982) *Resource Competition and Community Structure*. Princeton: Princeton University Press.

Tipper, J.C. (1979) Rarefaction and rarefiction – The use and abuse of a method of paleoecology. *Paleobiology*, **5**, 423–34.

Titterington, R.W., Crawford, H.S. & Burgason, B.N. (1979) Songbird responses to commercial clear-cutting in Maine spruce-fir forests. *Journal of Wildlife Management*, **43**, 602–9.

Toft, C.A. & Shea, P.J. (1983) Detecting community-wide patterns: estimating power strengthens statistical inference. *The American Naturalist*, **122**, 618–25.

Toft, C.A., Trauger, D.L. & Murdy, H.W. (1982) Tests for species interactions: breeding phenology and habitat use in subarctic ducks. *The American Naturalist*, **120**, 586–13.

Tomoff, C.W. (1974) Avian species diversity in desert scrub. *Ecology*, **55**, 396–403.

Tonkyn, D.W. & Cole, B.J. (1986) The statistical analysis of size ratios. *The American Naturalist*, **128**, 66–81.

Törmälä, T. (1980) The bird community of reserved fields in central Finland. *Ornis Fennica*, **57**, 161–66.

Török, J. (1986) Food segregation in three hole-nesting bird species during the breeding season. *Ardea*, **74**, 129–36.

Tracy, C.R. & Christian, K.A. (1986) Ecological relations among space, time, and thermal niche axes. *Ecology*, **67**, 609–15.

Tramer, E.J. (1969) Bird species diversity: components of Shannon's formula. *Ecology*, **50**, 927–29.

(1974) An analysis of the species density of US landbirds during winter using the 1971 Christmas Bird Count. *American Birds*, **28**, 563–67.

Travis, J. & Ricklefs, R.E. (1983) A morphological comparison of island and mainland assemblages of neotropical birds. *Oikos*, **41**, 434–41.

Trivelpiece, W.Z., Trivelpiece, S.G. & Volkman, N.J. (1987) Ecological segregation of Adelie, Gentoo, and Chinstrap penguins at King George Island, Antarctica. *Ecology*, **68**, 351–61.

Tukey, J.W. (1977) *Exploratory Data Analysis*. Reading, Massachusetts: Addison-Wesley Publishing Company.

Turelli, M. (1978) Does environmental variability limit niche overlap? *Proceedings of the National Academy of Sciences USA*, **75**, 5085–89.

(1981) Niche overlap and invasion of competitors in random environments. I. Models without demographic stochasticity. *Theoretical Population Biology*, **20**, 1–56.

Twomey, A.C. (1945) The bird population of an elm–maple forest with special reference to aspection, territorialism, and coactions. *Ecological Monographs*, **15**, 173–205.

Tye, A. (1981) Ground-feeding methods and niche separation in thrushes. *The Wilson Bulletin*, **93**, 112–14.

Udvardy, M.D.F. (1969) *Dynamic Zoogeography*. New York: Van Nostrand Reinhold Company.

Ugland, K.I. & Gray, J.S. (1982) Lognormal distributions and the concepts of community equilibrium. *Oikos*, **39**, 171–78.

Ulfstrand, S. (1975) Bird flocks in relation to vegetation diversification in South Swedish coniferous plantation during winter. *Oikos*, **26**, 65–73.

(1976) Feeding niches of some passerine birds in a South Swedish coniferous plantation in winter and summer. *Ornis Scandinavica*, **7**, 21–27.

(1977) Foraging niche dynamics and overlap in a guild of passerine birds in a south Swedish coniferous woodland. *Oecologia*, **27**, 23–45.

Ulfstrand, S., Alatalo, R.V., Carlson, A., & Lundberg, A. (1981) Habitat distribution and body size of the Great Tit *Parus major*. *The Ibis*, **123**, 494–99.

Underwood, A.J. (1980) The effects of grazing by gastropods and physical factors on the

upper limits of distribution of intertidal macroalgae. *Oecologia*, **46**, 201–13.
(1986a) What is a community? In *Patterns and Processes in the History of Life*, ed. Raup, D.M. & Jablonski, D., pp. 351–67. Berlin: Springer-Verlag.
(1986b) The analysis of competition by field experiments. In *Community Ecology. Patterns and Process*, ed. Kikkawa, J. & Anderson, D.J., pp. 240–68. Oxford: Blackwell Scientific Publications.

Underwood, A.J. & Denley, E.J. (1984) Paradigms, explanations, and generalizations in models for the structure of intertidal communities on rocky shores. In *Ecological Communities. Conceptual Issues and the Evidence*, ed. Strong, D.R. Jr, Simberloff, D., Abele, L.G. & Thistle, A.B., pp. 151–80. Princeton: Princeton University Press.

Utter, J.M. & Lefebvre, E.A. (1973) Daily energy expenditure of Purple Martins (*Progne subis*) during the breeding season: estimates using D_2O^{18} and time budget methods. *Ecology*, **54**, 597–603.

Väisänen, R.A. & Järvinen, O. (1977) Structure and fluctuation of the breeding bird fauna of a north Finnish peatland area. *Ornis Fennica*, **54**, 143–53.

Vandermeer, J. (1981) *Elementary Mathematical Ecology*. New York: John Wiley & Sons.

Van Horne, B. (1983) Density as a misleading indicator of habitat quality. *Journal of Wildlife Management*, **47**, 893–901.
(1986) Summary: When habitats fail as predictors – the researcher's viewpoint. In *Wildlife 2000. Modeling Habitat Relationships of Terrestrial Vertebrates*, ed. Verner, J., Morrison, M.L. & Ralph, C.J., pp. 257–58. Madison: University of Wisconsin Press.

Van Horne, B. & Ford, R.G. (1982) Niche breadth calculation based on discriminant analysis. *Ecology*, **63**, 1172–74.

Van Riper, C., III, Van Riper, S.G., Goff, M.L. & Laird, M. (1986) The epizootiology and ecological significance of malaria in Hawaiian land birds. *Ecological Monographs*, **56**, 327–44.

Van Valen, L. (1965) Morphological variation and the width of the ecological niche. *The American Naturalist*, **100**, 377–89.

Van Valen, L. & Pitelka, F.A. (1974) Commentary – intellectual censorship in ecology. *Ecology*, **55**, 925–26.

Vepsäläinen, K. & Pisarski, B. (1982) Assembly of island ant communities. *Annales Zoologi Fennici*, **19**, 327–35.

Verner, J. (1983) An integrated system for monitoring wildlife on Sierra National Forest. *Proceedings Forty-Eighth North American Wildlife Conference*, 355–66.
(1985) Assessment of counting techniques. *Current Ornithology*, **2**, 247–302.

Verner, J. & Ritter, L.V. (1985) Comparison of transect and point counts in oak-pine woodlands of California. *The Condor*, **87**, 47–68.

Verner, J., Morrison, M.L. & Ralph, C.J., eds. (1986) *Wildlife 2000. Modeling Habitat Relationships of Terrestrial Vertebrates*. Madison: University of Wisconsin Press.

Vézina, A.F. (1985) Empirical relationships between predator and prey size among terrestrial vertebrate predators. *Oecologia*, **67**, 555–65.

Vuilleumier, F. & Ewert, D.N. (1978) The distribution of birds in Venezuelan Paramos. *Bulletin of the American Museum of Natural History*, **162**, 51–90.

Vuilleumier, F. & Simberloff, D. (1980) Ecology versus history as determinants of patchy and insular distributions in high Andean birds. *Evolutionary Biology*, **12**, 235–379.

Wagner, J.L. (1981) Seasonal change in guild structure: oak woodland insectivorous birds. *Ecology*, **62**, 973–81.

Walsberg, G.E. (1977) Ecology and energetics of contrasting social systems in *Phainopepla nitens*, (Aves: Ptilogonatidae). *University of California Publications in Zoology*, **108**, 1–63.
(1980) Energy expenditure in free-living birds: patterns and diversity. *Acta XVII*

International Ornithological Congress, 300–5.

(1983) Avian ecological energetics. In *Avian Biology*, ed. Farner, D.S., King, J.R. & Parkes, K.C., pp. 161–220. New York: Academic Press.

Weathers, W.W. (1983) *Birds of Southern California's Deep Canyon*. Berkeley: University of California Press.

Weathers, W.W., Buttemer, W.A., Hayworth, A.M. & Nagy, K.A. (1984) An evaluation of time-budget estimates of daily energy expenditure in birds. *The Auk*, **101**, 459–72.

Weiner, J. & Głowacinski, Z. (1975) Energy flow through a bird community in a deciduous forest in southern Poland. *The Condor*, **77**, 233–42.

Werner, E.E. (1984) The mechanisms of species interactions and community organization in fish. In *Ecological Communities: Conceptual Issues and the Evidence*, ed. Strong, D.R. Jr, Simberloff, D., Abele, L.G. & Thistle, A.B., pp. 360–82. Princeton, NJ: Princeton University Press.

West, G.C. & DeWolfe, B.B. (1974) Populations and energetics of taiga birds near Fairbanks, Alaska. *The Auk*, **91**, 757–75.

Wheelwright, N.T. (1985) Fruit size, gape width, and the diets of fruit-eating birds. *Ecology*, **66**, 808–18.

Whitcomb, R.F., Lynch, J.F., Klimkiewicz, M.K., Robbins, C.S., Whitcomb, B.L. & Bystrak, D. (1981) Effects of forest fragmentation on avifauna of the eastern deciduous forest. In *Forest Island Dynamics in Man-dominated Landscapes*, ed. Burgess, R.L. & Sharpe, D.M., pp. 125–205. New York: Springer-Verlag.

Whitmore, R.C. (1975) Habitat ordination of the passerine birds of the Virgin River Valley, southwestern Utah. *The Wilson Bulletin*, **87**, 65–74.

(1977) Habitat partitioning in a community of passerine birds. *The Wilson Bulletin*, **89**, 253–65.

(1979) Temporal variation in the selected habitats of a guild of grassland sparrows. *The Wilson Bulletin*, **91**, 592–98.

(1981) Applied aspects of choosing variables in studies of bird habitats. In *The Use of Multivariate Statistics in Studies of Wildlife Habitat. USDA Forest Service General Technical Report RM-87*, ed. Capen, D.E., pp. 39–41. Fort Collins, Colorado: Rocky Mountain Forest and Range Experiment Station.

Whittaker, R.H. (1953) A consideration of climax theory: the climax as a population and pattern. *Ecological Monographs*, **23**, 41–78.

(1960) Vegetation of the Siskiyou Mountains, Oregon and California. *Ecological Monographs*, **30**, 279–338.

(1965) Dominance and diversity in land plant communities. *Science*, **147**, 250–60.

(1972) Evolution and measurement of species diversity. *Taxon*, **21**, 213–51.

(1975) *Communities and Ecosystems*, 2nd edn. New York: Macmillan.

(1977) Evolution of species diversity in land communities. *Evolutionary Biology*, **10**, 1–67.

Whittam, T.S. & Siegel-Causey, D. (1981a) Species incidence functions and Alaskan seabird colonies. *Journal of Biogeography*, **8**, 421–25.

(1981b) Species interactions and community structure in Alaskan seabird colonies. *Ecology*, **62**, 1515–24.

Wiens, J.A. (1969) An approach to the study of ecological relationships among grassland birds. *Ornithological Monographs*, **8**, 1–93.

(1973a) Interterritorial habitat variation in Grasshopper and Savannah sparrows. *Ecology*, **54**, 877–84.

(1973b) Pattern and process in grassland bird communities. *Ecological Monographs*, **43**, 237–70.

(1974a) Climatic instability and the "ecological saturation" of bird communities in

North American grasslands. *The Condor*, **76**, 385–400.
(1974b) Habitat heterogeneity and avian community structure in North American grasslands. *The American Midland Naturalist*, **91**, 195–213.
(1975) Avian communities, energetics, and functions in coniferous forest habitats. In *Proceedings of the Symposium on Management of Forest and Range Habitats for Nongame Birds. USDA Forest Service General Technical Report WO-1*, ed. Smith, D.R., pp. 226–65. Washington, DC: USDA Forest Service.
(1976a) Review of 'Competition and the Structure of Bird Communities' by M. Cody. *The Auk*, **93**, 396–400.
(1976b) Population responses to patchy environments. *Annual Review of Ecology and Systematics*, **7**, 81–120.
(1977a) Review of 'Ecology and Evolution of Communities', ed. by M. Cody and J. Diamond. *The Auk*, **94**, 792–94.
(1977b) On competition and variable environments. *The American Scientist*, **65**, 590–97.
(1977c) Model estimation of energy flow in North American grassland bird communities. *Oecologia*, **31**, 135–51.
(1981a) Scale problems in avian censusing. *Studies in Avian Biology*, **6**, 513–21.
(1981b) Single-sample surveys of communities: Are the revealed patterns real? *The American Naturalist*, **117**, 90–8.
(1982) On size ratios and sequences in ecological communities: Are there no rules? *Annales Zoologici Fennici*, **19**, 297–308.
(1983) Avian community ecology: an iconoclastic view. In *Perspectives in Ornithology*, ed. Brush, A.H. & Clark, G.A. Jr, pp. 355–403. Cambridge: Cambridge University Press.
(1984a) On understanding a non-equilibrium world: myth and reality in community patterns and processes. In *Ecological Communities. Conceptual Issues and the Evidence*, ed. Strong, D.R. Jr, Simberloff, D., Abele, L.G. & Thistle, A.B., pp. 439–57. Princeton: Princeton University Press.
(1984b) Resource systems, populations, and communities. In *A New Ecology. Novel Approaches to Interactive Systems*, ed. Price, P.W., Slobodchikoff, C.N. & Gaud, W.S., pp. 397–436. New York: John Wiley & Sons.
(1984c) Modeling the energy requirements of seabird populations. In *Seabird Energetics*, ed. Whittow, G.C. & Rahn, H., pp. 255–84. New York: Plenum Press.
(1985a) Vertebrate responses to environmental patchiness in arid and semiarid ecosystems. In *The Ecology of Natural Disturbance and Patch Dynamics*, ed. Pickett, S.T.A. & White, P.S., pp. 169–93. New York: Academic Press.
(1985b) Habitat selection in variable environments: shrub-steppe birds. In *Habitat Selection in Birds*, ed. Cody, M.L., pp. 227–51. New York: Academic Press.
(1986) Spatial scale and temporal variation in studies of shrubsteppe birds. In *Community Ecology*, ed. Diamond, J. & Case, T.J., pp. 154–72. New York: Harper & Row.
Wiens, J.A. & Dyer, M.I. (1975) Simulation modelling of Red-winged Blackbird impact on grain crops. *Journal of Applied Ecology*, **12**, 63–82.
(1977) Assessing the potential impact of granivorous birds in ecosystems. In *Granivorous Birds in Ecosystems*, ed. Pinowski, J. & Kendeigh, S.C., pp. 205–66. Cambridge: Cambridge University Press.
Wiens, J.A. & Innis, G.S. (1973) Estimation of energy flow in bird communities. II. A simulation model of activity budgets and population bioenergetics. *Proceedings 1973 Summer Computer Simulation Conference, Montreal*, 739–52.
(1974) Estimation of energy flow in bird communities: a population bioenergetics model. *Ecology*, **55**, 730–46.

Wiens, J.A. & Nussbaum, R.A. (1975) Model estimation of energy flow in northwestern coniferous forest bird communities. *Ecology*, **56**, 547–61.

Wiens, J.A. & Rotenberry, J.T. (1979) Diet niche relationships among North American grassland and shrubsteppe birds. *Oecologia*, **42**, 253–92.

(1980) Patterns of morphology and ecology in grassland and shrubsteppe bird populations. *Ecological Monographs*, **50**, 287–308.

(1981a) Habitat associations and community structure of birds in shrubsteppe environments. *Ecological Monographs*, **51**, 21–41.

(1981b) Morphological size ratios and competition in ecological communities. *The American Naturalist*, **117**, 592–99.

(1981c) Censusing and the evaluation of avian habitat occupancy. *Studies in Avian Biology*, **6**, 522–32.

(1985) Response of breeding passerine birds to rangeland alteration in a North American shrubsteppe locality. *Journal of Applied Ecology*, **22**, 655–68.

(1987) Shrub-steppe birds and the generality of community models: a response to Dunning. *The American Naturalist*, **129**, 920–27.

Wiens, J.A. & Scott, J.M. (1975) Model estimation of energy flow in Oregon coastal seabird populations. *The Condor*, **77**, 439–52.

Wiens, J.A., Crawford, C.S. & Gosz, J.R. (1985) Boundary dynamics: a conceptual framework for studying landscape ecosystems. *Oikos*, **45**, 421–27.

Wiens, J.A., Ford, F.G. & Heinemann, D. (1984) Information needs and priorities for assessing the sensitivity of marine birds to oil spills. *Biological Conservation*, **28**, 21–49.

Wiens, J.A., Rotenberry, J.T. & Van Horne, B. (1986) A lesson in the limitations of field experiments: shrubsteppe birds and habitat alternation. *Ecology*, **67**, 365–76.

(1987) Habitat occupancy patterns of North American shrubsteppe birds: the effects of spatial scale. *Oikos*, **48**, 132–47.

Wiens, J.A., Van Horne, B. & Rotenberry, J. (1987) Temporal and spatial variations in the behavior of shrubsteppe birds. *Oecologia*, **73**, 60–70.

Williams, A.B. (1936) The composition and dynamics of a beech–maple climax community. *Ecological Monographs*, **6**, 317–408.

Williams, B.K. (1981) Discriminant analysis in wildlife research: theory and applications. In *The Use of Multivariate Statistics in Studies of Wildlife Habitat. USDA Forest Service General Technical Report RM-87*, ed. Capen, D.E., pp. 59–71. Fort Collins, Colorado: Rocky Mountain Forest and Range Experiment Station.

(1983) Some observations on the use of discriminant analysis in ecology. *Ecology*, **64**, 1283–91.

Williams, C.B. (1964) *Patterns in the Balance of Nature*. London: Academic Press.

Williams, G.R. (1981) Aspects of avian island biogeography in New Zealand. *Journal of Biogeography*, **8**, 439–56.

Williams, J.B. & Nagy, K.A. (1984) Daily energy expenditure of Savannah Sparrows: comparison of time-energy budget and doubly-labeled water estimates. *The Auk*, **101**, 221–29.

Williamson, K. (1969) Habitat preferences of the Wren on English farmland. *Bird Study*, **16**, 53–9.

Williamson, M. (1981) *Island Populations*. Oxford: Oxford University Press.

Willis, E.O. (1974) Populations and local extinctions of birds on Barro Colorado Island, Panama. *Ecological Monographs*, **44**, 153–69.

Willson, M.F. (1969) Avian niche size and morphological variation. *The American Naturalist*, **103**, 531–42.

(1971) Seed selection in some North American finches. *The Condor*, **73**, 415–29.

(1974) Avian community organization and habitat structure. *Ecology*, **55**, 1017–29.

Willson, M.F., Anderson, S.H. & Murray, B.G. Jr (1973) Tropical and temperate bird species diversity: within-habitat and between-habitat comparisons. *Caribbean Journal of Science*, **13**, 81–90.

Willson, M.F., Karr, J.R., & Roth, R.R. (1975) Ecological aspects of avian bill-size variation. *The Wilson Bulletin*, **87**, 32–44.

Wilson, E.O. (1961) The nature of the taxon cycle in the Melanesian ant fauna. *The American Naturalist*, **95**, 169–93.

Winkler, H. & Leisler, B. (1985) Morphological aspects of habitat selection in birds. In *Habitat Selection in Birds*, ed. Cody, M.L., pp. 415–34. New York: Academic Press.

Winternitz, B.L. (1976) Temporal change and habitat preference of some montane breeding birds. *The Condor*, **78**, 383–93.

Woinarski, J.C.Z. (1985) Foliage gleaners of the tree tops, the pardalotes. In *Birds of Eucalypt Forests and Woodlands: Ecology, Conservation, Management*, ed. Keast, A., Recher, H.F., Ford, H. & Saunders, D., pp. 165–75. Chipping Norton, NSLO: Surrey Beatty & Sons.

Wolf, L.L., Stiles, F.G. & Hainsworth, F.R. (1976) Ecological organization of a tropical, highland hummingbird community. *Journal of Animal Ecology*, **45**, 349–79.

Wong, M. (1986) Trophic organization of understory birds in a Malaysian dipterocarp forest. *The Auk*, **103**, 100–16.

Woods, P.E. (1984) Woodpecker bills and their conformance to Hutchinsonian ratios. *Ohio Journal of Science*, **84**, 255–58.

Wooller, R.D. (1984) Bill shape and size in honeyeaters and other small insectivorous birds in Western Australia. *Australian Journal of Zoology*, **32**, 657–61.

Wooller, R.D. & Calver, M.C. (1981) Feeding segregation within an assemblage of small birds in the Karri Forest understorey. *Australian Wildlife Research*, **8**, 401–10.

Wooller, R.D., Saunders, D.A., Bradley, J.S. & de Rebeira, C.P. (1985) Geographical variation in size of an Australian honeyeater (Aves: Meliphagidae): an example of Bergmann's rule. *Biological Journal of the Linnean Society*, **25**, 355–63.

Worster, D. (1977) *Nature's Economy: The Roots of Ecology*. San Francisco: Sierra Club Books.

Wright, S.J. (1980) Density compensation in island avifaunas. *Oecologia*, **45**, 385–89.

(1981) Intra-archipelago vertebrate distributions: the slope of the species–area relation. *The American Naturalist*, **118**, 726–48.

Wright, S.J. & Biehl, C.C. (1982) Island biogeographic distributions: testing for random, regular, and aggregated patterns of species occurrence. *The American Naturalist*, **119**, 345–57.

Wykes, B.J. (1985) The Helmeted Honeyeater and related honeyeaters of Victorian woodlands. In *Birds of Eucalypt Forests and Woodlands: Ecology, Conservation, Management*, ed. Keast, A., Recher, H.F., Ford, H. & Saunders, D., pp. 205–17. Chipping Norton, NSW: Surrey Beatty & Sons.

Wyndham, E. (1986) Length of birds' breeding seasons. *The American Naturalist*, **128**, 155–64.

Wyndham, E. & Cannon, C.E. (1985) Parrots of eastern Australian forests and woodlands: the genera *Platycercus* and *Trichoglossus*. In *Birds of Eucalypt Forests and Woodlands: Ecology, Conservation, Management*, ed. Keast, A., Recher, H.F., Ford, H. & Saunders, D., pp. 141–50. Chipping Norton, NSW: Surrey Beatty & Sons.

Yeaton, R.I. (1974) An ecological analysis of chaparral and pine forest bird communities on Santa Cruz Island and mainland California. *Ecology*, **55**, 959–73.

Yeaton, R.I. & Cody, M.L. (1974) Competitive release in island Song Sparrow populations. *Theoretical Population Biology*, **5**, 42–57.

Yom–Tov, Y. & Hilborn, R. (1981) Energetic constraints on clutch size and time of breeding in temperate zone birds. *Oecologia*, **48**, 234–43.

Zach, R. & Falls, J.B. (1979) Foraging and territoriality of male Ovenbirds (Aves: Parulidae) in a heterogeneous habitat. *Journal of Animal Ecology*, **48**, 33–52.

Zwarts, L. (1980) Intra- and interspecific competition for space in estuarine bird species in a one-prey situation. In *Acta XVII Congressus Internationalis Ornithologici*, ed. Nöhring, R., pp. 1045–50. Berlin: Deutsche Ornithologen-Gesellschaft.

Zwarts, L. & Wanink, J. (1984) How Oystercatchers and Curlews successively deplete clams. In *Coastal Waders and Wildfowl in Winter*, ed. Evans, P.R., Goss-Custard, J.D. & Hale, W.G., pp. 69–83. Cambridge: Cambridge University Press.

Author index

Subject index